Birkhäuser

Jean-Michel Muller • Nicolas Brunie
Florent de Dinechin • Claude-Pierre Jeannerod
Mioara Joldes • Vincent Lefèvre
Guillaume Melquiond • Nathalie Revol
Serge Torres

Handbook of Floating-Point Arithmetic

Second Edition

Birkhäuser

Jean-Michel Muller
CNRS - LIP
Lyon, France

Nicolas Brunie
Kalray
Grenoble, France

Florent de Dinechin
INSA-Lyon - CITI
Villeurbanne, France

Claude-Pierre Jeannerod
Inria - LIP
Lyon, France

Mioara Joldes
CNRS - LAAS
Toulouse, France

Vincent Lefèvre
Inria - LIP
Lyon, France

Guillaume Melquiond
Inria - LRI
Orsay, France

Nathalie Revol
Inria - LIP
Lyon, France

Serge Torres
ENS-Lyon - LIP
Lyon, France

ISBN 978-3-030-09513-0 ISBN 978-3-319-76526-6 (eBook)
https://doi.org/10.1007/978-3-319-76526-6

Mathematics Subject Classification: 65Y99, 68N30

Printed on acid-free paper

This book is published under the imprint Birkhäuser, www.birkhauser-science.com by the registered company Springer International Publishing AG part of Springer Nature.
The registered company address is: Gewerbestrasse 11, 6330 Cham, Switzerland

Contents

List of Figures

List of Tables

Preface

FLOATING-POINT ARITHMETIC is by far the most widely used way of approximating real-number arithmetic for performing numerical calculations on modern computers. A rough presentation of floating-point arithmetic requires only a few words: a number x is represented in radix β floating-point arithmetic with a sign s, a significand m, and an exponent e, such that $x = s \times m \times \beta^e$. Making such arithmetic reliable, fast, and portable is however a very complex task. Although it could be argued that, to some extent, the concept of floating-point arithmetic (in radix 60) was invented by the Babylonians, or that it is the underlying arithmetic of the slide rule, its first modern implementation appeared in Konrad Zuse's Z1 and Z3 computers.

A vast quantity of very diverse arithmetics was implemented between the 1960s and the early 1980s. The radix (radices 2, 4, 8, 16, and 10, and even radix 3, were then considered), and the sizes of the significand and exponent fields were not standardized. The approaches for rounding and for handling underflows, overflows, or "forbidden operations" (such as $5/0$ or $\sqrt{-3}$) were significantly different from one machine to another. This lack of standardization made it difficult to write reliable and portable numerical software.

Pioneering scientists including Brent, Cody, Kahan, and Kuki highlighted the relevant key concepts for designing an arithmetic that could be both useful for programmers and practical for implementers. These efforts resulted in the IEEE 754-1985 standard for radix-2 floating-point arithmetic. The standardization process was expertly orchestrated by William Kahan. The IEEE 754-1985 standard was a key factor in improving the quality of the computational environment available to programmers. A new version, the IEEE 754-2008 standard, was released in August 2008. It specifies radix-2 and radix-10 floating-point arithmetic. At the time of writing these lines, a working group is preparing a new version of IEEE 754, which should be released in 2018. It will not be very different from the 2008 version.

By carefully specifying the behavior of the arithmetic operators, the IEEE 754-1985 and 754-2008 standards allowed researchers and engineers to design extremely powerful yet portable algorithms, for example, to compute very accurate sums and dot products, and to formally prove some critical parts of programs. Unfortunately, the subtleties of the standard are little known by the nonexpert user. Even more worrying, they are sometimes overlooked by compiler designers. As a consequence, floating-point arithmetic is sometimes conceptually misunderstood and is often far from being exploited to its full potential.

This led us to the decision to compile into a book selected parts of the vast knowledge of floating-point arithmetic. This book is designed for programmers of numerical applications, compiler designers, programmers of floating-point algorithms, designers of arithmetic operators (floating-point adders, multipliers, dividers, ...), and more generally students and researchers in numerical analysis who wish to more accurately understand a tool that they manipulate on an everyday basis. During the writing, we tried, whenever possible, to illustrate the techniques described by an actual program, in order to allow a more direct practical use for coding and design.

The first part of the book presents the history and basic concepts of floating-point arithmetic (formats, exceptions, correct rounding, etc.) and various aspects of the 2008 version of the IEEE 754 standard. The second part shows how the features of the standard can be used to develop effective and nontrivial algorithms. This includes summation algorithms, and division and square root relying on a fused multiply-add. This part also discusses issues related to compilers and languages. The third part then explains how to implement floating-point arithmetic, both in software (on an integer processor) and in hardware (VLSI or reconfigurable circuits). It also surveys the implementation of elementary functions. The fourth part presents some extensions: complex numbers, interval arithmetic, verification of floating-point arithmetic, and extension of the precision. In the Appendix, the reader will find an introduction to relevant number theory tools and a brief presentation of the standards that predated IEEE 754-2008.

Acknowledgments

Some of our colleagues around the world, in academia and industry, and several students from École normale supérieure de Lyon and Université de Lyon greatly helped us by discussing with us, giving advice and reading preliminary versions of this edition, or by giving comments on the previous one that (hopefully) helped us to improve it: Nicolas Brisebarre, Jean-Yves L'Excellent, Warren Ferguson, Christian Fleck, John Harrison, Nicholas Higham, Tim Leonard, Dan Liew, Nicolas Louvet, David Lutz, Peter Markstein, Marc Mezzarobba, Kai Torben Ohlhus, Antoine Plet, Valentina Popescu, Guillaume Revy, Siegfried Rump, Damien Stehlé, and Nick Trefethen. We thank them all for their suggestions and interest.

Our dear colleague and friend Peter Kornerup left us recently. The computer arithmetic community misses one of its pioneers, and we miss a very kind and friendly person.

Grenoble, France	Nicolas Brunie
Villeurbanne, France	Florent de Dinechin
Lyon, France	Claude-Pierre Jeannerod
Toulouse, France	Mioara Joldes
Lyon, France	Vincent Lefèvre
Orsay, France	Guillaume Melquiond
Lyon, France	Jean-Michel Muller
Lyon, France	Nathalie Revol
Lyon, France	Serge Torres

Part I

Introduction, Basic Definitions, and Standards

Chapter 1

Introduction

REPRESENTING AND MANIPULATING real numbers efficiently is required in many fields of science, engineering, finance, and more. Since the early years of electronic computing, many different ways of approximating real numbers on computers have been introduced. One can cite (this list is far from being exhaustive): fixed-point arithmetic, logarithmic [337, 585] and semi-logarithmic [444] number systems, intervals [428], continued fractions [349, 622], rational numbers [348] and possibly infinite strings of rational numbers [418], level-index number systems [100, 475], fixed-slash and floating-slash number systems [412], tapered floating-point arithmetic [432, 22], 2-adic numbers [623], and most recently unums and posits [228, 229].

And yet, floating-point arithmetic is by far the most widely used way of representing real numbers in modern computers. Simulating an infinite, continuous set (the real numbers) with a finite set (the "machine numbers") is not a straightforward task: preserving all properties of real arithmetic is not possible, and clever compromises must be found between, e.g., speed, accuracy, dynamic range, ease of use and implementation, and memory cost. It appears that floating-point arithmetic, with adequately chosen parameters (radix, precision, extremal exponents, etc.), is a very good compromise for most numerical applications.

We will give a complete, formal definition of floating-point arithmetic in Chapter 3, but roughly speaking, a radix-β, precision-p, floating-point number is a number of the form

$$\pm m_0.m_1m_2\cdots m_{p-1} \times \beta^e,$$

where e, called the *exponent*, is an integer, and $m_0.m_1m_2\cdots m_{p-1}$, called the *significand*, is represented in radix β. The major purpose of this book is to explain how these numbers can be manipulated efficiently and safely.

J.-M. Muller et al., *Handbook of Floating-Point Arithmetic*,
https://doi.org/10.1007/978-3-319-76526-6_1

1.1 Some History

Even if the implementation of floating-point arithmetic on electronic computers is somewhat recent, floating-point arithmetic itself is an old idea. In *The Art of Computer Programming* [342], Knuth presents a short history. He views the radix-60 number system of the Babylonians as some kind of early floating-point system. That system allowed the Babylonians to perform arithmetic operations rather efficiently [498]. Since the Babylonians did not invent the number zero, if the ratio of two numbers is a power of 60, then their representation in the Babylonian system is the same. In that sense, the number represented is the *significand* of a radix-60 floating-point representation.

A famous tablet from the Yale Babylonian Collection (YBC 7289) gives an approximation to $\sqrt{2}$ with four sexagesimal places (the digits represented on the tablet are 1, 24, 51, 10). A photo of that tablet can be found in [633], and a very interesting analysis of the Babylonian mathematics used for computing square roots, related to YBC 7289, was carried out by Fowler and Robson [205].

Whereas the Babylonians invented the *significands* of our floating-point numbers, one may reasonably argue that Archimedes invented the *exponents*: in his famous treatise Arenarius (The Sand Reckoner) he invents a system of naming very large numbers that, in a way, "contains" an exponential representation [259]. The notation a^n for $a \times a \times a \times \cdots \times a$ seems to have been coined much later by Descartes (it first appeared in his book *La Géométrie* [169]).

The arithmetic of the slide rule, invented around 1630 by William Oughtred [632], can be viewed as another kind of floating-point arithmetic. Again, as with the Babylonian number system, we only manipulate significands of numbers (in that case, radix-10 significands).

The two modern co-inventors of floating-point arithmetic are probably Quevedo and Zuse. In 1914 Leonardo Torres y Quevedo described an electro-mechanical implementation of Babbage's Analytical Engine with floating-point arithmetic [504]. Yet, the first real, modern implementations of floating-point arithmetic were in Konrad Zuse's Z1 [514] and Z3 [92] computers. The Z3, built in 1941, used a radix-2 floating-point number system, with 15-bit significands (stored on 14 bits, using the *leading bit convention*, see Section 2.1.2), 7-bit exponents, and a 1-bit sign. It had special representations for infinities and indeterminate results. These characteristics made the real number arithmetic of the Z3 much ahead of its time.

The Z3 was rebuilt recently [515]. Photographs of Konrad Zuse and the Z3 can be viewed at http://www.konrad-zuse.de/.

Readers interested in the history of computing devices should have a look at the excellent books by Aspray et al. [20] and Ceruzzi [93].

When designing a floating-point system, the first thing one must think about is the choice of the radix β. Radix 10 is what humans use daily for representing numbers and performing paper and pencil calculations. Therefore, to avoid input and output radix conversions, the first idea that springs to mind for implementing automated calculations is to use the same radix.

And yet, since most of our computers are based on two-state logic, radix 2 (and, more generally, radices that are a power of 2) is by far the easiest to implement. Hence, choosing the right radix for the internal representation of floating-point numbers was not obvious. Indeed, several different solutions were explored in the early days of automated computing.

Various early machines used a radix-8 floating-point arithmetic: the PDP-10 and the Burroughs 570 and 6700 for example. The IBM 360 had a radix-16 floating-point arithmetic (and on its mainframe computers, IBM still offers hexadecimal floating-point, along with more conventional radix-2 and radix-10 arithmetics [213, 387]). Radix 10 has been extensively used in financial calculations[1] and in pocket calculators, and efficient implementation of radix-10 floating-point arithmetic is still a very active domain of research [87, 89, 116, 121, 122, 124, 191, 614, 613, 627, 628]. The computer algebra system Maple also uses radix 10 for its internal representation of the "software floats." It therefore seems that the various radices of floating-point arithmetic systems that have been implemented so far have almost always been either 10 or a power of 2.

There was a very odd exception. The Russian SETUN computer, built at Moscow State University in 1958, represented numbers in radix 3, with digits -1, 0, and 1 [630]. This "balanced ternary" system has several advantages. One of them is the fact that rounding to a nearest number the sum or product of two numbers is equivalent to truncation [342]. Another one [250] is the following. Assume you use a radix-β fixed-point system, with p-digit numbers. A large value of β makes the implementation complex: the system must be able to "recognize" and manipulate β different symbols. A small value of β means that more digits are needed to represent a given number: if β is small, p has to be large. To find a compromise, we can try to minimize $\beta \times p$, while having the largest representable number $\beta^p - 1$ (almost) constant. The optimal solution[2] will almost always be $\beta = 3$. See http://www.computer-museum.ru/english/setun.htm for more information on the SETUN computer.

Johnstone and Petry have argued [306] that radix 210 could be a sensible choice because it would allow exact representation of many rational numbers.

Various studies (see references [63, 104, 352] and Chapter 2) have shown that radix 2 with the *implicit leading bit convention* gives better worst-case and average accuracy than all other radices. This and the ease of implementation explain the current prevalence of radix 2.

---

[1]For legal reasons, financial calculations frequently require special rounding rules that are very tricky to implement if the underlying arithmetic is binary: this is illustrated in [320, Section 2].

[2]If p and β were real numbers, the value of β that would minimize $\beta \times p$ while letting β^p be constant would be $e = 2.7182818\cdots$.

The world of numerical computation changed a great deal in 1985, when the IEEE 754-1985 Standard for Binary Floating-Point Arithmetic was published [12]. This standard specifies various formats, the behavior of the basic operations and conversions, and exception conditions. As a matter of fact, the Intel 8087 mathematics co-processor, built a few years before (in 1980) to be paired with the Intel 8088 and 8086 processors, was already extremely close to what would later become the IEEE 754-1985 standard. And the HP35 pocket calculator (a landmark: it was the machine that killed the slide rule!), launched in 1972, already implemented related ideas. Now, most systems of commercial significance offer compatibility[3] with IEEE 754-1985 or its successor IEEE 754-2008. This has resulted in significant improvements in terms of accuracy, reliability, and portability of numerical software. William Kahan [553] played a leading role in the conception of the IEEE 754-1985 standard and in the development of smart algorithms for floating-point arithmetic. His website[4] contains much useful information. He received the Turing award in 1989.

IEEE 754-1985 only dealt with radix-2 arithmetic. Another standard, released in 1987, the IEEE 854-1987 Standard for Radix Independent Floating-Point Arithmetic [13], was devoted to both binary (radix-2) and decimal (radix-10) arithmetic.

In 1994, a number theorist, Thomas Nicely, who was working on the twin prime conjecture,[5] noticed that his Pentium-based computer delivered very inaccurate results when performing some divisions. The reason was a flaw in the choice of tabulated constants needed by the division algorithm [108, 183, 437]. For instance, when dividing 4195835.0 by 3145727.0, one would get 1.333739068902 instead of 1.3338204491. This announcement of a "Pentium FDIV bug" provoked a great deal of discussion at the time but has had very positive long-term effects: most arithmetic algorithms used by the manufacturers are now published (for instance a few years after, the division algorithm used on the Intel/HP Itanium [120, 242] was made public) so that everyone can check them, and everybody understands that a particular effort must be made to build formal proofs of the arithmetic algorithms and their implementation [240, 243].

A revision of the standard, which replaced both IEEE 754-1985 and 854-1987, was adopted in 2008 [267]. That IEEE 754-2008 standard brought significant improvements. It specified the Fused Multiply-Add (FMA) instruction—which makes it possible to evaluate $ab + c$, where a, b, and c are floating-point numbers, with one final rounding only. It resolved some ambiguities in IEEE 754-1985, especially concerning expression evaluation and exception

[3]Even if sometimes you need to dive into the compiler documentation to find the right options; see Chapter 6.

[4]http://www.cs.berkeley.edu/~wkahan/.

[5]That conjecture asserts that there are infinitely many pairs of prime numbers that differ by 2.

handling. It also included "quadruple precision" (now called binary128 or decimal128), and recommended (yet did not mandate) that some elementary functions should be correctly rounded (see Chapter 10).

At the time of writing these lines, the IEEE 754 Standard is again under revision: the new standard is to be released in 2018.

1.2 Desirable Properties

Specifying a floating-point arithmetic (formats, behavior of operators, etc.) demands that one find compromises between requirements that are seldom fully compatible. Among the various properties that are desirable, one can cite:

- **Speed**: Tomorrow's weather must be computed in less than 24 hours;
- **Accuracy**: Even if speed is important, getting a wrong result right now is not better than getting the correct one too late;
- **Range**: We may need to represent large as well as tiny numbers;
- **Portability**: The programs we write on a given machine should run on different machines without requiring modifications;
- **Ease of implementation and use**: If a given arithmetic is too arcane, almost nobody will use it.

With regard to accuracy, the most accurate current physical measurements allow one to check some predictions of quantum mechanics or general relativity with a relative accuracy better than 10^{-15} [99].

This means that in some cases, we must be able to represent numerical data with a similar accuracy (which is easily done, using formats that are implemented on almost all current platforms: for instance, with the binary64 format of IEEE 754-2008, one represents numbers with relative error less than $2^{-53} \approx 1.11 \times 10^{-16}$). But this also means that we must sometimes be able to carry out long computations that end up with a relative error less than or equal to 10^{-15}, which is much more difficult. Sometimes, one will need a significantly larger floating-point format or clever "tricks" such as those presented in Chapter 4.

An example of a huge calculation that requires great care was carried out by Laskar's team at the Paris Observatory [369]. They computed long-term numerical solutions for the insolation quantities of the Earth (very long-term, ranging from -250 to $+250$ millions of years from now).

In other domains, such as number theory, some multiple-precision computations are indeed carried out using a very large precision. For instance, in

2002, Kanada's group computed 1.241 trillion decimal digits of π [25], using the two formulas

$$\begin{aligned}\pi &= 48\arctan\frac{1}{49}+128\arctan\frac{1}{57}-20\arctan\frac{1}{239}+48\arctan\frac{1}{110443}\\ &= 176\arctan\frac{1}{57}+28\arctan\frac{1}{239}-48\arctan\frac{1}{682}+96\arctan\frac{1}{12943},\end{aligned}$$

and in 2013 Lee and Kondo computed 12.1 trillion digits (see http://www.numberworld.org/misc_runs/pi-12t/).

These last examples are extremes. One should never forget that with 50 bits, one can express the distance from the Earth to the Moon with an error less than the thickness of a bacterium. It is very uncommon to need such an accuracy on a final result and, actually, very few physical quantities are defined that accurately. Choices made in the definition of the usual floating-point formats target the common cases.

1.3 Some Strange Behaviors

Designing efficient and reliable hardware or software floating-point systems is a difficult and somewhat risky task. Some famous bugs have been widely discussed; we review some of them below. Also, even when the arithmetic is not flawed, some strange behaviors can sometimes occur, just because they correspond to a numerical problem that is intrinsically difficult. All this is not surprising: mapping the continuous real numbers on a finite structure (the floating-point numbers) cannot be done without any trouble.

1.3.1 Some famous bugs

Let us start with some famous arithmetic bugs.

- In Section 1.1, we have already discussed the bug in the floating-point division instruction in the early Intel Pentium processors. In extremely rare cases, one would get only four correct decimal digits.
- With release 7.0 of the computer algebra system Maple (distributed in 2001), when computing
$$\frac{1001!}{1000!},$$
we would get 1 instead of 1001.
- With the previous release (6.0, distributed in 2000) of the same system, when entering
$$21474836480413647819643794$$

we would get

$$413647819643790)+'--.(--.($$

- Kahan [320, Section 2] mentions some strange behavior of some versions of the Excel spreadsheet. They seem to be due to an attempt to mimic decimal arithmetic with an underlying binary one.

  Another strange behavior happens with some early versions of Excel 2007: When you try to compute

  $$65536 - 2^{-37}$$

  the result displayed is 100001. This is an error in the binary-to-decimal conversion used for displaying that result: the internal binary value is correct,[6] and if you add 1 to the result you get 65537.

- Some bugs do not require any programming error: they are due to poor specifications. For instance, the Mars Climate Orbiter probe crashed on Mars in September 1999 because of an astonishing mistake: one of the teams that designed the numerical software assumed the unit of distance was the meter, while another team assumed it was the foot [9, 466].

  Very similarly, in June 1985, a Space Shuttle positioned itself to receive a laser beamed from the top of a mountain that was supposed to be 10,000 miles high, instead of the correct 10,000 feet [9].

  Also, in January 2004, a bridge between Germany and Switzerland did not fit at the border because the two countries use a different definition of sea level.[7]

1.3.2 Difficult problems

Sometimes, even with a correctly implemented floating-point arithmetic, the result of a computation is far from what could be expected.

1.3.2.1 A sequence that seems to converge to a wrong limit

Consider the following example, due to one of us [436] and analyzed by Kahan [320]. Let (u_n) be the sequence defined as

$$\begin{cases} u_0 &=& 2, \\ u_1 &=& -4, \\ u_n &=& 111 - \dfrac{1130}{u_{n-1}} + \dfrac{3000}{u_{n-1}u_{n-2}}, \quad n \geq 2. \end{cases} \tag{1.1}$$

[6] A detailed analysis of this bug can be found at http://www.lomont.org/Math/Papers/2007/Excel2007/Excel2007Bug.pdf.

[7] See http://www.science20.com/news_articles/what_happens_bridge_when_one_side_uses_mediterranean_sea_level_and_another_north_sea-121600.

One can easily show that this sequence tends to 6 as $n \to \infty$. Yet, on any system with any precision, and with any order of evaluation of the sums in (1.1), it will seem to tend to 100.

For example, Table 1.1 gives the results obtained by compiling Program 1.1 and running it on a Macintosh with an Intel Core i5 processor, using the GNU Compiler Collection (GCC).

```
#include <stdio.h>

int main(void)
{
  double u, v, w;
  int i, max;

  printf("n =");
  scanf("%d",&max);
  printf("u0 = ");
  scanf("%lf",&u);
  printf("u1 = ");
  scanf("%lf",&v);
  printf("Computation from 3 to n:\n");
  for (i = 3; i <= max; i++)
    {
      w = 111. - 1130./v + 3000./(v*u);
      u = v;
      v = w;
      printf("u%d = %1.17g\n", i, v);
    }
  return 0;
}
```

Program 1.1: A C program that is supposed to compute the sequence (u_n) using double-precision arithmetic. The results obtained are given in Table 1.1.

The explanation of this weird phenomenon is quite simple. The general solution for the recurrence

$$u_n = 111 - \frac{1130}{u_{n-1}} + \frac{3000}{u_{n-1}u_{n-2}}$$

is

$$u_n = \frac{\alpha \cdot 100^{n+1} + \beta \cdot 6^{n+1} + \gamma \cdot 5^{n+1}}{\alpha \cdot 100^n + \beta \cdot 6^n + \gamma \cdot 5^n},$$

where α, β, and γ depend on the initial values u_0 and u_1. Therefore, if $\alpha \neq 0$ then the limit of the sequence is 100, otherwise (assuming $\beta \neq 0$), it is 6. In the present example, the starting values $u_0 = 2$ and $u_1 = -4$ were chosen so that $\alpha = 0$, and $\beta/\gamma = -3/4$. Therefore, the "exact" limit of u_n is 6. Yet, when computing the values u_n in floating-point arithmetic using (1.1), due to the various rounding errors, even the very first computed terms become

n	Computed value	Exact value
3	**18.5**	18.5
4	**9.37837837837837**9	$9.3783783783783783783\cdots$
5	**7.80115273775216**88	$7.8011527377521613833\cdots$
6	**7.15441448097533**34	$7.1544144809752493535\cdots$
11	**6.2744386**627281159	$6.2744385982163279138\cdots$
12	**6.21869**67685821628	$6.2186957398023977883\cdots$
16	**6.1**660865595980994	$6.0947394393336811283\cdots$
17	7.2356654170119432	$6.0777223048472427363\cdots$
18	22.069559154531031	$6.0639403224998087553\cdots$
19	78.58489258126825	$6.0527217610161521934\cdots$
20	98.350416551346285	$6.0435521101892688678\cdots$
21	99.898626342184102	$6.0360318810818567800\cdots$
22	99.993874441253126	$6.0298473250239018567\cdots$
23	99.999630595494608	$6.0247496523668478987\cdots$
30	99.999999999998948	$6.0067860930312057585\cdots$
31	99.999999999999943	$6.0056486887714202679\cdots$

Table 1.1: Results obtained by running Program 1.1 on a Macintosh with an Intel Core i5 processor, using GCC, compared to the exact values of the sequence (u_n).

slightly different from the exact terms. Hence, the value α corresponding to these computed terms is tiny, but nonzero. This suffices to make the computed sequence converge to 100.

1.3.2.2 The Chaotic Bank Society

Recently, Mr. Gullible went to the Chaotic Bank Society, to learn more about the new kind of account they offer to their best customers. He was told:

> You first deposit \$$e - 1$ on your account, where $e = 2.7182818\ldots$ is the base of the natural logarithms. The first year, we take \$1 from your account as banking charges. The second year is better for you: We multiply your capital by 2, and we take \$1 of banking charges. The third year is even better: We multiply your capital by 3, and we take \$1 of banking charges. And so on: The n-th year, your capital is multiplied by n and we just take \$1 of charges. Interesting, isn't it?

Mr. Gullible wanted to secure his retirement. So before accepting the offer, he decided to perform some simulations on his own computer to see what his capital would be after 25 years. Once back home, he wrote a C program (Program 1.2).

```
#include <stdio.h>

int main(void)
{
  double account = 1.71828182845904523536028747135;
  int i;
  for (i = 1; i <= 25; i++)
    {
      account = i*account - 1;
    }
  printf("You will have $%1.17e on your account.\n", account);
}
```

Program 1.2: Mr. Gullible's C program.

On his computer (with an Intel Xeon processor, and GCC on Linux, but strange things would happen with any other equipment), he got the following answer:

```
You will have $1.20180724741044855e+09 on your account.
```

So he immediately decided to accept the offer. He will certainly be sadly disappointed, 25 years later, when he realizes that he actually has around \$0.0399 in his account.

What happens in this example is easy to understand. If you call a_0 the amount of the initial deposit and a_n the capital after the end of the n-th year, then

$$\begin{aligned} a_n &= n! \times \left(a_0 - 1 - \frac{1}{2!} - \frac{1}{3!} - \cdots - \frac{1}{n!}\right) \\ &= n! \times \left(a_0 - (e-1) + \frac{1}{(n+1)!} + \frac{1}{(n+2)!} + \frac{1}{(n+3)!} + \cdots\right), \end{aligned}$$

so that:

- if $a_0 < e - 1$, then a_n goes to $-\infty$;
- if $a_0 = e - 1$, then a_n goes to 0;
- if $a_0 > e - 1$, then a_n goes to $+\infty$.

In our example, $a_0 = e-1$, so the *exact* sequence a_n goes to zero. This explains why the exact value of a_{25} is so small. And yet, even if the arithmetic operations were errorless (which of course is not the case), since $e-1$ is not exactly representable in floating-point arithmetic,[8] the *computed* sequence will go to $+\infty$ or $-\infty$, depending on rounding directions.

[8]Interestingly enough, the decimal value $a_0 = 1.71828182845904523536028747135$ given in Program 1.2 is less than $e - 1$, but when rounded to the nearest binary64/double precision floating-point number, it becomes larger than $e - 1$.

1.3.2.3 Rump's example

Consider the following function, designed by Rump in 1988 [518] and analyzed by various authors [125, 393]: given

$$f(a,b) = 333.75b^6 + a^2\left(11a^2b^2 - b^6 - 121b^4 - 2\right) + 5.5b^8 + \frac{a}{2b}, \quad (1.2)$$

let us try to compute $f(a,b)$ for $a = 77617$ and $b = 33096$.

Note that a and b, as well as the coefficients 333.75, 11, 121, and 5.5 that appear in (1.2) are all exactly representable in binary floating-point arithmetic as soon as the precision is larger than 17 bits. On an IBM S/370 computer, the results obtained by Rump were

- 1.172603 in single precision;
- 1.1726039400531 in double precision; and
- 1.172603940053178 in extended precision.

Anybody looking at these figures would feel that the single precision result is certainly very accurate. And yet, the exact result is $-0.8273960599\ldots$. On more recent systems, we do not see the same behavior exactly. For instance, on a Macintosh with an Intel Core i5 processor, using GCC, the C program (Program 1.3) which uses double-precision computation, will return $-1.18059162 \times 10^{21}$, whereas its single-precision equivalent will return -9.87501×10^{29}. We still get totally wrong results, but at least, the clear differences between them show that something weird is going on. Notice that if we change the order of evaluation of the sums in (1.2), we get different results, but we still observe a strange behavior.

```
#include <stdio.h>
int main(void)
{
  double a = 77617.0;
  double b = 33096.0;
  double b2,b4,b6,b8,a2,firstexpr,f;
  b2 = b*b;
  b4 = b2*b2;
  b6 = b4*b2;
  b8 = b4*b4;
  a2 = a*a;
  firstexpr = 11*a2*b2-b6-121*b4-2;
  f = 333.75*b6 + a2 * firstexpr + 5.5*b8 + (a/(2.0*b));
  printf("Double precision result: $ %1.17e \n",f);
}
```

Program 1.3: Rump's example.

These examples are not meant to frighten the reader. The floating-point arithmetic that is available on our modern computers is a very powerful tool. As with other tools, making sure that we always correctly manipulate it requires some care and knowledge. The major aim of this book is to provide the reader with the necessary knowledge.

Chapter 2

Definitions and Basic Notions

As stated in the Introduction, roughly speaking, a radix-β floating-point number x is a number of the form

$$m \cdot \beta^e,$$

where β is the *radix* of the floating-point system, m such that $|m| < \beta$ is the *significand* of x, and e is its *exponent*. However, portability, accuracy, and the ability to prove interesting and useful properties as well as to design effective algorithms require more rigorous definitions, and much care in the specifications. This is the first purpose of this chapter. The second one is to deal with basic problems: rounding, exceptions, properties of real arithmetic that cease to hold in floating-point arithmetic, best choices for the radix, and radix conversions.

2.1 Floating-Point Numbers

Let us now give a more formal definition of the floating-point numbers. Although we try to be somewhat general, the definition is largely inspired by the IEEE 754-2008 standard for floating-point arithmetic (see Chapter 3).

2.1.1 Main definitions

A floating-point format is (partially)[1] characterized by four integers:

- a *radix* (or *base*) $\beta \geq 2$;

[1]A full definition of a floating-point format also specifies the binary encoding of the significands and exponents. It also deals with special values such as infinities and results of invalid operations.

J.-M. Muller et al., *Handbook of Floating-Point Arithmetic*,
https://doi.org/10.1007/978-3-319-76526-6_2

- a *precision* $p \geq 2$ (roughly speaking, p is the number of "significant digits" of the representation);

- two *extremal exponents* $e_{\min}$ and $e_{\max}$ such that $e_{\min} < e_{\max}$. In all practical cases, $e_{\min} < 0 < e_{\max}$, and with all formats specified by the IEEE 754 standard, $e_{\min} = 1 - e_{\max}$.

A finite floating-point number in such a format is a number x for which there exists at least one representation (M, e) such that

$$x = M \cdot \beta^{e-p+1}, \tag{2.1}$$

where

- M is an integer such that $|M| \leq \beta^p - 1$, called the *integral significand* of the representation of x;

- e is an integer such that $e_{\min} \leq e \leq e_{\max}$, called the *exponent* of the representation of x.

The representation (M, e) of a floating-point number is not necessarily unique. For instance, with $\beta = 10$ and $p = 3$, the number 17 can be represented either by 17×10^0 or by 170×10^{-1}, since both 17 and 170 are less than $\beta^p = 1000$. The set of these equivalent representations is called a *cohort*.

The number

$$\beta^{e-p+1}$$

from Equation (2.1) is called the *quantum* of the representation of x. We will call the *quantum exponent* the number

$$q = e - p + 1.$$

The notion of quantum is closely related to the notion of *ulp* (*unit in the last place*); see Section 2.3.2.

Another way to express the same floating-point number x is by using the triple (s, m, e), so that

$$x = (-1)^s \cdot m \cdot \beta^e,$$

where

- e is the same as before;

- $m = |M| \cdot \beta^{1-p}$ is called the *normal significand* (or, more simply, the *significand* of the representation). It has one digit before the radix point, and at most $p - 1$ digits after (notice that $0 \leq m < \beta$); and

- $s \in \{0, 1\}$ is the sign bit of x.

The significand is also frequently (and slightly improperly) called the *mantissa* in the literature.[2] According to Goldberg [214], the term "significand" was coined by Forsythe and Moler in 1967 [202].

Consider the following toy system. We assume radix $\beta = 2$, precision $p = 4$, $e_{\min} = -7$, and $e_{\max} = +8$. The number[3] $416_{10} = 110100000_2$ is a floating-point number. It has one representation only, with integral significand 13_{10} and exponent 8_{10}, since

$$416 = 13 \cdot 2^{8-4+1}.$$

The quantum of this representation is $2^5 = 32$. Note that a representation such as $26 \cdot 2^{7-4+1}$ is excluded because $26 > 2^p - 1 = 15$. In the same toy system, the number $4.25_{10} = 17 \cdot 2^{-2}$ is not a floating-point number, as it cannot be exactly expressed as $M \cdot \beta^{e-p+1}$ with $|M| \leq 2^4 - 1$.

When x is a nonzero arbitrary real number (i.e., x is not necessarily exactly representable in a given floating-point format), we will call *infinitely precise significand* of x (in radix β) the number

$$\frac{x}{\beta^{\lfloor \log_\beta |x| \rfloor}},$$

where $\beta^{\lfloor \log_\beta |x| \rfloor}$ is the largest integer power of β smaller than or equal to $|x|$.

2.1.2 Normalized representations, normal and subnormal numbers

As explained above, some floating-point numbers may have several representations (M, e). Nevertheless, it is frequently desirable to require unique representations. In order to have a unique representation, one may want to *normalize* the finite nonzero floating-point numbers by choosing the representation for which the exponent is minimum (yet greater than or equal to $e_{\min}$). The representation obtained will be called a *normalized representation*. Requiring normalized representations allows for easier expression of error bounds, it somewhat simplifies the implementation, and it allows for a one-bit saving in radix 2.[4] Two cases may occur.

- In general, such a representation satisfies $1 \leq |m| < \beta$, or, equivalently, $\beta^{p-1} \leq |M| < \beta^p$. In such a case, one says that the corresponding value x is a *normal* number. When x is a normal floating-point number, its infinitely precise significand is equal to its significand.

---

[2]The *mantissa* of a nonnegative number is the fractional part of its logarithm.

[3]We will frequently write xxx...xx$_\beta$ to denote the number whose radix-β representation is xxx...xx. To avoid complicated notation, we will tend to omit β whenever its value is clear from the context.

[4]We will see in Chapter 3 that the IEEE 754-2008 standard requires normalized representations in radix 2, but not in radix 10.

- Otherwise, one necessarily has $e = e_{\min}$, and the corresponding value x is said to be a *subnormal* number (the term *denormal number* is often used too). In that case, $|m| < 1$ or, equivalently, $|M| \leq \beta^{p-1} - 1$. The special case of zero will be dealt with later.

With such a normalization, as the representation of a finite nonzero number x is unique, the values M, q, m, and e only depend on the value of x. We therefore call e the exponent of x, q its quantum exponent ($q = e - p + 1$), M its integral significand, and m its significand.

For instance, in radix-2 normalized floating-point arithmetic, with[5] $p = 24$, $e_{\min} = -126$, and $e_{\max} = 127$, the floating-point number f that is nearest to $1/3$ is a normal number. Its exponent is -2, and its integral significand is 11184811. Therefore,

$$f = 11184811 \times 2^{-2-24+1} = 0.3333333432674407958984375_{10}$$

and the quantum of f is $2^{-2-24+1} \approx 2.98 \times 10^{-8}$, while its quantum exponent is -25.

In the same floating-point system, the number 3×2^{-128} is a subnormal number. Its exponent is $e_{\min} = -126$, its quantum is 2^{-149}, and its integral significand is

$$3 \times 2^{149-128} = 6291456.$$

In radix 2, the first digit of the significand of a normal number is a 1, and the first digit of the significand of a subnormal number is a 0. If we have a special encoding (in practice, in the exponent field, see Section 3.1.1.1) that tells us if a number is normal or subnormal, there is no need to store the first bit of its significand. This *leading bit convention*, or *implicit bit convention*, or *hidden bit convention* is frequently used. It was already used by Zuse for his Z3 computer [92].

Hence, in radix 2, the significand of a normal number always has the form

$$1.m_1m_2m_3 \cdots m_{p-1},$$

whereas the significand of a subnormal number always has the form

$$0.m_1m_2m_3 \cdots m_{p-1}.$$

In both cases, the digit sequence $.m_1m_2m_3 \cdots m_{p-1}$ is called the *trailing significand* of the number. It is also sometimes called the *fraction*.

Some "extremal" floating-point numbers are important and will be used throughout this book:

- the smallest positive subnormal number is

$$\alpha = \beta^{e_{\min}-p+1};$$

[5]This is the *binary32* format of IEEE 754-2008, formerly called *single precision*. See Chapter 3 for details.

- the smallest positive normal number is $\beta^{e_{\min}}$;
- the largest finite floating-point number is

$$\Omega = (\beta - \beta^{1-p}) \cdot \beta^{e_{\max}}.$$

For example, still using the format ($\beta = 2$, $p = 24$, $e_{\min} = -126$, and $e_{\max} = +127$) of the example given above, the smallest positive subnormal number is

$$\alpha = 2^{-149} \approx 1.401 \times 10^{-45};$$

the smallest positive normal number is

$$2^{-126} \approx 1.175 \times 10^{-38};$$

and the largest finite floating-point number is

$$\Omega = \left(2 - 2^{-23}\right) \cdot 2^{127} \approx 3.403 \times 10^{+38}.$$

Subnormal numbers have probably been the most controversial part of IEEE 754-1985 [214, 553]. As stated by Schwarz, Schmookler, and Trong [549] (who present solutions for implementing subnormal numbers on FPUs), they are the most difficult type of numbers to implement in floating-point units. As a consequence, they are often implemented in software rather than in hardware, which may result in huge execution times when such numbers appear in a calculation.

One can of course define floating-point systems without subnormal numbers. And yet, the availability of these numbers allows for what Kahan calls *gradual underflow*: the loss of precision is slow instead of being abrupt. For example [249, 318], the availability of subnormal numbers implies the following interesting property: if a and b are floating-point numbers such that $a \neq b$, then the computed value of $b - a$ is necessarily nonzero.[6] This is illustrated by Figure 2.1. Gradual underflow is also sometimes called *graceful underflow*. In 1984, Demmel [160] analyzed various numerical programs, including Gaussian elimination, polynomial evaluation, and eigenvalue calculation, and concluded that the availability of gradual underflow significantly eases the writing of stable numerical software.

Many properties presented in Chapter 4, such as Sterbenz's lemma (Lemma 4.1), are true only if subnormal numbers are available (otherwise, we must add the condition *if no underflow occurs then...* in these properties).

2.1.3 A note on underflow

The word "underflow"can be ambiguous. For many naive users, it may mean that the exact result of an arithmetic operation has an absolute value below

[6]Note that this property is also true if a and b are subnormal numbers.

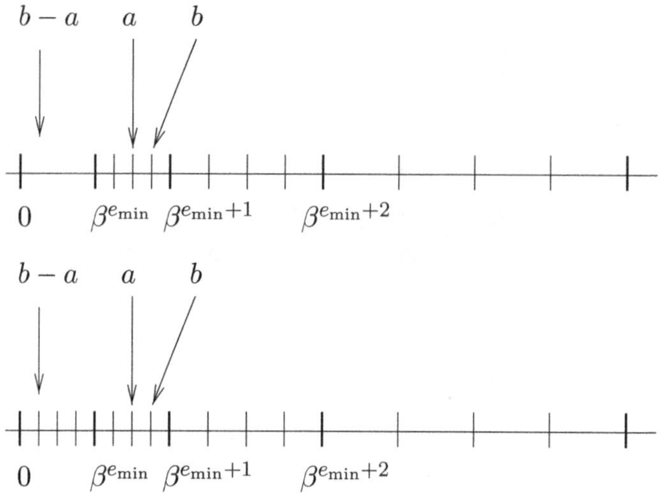

Figure 2.1: The positive floating-point numbers in the "toy system" $\beta = 2$ and $p = 3$. Above: normal floating-point numbers only. In that set, $b - a$ cannot be represented, so the computation of $b - a$ in round-to-nearest mode (see Section 2.2) will return 0. Below: the subnormal numbers are included in the set of floating-point numbers, and $b - a$ is one of them (as we shall see in Section 4.2.1)

the smallest nonzero *representable* number (that is, when subnormal numbers are available, $\alpha = \beta^{e_{\min}-p+1}$). *This is not the definition chosen in the context of floating-point arithmetic.* As a matter of fact, there are two slightly different definitions.

Definition 2.1 (Underflow before rounding). *In radix-β arithmetic, an arithmetic operation or function underflows if the exact result of that operation or function is of absolute value strictly less than $\beta^{e_{\min}}$.*

Definition 2.2 (Underflow after rounding). *In radix-β, precision-p arithmetic, an arithmetic operation or function underflows if the result we would compute with an* unbounded exponent range *and precision p would be nonzero and of absolute value strictly less than $\beta^{e_{\min}}$.*

For binary formats, unfortunately, the IEEE 754-1985 and 754-2008 standards did not make a choice between these two definitions (see Chapter 3). As a consequence, for the very same sequence of calculations, the underflow exception may be signaled on one "conforming" platform, and not signaled on another one. For decimal formats, however, the IEEE 754-2008 standard requires that underflow be detected *before* rounding.

Figure 2.2 illustrates Definitions 2.1 and 2.2 by pointing out (assuming round-to-nearest—see Section 2.2—and radix 2) the tiny domain in which they lead to a different conclusion. For instance in radix-2, precision-p arithmetic, as soon as $2e_{\min} < -p+1$ (which holds in all practical cases), $\sin(2^{e_{\min}})$ should underflow according to Definition 2.1, and should not underflow according to Definition 2.2.

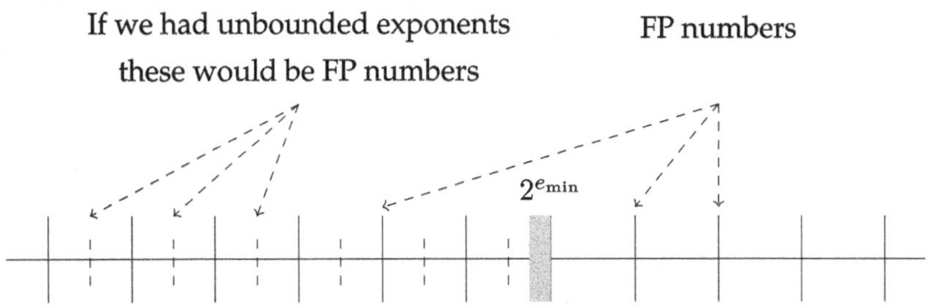

Figure 2.2: In this radix-2 example, if the exact result is in the grey area, then there is an underflow before rounding (Definition 2.1), and no underflow after rounding (Definition 2.2).

One should not worry too much about this ambiguity, as the two definitions disagree extremely rarely. What really matters is that when the result returned is subnormal and inexact,[7] an underflow is signaled. This is useful since it warns the user that the arithmetic operation that returned this result might be significantly less accurate (in terms of relative error) than usual. Indeed, we will see throughout this book that many algorithms and properties hold "provided that no underflow occurs."

2.1.4 Special floating-point data

Some data cannot be expressed as normal or subnormal numbers. An obvious example is the number zero, which requires a special encoding. There are also other examples that are not fully "numeric."

- It is in general highly desirable to have a closed system so that any machine operation can be well specified (without generating any trap[8]), even if it is an invalid operation over the real numbers (e.g., $\sqrt{-5}$ or $0/0$). A special datum, called NaN (*Not a Number*) in IEEE 754-1985 and its successors, can be introduced for this purpose. Any invalid operation will return a NaN.[9]

- Moreover, due to the limited exponent range, one needs to introduce more special data. To cope with values whose magnitude is larger than

[7] As noted by the working group who designed IEEE 754-1985, there is no point in warning the user when an exact result is returned.

[8] A trap is a transfer of control to a special handler routine.

[9] The IEEE 754-1985 standard and its followers actually define two kinds of NaN: *quiet* and *signaling* NaNs. See Section 3.1.7 for an explanation.

the maximum representable number, either an unsigned infinity (∞) or two signed infinities ($+\infty$ and $-\infty$) can be added to the floating-point system. If signed infinities are chosen, one may want signed zeros (denoted $+0$ and -0) too, for symmetry. The IEEE standards for floating-point arithmetic have two signed zeros and two signed infinities. As noted by Kahan [317], signed zeros also help greatly in dealing with branch cuts of complex elementary functions.

This choice yields an asymmetry for representing the exact zero and the result of $(+0) + (-0)$. To avoid this, the first solution that springs in mind is the introduction of a third, unsigned zero. As an example, the system of the Texas Instruments pocket calculators[10] has 3 zeros and 3 infinities: positive, negative, and with indeterminate sign. It is not certain that the benefit is worth the additional complexity.

Konrad Zuse's Z3 computer, built in 1941, already had mechanisms for dealing with floating-point exceptions [513]. The current choices will be detailed in Chapter 3.

2.2 Rounding

2.2.1 Rounding functions

In general, the result of an operation (or function) on floating-point numbers is not exactly representable in the floating-point system being used, so it has to be *rounded*. In the first floating-point systems, the way results were rounded was not always fully specified. One of the most interesting ideas brought out by the IEEE 754 standard is the concept of *correct rounding*. How a numerical value is rounded to a finite (or, possibly, infinite) floating-point number is specified by what is called a *rounding mode* in IEEE 754-1985, and *rounding direction attribute* in IEEE 754-2008, that defines a *rounding function* $\circ$. For example, when computing $a + b$, where a and b are floating-point numbers, the result returned is $\circ(a + b)$. One can define many possible rounding functions. The IEEE 754-2008 standard specifies 5 different rounding functions:

- round toward $-\infty$ (or "round downwards"): the rounding function RD is such that $\mathrm{RD}(x)$ is the largest floating-point number (possibly $-\infty$) less than or equal to x;

- round toward $+\infty$ (or "round upwards"): the rounding function RU is such that $\mathrm{RU}(x)$ is the smallest floating-point number (possibly $+\infty$) greater than or equal to x;

[10]http://tigcc.ticalc.org/doc/timath.html.

- round toward zero: the rounding function RZ is such that $\mathrm{RZ}(x)$ is the closest floating-point number to x that is no greater in magnitude than x (it is equal to $\mathrm{RD}(x)$ if $x \geq 0$, and to $\mathrm{RU}(x)$ if $x \leq 0$);
- two round to nearest functions $\mathrm{RN}_{\mathrm{even}}$ (round to nearest *ties to even*) and $\mathrm{RN}_{\mathrm{away}}$ (round to nearest *ties to away*) : for any floating-point number t, $|\,\mathrm{RN}_{\mathrm{even}}(x) - x|$ and $|\,\mathrm{RN}_{\mathrm{away}}(x) - x|$ are less than or equal to $|t - x|$. A tie-breaking rule must be chosen when x falls exactly halfway between two consecutive floating-point numbers:
  - $\mathrm{RN}_{\mathrm{even}}(x)$ is the only one of these two consecutive floating-point numbers whose integral significand is even. This is the default rounding function in the IEEE 754-2008 standard (see Chapter 3);
  - $\mathrm{RN}_{\mathrm{away}}(x)$ is the one of these two consecutive floating-point numbers whose magnitude is largest.

  When the tie-breaking rule does not matter (which is the case in general), we just write "RN" for designing one of these two round to nearest functions.

Figure 2.3 illustrates these rounding functions.

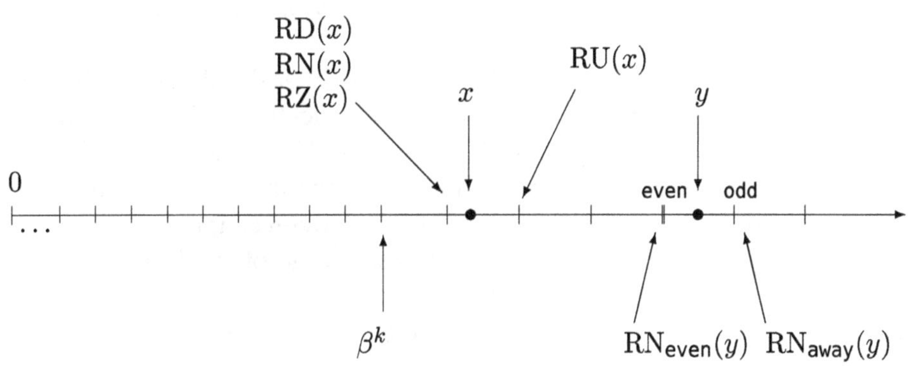

Figure 2.3: The standard rounding functions. Here we assume that the real numbers x and y are positive.

When the computed result of a function is always the same as if the function was first computed with infinite precision and unlimited range, then rounded according to the chosen rounding function, one says that the function is *correctly rounded*. The term *exactly rounded* is sometimes used [214].

In radix 2 and precision p, how a positive real value x greater than or equal to $2^{e_{\min}}$, whose infinitely precise significand is $1.m_1m_2m_3\ldots$, is rounded can be expressed as a function of the bit $round = m_p$ (*round bit*) and the bit $sticky = m_{p+1} \vee m_{p+2} \vee \cdots$ (*sticky bit*), as summarized in Table 2.1 (see Chapter 7).

round / sticky	RD	RU	RN
0 / 0	$-$	$-$	$-$
0 / 1	$-$	$+$	$-$
1 / 0	$-$	$+$	$-/+$
1 / 1	$-$	$+$	$+$

Table 2.1: Rounding a radix-2 infinitely precise significand, depending on the "round" and "sticky" bits. Let $\circ \in \{\mathrm{RN}, \mathrm{RD}, \mathrm{RU}\}$ be the rounding mode we wish to implement. A "$-$" in the table means that the significand of $\circ(x)$ is the truncated exact significand $1.m_1m_2m_3\ldots m_{p-1}$. A "$+$" in the table means that one needs to add 2^{-p+1} to this truncated significand (possibly leading to an exponent change if all the m_i's up to m_{p-1} are equal to 1). The "$-/+$" corresponds to the halfway cases for the round-to-nearest (RN) mode, where the rounded result depends on the chosen tie-breaking rule.

In the following, we will call a *rounding breakpoint* a value where the rounding function changes. In round-to-nearest mode, the rounding breakpoints are the exact middles of consecutive floating-point numbers. In the other rounding modes, called *directed rounding modes*, they are the floating-point numbers themselves.

Returning a correctly rounded result is fairly easily done for the arithmetic functions (addition, subtraction, multiplication, division) and the square root, as Chapters 7, 8, and 9 will show. This is why the IEEE 754-1985 and 754-2008 standards for floating-point arithmetic require that these functions be correctly rounded (see Chapter 3). However, it may be extremely difficult for some other functions: if the exact value of the function is very close to a rounding breakpoint, the function must be approximated with great accuracy to make it possible to decide which value must be returned. This problem, called the *Table maker's dilemma*, is addressed in Section 10.5.

When correct rounding is not guaranteed, if for any exact result y, one always returns either $\mathrm{RD}(y)$ or $\mathrm{RU}(y)$, one says that the returned value is a *faithful* result and that the arithmetic is faithful [158].

Caution: sometimes, this is called a "faithful rounding" in the literature, but this is not a *rounding function* as defined above, since the result obtained is not a fully specified function of the input value. With a faithful arithmetic, if the exact result of an arithmetic operation is a floating-point number, then that very result is returned. A correctly rounded arithmetic is always faithful.

2.2.2 Useful properties

As shown later (especially in Chapters 4 and 5), correct rounding is useful to design and prove algorithms and to find tight and verified error bounds.

An important and helpful property is that any of the five rounding functions presented above is *nondecreasing function*; i.e., if $\circ$ is the rounding function, if $x \le y$, then $\circ(x) \le \circ(y)$. Moreover, if y is a floating-point number, then $\circ(y) = y$, which means that when the exact result of a correctly rounded function is a floating-point number, we obtain that result exactly.

Also, if the rounding function is *symmetric* (i.e., it is RZ or RN with a symmetrical choice in case of a tie), then a correctly rounded implementation preserves the symmetries of the function being evaluated. With the other rounding functions, properties such as

$$\mathrm{RU}(a+b) = -\,\mathrm{RD}(-a-b)$$

or

$$\mathrm{RD}(a \times b) = -\,\mathrm{RU}((-a) \times b)$$

can sometimes be used for saving a change of rounding mode if this is a complicated or costly operation, as discussed in Section 12.3.3.

Finally, we note that in the case of tiny or huge values, the rounding functions of the IEEE 754-2008 standard behave as shown by the following property. This property is useful in particular when optimizing the implementation of correctly rounded arithmetic operators (see Chapters 7, 8, and 9).

Property 2.1. *With $\alpha = \beta^{e_{\min}-p+1}$ (smallest positive subnormal number) and $\Omega = (\beta - \beta^{1-p}) \cdot \beta^{e_{\max}}$ (largest finite floating-point number), one has the following values when rounding a real x:*

- $\mathrm{RN}_{even}(x) = \begin{cases} +0 & \text{if } 0 < x \le \alpha/2, \\ +\infty & \text{if } x \ge (\beta - \beta^{1-p}/2) \cdot \beta^{e_{\max}}; \end{cases}$

- $\mathrm{RN}_{away}(x) = \begin{cases} +0 & \text{if } 0 < x < \alpha/2, \\ +\infty & \text{if } x \ge (\beta - \beta^{1-p}/2) \cdot \beta^{e_{\max}}; \end{cases}$

- $\mathrm{RD}(x) = \begin{cases} +0 & \text{if } 0 < x < \alpha, \\ \Omega & \text{if } x \ge \Omega; \end{cases}$

- $\mathrm{RU}(x) = \begin{cases} \alpha & \text{if } 0 < x \le \alpha, \\ +\infty & \text{if } x > \Omega. \end{cases}$

2.3 Tools for Manipulating Floating-Point Errors

2.3.1 Relative error due to rounding

In the following we call the set of real numbers x such that $\beta^{e_{\min}} \le |x| \le \Omega$ the *normal range*, and the set of real numbers x such that $|x| < \beta^{e_{\min}}$ the *subnormal range*.

For a given rounding function $\circ$, when approximating a nonzero real number x by $\circ(x)$, the *relative error*

$$\epsilon(x) = \left|\frac{x - \circ(x)}{x}\right|$$

is introduced. This relative error is plotted in Figure 2.4 in a simple case. When $\circ(x) = x = 0$, we consider that the relative error is 0.

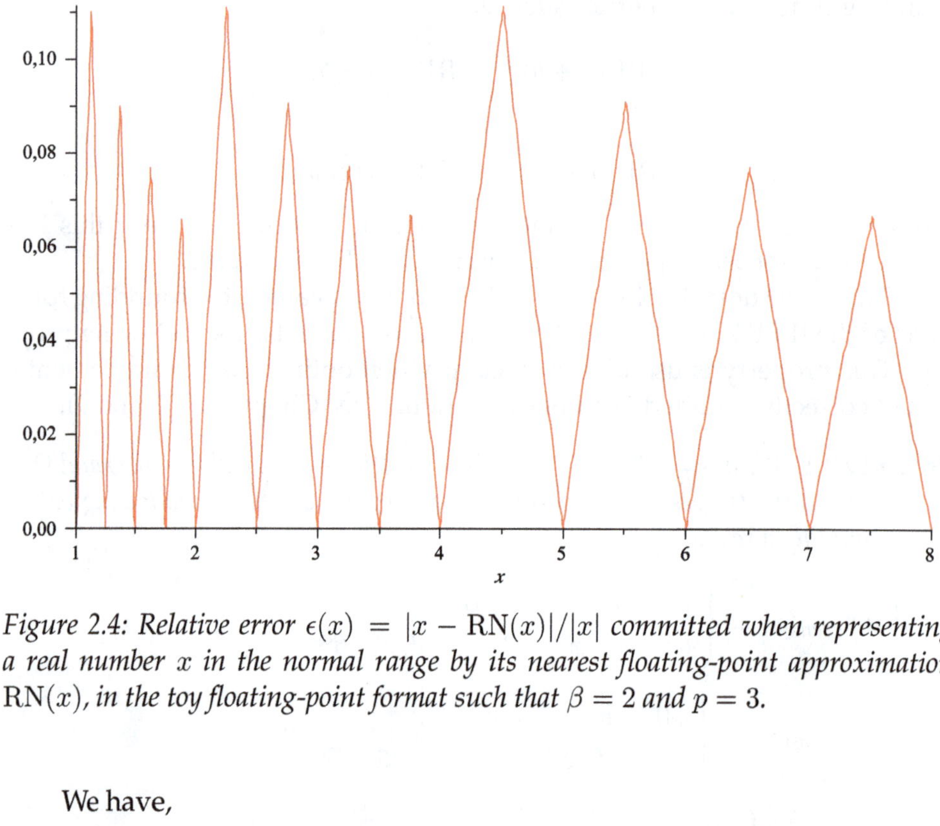

Figure 2.4: Relative error $\epsilon(x) = |x - \mathrm{RN}(x)|/|x|$ committed when representing a real number x in the normal range by its nearest floating-point approximation $\mathrm{RN}(x)$, in the toy floating-point format such that $\beta = 2$ and $p = 3$.

We have,

Theorem 2.2. *If x is in the normal range, then the relative error $\epsilon(x)$ satisfies*

$$\epsilon(x) < \begin{cases} \dfrac{1}{2}\beta^{1-p} & \text{if } \circ = \mathrm{RN} \text{ (rounding to nearest, with any choice in case of a tie)}, \\ \beta^{1-p} & \text{if } \circ \in \{\mathrm{RD}, \mathrm{RU}, \mathrm{RZ}\} \text{ (directed rounding)}. \end{cases}$$

These (strict) inequalities are well known and we give a proof for completeness only.

Proof. Since x is in the normal range, it is nonzero and there is a unique integer, say e, such that $\beta^e \le |x| < \beta^{e+1}$. If $|x| = \beta^e$ then x is a floating-point number and thus $\epsilon(x) = 0$. Assume now that $|x| > \beta^e$. The distance between two consecutive floating-point numbers in either $[\beta^e, \beta^{e+1}]$ or $[-\beta^{e+1}, -\beta^e]$ is exactly β^{e-p+1}, so the absolute error due to rounding satisfies

$$|x - \mathrm{RN}(x)| \le \frac{1}{2}\beta^{e-p+1} \tag{2.2}$$

for rounding to nearest, and

$$|x - \circ(x)| \le \beta^{e-p+1}$$

otherwise. Dividing by $|x|$ such that $|x| > \beta^e$ then yields the announced bounds on the relative error $\epsilon(x)$. □

The value $\frac{1}{2}\beta^{1-p}$ or β^{1-p} (depending on the rounding function) is so frequently used that a notation has been widely adopted for representing it:

Definition 2.3 (Unit roundoff). *The* unit roundoff $\mathbf{u}$ *of a radix-β, precision-p, floating-point system is defined as*

$$\mathbf{u} = \begin{cases} \frac{1}{2}\,\mathrm{ulp}(1) = \frac{1}{2}\beta^{1-p} & \text{in round-to-nearest mode,} \\ \mathrm{ulp}(1) = \beta^{1-p} & \text{in directed rounding modes.} \end{cases}$$

This notion is widespread in the analysis of numerical algorithms. See for instance the excellent book by Higham [258] (for a more general definition, we also refer to [530]).

Note that for rounding to nearest one can show the following slightly smaller bound on $|(x - \mathrm{RN}(x))/x|$ (see [342]):

Theorem 2.3. *If x is in the normal range, then the relative error $\epsilon(x)$ introduced when rounding x to nearest satisfies*

$$\epsilon(x) \le \frac{\beta^{1-p}}{2+\beta^{1-p}} = \frac{\mathbf{u}}{1+\mathbf{u}}, \tag{2.3}$$

Proof. Using the same notation as in the proof of Theorem 2.2, we can show (2.3) easily as follows. If $|x| \ge (1 + \frac{1}{2}\beta^{1-p})\beta^e$, then it suffices to apply (2.2). Otherwise, $\beta^e \le |x| < (1 + \frac{1}{2}\beta^{1-p})\beta^e$ and the two consecutive floating-point numbers surrounding $|x|$ are $f = \beta^e$ and $g = (1 + \beta^{1-p})\beta^e$, with f being closest to $|x|$. Hence $\mathrm{RN}(x) = \mathrm{sign}(x) \cdot f$ and the absolute error is given by $|x - \mathrm{RN}(x)| = |x| - f = |x| - \beta^e$. This implies

$$\epsilon(x) = 1 - \frac{\beta^e}{|x|} < 1 - \frac{\beta^e}{(1+\frac{1}{2}\beta^{1-p})\beta^e} = \frac{\beta^{1-p}}{2+\beta^{1-p}},$$

which concludes the proof. □

The bound given by Theorem 2.3 is valid for any tie-breaking rule. It is attained for example at the rounding breakpoint $x = 1 + \mathbf{u}$.

If x is in the subnormal range (assuming subnormal numbers are available), then the relative error $|(x - \circ(x))/x|$ introduced when rounding x can become very large (it can be close to 1). In that case, we have a bound on the *absolute* error due to rounding:

$$|x - \mathrm{RN}(x)| \leq \frac{1}{2}\beta^{e_{\min}-p+1} \tag{2.4}$$

in round-to nearest mode, and

$$|x - \circ(x)| < \beta^{e_{\min}-p+1} \tag{2.5}$$

if $\circ \in \{\mathrm{RD}, \mathrm{RU}, \mathrm{RZ}\}$.

The "standard models" of floating-point arithmetic are a consequence of Theorem 2.2: for any arithmetic operation $\top \in \{+, -, \times, \div\}$, for any rounding function $\circ \in \{\mathrm{RN}, \mathrm{RD}, \mathrm{RU}, \mathrm{RZ}\}$, and for all floating-point numbers a, b such that $a \top b$ does not underflow[11] nor overflow, we have

$$\begin{aligned} \circ(a \top b) &= (a \top b)(1 + \epsilon_1), \qquad |\epsilon_1| \leq \mathbf{u}, &(2.6)\\ &= (a \top b)/(1 + \epsilon_2), \qquad |\epsilon_2| \leq \mathbf{u}. &(2.7) \end{aligned}$$

These two equalities, which also apply to the square root and fused multiply-add (FMA) operations, are called the first and second *standard models* of floating-point arithmetic. They are the most classically used properties in the derivation of rounding error bounds [258], and we shall give a few examples of such analyses in Chapter 5.

In the case of rounding to nearest, that is, when $\circ = \mathrm{RN}$ and $\mathbf{u} = \frac{1}{2}\beta^{1-p}$, it follows from Theorem 2.3 that the first standard model in (2.6) can be refined slightly:

$$\mathrm{RN}(a \top b) = (a \top b)(1 + \epsilon_1), \qquad |\epsilon_1| \leq \frac{\mathbf{u}}{1 + \mathbf{u}}.$$

As noted before, this refined bound is sometimes attained (for example when $a \top b$ can be equal to $1 + \mathbf{u}$). However, for some operations, this may not be the case any longer: for example, for division in radix 2, the best possible bound on $|\epsilon_1|$ is not $\mathbf{u}/(1 + \mathbf{u}) = \mathbf{u} - \mathbf{u}^2 + \mathbf{u}^3 - \cdots$ but $\mathbf{u} - 2\mathbf{u}^2$. We refer to [303] for a proof of this fact and for similar optimal bounds for all the basic operations.

The standard model and the absolute error bounds 2.4 and 2.5 can be combined: we find that if z is the result of the correctly rounded operation $a \top b$ (that is, if $z = \circ(a \top b)$), and if no overflow occurs, then

$$z = (a \top b)(1 + \epsilon) + \epsilon',$$

with $|\epsilon| \leq \mathbf{u}$ and $|\epsilon'| \leq \beta^{e_{\min}} \cdot \mathbf{u}$. Moreover, ϵ and ϵ' cannot both be nonzero [160, 318]. One should note that

[11] We consider that $a \top b$ underflows when it is in the subnormal range and the operation is inexact (i.e., $a \top b$ is not equal to a floating-point number).

- if z is in the normal range (i.e., if no underflow occurred) then $\epsilon' = 0$;
- if z is in the subnormal range, then $\epsilon = 0$. Moreover, in this case, if the arithmetic operation being performed is addition or subtraction ($\top$ is $+$ or $-$), then we will see (Chapter 4, Theorem 4.2) that the result is *exact*, so that $z = a \top b$ (i.e., $\epsilon' = 0$).

The bounds given here on the errors due to rounding will be used in particular in Chapters 5 and 11.

2.3.2 The ulp function

In numerical analysis, errors are very often expressed in terms of relative errors. And yet, when we want to express the errors of "nearly atomic" functions (arithmetic operations, elementary functions, small polynomials, sums, and inner products, etc.), it is more appropriate (and frequently more accurate) to express errors in terms of what we would intuitively define as the "weight of the last digit of the significand." Let us define that notion more precisely. The term ulp, which is an acronym for *unit in the last place*, was coined by William Kahan in 1960. The original definition was as follows [321]:

> $\mathrm{ulp}(x)$ is the gap between the two floating-point numbers nearest to x, even if x is one of them.

When x is a floating-point number (except, possibly, when x is a power of the radix; see below), we would like $\mathrm{ulp}(x)$ to be equal to the *quantum* of x. However, it is frequently useful to define that function for other numbers too.

Several slightly different definitions of $\mathrm{ulp}(x)$ appear in the literature [214, 240, 406, 274, 477, 321]. They all coincide whenever x is not extremely close to a power of the radix. They have properties that differ to a small degree. A good knowledge of these properties may be important, for instance, for anyone who wants to prove sure yet tight bounds on the errors of atomic computations. For instance, we frequently hear or read that correctly rounding to nearest is equivalent to having an error less than 0.5 ulp. This might be true, depending on the radix, on what we define as an ulp (especially near the powers of the radix), and on whether we consider the ulp function to be taken at the real value being approximated, or at the floating-point value that approximates it.

Consider first the following definition of the ulp function, due to John Harrison [240, 243].

Definition 2.4 (Harrison). $\mathrm{ulp}(x)$ *is the distance between the closest straddling floating-point numbers* a *and* b *(i.e., those with* $a \le x \le b$ *and* $a \ne b$*), assuming that the exponent range is not upper bounded.*

Figure 2.5 shows the values of Harrison's ulp near 1. One can easily confirm that, in radix β floating-point arithmetic, if x is a floating-point number

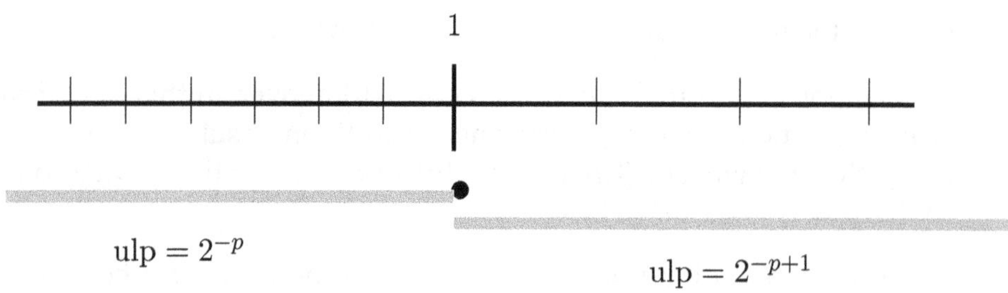

Figure 2.5: The values of $\mathrm{ulp}(x)$ *near* 1*, assuming a binary floating-point system with precision* p*, according to Harrison's definition.*

then Harrison's ulp of x and the quantum of x have the same value, except if x is an integer power of β. Goldberg [214] gives another definition the ulp function.

Definition 2.5 (Goldberg). *If the FP number* $d_0.d_1d_2d_3d_4\ldots d_{p-1}\beta^e$ *is used to represent* x*, it is in error by*

$$\left| d_0.d_1d_2d_3d_4\ldots d_{p-1} - \frac{x}{\beta^e} \right|$$

units in the last place.

This definition does not define ulp as a function of x, since the value depends on which floating-point number approximates x. However, it clearly defines a function $\mathrm{ulp}(X)$, for a *floating-point* number $X \in [\beta^e, \beta^{e+1})$, as β^{e-p+1} (or, more precisely, as $\beta^{\max(e,e_{\min})-p+1}$, if we want to handle the subnormal numbers properly). Hence, a natural generalization to real numbers is the following, which is equivalent to the one given by Cornea, Golliver, and Markstein[12] [120, 406].

Definition 2.6 (Goldberg's definition, extended to reals). *If* $|x| \in [\beta^e, \beta^{e+1})$*, then* $\mathrm{ulp}(x) = \beta^{\max(e,e_{\min})-p+1}$.

When x is a floating-point number, this definition coincides with the *quantum* of x. Figure 2.6 shows the values of ulp near 1 according to Definition 2.6.

Let us now examine some properties of these definitions (see [439] for comments and some proofs). In the following, x is a real number and X is a radix-β, precision-p floating-point number, $\mathrm{HarrisonUlp}(x)$ is $\mathrm{ulp}(x)$ according to Harrison's definition, and $\mathrm{GenGoldbergUlp}(x)$ is $\mathrm{ulp}(x)$ according to Goldberg's definition extended to the reals. We assume that $|x|$ is less than or equal to the largest representable number, $\Omega = (\beta - \beta^{1-p}) \cdot \beta^{e_{\max}}$.

[12]They stated the definition in radix 2, but generalization to radix β is straightforward.

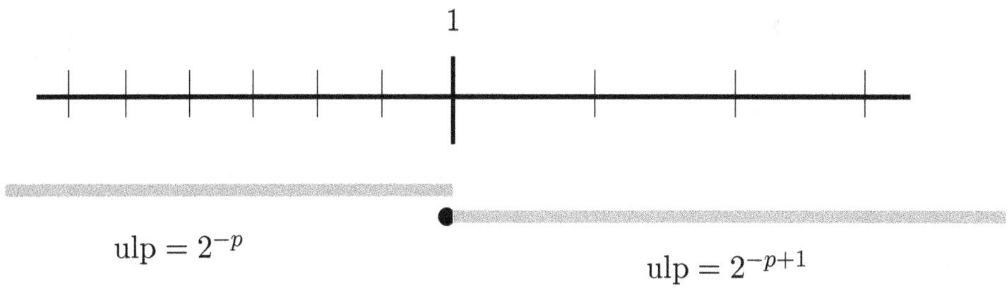

Figure 2.6: The values of $\mathrm{ulp}(x)$ *near* 1*, assuming a binary floating-point system with precision* p*, according to Definition 2.6 (Goldberg's definition extended to the reals). Notice that this definition and Harrison's only differ when* x *is a power of the radix.*

Property 2.4. *In radix 2,*

$$|X - x| < \frac{1}{2}\,\mathrm{HarrisonUlp}(x) \Rightarrow X = \mathrm{RN}(x).$$

It is important to note that Property 2.4 is not true in radices greater than or equal to 3. Figure 2.7 gives a counterexample in radix 3.

If, instead of considering ulps of the "exact" value x, we consider ulps of the floating-point value X, we have a property that is very similar to Property 2.4, with the interesting difference that now it holds for any value of the radix.

Property 2.5. *For any value of the radix* β*,*

$$|X - x| < \frac{1}{2}\,\mathrm{HarrisonUlp}(X) \Rightarrow X = \mathrm{RN}(x).$$

Now, still with Harrison's definition, we might be interested in knowing if the converse property holds; that is, if $X = \mathrm{RN}(x)$ implies that X is within $\frac{1}{2}\,\mathrm{HarrisonUlp}(x)$ or $\frac{1}{2}\,\mathrm{HarrisonUlp}(X)$ of x. For the first case, we have the following.

Property 2.6. *For any radix,*

$$X = \mathrm{RN}(x) \Rightarrow |X - x| \le \frac{1}{2}\,\mathrm{HarrisonUlp}(x).$$

On the other hand, there is no similar property for the second case: $X = \mathrm{RN}(x)$ does not imply $|X - x| \le \frac{1}{2}\,\mathrm{HarrisonUlp}(X)$. For example, assume radix 2. Any number x strictly between $1 + 2^{-p-1}$ and $1 + 2^{-p}$ will round to 1, but it will be at a distance from 1 larger than $\frac{1}{2}\,\mathrm{HarrisonUlp}(1) = 2^{-p-1}$.

Concerning Goldberg's definition extended to the reals, we have very similar properties.

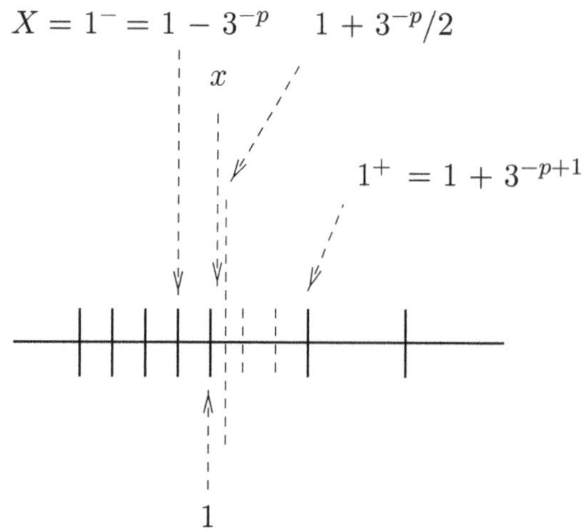

Figure 2.7: This example shows that Property 2.4 is not true in radix 3. Here, x satisfies $1 < x < 1 + \frac{1}{2}3^{-p}$ and $X = 1^- = 1 - 3^{-p}$ (if v is a floating-point number, v^- denotes its predecessor, namely, the largest floating-point number less than v, and v^+ denotes its successor). We have $\mathrm{HarrisonUlp}(x) = 3^{-p+1}$, and $|x - X| < 3^{-p+1}/2$, so that $|x - X| < \frac{1}{2}\,\mathrm{HarrisonUlp}(x)$. However, $X \neq \mathrm{RN}(x)$.

Property 2.7. *In radix 2,*

$$|X - x| < \frac{1}{2}\mathrm{GenGoldbergUlp}(x) \Rightarrow X = \mathrm{RN}(x).$$

Property 2.7 is not true in higher radices: The example of Figure 2.7, designed as a counterexample to Property 2.4, is also a counterexample to Property 2.7.

Also, Property 2.7 does not hold if we replace $\mathrm{GenGoldbergUlp}(x)$ by $\mathrm{GenGoldbergUlp}(X)$. Indeed, $|X - x| < \frac{1}{2}\mathrm{GenGoldbergUlp}(X)$ does not imply $X = \mathrm{RN}(x)$ (it suffices to consider x very slightly above $1^- = 1 - \beta^{-p}$, the floating-point predecessor of 1: x will be within $\frac{1}{2}\mathrm{GenGoldbergUlp}(1)$ from 1, and yet $\mathrm{RN}(x) = 1^-$). In a way, this kind of counterexample is the only one; see Property 2.8.

Property 2.8. *For any radix, if X is not an integer power of β,*

$$|X - x| < \frac{1}{2}\mathrm{GenGoldbergUlp}(X) \Rightarrow X = \mathrm{RN}(x).$$

We also have the following.

Property 2.9. *For any radix,*

$$X = \mathrm{RN}(x) \Rightarrow |X - x| \leq \frac{1}{2}\mathrm{GenGoldbergUlp}(x).$$

Property 2.10. *For any radix,*

$$X = \mathrm{RN}(x) \Rightarrow |X - x| \le \frac{1}{2}\mathrm{GenGoldbergUlp}(X).$$

After having considered properties linked to the round-to-nearest mode, we can try to consider properties linked to the directed rounding modes (i.e., rounding toward $\pm\infty$ and rounding toward zero). One can show the following properties (still assuming $|x|$ is less than or equal to the largest representable number).

Property 2.11. *For any value of the radix β,*

$$X \in \{\mathrm{RD}(x), \mathrm{RU}(x)\} \Rightarrow |X - x| < \mathrm{HarrisonUlp}(x).$$

Note that the converse is not true. There are values X and x for which $|X - x| < \mathrm{HarrisonUlp}(x)$, and X is not in $\{\mathrm{RD}(x), \mathrm{RU}(x)\}$. It suffices to consider the case x slightly above 1 and X equal to $1^- = 1 - \beta^{-p}$, the floating-point predecessor of 1. However, if we consider $\mathrm{HarrisonUlp}(X)$ instead of $\mathrm{HarrisonUlp}(x)$, we have

Property 2.12. *For any value of the radix β,*

$$|X - x| < \mathrm{HarrisonUlp}(X) \Rightarrow X \in \{\mathrm{RD}(x), \mathrm{RU}(x)\}.$$

But the converse is not true: $X \in \{\mathrm{RD}(x), \mathrm{RU}(x)\}$ does not imply $|X - x| \le \mathrm{HarrisonUlp}(X)$.

Property 2.13.

$$X \in \{\mathrm{RD}(x), \mathrm{RU}(x)\} \Rightarrow |X - x| \le \mathrm{GenGoldbergUlp}(x).$$

The converse is not true: $|X - x| < \mathrm{GenGoldbergUlp}(x)$ does not imply $X \in \{\mathrm{RD}(x), \mathrm{RU}(x)\}$. It suffices to consider $X = 1^- = 1 - \beta^{-p}$, the floating-point predecessor of 1, and x slightly above 1.

Property 2.14.

$$X \in \{\mathrm{RD}(x), \mathrm{RU}(x)\} \Rightarrow |X - x| \le \mathrm{GenGoldbergUlp}(X).$$

Again, the converse is not true: $|X - x| < \mathrm{GenGoldbergUlp}(X)$ does not imply $X \in \{\mathrm{RD}(x), \mathrm{RU}(x)\}$.

After this examination of the properties of these two definitions of the ulp function, which one is to be chosen? A good definition of function ulp:

- should (of course) agree with the "intuitive" notion when x is not in an "ambiguous area" (i.e., x is not very close to a power of the radix);

- should be *useful*: after all, for a binary format with precision p, defining $\mathrm{ulp}(1)$ as 2^{-p} (i.e., $1 - 1^-$) or 2^{-p+1} (i.e., $1^+ - 1$) are equally legitimate from a theoretical point of view. What matters is which choice is helpful (i.e., which choice will preserve in "ambiguous areas" properties that are true when we are far enough from them).

From this point of view, it is still not very easy to decide between Definitions 2.4 and 2.6. Both preserve interesting properties, yet also set some traps (e.g., the fact that $X = \mathrm{RN}(x)$ does not imply $|X - x| \le \frac{1}{2}\,\mathrm{HarrisonUlp}(X)$, or the fact that $|X - x| < \mathrm{GenGoldbergUlp}(x)$ does not imply $X \in \{\mathrm{RD}(x), \mathrm{RU}(x)\}$). These traps sometimes make the task of proving properties of arithmetic algorithms a difficult job when some operand can be very near a power of the radix.

In the remainder of this book, $\mathrm{ulp}(x)$ *will be* $\mathrm{GenGoldbergUlp}(x)$. That is, we will follow Definition 2.6, not because it is the best (as we have seen, it is difficult to tell which one is the best), but because it is the most used.

The quantity $\mathrm{ulp}(1) = \beta^{1-p}$, which gives the distance between 1 and its floating-point successor, is also called *machine epsilon*. In particular, this quantity is the one returned by MATLAB's function `eps`; see [256]. It thus coincides with the above notion of unit roundoff in the case of the directed rounding modes, but not in the case of rounding to nearest.

Note finally that rounding error analyses can also be carried out using the related notion of *unit in the first place* (ufp), introduced in [531] and defined for any real number x as follows:

$$\mathrm{ufp}(x) = \begin{cases} 0 & \text{if } x = 0, \\ \beta^{\lfloor \log_\beta |x| \rfloor} & \text{if } x \neq 0. \end{cases}$$

Thus, if x is nonzero, its unit in the first place is the largest integral power of the radix that is less than or equal to $|x|$. In particular, if x is a real lying in the normal range of a given floating-point number set (with radix β and precision p), then its ufp and its ulp are related to each other via the equality

$$\mathrm{ufp}(x) = \beta^{p-1} \cdot \mathrm{ulp}(x).$$

2.3.3 Link between errors in ulps and relative errors

It is important to be able to establish links between errors expressed in ulps, and relative errors. Inequalities (2.8) and (2.10) below exhibit such links when X is a normal floating-point number and x is a real number of the same sign.

2.3.3.1 Converting from errors in ulps to relative errors

First, let us convert from errors in ulps to relative errors. Assume that $|x - X| = \alpha\,\mathrm{ulp}(x)$, and that x is in the normal range. Assuming no underflow, we easily get

$$\left|\frac{x-X}{x}\right| \le \alpha \times \beta^{1-p}. \tag{2.8}$$

2.3.3.2 Converting from relative errors to errors in ulps

Now, let us convert from relative errors to errors in ulps. A relative error

$$\epsilon_r = \left|\frac{x-X}{x}\right| \tag{2.9}$$

implies an error in ulps bounded by

$$|x-X| \le \epsilon_r \beta^p \,\mathrm{ulp}(x). \tag{2.10}$$

Hence, one can easily switch from an error in ulps to a relative error, and conversely. This is convenient, since for the correctly rounded arithmetic operations and functions, we have an error in ulps, whereas it is generally much easier to deal with relative errors for performing error calculations. A typical example of this is detailed in the Section 2.3.4.

2.3.3.3 Loss of information during these conversions

One must keep in mind that each time we switch from one form of error to the other, we lose some information. For instance, a correctly rounded-to-nearest operation returns a result X within $0.5\,\mathrm{ulp}$ of the exact value x. This implies a relative error bounded by 2^{-p}. Figure 2.8 shows, in the interval $[-1/2, 8]$, the bound on the distance between x and X one can infer from the information in terms of ulps and the information in terms of relative errors. We immediately see that we have lost some information in the conversion.

When converting from relative errors to errors in ulps, some information is lost too. This is illustrated by Figure 2.9.

2.3.4 An example: iterated products

Assume that we iteratively compute the product of the floating-point numbers $x_1, x_2, \ldots, x_n$ in binary, precision-p, rounded-to-nearest floating-point arithmetic. That is, we perform

$P \leftarrow x_1$
for $i = 2$ to n **do**
  $P \leftarrow \mathrm{RN}(P \times x_i)$
end for
return P

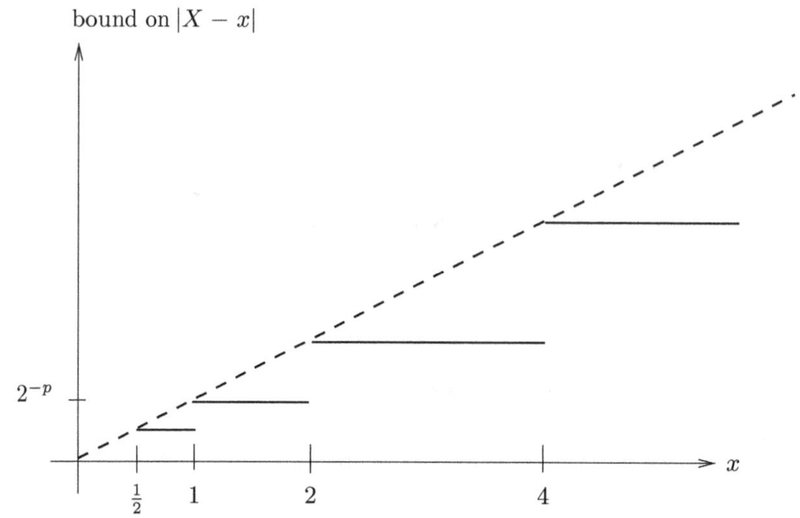

Figure 2.8: Conversion from ulps to relative errors. Assume we know that an operation is correctly rounded: the computed result X is within $0.5\,\mathrm{ulp}$ of the exact result x. This implies that $|x - X|$ is below the bold (noncontinuous) curve. Converted in terms of relative errors, this information becomes $X = x(1+\epsilon)$, with $|\epsilon| \leq 2^{-p}$, i.e., $|x - X|$ is below the dashed curve. This last property is less accurate.

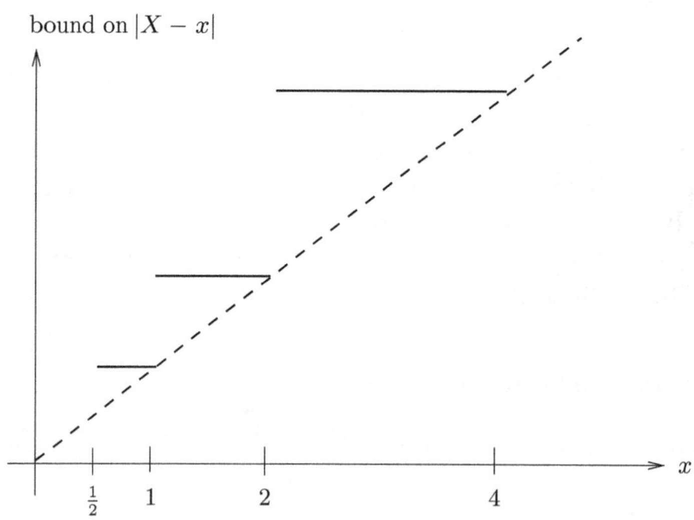

Figure 2.9: Conversion from relative errors to ulps. Assume we have a bound on the relative error between an exact value x and a floating-point approximation X (dashed curve). From it, we can infer an error in ulps that implies that X is below the bold curve. This last property is less accurate.

Each multiplication is correctly rounded, leading to an error less than or equal to $(1/2)$ ulp. However, reasoning in terms of ulps will lead to calculations that are much too complex. By contrast, if we assume that no underflows or overflows occur, a simple reasoning with relative errors shows that the final value of P satisfies

$$P = x_1 x_2 \cdots x_n \times K,$$

where

$$\left(1 - 2^{-p}\right)^{n-1} \leq K \leq \left(1 + 2^{-p}\right)^{n-1},$$

which means that the relative error of the result is upper bounded by

$$\left(1 + 2^{-p}\right)^{n-1} - 1,$$

which is close to (and larger than) $(n-1) \times 2^{-p}$ as long as $n \ll 2^p$. Note that it was shown recently in [527] that if $n - 1 < 2^{p/2}$, then the relative error of the result is always upper bounded by $(n-1) \times 2^{-p}$.

2.4 The Fused Multiply-Add (FMA) Instruction

The FMA instruction was introduced in 1990 on the IBM RS/6000 processor to facilitate correctly rounded software division and to make some calculations (especially dot products and polynomial evaluations) faster and in general more accurate. It also facilitates the calculation of the error of a floating-point multiplication.

Definition 2.7 (FMA instruction). *Assume that the rounding function is* $\circ$, *and that* a, b, *and* c *are floating-point numbers. Then* FMA(a, b, c) *is* $\circ(a \cdot b + c)$.

Some algorithms facilitated by the availability of this instruction are presented in Chapters 4 and 11. A brief discussion on current implementations is given in Section 3.4.2.

The IEEE 754-2008 standard for floating-point arithmetic specifies the FMA instruction. This was not the case in the previous version of the standard.

2.5 Exceptions

In IEEE 754 arithmetic, an *exception* can be signaled along with the result of an operation. The default is to raise a status flag (which must be "sticky," so that the user does not need to check it immediately, but can do so after a sequence of operations, for instance at the end of the evaluation of a function).

Invalid: This exception is signaled when an input is invalid for the function. The result is a NaN (when supported). Examples: $(+\infty) - (+\infty)$, $0/0$, $\sqrt{-1}$.

DivideByZero: This exception is signaled when an exact infinite result is defined for a function on finite inputs, e.g., at a pole. Examples: $1/0$, $\log(+0)$.

Overflow: This exception is signaled when the rounded result with an unbounded exponent range would have an exponent larger than $e_{\max}$.

Underflow: This exception is signaled when a tiny nonzero result is detected. This can be according to either Definition 2.1 (i.e., before rounding) or Definition 2.2 (i.e., after rounding).

Furthermore, underflow handling can be different depending on whether the exact result is exactly representable or not, which makes sense: the major interest in signaling the underflow exception is to warn the user that the result obtained might not be very accurate (in terms of relative error). Of course, this is not at all the case when the result obtained is exact. This is why, in the IEEE 754-2008 standard, if the result of an operation is exact, the underflow flag is *not* raised (see Chapter 3).

Inexact: This exception is signaled when the exact result y is not exactly representable, that is, when $\circ(y) \neq y$ with y not being a NaN.

In general, the values of $e_{\min}$ and $e_{\max}$ are chosen to be almost symmetrical: $e_{\min} \approx -e_{\max}$. One of the reasons for that is that we expect an acceptable behavior of the reciprocal function $1/x$. The minimum positive normal number is $\beta^{e_{\min}}$. Its reciprocal is $\beta^{-e_{\min}}$, which is below the maximum normal number threshold as long as $-e_{\min} \leq e_{\max}$, i.e., $e_{\min} \geq -e_{\max}$. So, if one chooses $e_{\min} = -e_{\max}$ or $e_{\min} = 1 - e_{\max}$ (which is the choice in IEEE 754-2008, probably for parity reasons: the number of different exponents that can be represented in a given binary integer format is an even number, whereas if $e_{\min}$ were exactly equal to $-e_{\max}$, we would have an odd number of exponents), one has the following properties.

- If x is a normal number, $1/x$ never produces an overflow.
- If x is a finite floating-point number, $1/x$ can underflow, but the rounded result is not zero (for common values of p), if subnormal numbers are available.

Referring to [164] and [264], where algorithms take advantage of the available exception-handling system, Hauser [249] explains that "it is often both easier and cheaper to respond to an exception after-the-fact than to prevent the exception from occurring in the first place." He also notes that, on the other hand, when exception handling is not available, avoiding exceptional cases in programs requires artful numerical tricks that can make programs slower and much less clear. Hauser gives the example of the calculation of the Euclidean norm

$$\mathcal{N} = \sqrt{\sum_{i=1}^{N} x_i^2},$$

where the x_i's are floating-point numbers. Consider computing $\mathcal{N}$ using a straightforward algorithm (Algorithm 2.1).

Algorithm 2.1 Straightforward calculation of $\sqrt{\sum_{i=1}^{N} x_i^2}$.

```
Inputs N, x_1, x_2, ..., x_N
S ← 0.0
for i = 1 to N do
  S ← RN(S + RN(x_i × x_i))
end for
return RN(√S)
```

Even when the exact value of $\mathcal{N}$ lies in the normal range of the floating-point format being used, Algorithm 2.1 may return $+\infty$ or a very inaccurate result, due to underflow or overflow when evaluating the square of one or several of the x_i's, or when evaluating their sum. There are several solutions for avoiding this. We may for instance emulate an extended range arithmetic or scale the operands (i.e., first examine the input operands, then multiply them by an adequate factor K, so that Algorithm 2.1 can be used with the scaled operands[13] without underflow or overflow, and finally divide the result obtained by K).

These solutions would lead to reliable programs, but the extended range arithmetic as well as the scaling would significantly slow down the calculations. This is unfortunate since, in the vast majority of practical cases, Algorithm 2.1 would have behaved satisfactorily. A possibly better solution, when alternate exception handling is available (see Section 3.1.2.2 and [267]), is to first use Algorithm 2.1, and then to resort to extended range arithmetic or a scaling technique only if overflows or underflows occurred during that preliminary computation.

[13]Notice that Hammarling's algorithm for routine xNRM2 in LAPACK [258, Problem 27.5] goes only once through the x_i's but nevertheless avoids overflow. This is done by computing the scaling factor K on the fly while computing the norm.

2.6 Lost and Preserved Properties of Real Arithmetic

The arithmetic of real numbers has several well-known properties. Among them:

- addition and multiplication are commutative: $a + b = b + a$ and $a \times b = b \times a$ for all a and b;
- addition and multiplication are associative: $a + (b + c) = (a + b) + c$ and $a \times (b \times c) = (a \times b) \times c$ for all a, b, and c;
- distributivity: $a \times (b + c) = a \times b + a \times c$.

When the arithmetic operations are correctly rounded, with any of the rounding functions presented in Section 2.2, floating-point addition and multiplication remain commutative[14]: if $\circ$ is the rounding function then, of course, $\circ(a + b) = \circ(b + a)$ and $\circ(a \times b) = \circ(b \times a)$ for all floating-point numbers a and b. However, associativity and distributivity are lost. More precisely, concerning associativity, the following can occur.

- In some extreme cases, $\circ(a + \circ(b + c))$ can be *drastically* different from $\circ(\circ(a+b)+c)$. A simple example, in radix-β, precision-p arithmetic with round-to-nearest mode is $a = \beta^{p+1}$, $b = -\beta^{p+1}$, and $c = 1$, since

  $$\mathrm{RN}(a + \mathrm{RN}(b + c)) = \mathrm{RN}(\beta^{p+1} - \beta^{p+1}) = 0,$$

  whereas

  $$\mathrm{RN}(\mathrm{RN}(a + b) + c) = \mathrm{RN}(0 + 1) = 1.$$

  Many studies have been devoted to finding good ways of reordering the operands when one wants to evaluate the sum of several floating-point numbers. See Chapter 5 for more details.

- If no overflow or underflow occurs, then $P_1 = \circ(\circ(a \times b) \times c)$ can differ from $P_2 = \circ(a \times \circ(b \times c))$ but only *slightly*. More precisely, in radix-β, precision-p arithmetic with round-to-nearest mode,

  $$\mathrm{RN}(a \times b) = a \times b \times (1 + \epsilon_1),$$

  with $|\epsilon_1| \leq \frac{1}{2}\beta^{1-p}$, so that

  $$P_1 = \mathrm{RN}(\mathrm{RN}(a \times b) \times c) = a \times b \times c \times (1 + \epsilon_1)(1 + \epsilon_2),$$

  with $|\epsilon_2| \leq \frac{1}{2}\beta^{1-p}$. Similarly,

  $$P_2 = \mathrm{RN}(a \times \ \mathrm{RN}(b \times c)) = a \times b \times c \times (1 + \epsilon_3)(1 + \epsilon_4),$$

[14]However, in a programming language, swapping the terms may yield a different result in practice. This can be noticed in the expression `a * b + c * d` when a fused multiply-add (FMA) instruction is used; see, e.g., Section 6.2.3.2.

with $|\epsilon_3|, |\epsilon_4| \leq \frac{1}{2}\beta^{1-p}$. Therefore,

$$\left(\frac{1-\frac{1}{2}\beta^{1-p}}{1+\frac{1}{2}\beta^{1-p}}\right)^2 \leq \frac{P_1}{P_2} \leq \left(\frac{1+\frac{1}{2}\beta^{1-p}}{1-\frac{1}{2}\beta^{1-p}}\right)^2,$$

which gives

$$\frac{P_1}{P_2} = 1 + \epsilon,$$

with

$$|\epsilon| \leq \sum_{i=1}^{+\infty} \frac{i}{2^{i-2}} \cdot \beta^{i-ip} = 2\,\beta^{1-p} + 2\,\left(\beta^{1-p}\right)^2 + 3/2\,\left(\beta^{1-p}\right)^3 + \cdots. \quad (2.11)$$

One should note that this bound is rather tight. For instance, in the binary32 arithmetic of the IEEE 754-2008 standard ($\beta = 2$, $p = 24$, $e_{\min} = -126$, $e_{\max} = 127$; see Section 3.1), the bound on ϵ given by (2.11) is $4.00000048 \times 2^{-24}$, whereas, if $a = 8622645$, $b = 16404663$, and $c = 8647279$, then

$$\frac{P1}{P2} = 1 + 3.86\ldots \times 2^{-24}.$$

- In case of overflow or underflow, $\circ(a \times \circ(b \times c))$ can be *drastically* different from $\circ(\circ(a \times b) \times c)$. For instance, in binary64 arithmetic ($\beta = 2, p = 53$, $e_{\min} = -1022$, $e_{\max} = 1023$; see Section 3.1), with $a = b = 2^{513}$ and $c = 2^{-1022}$, $\mathrm{RN}(\mathrm{RN}(a \times b) \times c)$ will be $+\infty$, whereas $\mathrm{RN}(a \times \mathrm{RN}(b \times c))$ will be 16.

2.7 Note on the Choice of the Radix

2.7.1 Representation errors

As stated in Chapter 1, various different radices were chosen in the early days of electronic computing, and several studies [413, 77, 63, 104, 352] have been devoted to the best radix choice, in terms of maximal or average representation error. These studies have shown that radix 2 with the implicit leading bit convention gives better worst-case and average accuracy than all other radices.

As an illustration, let us briefly present the analysis done by McKeeman. Assume that floating-point numbers are represented in radix β and that their significands and exponents are stored on w_s and w_e bits, respectively. As explained in Section 2.3.1, when a nonzero real number x in the normal range is represented by a nearest floating-point number $\mathrm{RN}(x)$, the relative representation error

$$\left|\frac{x - \mathrm{RN}(x)}{x}\right|$$

is introduced. Given a fixed value of $w_s + w_e$, we want to approximate the maximum and average values of this error by functions of β and w_s.

First, notice that for any integer k, if $\beta^k|x|$ remains between the smallest positive normal floating-point number $\beta^{e_{\min}}$ and the largest one $\Omega = (\beta - \beta^{1-p}) \cdot \beta^{e_{\max}}$, then $|x - \mathrm{RN}(x)|/|x|$ is equal to $\big||x| \cdot \beta^k - \mathrm{RN}(|x| \cdot \beta^k)\big|/(|x| \cdot \beta^k)$, so we can restrict x to be between two consecutive powers of β, say, $1/\beta$ and 1.

Second, for evaluating average values, one must choose a probability distribution for the significands of floating-point numbers. For various reasons [231, 342], the most sensible choice is the *logarithmic distribution*, also called *Benford's law* [35] and obtained from the density distribution

$$P(s) = \frac{1}{s \ln \beta}, \qquad \frac{1}{\beta} \leq s < 1.$$

We then easily obtain the following results.

- If $\beta > 2$, or if $\beta = 2$ and we do not use the hidden bit convention (that is, the first "1" of the significand of a normal number is actually stored), then the maximum relative representation error (MRRE) is, approximately,
$$\mathrm{MRRE}(w_s, \beta) \approx 2^{-w_s-1}\beta.$$
- If $\beta = 2$ and we use the hidden bit convention, we have
$$\mathrm{MRRE}(w_s, 2) \approx 2^{-w_s-1}.$$
- If $\beta > 2$, or if $\beta = 2$ and we do not use the hidden bit convention, then the average relative representation error (ARRE) is
$$\mathrm{ARRE}(w_s, \beta) = \int_{1/\beta}^{1} \left(\frac{1}{s \ln \beta}\right) \frac{2^{-w_s} ds}{4s} = \frac{\beta - 1}{4 \ln \beta} 2^{-w_s}.$$
- If $\beta = 2$ and we use the hidden bit convention, that value is halved and we get
$$\mathrm{ARRE}(w_s, 2) = \frac{1}{8 \ln 2} 2^{-w_s}.$$

These values seem much in favor of small radices, and yet, we must take into account the following. To achieve the same dynamic range (i.e., to have a similar order of magnitude of the extremal values) as in binary, in radix 2^k, we need around $\log_2(k)$ fewer bits for representing the exponent. These saved bits can, however, be used for the significands. Hence, for a similar total number of bits (sign, exponent, and significand) for the representation of floating-point numbers, a fair comparison between radices 2, 4, and 16

Format	MRRE	ARRE
$\beta = 2, w_s = 64$ first bit stored	5.42×10^{-20}	1.95×10^{-20}
$\beta = 2, w_s = 64$ first bit hidden	2.71×10^{-20}	9.77×10^{-21}
$\beta = 4, w_s = 65$	5.42×10^{-20}	1.46×10^{-20}
$\beta = 8, w_s = 65$	1.08×10^{-19}	2.28×10^{-20}
$\beta = 8, w_s = 66$	5.42×10^{-20}	1.14×10^{-20}
$\beta = 16, w_s = 66$	1.08×10^{-19}	1.83×10^{-20}

Table 2.2: Approximate values of ARRE and MRRE for various formats of comparable dynamic range. The cases $\beta = 2, 4,$ and 16 can be directly compared. In the case $\beta = 8$, one cannot get the same dynamic range exactly.

is obtained by taking a value of w_s larger by one unit for radix 4, and two units for radix 16, so that we compare number systems with similar dynamic ranges, and the same value of $w_s + w_e$.

Table 2.2 gives some values of MRRE and ARRE for various formats. From this table, one can infer that radix 2 with implicit bit convention is the best choice from the point of view of the relative representation error. Since radix-2 floating-point arithmetic is also the easiest to implement using digital logic, this explains why it is predominant on current systems.

2.7.2 A case for radix 10

Representation error, however, is not the only characteristic to consider. A strong argument in favor of radix 10 is that we humans work, read, and write in that radix. Section 4.9 will show how to implement the best possible conversions between radices 2 and 10, but such conversions may sometimes introduce errors that, in some applications, are unacceptable. A typical example is banking. An interest rate is written in a contract in decimal, and the computer at the bank is legally mandated to carry out interest computations using the exact decimal value of this rate, not a binary approximation to it.

As another example, when some European countries abandoned their local currencies in favor of the euro, the conversion rate was defined by law as a decimal number (e.g., 1 euro = 6.55957 French francs), and the way this conversion had to be implemented was also defined by law. Using any other conversion value, such as 6.559569835662841796875 (the binary32 number nearest to the legal value), was simply illegal.

Colishaw [121] gives other examples and also shows how pervasive decimal computing is in business applications.

As the current technology is fundamentally binary, radix 10 will intrinsically be less efficient than radix 2, and indeed even hardware implementations of decimal floating-point are much slower than their binary counterparts (see Table 3.18 for an example).

However, this is at least partially compensated by other specifics of the financial application domain. In particular, in accounting applications, the floating point in most of the additions and subtractions is actually fixed. Indeed, one adds cents to cents. The good news is that this common case of addition is much easier to implement than the general case:

- first, the significands need not be shifted as in the general case [121] (see Chapters 7 and 8);
- second, such fixed-point additions and subtractions will be exact (entailing no rounding error).

Rounding does occur in financial applications; for instance, when applying a sales tax or an interest rate. However, from an accounting point of view, it is best managed in such a way that one eventually adds only numbers which have already been rounded to the nearest cent.

The reader should have in mind these peculiarities when reading about decimal formats and operations in this book. Most of the intricacies of the decimal part of the IEEE 754-2008 floating-point standard are directly motivated by the needs of accounting applications.

2.8 Reproducibility

In recent years, numerical reproducibility has become an important topic [161, 508, 162, 165]. The goal is to always get bitwise identical results when running the same program. As explained by Demmel, Ahrens, and Nguyen [165], this presents a number of advantages:

- it helps to debug and test a program, since errors can be reproduced;
- in massively parallel computations, it is becoming more and more common to have the same quantity computed on different processors (because it is often cheaper to recompute it locally than to transmit it): in such a case, for elementary reasons of consistency (e.g., the same branches must be taken), it must be identical;
- one may wish to check previously published results;
- there may even be legal reasons, when some critical decision is based on the result of some computation: one may wish to be able to check later on what result was obtained.

It is therefore important to have algorithms, languages, compilation options, instructions sets, etc. that allow one to obtain reproducible results when needed. However, we must warn the reader that this approach also has drawbacks. When the same program returns results that differ significantly, either when run on different platforms, or when run twice on the same platform, *this is a clear sign that something worrying is going on*: it can be related to the problem being solved (e.g., high sensitivity to initial conditions), or it can be a "numerical artifact" (e.g., programming bug, poor algorithm). In any case, this is important information that is lost if reproducibility is enforced.

Chapter 3

Floating-Point Formats and Environment

OUR MAIN FOCUS IN THIS CHAPTER is the IEEE[1] 754-2008 Standard for Floating-Point Arithmetic [267], a revision and merge of the earlier IEEE 754-1985 [12] and IEEE 854-1987 [13] standards. A paper written in 1981 by Kahan, *Why Do We Need a Floating-Point Standard?* [315], depicts the rather messy situation of floating-point arithmetic before the 1980s. Anybody who takes the view that the current standard is too constraining and that circuit and system manufacturers could build much more efficient machines without it should read that paper and think about it. Even if there were at that time a few reasonably good environments, the various systems available then were so different that writing portable yet reasonably efficient numerical software was extremely difficult. For instance, as pointed out in [553], sometimes a programmer had to insert multiplications by 1.0 to make a program work reliably.

The IEEE 754-1985 Standard for Binary Floating-Point Arithmetic was released in 1985, but the first meetings of the working group started more than eight years earlier [553]. William Kahan, a professor at the University of California at Berkeley, played a leading role in the development of the standard.

IEEE 754-1985 drastically changed the world of numerical computing, by clearly specifying formats and the way exceptions must be handled, and by requiring correct rounding of the arithmetic operations and the square root. Two years later, another standard, the IEEE 854-1987 Standard for "Radix-Independent" (in fact, radix 2 or 10) Floating-Point Arithmetic was released. It generalized to radix 10 the main ideas of IEEE 754-1985.

[1]IEEE is an acronym for the Institute of Electrical and Electronics Engineers. For more details, see https://www.ieee.org/about/index.html.

J.-M. Muller et al., *Handbook of Floating-Point Arithmetic*,
https://doi.org/10.1007/978-3-319-76526-6_3

IEEE 754-1985 was under revision between 2000 and 2008, and the new standard was adopted in June 2008. In the following, it will be called IEEE 754-2008. In the literature published before its official release (e.g., [116]), IEEE 754-2008 is sometimes called IEEE 754R. IEEE 754-2008 is also known as ISO/IEC/IEEE 60559:2011 *Information technology – Microprocessor Systems – Floating-Point arithmetic* [277]. At the time of writing these lines, IEEE 754-2008 is under revision, and a new version (which will not be much different) is to be released in 2018.

Some languages, such as Java and ECMAScript, are based on IEEE 754-1985. The ISO C11 standard (released in 2011) for the C language has optional support for IEEE 754-1985 in its normative annex F; support for IEEE 754-2008 is planned for the C2x standard. Details will be given in Chapter 6.

The description of the IEEE standard given in this chapter is not exhaustive: standards are big documents that contain many details. Anyone who wants to implement a floating-point arithmetic function compliant to IEEE 754-2008 must carefully read that standard. Also, readers interested in the older IEEE 754-1985 and IEEE 854-1987 standards will find brief descriptions of them in Appendix B.

3.1 The IEEE 754-2008 Standard

3.1.1 Formats

The standard requires that the radix β be 2 or 10, and that for all formats, the minimum and maximum exponents obey the following relation:

$$e_{\min} = 1 - e_{\max} .$$

It defines several *interchange formats*, whose encodings are fully specified as bit strings, and that enable lossless data interchange between different platforms.[2] The main parameters of the interchange formats of size up to 128 bits are given in Tables 3.1 and 3.2.

A format is said to be an *arithmetic format* if all the mandatory operations defined by the standard are supported by the format.

Among the interchange formats, the standard defines five *basic formats* which must also be arithmetic formats: the three binary formats on 32, 64, and 128 bits, and the two decimal formats on 64 and 128 bits. *A conforming implementation must implement at least one of them.*

To implement a function that must return a result in a basic format, it may be convenient to have the intermediate calculations performed in a somewhat wider format:

[2]Platforms may exchange character strings, or binary data. In the latter case, endianness problems (see Section 3.1.1.5) must be considered.

Name	binary16	binary32 (basic)	binary64 (basic)	binary128 (basic)
Former name	N/A	single precision	double precision	N/A
p	11	24	53	113
$e_{\max}$	+15	+127	+1023	+16383
$e_{\min}$	−14	−126	−1022	−16382

Table 3.1: Main parameters of the binary interchange formats of size up to 128 *bits specified by the 754-2008 standard [267]. In some articles and software libraries, 128-bit formats were sometimes called "quad precision." However, quad precision was not specified by IEEE 754-1985.*

Name	decimal32	decimal64 (basic)	decimal128 (basic)
p	7	16	34
$e_{\max}$	+96	+384	+6144
$e_{\min}$	−95	−383	−6143

Table 3.2: Main parameters of the decimal interchange formats of size up to 128 *bits specified by the 754-2008 standard [267].*

- the wider precision makes it possible to get a result that will almost always be significantly more accurate than that obtained with the basic format only;

- the wider range will drastically limit the occurrences of "apparent" or "spurious" under/overflow (that is, cases where there is an underflow or overflow in an intermediate result, whereas the final value would have been in the range of the basic format).

For this purpose, the standard also defines an *extended precision* format as a format that extends a basic format with a wider precision and range, in a language-defined or implementation-defined way. Also, an *extendable precision* format is similar to an extended format, but its precision and range are defined under program control. For both extended and extendable formats, the standard does not specify the binary encoding: it may be a straightforward generalization of the encoding of an interchange format, or not. Offering extended or extendable formats is optional.

Finally, the standard requires that conversions between any two supported formats be implemented.

Let us now describe the interchange formats in more detail.

3.1.1.1 Binary interchange format encodings

The binary interchange formats, whose main parameters are given in Table 3.1, are very much like the formats of the older IEEE 754-1985 standard. Actually, the binary32 and binary64 formats correspond to the single- and double-precision formats of IEEE 754-1985: the encodings are exactly the same. However, IEEE 754-1985 did not specify any 128-bit format. What was called "quad precision" in some articles and software libraries was sometimes slightly different from the new binary128 format of IEEE 754-2008. Some authors call "half precision" the binary16 format.

As illustrated in Figure 3.1, a floating-point number is encoded using a 1-bit sign, a W_E-bit exponent field, and a $(p-1)$-bit field for the trailing significand. Let us remind what we mean by "trailing significand." As we are going to see, the information according to which the number is normal or subnormal is encoded in the exponent field. As already explained in Section 2.1.2, since the first bit of the significand of a binary floating-point number is necessarily a "1" if the number is normal (i.e., larger than or equal to $2^{e_{\min}}$) and a "0" if the number is subnormal, there is no need to store that bit. We only store the $p-1$ other bits: these bits constitute the trailing significand (also known as "fraction"). That choice of not storing the leftmost bit of the significand is sometimes called the "hidden bit convention" or the "leading bit convention."

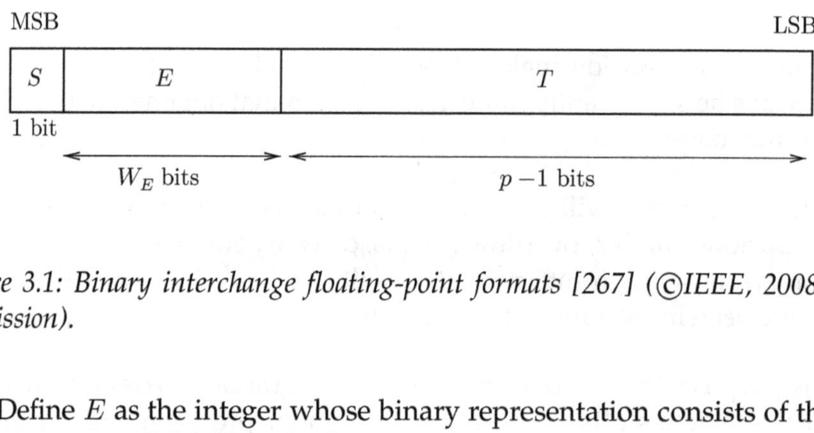

Figure 3.1: Binary interchange floating-point formats [267] (©IEEE, 2008, with permission).

Define E as the integer whose binary representation consists of the bits of the exponent field, T as the integer whose representation consists of the bits of the trailing significand field, and S as the sign bit. The binary encoding (S, E, T) should be interpreted as follows [267]:

- if $E = 2^{W_E} - 1$ (i.e., E is a string of ones) and $T \neq 0$, then a NaN, either quiet (qNaN) or signaling (sNaN), is represented. A quiet NaN is the default result of an invalid operation (e.g., $\sqrt{-5.0}$), and for most operations, a signaling NaN will signal the invalid operation exception whenever it appears as an operand (see Section 3.1.7.1);

- if $E = 2^{W_E} - 1$ and $T = 0$, then $(-1)^S \times (+\infty)$ is represented;
- if $1 \leq E \leq 2^{W_E} - 2$, then the (normal) floating-point number being represented is
$$(-1)^S \times 2^{E-b} \times \left(1 + T \cdot 2^{1-p}\right),$$
where the *bias* b is defined as $b = e_{\max} = 2^{W_E - 1} - 1$ (see Table 3.4 for the actual values);
- if $E = 0$ and $T \neq 0$, then the (subnormal) number being represented is
$$(-1)^S \times 2^{e_{\min}} \times \left(0 + T \cdot 2^{1-p}\right);$$
- if $E = 0$ and $T = 0$, then the number being represented is the signed zero $(-1)^S \times (+0)$.

Biased exponent N_e	Trailing significand $t_1 t_2 \dots t_{p-1}$	Value represented
$111 \cdots 1_2$	$\neq 000 \cdots 0_2$	NaN
$111 \cdots 1_2$	$000 \cdots 0_2$	$(-1)^s \times \infty$
$000 \cdots 0_2$	$000 \cdots 0_2$	$(-1)^s \times 0$
$000 \cdots 0_2$	$\neq 000 \cdots 0_2$	$(-1)^s \times 0.t_1 t_2 \dots t_{p-1} \times 2^{e_{\min}}$
$0 < N_e < 2^{W_E} - 1$	any	$(-1)^s \times 1.t_1 t_2 \dots t_{p-1} \times 2^{N_e - b}$

Table 3.3: How to interpret the encoding of an IEEE 754 binary floating-point number.

format	binary16	binary32	binary64	binary128
former name	N/A	single precision	double precision	N/A
storage width	16	32	64	128
$p - 1$, trailing significand width	10	23	52	112
W_E, exponent field width	5	8	11	15
$b = e_{\max}$	15	127	1023	16383
$e_{\min}$	−14	−126	−1022	−16382

Table 3.4: Parameters of the encodings of binary interchange formats [267]. As stated above, in some articles and software libraries, 128-bit formats were called "quad precision." However, quad precision was not specified by IEEE 754-1985.

	Smallest positive subnormal $2^{e_{\min}+1-p}$	Smallest positive normal $2^{e_{\min}}$	Largest finite $2^{e_{\max}}(2-2^{1-p})$
binary16	2^{-14-10} $\approx 5.96 \times 10^{-8}$	2^{-14} $\approx 6.10 \times 10^{-5}$	$(2-2^{-10}) \times 2^{15}$ $= 65504$
binary32	$2^{-126-23}$ $\approx 1.401 \times 10^{-45}$	2^{-126} $\approx 1.175 \times 10^{-38}$	$(2-2^{-23}) \times 2^{127}$ $\approx 3.403 \times 10^{38}$
binary64	$2^{-1022-52}$ $\approx 4.941 \times 10^{-324}$	2^{-1022} $\approx 2.225 \times 10^{-308}$	$(2-2^{-52}) \times 2^{1023}$ $\approx 1.798 \times 10^{308}$
binary128	$2^{-16382-112}$ $\approx 6.68 \times 10^{-4966}$	2^{-16382} $\approx 3.36 \times 10^{-4932}$	$(2-2^{-112}) \times 2^{16383}$ $\approx 1.19 \times 10^{4932}$

Table 3.5: Extremal values of the IEEE 754-2008 binary interchange formats.

Datum	Sign	Biased exponent	Trailing significand
-0	1	00000000	00000000000000000000000
$+0$	0	00000000	00000000000000000000000
$-\infty$	1	11111111	00000000000000000000000
$+\infty$	0	11111111	00000000000000000000000
NaN	0	11111111	nonzero string
5	0	10000001	01000000000000000000000

Table 3.6: Binary encoding of various floating-point data in the binary32 format.

Table 3.3 summarizes this encoding. The sizes of the various fields are given in Table 3.4, and extremal values are given in Table 3.5.

Table 3.6 gives examples of the binary encoding of various floating-point values in the binary32 format. Let us now detail two examples.

Example 3.1 (Binary encoding of a normal number). *Consider the binary32 number x whose binary encoding is*

sign	exponent	trailing significand
0	01101011	01010101010101010101010

- *the bit sign of x is a zero, which indicates that $x \geq 0$;*
- *the biased exponent is neither 00000000_2 nor 11111111_2, which indicates that x is a normal number. It is $01101011_2 = 107_{10}$, hence, since the bias in binary32 is 127, the actual exponent of x is $107 - 127 = -20$;*

- *by placing the hidden bit (which is a* 1*, since* x *is not subnormal) at the left of the trailing significand, we get the significand of* x*:*

$$1.0101010101010101010101_2 = \frac{5592405}{2^{22}};$$

- *hence,* x *is equal to*

$$\begin{aligned} \frac{5592405}{2^{22}} \times 2^{-20} &= \frac{5592405}{2^{42}} \\ &= 0.000001271565679417108185589313507080078125. \end{aligned}$$

Example 3.2 (Binary encoding of a subnormal number). *Consider the binary32 number* x *whose binary encoding is*

sign	exponent	trailing significand
1	00000000	01100000000000000000000

- *the bit sign of* x *is a one, which indicates that* $x \leq 0$*;*
- *the biased exponent is* 00000000*, which indicates that* x *is a subnormal number. It is not a zero, since the significand field is not a string of zeros. Hence, the real exponent of* x *is* $e_{\min} = -126$*;*
- *by placing the hidden bit (which is a* 0*, since* x *is subnormal) at the left of the trailing significand, we get the significand of* x*:*

$$0.01100000000000000000000_2 = \frac{3}{8};$$

- *hence,* x *is equal to*

$$\begin{aligned} -\frac{3}{8} \times 2^{-126} &= -4.408103815583578154882762014583421291819995837895 \\ &\quad 32820565781889854406472295522689819335937 5 \times 10^{-39}. \end{aligned}$$

3.1.1.2 Decimal interchange format encodings

The decimal format encodings are more complex than the binary ones, for several reasons.

- *Two* encoding systems are specified, called the *decimal* and *binary* encodings: the members of the revision committee could not agree on a single encoding system. The reason for that is that the binary encoding makes a *software* implementation of decimal arithmetic easier, whereas the decimal encoding is more suited for a *hardware* implementation. And yet, despite this problem, one must understand that the set of representable floating-point numbers is *the same* for both encoding systems, so that this additional complexity will be transparent for most users. Also, a conforming implementation must provide conversions between these two encoding systems [267, §5.5.2].

- Contrary to the binary interchange formats, the sign, exponent, and (trailing) significand fields are not fully separated: to preserve as much accuracy as possible, some information on the significand is partly encoded in what used to be the exponent field and is hence called the *combination* field.
- In the decimal formats, the representations (M, e) are not normalized (i.e., it is not required that e should be minimal), so that a decimal floating-point number may have multiple valid representations. The set of the various representations of a same number is called a *cohort*. As a consequence, we will have to explain which exponent is *preferred* for the result of an arithmetic operation.
- Even if the representation itself (that is, values of the sign, exponent, and significand) of a number x (or an infinite, or a NaN) and the type of encoding (binary or decimal) are chosen, a same number (or infinite, or NaN) can still be encoded by different bit strings. One of them will be said to be *canonical*.

Roughly speaking, the difference between the decimal and binary encodings of decimal floating-point numbers originates from a choice in the encoding of the significand. The integral significand is a nonnegative integer less than or equal to $10^p - 1$. One can encode it either in binary (which gives the binary encoding) or in decimal (which gives the decimal encoding).

Concerning the decimal encoding, in the early days of computer arithmetic, people would use binary coded decimal (BCD) encoding, where each decimal digit was encoded by four bits. That encoding was quite wasteful, since among the 16 possible values representable on four bits, only 10 were actually used. Since $2^{10} = 1024$ is very close to 10^3 (and larger), one can design a much denser encoding by encoding three consecutive decimal digits by a 10-bit *declet* [95]. Many possible ways of performing this encoding are possible. The one chosen by the standard committee for the decimal encoding of decimal numbers is given in Tables 3.10 (declet to decimal) and 3.11 (decimal to declet). It was designed to facilitate conversions: all these tables have a straightforward hardware implementation and can be implemented in three gate levels [184]. Note that Table 3.10 has 1024 possible inputs and 1000 possible outputs (hence, there is some redundancy), and Table 3.11 has 1000 possible inputs and outputs. This implies that there are 24 "noncanonical" bit patterns,[3] which are accepted as input values but cannot result from an arithmetic operation. An encoding that contains a noncanonical bit pattern is called *noncanonical*.

Let us explain more precisely why there is no clear separation between an exponent field and a significand field (as is the case in the binary formats). Consider as an example the decimal64 format (see Table 3.2). In that format, $e_{\max} = 384$ and $e_{\min} = -383$; therefore, there are 768 possible values of the exponent. Storing all these values in binary in an exponent field would require 10 bits. Since we can store 1024 possible values in a 10-bit field, that would be wasteful. This explains why it was decided to put all the information about the exponent plus some other information in a "combination field," where will be stored:

- "classification" information: Does the datum represent a finite number, or $\pm\infty$, or a NaN (see Section 3.1.7)?
- the exponent (if the datum represents a finite number);
- the leading part of the significand (if the datum represents a finite number); more precisely, the leading decimal digit (if the *decimal* encoding is used) or 3 to 4 leading bits (if the *binary* encoding is used). The remaining significand bits/digits are stored in the trailing significand field.

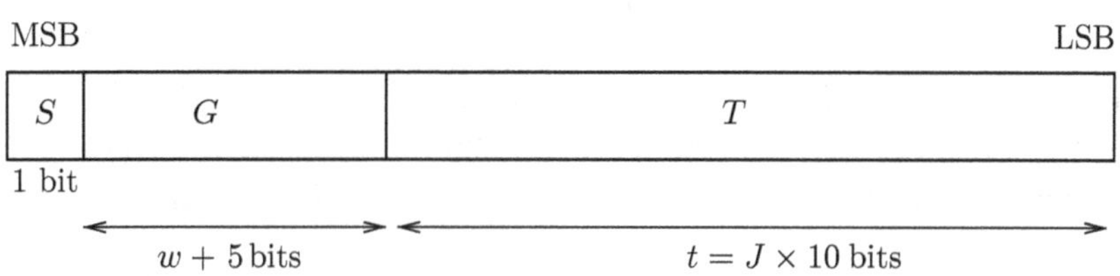

Figure 3.2: Decimal interchange floating-point formats [267] (©IEEE, 2008, with permission).

The widths of the various fields are given in Table 3.7. It is important to note that in this table the bias b is related to the *quantum exponent* (see Section 2.1.1), which means that if e is the exponent of x, if $q = e - p + 1$ is its quantum exponent, then the biased exponent E is

$$E = q + b = e - p + 1 + b.$$

The floating-point format illustrated in Figure 3.2, with a 1-bit sign, a $(w + 5)$-bit *combination* field, and a $t = (J \times 10)$-bit *trailing significand* field must be interpreted as follows [267]:

[3]Those of the form $01x11x111x$, $10x11x111x$, or $11x11x111x$.

	decimal32	decimal64	decimal128
storage width	32	64	128
$t = 10J$, trailing significand width	20	50	110
$w + 5$, combination field width	11	13	17
$b = E - (e - p + 1)$, bias	101	398	6176

Table 3.7: Width (in bits) of the various fields in the encodings of the decimal interchange formats of size up to 128 *bits [267].*

- if the most significant five bits of G (numbered from the left, G_0 to G_4) are all ones, then the datum being represented is a NaN (see Section 3.1.7.1). Moreover, if G_5 is 1, then it is an sNaN, otherwise it is a qNaN. In a canonical encoding of a NaN, the bits G_6 to G_{w+4} are all zeros;

- if the most significant five bits of G are 11110, then the value being represented is $(-1)^S \times (+\infty)$. Moreover, the canonical encodings of infinity have bits G_5 to G_{w+4} as well as trailing significand T equal to 0;

- if the most significant four bits of G, i.e., G_0 to G_3, are not all ones, then the value being represented is a finite number, equal to

$$(-1)^S \times 10^{E-b} \times C. \tag{3.1}$$

  Here, the value $E - b$ is the *quantum exponent* (see Section 2.1.1), where b, the *exponent bias*, is equal to 101, 398, and 6176 for the decimal32, decimal64, and decimal128 formats, respectively. E and C are obtained as follows.

  1. If the *decimal encoding* is used for the significand, then the least significant w bits of the biased exponent E are made up of the bits G_5 to G_{w+4} of G, whereas the most significant two bits of E and the most significant two digits of C are obtained as follows:
     - if the most significant five bits $G_0G_1G_2G_3G_4$ of G are of the form $110xx$ or $1110x$, then the leading significand digit C_0 is $8 + G_4$ (which equals 8 or 9), and the leading biased exponent bits are G_2G_3;
     - if the most significant five bits of G are of the form $0xxxx$ or $10xxx$, then the leading significand digit C_0 is $4G_2 + 2G_3 + G_4$ (which is between 0 and 7), and the leading biased exponent bits are G_0G_1.

The $p-1=3J$ decimal digits $C_1,\ldots,C_{p-1}$ of C are encoded by T, which contains J declets encoded in densely packed decimal (see Tables 3.10 and 3.11). Note that if the most significant five bits of G are 00000, 01000, or 10000, and $T=0$, then the significand is 0 and the number represented is $(-1)^S \times (+0)$.

Table 3.8 summarizes these rules.

2. If the *binary encoding* is used for the significand, then
   - if G_0G_1 is 00, 01, or 10, then E is made up of the bits G_0 to G_{w+1}, and the binary encoding of the significand C is obtained by prefixing the last 3 bits of G (i.e., $G_{w+2}G_{w+3}G_{w+4}$) to T;
   - if G_0G_1 is 11 and G_2G_3 is 00, 01 or 10, then E is made up of the bits G_2 to G_{w+3}, and the binary encoding of the significand C is obtained by prefixing $100G_{w+4}$ to T.

   Remember that the maximum value of the integral significand is $10^p-1=10^{3J+1}-1$. If the value of C computed as above is larger than that maximum value, then the value used for C will be zero [267], and the encoding will not be canonical. Table 3.9 summarizes these rules.

A decimal software implementation of IEEE 754-2008, based on the binary encoding of the significand, is presented in [116, 117]. Interesting information on decimal arithmetic can be found in [121]. A decimal floating-point multiplier that uses the decimal encoding of the significand is presented in [192].

Example 3.3 (Finding the encoding of a decimal number assuming decimal encoding of the significands). *Consider the number*

$$x = 3.141592653589793 \times 10^0 = 3141592653589793 \times 10^{-15}.$$

This number is exactly representable in the decimal64 format. Let us find its encoding, assuming decimal encoding of the significands.

- *First, the sign bit is* 0*;*
- *since the quantum exponent is* -15*, the biased exponent will be* 383 *(see Table 3.7), whose 10-bit binary representation is* 0101111111_2*. One should remember that the exponent is not directly stored in an exponent field, but combined with the most significant digit of the significand in a* combination field G*. Since the leading significand digit is* 3*, we are in the case*

G	T	Datum being represented
$11111\,0xxx\cdots x$	any	qNaN
$11111\,1xxx\cdots x$	any	sNaN
$11110\,xxxx\cdots x$	any	$(-1)^S \times (+\infty)$
$110xx\cdots$ or $1110x\cdots$	$T_0T_1\cdots T_{10J-1}$	$(-1)^S \times 10^{\overbrace{G_2G_3G_5G_6\cdots G_{w+4}}^{\text{binary}}-b} \times \underbrace{(8+G_4)C_1C_2\cdots C_{p-1}}_{\text{decimal}}$ with $C_{3j+1}C_{3j+2}C_{3j+3}$ deduced from $T_{10j}T_{10j+1}T_{10j+2}\cdots T_{10j+9}$ for $0 \le j < J$ using Table 3.10.
$0xxxx\cdots$ or $10xxx\cdots$	$T_0T_1\cdots T_{10J-1}$	$(-1)^S \times 10^{\overbrace{G_0G_1G_5G_6\cdots G_{w+4}}^{\text{binary}}-b} \times \underbrace{(4G_2+2G_3+G_4)C_1C_2\cdots C_{p-1}}_{\text{decimal}}$ with $C_{3j+1}C_{3j+2}C_{3j+3}$ deduced from $T_{10j}T_{10j+1}T_{10j+2}\cdots T_{10j+9}$ for $0 \le j < J$ using Table 3.10.

Table 3.8: Decimal encoding of a decimal floating-point number (IEEE 754-2008).

G	T	Datum being represented
$11111\,0xxx\cdots x$	any	qNaN
$11111\,1xxx\cdots x$	any	sNaN
$11110\,xxxx\cdots x$	any	$(-1)^S \times (+\infty)$
$00xxx\cdots$ or $01xxx\cdots$ or $10xxx\cdots$	$T_0T_1\cdots T_{10J-1}$	$(-1)^S \times 10^{\overbrace{G_0G_1G_2\cdots G_{w+1}}^{\text{binary}}-b} \times \underbrace{G_{w+2}G_{w+3}G_{w+4}T_0T_1\cdots T_{10J-1}}_{\text{binary}}$ if $G_{w+2}G_{w+3}G_{w+4}T_0T_1\cdots T_{10J-1} \le 10^p-1$, otherwise $(-1)^S \times (+0)$.
$1100xxx\cdots$ or $1101xxx\cdots$ or $1110xxx\cdots$	$T_0T_1\cdots T_{10J-1}$	$(-1)^S \times 10^{\overbrace{G_2G_3G_4\cdots G_{w+3}}^{\text{binary}}-b} \times \underbrace{100G_{w+4}T_0T_1\cdots T_{10J-1}}_{\text{binary}}$ if $100G_{w+4}T_0T_1\cdots T_{10J-1} \le 10^p-1$, otherwise $(-1)^S \times (+0)$.

Table 3.9: Binary encoding of a decimal floating-point number (IEEE 754-2008).

$b_6b_7b_8b_3b_4$	d_0	d_1	d_2
$0xxxx$	$4b_0+2b_1+b_2$	$4b_3+2b_4+b_5$	$4b_7+2b_8+b_9$
$1\,0\,0xx$	$4b_0+2b_1+b_2$	$4b_3+2b_4+b_5$	$8+b_9$
$1\,0\,1xx$	$4b_0+2b_1+b_2$	$8+b_5$	$4b_3+2b_4+b_9$
$1\,1\,0xx$	$8+b_2$	$4b_3+2b_4+b_5$	$4b_0+2b_1+b_9$
$1\,1\,1\,0\,0$	$8+b_2$	$8+b_5$	$4b_0+2b_1+b_9$
$1\,1\,1\,0\,1$	$8+b_2$	$4b_0+2b_1+b_5$	$8+b_9$
$1\,1\,1\,1\,0$	$4b_0+2b_1+b_2$	$8+b_5$	$8+b_9$
$1\,1\,1\,1\,1$	$8+b_2$	$8+b_5$	$8+b_9$

Table 3.10: Decoding the declet $b_0b_1b_2\cdots b_9$ of a densely packed decimal encoding to three decimal digits $d_0d_1d_2$ [267] (©IEEE, 2008, with permission).

$d_0^0\ d_1^0\ d_2^0$	$b_0\ b_1\ b_2$	$b_3\ b_4\ b_5$	b_6	$b_7\ b_8\ b_9$
0 0 0	$d_0^1\ d_0^2\ d_0^3$	$d_1^1\ d_1^2\ d_1^3$	0	$d_2^1\ d_2^2\ d_2^3$
0 0 1	$d_0^1\ d_0^2\ d_0^3$	$d_1^1\ d_1^2\ d_1^3$	1	$0\ 0\ d_2^3$
0 1 0	$d_0^1\ d_0^2\ d_0^3$	$d_2^1\ d_2^2\ d_1^3$	1	$0\ 1\ d_2^3$
0 1 1	$d_0^1\ d_0^2\ d_0^3$	$1\ 0\ d_1^3$	1	$1\ 1\ d_2^3$
1 0 0	$d_2^1\ d_2^2\ d_0^3$	$d_1^1\ d_1^2\ d_1^3$	1	$1\ 0\ d_2^3$
1 0 1	$d_1^1\ d_1^2\ d_0^3$	$0\ 1\ d_1^3$	1	$1\ 1\ d_2^3$
1 1 0	$d_2^1\ d_2^2\ d_0^3$	$0\ 0\ d_1^3$	1	$1\ 1\ d_2^3$
1 1 1	$0\ 0\ d_0^3$	$1\ 1\ d_1^3$	1	$1\ 1\ d_2^3$

Table 3.11: Encoding the three consecutive decimal digits $d_0d_1d_2$, each of them being represented in binary by four bits (e.g., d_0 is written in binary $d_0^0d_0^1d_0^2d_0^3$), into a 10-bit declet $b_0b_1b_2\cdots b_9$ of a densely packed decimal encoding [267] (©IEEE, 2008, with permission).

> *"if the most significant five bits of G are of the form* $0xxxx$ *or* $10xxx$*, then the leading significand digit* C_0 *is* $4G_2 + 2G_3 + G_4$ *(which is between* 0 *and* 7*), and the leading biased exponent bits are* G_0G_1*."*

Hence,

- G_0 *and* G_1 *are the leading biased exponent bits, namely* 0 *and* 1*;*
- G_2*,* G_3*, and* G_4 *are the binary encoding of the first significand digit* 3*, i.e.,* $G_2 = 0$*, and* $G_3 = G_4 = 1$*; and*
- *the bits* G_5 *to* G_{12} *are the least significant bits of the biased exponent, namely* 01111111.

- *Now, the trailing significand field* T *is made up of the five declets of the densely packed decimal encoding of the trailing significand* 141592653589793*:*
  - *the 3-digit chain* 141 *is encoded by the declet* 0011000001*, according to Table 3.11;*
  - 592 *is encoded by the declet* 1010111010*;*
  - 653 *is encoded by the declet* 1101010011*;*
  - 589 *is encoded by the declet* 1011001111*;*
  - 793 *is encoded by the declet* 1110111011.
- *Therefore, the encoding of* x *is*

$$\underbrace{0}_{\text{sign}}\ \underbrace{01011011111111}_{\text{combination field}}\ \ldots$$

$$\underbrace{00110000011010111010110101001110110011111110111011}_{\text{trailing significand field}}.$$

Example 3.4 (Finding an encoding of a decimal number assuming binary encoding of the significands). *Consider the number*

$$x = 3.141592653589793 \times 10^0 = 3141592653589793 \times 10^{-15}.$$

(This is the same number as in Example 3.3, but now we consider binary encoding, in the decimal64 format.) The sign bit will be zero. Since 3141592653589793 *is a 16-digit integer that does not end with a* 0*, the quantum exponent can only be* -15*; therefore, the biased exponent* E *will be* $398 - 15 = 383$*, whose binary representation is* 101111111*. The binary representation of the integral significand of* x *is*

$$10\,\underbrace{11001010010100001100001010001001010110110100100001}_{t = 50 \text{ bits (trailing significand)}}.$$

The length of this bit string is 52*, which is less than* $t + 4 = 54$*, hence we are not in the case*

> *"if G_0G_1 is 11 and G_2G_3 is 00, 01 or 10, then E is made up of the bits G_2 to G_{w+3}, and the binary encoding of the significand C is obtained by prefixing $100G_{w+4}$ to T,"*

which means that we are in the case

> *"if G_0G_1 is 00, 01, or 10, then E is made up of the bits G_0 to G_{w+1}, and the binary encoding of the significand C is obtained by prefixing the last 3 bits of G (i.e., $G_{w+2}G_{w+3}G_{w+4}$) to T."*

Therefore, $G_0G_1 \ldots G_9 = 0101111111$, $G_{10}G_{11}G_{12} = 010$ and T is made up with the 50 rightmost bits of t, resulting in an encoding of x as

$$T = 11001010010100001100001010001001010110110100100001.$$

Example 3.5 (Finding the value of a decimal floating-point number from its encoding, assuming decimal encoding of the significand). *Consider the decimal32 number x whose encoding is*

$$\underbrace{1}_{\textit{sign}} \quad \underbrace{11101101101}_{\textit{combination field } G} \quad \underbrace{01101001101111000011}_{\textit{trailing significand field } T}.$$

- *Since the bit sign is 1, we have $x \leq 0$;*
- *since the most significant four bits of G are not all ones, x is not an infinity or a NaN;*
- *by looking at the most significant four bits of G, we deduce that we are in the case*

  > *if the most significant five bits $G_0G_1G_2G_3G_4$ of G are of the form $110xx$ or $1110x$, then the leading significand digit C_0 is $8 + G_4$ (which equals 8 or 9), and the leading biased exponent bits are G_2G_3.*

  Therefore, the leading significand bit C_0 is $8 + G_4 = 9$, and the leading biased exponent bits are 10. The least significant bits of the exponent are 101101; therefore, the biased exponent is $10101101_2 = 173_{10}$. Hence, the (unbiased) quantum exponent of x is $173 - 101 = 72$;
- *the trailing significand field T is made up of two declets, 0110100110 and 1111000011. According to Table 3.10,*
  - *the first declet encodes the 3-digit chain 326;*
  - *the second declet encodes 743.*
- *Therefore, x is equal to*

$$-9326743 \times 10^{72} = -9.326743 \times 10^{78}.$$

Example 3.6 (Finding the value of a decimal floating-point number from its encoding, assuming binary encoding of the significand). *Consider the decimal32 number x whose encoding is*

$$\underbrace{1}_{\textit{sign}}\quad\underbrace{11101101101}_{\textit{combination field } G}\quad\underbrace{01101001101111000011}_{\textit{trailing significand field } T}.$$

(It is the same bit string as in Example 3.5, but now we consider binary encoding.)

- *Since the bit sign is 1, we have $x \le 0$;*
- *since the most significant four bits of G are not all ones, x is not an infinity or a NaN;*
- *since $G_0G_1 = 11$ and $G_2G_3 = 10$, we are in the case*

  > *if G_0G_1 is 11 and G_2G_3 is 00, 01, or 10, then E is made up of the bits G_2 to G_{w+3}, and the binary encoding of the significand C is obtained by prefixing $100G_{w+4}$ to T.*

  Therefore, the biased exponent E is $10110110_2 = 182_{10}$, which means that the quantum exponent of x is $182 - 101 = 81$, and the integral significand of x is

  $$100101101001101111000011_2 = 9870275_{10}.$$

- *Therefore, x is equal to*

  $$-9870275 \times 10^{81} = -9.870275 \times 10^{87}.$$

3.1.1.3 Larger formats

The IEEE 754-2008 standard also specifies larger interchange formats for widths of at least 128 bits that are multiples of 32 bits. Their parameters are given in Table 3.12, and examples are given in Tables 3.13 and 3.14. This allows one to define "big" (yet, fixed) precisions. A format is fully defined from its radix (2 or 10) and size: the various parameters (precision, $e_{\min}$, $e_{\max}$, bias, etc.) are derived from them, using the formulas given in Table 3.12. Hence, binary1024 or decimal512 will mean the same thing on all platforms.

3.1.1.4 Extended and extendable precisions

Beyond the interchange formats, the IEEE 754-2008 standard partially specifies the parameters of possible *extended precision* and *extendable precision* formats. These formats are optional, and their binary encoding is not specified.

Parameter	Binary$k$ format (k is a multiple of 32)	Decimal$k$ format
k	≥ 128	≥ 32
p	$k - \lfloor 4\log_2(k) \rceil + 13$	$9 \times \frac{k}{32} - 2$
t	$p - 1$	$(p-1) \times 10/3$
w	$k - t - 1$	$k - t - 6$
$e_{\max}$	$2^{w-1} - 1$	$3 \times 2^{w-1}$
$e_{\min}$	$1 - e_{\max}$	$1 - e_{\max}$
b	$e_{\max}$	$e_{\max} + p - 2$

Table 3.12: Parameters of the interchange formats. $\lfloor u \rceil$ *is* u *rounded to the nearest integer,* t *is the trailing significand width,* w *is the width of the exponent field for the binary formats, and the width of the combination field minus* 5 *for the decimal formats, and* b *is the exponent bias [267], (©IEEE, 2008, with permission).*

Format	p	t	w	$e_{\min}$	$e_{\max}$	b
binary256	237	236	19	−262142	+262143	262143
binary1024	997	996	27	−67108862	+67108863	67108863

Table 3.13: Parameters of the binary256 and binary1024 interchange formats deduced from Table 3.12. Variables p, t, w, $e_{\min}$, $e_{\max}$, *and* b *are the precision, the trailing significant field length, the exponent field length, the minimum exponent, the maximum exponent, and the exponent bias, respectively.*

- An *extended precision format* extends a basic format with a wider precision and range, and is either language defined or implementation defined. The constraints on these wider precisions and ranges are listed in Table 3.15. The basic idea behind these formats is that they should be used to carry out intermediate computations, in order to return a final result in the associated basic formats. The wider precision makes it possible to get a result that will generally be more accurate than that obtained with the basic formats only, and the wider range will drastically limit the cases of "apparent under/overflow" (that is, cases where there is an underflow or overflow in an intermediate result, whereas the final value would have been representable). An example of extended precision format, still of importance, is the 80-bit "double-extended format" (radix 2, precision 64, extremal exponents -16382 and 16383) specified by the IA-32 instruction set.

Format	p	t	$w+5$	$e_{\max}$	b
decimal256	70	230	25	+1572864	1572932
decimal512	142	470	41	+103079215104	103079215244

Table 3.14: Parameters of the decimal256 and decimal512 interchange formats deduced from Table 3.12. $e_{\min}$ (not listed in the table) equals $1 - e_{\max}$. Variables p, t, w, $e_{\min}$, $e_{\max}$, and b are the precision, the combination field length, the exponent field length, the minimum exponent, the maximum exponent, and the exponent bias, respectively.

	Extended formats associated with:				
Parameter	binary32	binary64	binary128	decimal64	decimal128
$p \geq$	32	64	128	22	40
$e_{\max} \geq$	1023	16383	65535	6144	24576
$e_{\min} \leq$	−1022	−16382	−65534	−6143	−24575

Table 3.15: Extended format parameters in the IEEE 754-2008 standard [267] (©IEEE, 2008, with permission).

- An *extendable precision format* is a format whose precision and range are defined under user or program control. The standard says that language standards supporting extendable precision shall allow users to specify p and $e_{\max}$ (or, possibly, p only with constraints on $e_{\max}$), and define $e_{\min} = 1 - e_{\max}$.

3.1.1.5 Little-endian, big-endian

The IEEE 754 standard specifies how floating-point data are encoded, but only as a sequence of bits. How such a sequence of bits is ordered in the memory depends on the platform. In general, the bits are grouped into bytes, and these bytes are ordered according to what is called the *endianness* of the platform. On big-endian platforms the most significant byte has the lowest address. This is the opposite on little-endian platforms.

Some architectures, such as IA-64, ARM, and PowerPC are *bi-endian*, i.e., they may be either little-endian or big-endian depending on their configuration.

For instance, the binary64 number that is closest to $-7.0868766365730135 \times 10^{-268}$ is encoded by the sequence of bytes 11 22 33 44 55 66 77 88 on little-endian platforms and by 88 77 66 55 44 33 22 11 on big-endian platforms.

When a format fits into several words (for instance, a 128-bit format on a 64-bit processor), the order of the words does not necessarily follow the endianness of the platform.

Endianness must be taken into account by users who want to exchange data in binary between different platforms.

3.1.2 Attributes and rounding

The IEEE 754-2008 standard defines *attributes* as parameters, attached to a program block, that specify some of its numerical and exception semantics. The availability of *rounding direction attributes* is mandatory, whereas the availability of *alternate exception-handling attributes, preferred width attributes, value-changing optimization attributes,* and *reproducibility attributes* is recommended only. Language standards must provide for constant specification of the attributes, and should also allow for dynamic-mode specification of them.

3.1.2.1 Rounding direction attributes

IEEE 754-2008 requires that the following operations and functions (among others), called "General-computational operations" in [267], be correctly rounded:

- arithmetic operations: addition, subtraction, multiplication, division, fused multiply-add (FMA);
- unary functions: square root, conversion from a supported format to another supported format.

Also, most conversions from a variable in a binary format to a decimal character sequence (typically for printing), or from a decimal character sequence to a binary format, must be correctly rounded: see Section 3.1.5.

Let us now describe the various *directed rounding attributes*.

- The roundTowardPositive attribute corresponds to what was called the round-toward $+\infty$ mode in IEEE 754-1985. The rounding function (see Section 2.2) is RU.
- The roundTowardNegative attribute corresponds to what was called the round-toward $-\infty$ mode in IEEE 754-1985. The rounding function is RD.
- The roundTowardZero attribute corresponds to what was called the round-toward-zero mode in IEEE 754-1985. The rounding function is RZ.

Concerning rounding to nearest, the situation is somewhat different. IEEE 754-1985 had one round-to-nearest mode only, named *round-to-nearest even*. The IEEE 754-2008 standard specifies two *rounding direction attributes to nearest*, which differ in the way of handling the case when an exact result is halfway between two consecutive floating-point numbers:

- roundTiesToEven attribute: if the two nearest floating-point numbers bracketing the exact result are equally near, the one whose least significant significand digit is even is delivered. This corresponds to the *round-to-nearest-even mode* of IEEE 754-1985 (in binary) and IEEE 854-1987[4];

- roundTiesToAway attribute: in the same case as above, the value whose magnitude is larger is delivered.

For instance, in the decimal64 format ($p = 16$), if the exact result of some arithmetic operation is 1.2345678901234565, then the result returned should be 1.234567890123456 with the roundTiesToEven attribute, and 1.234567890123457 with the roundTiesToAway attribute.

There is another important issue with rounding to nearest: In radix-β, precision-p arithmetic, a number of absolute value greater than or equal to $\beta^{e_{\max}}(\beta - \frac{1}{2}\beta^{-p+1})$ will be rounded to infinity (with the appropriate sign). This of course is not what one would infer from a naive understanding of the words *round to nearest*, but the advantage is clear: when the result of an arithmetic operation is a normal number (including the largest one, $\Omega = \beta^{e_{\max}}(\beta - \beta^{-p+1})$), we know that the relative error induced by that operation is small. If huge numbers were rounded to the floating-point value that is really closest to them (namely, $\pm\Omega$), we would have no bound on the relative error induced by an arithmetic operation whose result is $\pm\Omega$.

The standard requires that an implementation (be it binary or decimal) provide the roundTiesToEven and the three directed rounding attributes. A *decimal* implementation must also provide the roundTiesToAway attribute (this is not required for binary implementations).

Having roundTiesToEven as the default rounding direction attribute is mandatory for binary implementations and recommended for decimal implementations. Whereas roundTiesToEven has several advantages (see [342]), roundTiesToAway is useful for some accounting calculations. This is why it is required for radix-10 implementations only, the main use of radix 10 being financial calculations. For instance, the European Council Regulation No. 1103/97 of June 17th 1997 on certain provisions relating to the introduction of the Euro sets out a number of rounding and conversion rules. Among them,

[4]The case where these floating-point numbers both have an odd least significant significand digit (this can occur in precision 1 only, possibly when converting a number such as 9.5 into a decimal string for instance) has been forgotten in the standard, but for the next revision, it has been proposed—See http://speleotrove.com/misc/IEEE754-errata.html—to deliver the one larger in magnitude.

If the application of the conversion rate gives a result which is exactly half-way, the sum shall be rounded up.

3.1.2.2 Alternate exception-handling attributes

It is recommended (but not required) that language standards define means for programmers to be able to associate alternate exception-handling attributes with a block. The alternate exception handlers provide lists of exceptions (invalid operation, division by zero, overflow, underflow, inexact, all exceptions) and specify what should be done when each of these exceptions is signaled. If no alternate exception-handling attribute is associated with a block, the exceptions are treated as explained in Section 3.1.6 (default exception handling).

3.1.2.3 Preferred width attributes

Consider an expression of the form

$$((a+b) \times c + (d+e)) \times f,$$

where a, b, c, d, e, and f are floating-point numbers, represented in the same radix, but possibly with different formats. Variables a, b, c, d, e, and f are *explicit*, but during the evaluation of that expression, there will also be *implicit* variables; for instance, the result r_1 of the calculation of $a+b$, and the result r_2 of the calculation of $r_1 \times c$. When more than one format is available on the system under consideration, an important question arises: *In which format should these intermediate values be represented?* That point was not very clear in IEEE 754-1985. Many choices are possible for the "destination width" of an implicit variable. For instance:

- one might prefer to always have these implicit variables in the largest format provided in hardware. This choice will generally lead to more accurate computations (although it is quite easy to construct counterexamples for which this is not the case);
- one might prefer to clearly specify a destination format. This will increase the portability of the program being written;
- one might require the implicit variables to be of the same format as the operands (and, if the operands are of different formats, to be of the widest format among the operands). This also will improve the portability of programs and will ease the use of smart algorithms such as those presented in Chapters 4, 5, and 11.

The standard recommends (but does not require) that the following preferredWidthNone and preferredWidthFormat attributes should be defined by language standards.

preferredWidthNone attribute: When the user specifies a preferredWidthNone attribute for a block, the destination width of an operation is the maximum of the operand widths.

preferredWidthFormat attributes: When the user specifies a preferredWidthFormat attribute for a block, the destination width is the maximum of the width of the preferredWidthFormat and the operand widths.

3.1.2.4 Value-changing optimization attributes

Some optimizations (e.g., generation of FMAs, use of distributive and associative laws) can enhance performance in terms of speed, and yet seriously hinder the portability and reproducibility of results. Therefore, it makes sense to let the programmer decide whether to allow them or not. The value-changing optimization attributes are used in this case. The standard recommends that language standards should clearly define what is called the "literal meaning" of the source code of a program (that is, the order of the operations and the destination formats of the operations). By default, the implementations should preserve the literal meaning. Language standards should define attributes for allowing or disallowing value-changing optimizations such as:

- applying relations such as $x \cdot y + x \cdot z = x \cdot (y + z)$ (distributivity), or $x + (y + z) = (x + y) + z$ (associativity);
- using FMAs for replacing, e.g., an expression of the form $a \cdot b + c \cdot d$ by $\text{FMA}(a, b, c \cdot d)$;
- using larger formats for storing intermediate results.

3.1.2.5 Reproducibility attributes

The standard requires that conforming language standards should define ways of expressing when reproducible results are required. To get reproducible results, the programs must be translated into an unambiguous sequence of reproducible operations in reproducible formats. As explained in the standard [267], when the user requires reproducible results:

- the execution behavior must preserve what the standard calls the *literal meaning* of the source code[5];
- conversions from and to external character strings must not bound the value of the maximum precision H (see Section 3.1.5) of these strings;

[5]This implies that the language standards must specify what that literal meaning is: order of operations, destination formats of operations, etc.

- when the reproducibility of some operation is not guaranteed, the user must be warned;
- only default exception handling is allowed.

3.1.3 Operations specified by the standard

3.1.3.1 Arithmetic operations and square root

The IEEE 754-2008 standard requires that addition, subtraction, multiplication, FMA, division, and square root of operands of any supported format be provided, with correct rounding (according to the supported rounding direction attributes) to any of the supported formats with same radix.

In other words, it not only mandates support of these basic operations with identical input and output formats (*homogeneous* operations in the standard's terminology), but also with different input and output formats (*formatOf*-operations in the standard). The hardware typically supports the homogeneous case, which is the most common, and heterogeneous operations can be provided at low cost in software [400].

When the sum or difference of two numbers is exactly zero, the returned result is zero, with a $+$ sign in the round-to-nearest, round-toward-zero, and round-toward $+\infty$ modes, and with a $-$ in the round-toward $-\infty$ mode, except for $x + x$ and $x - (-x)$ with x being ± 0, in which case the result has the same sign as x. As noticed by Boldo et al. [51], this means that $x + 0$ cannot be blindly replaced by x: when x is -0, assuming round-to-nearest, $+0$ must be returned. Compiler designers should be aware of such subtleties (see Section 6.2.3.4).

Concerning square root, the result is defined and has a positive sign for all input values greater than or equal to zero, with the exception[6] that $\sqrt{-0} = -0$.

3.1.3.2 Remainders

The remainder must also be provided, but only in homogeneous variants. There are several different definitions of remainders [62]; here is the one chosen for the standard. If x is a finite floating-point number and y is a finite, nonzero floating-point number, then the remainder $r = x \text{ REM } y$ is defined as

1. $r = x - y \times n$, where n is the integer nearest to the exact value x/y;

[6]This rule (which may help in implementing complex functions [317]) may seem strange, but the most important point is that any sequence of exact computations on real numbers will give the correct result, even when $\sqrt{-0}$ is involved. Also let us recall that -0 is regarded as a null value, not a negative number.

2. if x/y is an odd multiple of $1/2$ (i.e., there are two integers nearest to x/y), then n is even;

3. if $r = 0$, its sign is that of x.

A consequence of this definition is that remainders are always exactly representable, which implies that the result returned does not depend on the rounding function.

The result of x REM ∞ is x.

3.1.3.3 Preferred exponent for arithmetic operations in the decimal format

Let $Q(x)$ be the quantum exponent of a floating-point number x. Since some numbers in the decimal format have several possible representations (as mentioned in Section 3.1.1.2, the set of their representations is a *cohort*), the standard specifies for each operation which exponent is preferred for representing the result of a calculation. The rule to be followed is:

- if the result of an operation is inexact, the cohort member of smallest exponent is used;

- if the result of an operation is exact, then if the result's cohort includes a member with the preferred exponent (see below), that member is returned; otherwise, the member with the exponent closest to the preferred exponent is returned.

The preferred quantum exponents for the most common operations are:

- $x + y$ and $x - y$: $\min(Q(x), Q(y))$;

- $x \times y$: $Q(x) + Q(y)$;

- x/y: $Q(x) - Q(y)$;

- FMA(x, y, z) (i.e., $xy + z$ using an FMA): $\min(Q(x) + Q(y), Q(z))$;

- $\sqrt{x}$: $\lfloor Q(x)/2 \rfloor$.

3.1.3.4 scaleB and logB

When designing fast software for evaluating elementary functions, or for efficiently scaling variables (for instance, to write robust code for computing functions such as $\sqrt{x^2 + y^2}$), it is sometimes very useful to have functions $x \cdot \beta^n$ and $\lfloor \log_\beta |x| \rfloor$, where β is the radix of the floating-point system, n is an integer, and x is a floating-point number. This is the purpose of the functions scaleB and logB:

- scaleB(x, n) is equal to $x \cdot \beta^n$, correctly rounded[7] (following the rounding direction attribute);
- when x is finite and nonzero, logB(x) equals $\lfloor \log_\beta |x| \rfloor$. When the output format of logB is a floating-point format, logB(NaN) is NaN, logB($\pm\infty$) is $+\infty$, and logB(± 0) is $-\infty$.

3.1.3.5 Miscellaneous

The standard defines many useful operations, see [267]. Some examples are

- nextUp(x), which returns the smallest floating-point number in the format of x that is greater than x;
- maxNum(x, y), which returns the maximum of x and y (the next revision plans to replace the current minimum/maximum operations to solve an issue as explained in Section 3.1.7.1);
- class(x), which tells whether x is a signaling NaN, a quiet NaN, $-\infty$, a negative normal number, a negative subnormal number, -0, $+0$, a positive subnormal number, a positive normal number, or $+\infty$.

3.1.4 Comparisons

It must be possible to compare two floating-point numbers, in all formats specified by the IEEE 754-2008 standard, even if their formats differ, provided that they have the same radix. This can be done either by means of a condition code identifying one of the four mutually exclusive following conditions: *less than, equal, greater than,* and *unordered*; or as a Boolean response to a predicate that gives the desired comparison. The *unordered* condition arises when at least one of its operands is a NaN: a NaN compares unordered with everything *including itself*. A consequence of this is that the test

$$x \neq x$$

returns **true** when x is a NaN. As pointed out by Kahan [318], this provides a way of checking if a floating-point datum is a NaN in languages that lack an instruction for doing that (assuming the test is not optimized out). The other tests involving a NaN will return **false**. Hence, the test

$$x \leq y$$

is not always equivalent to the test

$$\text{not}(x > y).$$

[7] In most cases, $x \cdot \beta^n$ is exactly representable so that there is no rounding at all, but requiring correct rounding is the simplest way of defining what should be returned if the result is outside the normal range.

If at least one of the two operands is a NaN, the first test will return *false* whereas the second one will return *true*.

Also, the test $+0 = -0$ must return *true*.

Again, users and especially compiler designers should be aware of these subtleties.

As mentioned above, floating-point data represented in different formats specified by the standard must be comparable, but only *if these formats have the same radix*; the standard does not require that comparing a decimal and a binary number should be possible without a preliminary conversion. Such mixed-radix comparisons appear extremely rarely in programs written by good programmers and, at the time the standard was released, it seemed very tricky to implement them without preliminary conversion. Performing correct comparisons, however, is presumably easier than what was believed (see [73, 40]).

3.1.5 Conversions to/from string representations

Concerning conversions between an external decimal or hexadecimal character sequence and an internal binary or decimal format, the requirements of IEEE 754-2008 are much stronger than those of IEEE 754-1985. They are described as follows.

1. Conversions between an external decimal character sequence and a supported decimal format: Input and output conversions are correctly rounded (according to the applicable rounding direction).

2. Conversions between an external hexadecimal character sequence and a supported binary format: Input and output conversions are also correctly rounded (according to the applicable rounding direction). They have been specified to allow any binary number to be represented exactly by a finite character sequence.

3. Conversions between an external decimal character sequence and a supported binary format: first, for each supported binary format, define a value p_{10} as the minimum number of decimal digits in the decimal external character sequence that allows for an error-free write-read cycle, as explained in Section 4.9. Table 3.16, which gives the value of p_{10} from the various basic binary formats of the standard, is directly derived from Table 4.2.

   The standard requires that there should be an implementation-defined value H, preferably unbounded, and in any case larger than or equal to 3 plus the largest value of p_{10} for all supported binary formats, such that the conversions are correctly rounded to and from external decimal character sequences with any number of significant digits between 1

format	binary32	binary64	binary128
p_{10}	9	17	36

Table 3.16: Minimum number of decimal digits in the decimal external character sequence that allows for an error-free write-read cycle, for the various basic binary formats of the standard. See Section 4.9 for further explanation.

and H. This implies that these conversions must always be correctly rounded if H is unbounded.

For output conversions, if the external decimal format has more than H significant digits, then the binary value is correctly rounded to H decimal digits and trailing zeros are appended to fill the output format. For input conversions, if the external decimal format has more than H significant digits, then the internal binary number is obtained by first correctly rounding the value to H significant digits (according to the applicable rounding direction), then by correctly rounding the resulting decimal value to the target binary format (with the applicable rounding direction). In the directed rounding directions, these rules allow intervals to be respected (interval arithmetic is dealt with in Chapter 12).

More details are given in the standard [267].

3.1.6 Default exception handling

The IEEE 754-2008 standard supports the five exceptions already listed in Section 2.5. Let us examine them.

3.1.6.1 Invalid operation

This exception is signaled each time there is no satisfactory way of defining the numeric result of some operation. The default result of such an operation is a quiet NaN (see Section 3.1.7.1), and it is recommended that its payload contains some diagnostic information. The operations that lead to an invalid operation exception are:

- an operation on a signaling NaN (see Section 3.1.7.1), for most operations;
- a multiplication of the form $0 \times \infty$ or $\infty \times 0$;
- an FMA of the form $\text{FMA}(0, \infty, x)$ (i.e., $0 \times \infty + x$) or $\text{FMA}(\infty, 0, x)$, unless x is a quiet NaN (in that last case, whether the invalid operation exception is signaled is implementation defined);
- additions/subtractions of the form $(-\infty) + (+\infty)$ or $(+\infty) - (+\infty)$;

- FMAs that lead to the subtraction of infinities of the same sign (e.g., FMA$(+\infty, -1, +\infty)$);
- divisions of the form $0/0$ or ∞/∞;
- remainder$(x, 0)$, where x is not a NaN;
- remainder(∞, y), where y is not a NaN;
- $\sqrt{x}$ where $x < 0$;
- conversion of a floating-point number x to an integer, where x is $\pm\infty$, or a NaN, or when the result would lie outside the range of the chosen integer format;
- comparison using unordered-signaling predicates (called in the standard compareSignalingEqual, compareSignalingGreater, compareSignalingGreaterEqual, compareSignalingLess, compareSignalingLessEqual, compareSignalingNotEqual, compareSignalingNotGreater, compareSignalingLessUnordered, compareSignalingNotLess, and compareSignalingGreaterUnordered), when the operands are unordered;
- logB(x) where x is NaN or ∞;
- logB(0) when the output format of logB is an integer format (when it is a floating-point format, the value to be returned is $-\infty$).

3.1.6.2 Division by zero

The words "division by zero" are misleading, since this exception is signaled whenever an *exact* infinite result is obtained from an operation on *finite* operands. The most frequent case, of course, is the case of a division by zero, but this can also appear, e.g., when computing the logarithm of zero or the arctanh of 1. An important case is logB(0) when the output format of logB is a floating-point format. The result returned is infinity, with the correct sign.

3.1.6.3 Overflow

Let us call an *intermediate* result what would have been the rounded result if the exponent range were unbounded. The overflow exception is signaled when the absolute value of the intermediate result is finite and strictly larger than the largest finite number $\Omega = (\beta - \beta^{1-p}) \cdot \beta^{e_{\max}}$, or equivalently, when it would have an exponent strictly larger than $e_{\max}$. When an overflow occurs, the result returned depends on the rounding direction attribute:

- it will be $\pm\infty$ with the two "round-to-nearest" attributes, namely roundTiesToEven and roundTiesToAway, with the sign of the intermediate result;
- it will be $\pm\Omega$ with the roundTowardZero attribute, with the sign of the intermediate result;
- it will be $+\Omega$ for a positive intermediate result and $-\infty$ for a negative one with the roundTowardNegative attribute;
- it will be $-\Omega$ for a negative intermediate result and $+\infty$ for a positive one with the roundTowardPositive attribute.

Furthermore, the overflow flag is raised and the inexact exception is signaled.

It is important to understand three consequences of these rules:

- as we have already seen, with the two "round-to-nearest" attributes, if the absolute value of the exact result of an operation is greater than or equal to
$$\beta^{e_{\max}} \cdot \left(\beta - \frac{1}{2}\beta^{1-p}\right) = \Omega + \frac{1}{2}\,\mathrm{ulp}(\Omega),$$
then an infinite result is returned, which is not what one could expect from a naive interpretation of the words "round to nearest";
- "overflow" is *not* equivalent to "infinite result returned";
- with the roundTowardZero attribute, "overflow" is *not* equivalent to "$\pm\Omega$ is returned": if the absolute value of the exact result of some operation is larger than or equal to Ω, and strictly less than $\beta^{e_{\max}}$, then $\pm\Omega$ is returned, and yet there is no overflow.

3.1.6.4 Underflow

The underflow exception is signaled when a nonzero result whose absolute value is strictly less than $\beta^{e_{\min}}$ is computed.

- For *binary formats*, unfortunately, there is some ambiguity in the standard.[8] See Section 2.1.3 for more explanation. The underflow can be signaled either *before rounding*, that is, when the absolute value of the exact result is nonzero and strictly less than $2^{e_{\min}}$, or *after rounding*, that is, when the absolute value of a nonzero result computed as if the exponent range were unbounded is strictly less than $2^{e_{\min}}$. In rare cases, this can make a difference, for instance, when computing
$$\mathrm{FMA}\left(-2^{e_{\min}}, 2^{-p-1}, 2^{e_{\min}}\right)$$

[8] This problem was already there in the 1985 version of the standard. The choice of not giving a clearer specification in IEEE 754-2008 probably results from the desire to keep existing implementations conforming to the standard.

in rounding to nearest, an underflow will be signaled if this is done before rounding, but not if it is done after rounding.

- For *decimal formats*, there is no ambiguity and the underflow result is signaled *before rounding*, i.e., when the absolute value of the exact result is nonzero and strictly less than $10^{e_{\min}}$.

The result is always correctly rounded: the choice (in the binary case) of how the underflow is detected (that is, before or after rounding) has no influence on the result delivered.

In case of underflow, if the result is inexact, then the underflow flag is raised and the inexact exception is signaled. If the result is exact, then the underflow flag is *not* raised. This might sound strange, but this was an adroit choice of the IEEE working group: the major use of the underflow flag is for warning that the result of some operation might not be very accurate—in terms of relative error. Thus, raising it when the operation is *exact* would be a needless warning. *This should not be thought of as an extremely rare event*: indeed, Theorem 4.2 shows that with any of the two round-to-nearest rounding direction attributes, whenever an addition or subtraction underflows, it is performed exactly.

3.1.6.5 Inexact

If the result of an operation differs from the exact result, then the inexact exception is signaled. The correctly rounded result is returned.

3.1.7 Special values

3.1.7.1 NaN: Not a Number

The standard defines two types of NaNs:

- *signaling NaNs* (sNaNs) do not appear, in default mode, as the result of arithmetic operations. Except for the sign-bit operations (such as copy, negate, and absolute value), decimal re-encoding operations, and some conversions, they signal the invalid operation exception whenever they appear as operands. For instance, they can be used for uninitialized variables;

- *quiet NaNs* (qNaNs) propagate through almost all operations[9] without signaling exceptions. They can be used for debugging and diagnostic purposes. As stated above, for operations that deliver a floating-point result, the default exception handling is that a quiet NaN is returned

[9]There are a few exceptions to this rule (but not with the basic arithmetic operations): for instance it is recommended that $\mathrm{pow}(+1, y)$ should be 1 even if y is a quiet NaN (this partial function being constant).

whenever an invalid operation exception occurs; this rule is valid for all operations of IEEE 754-2008, but the new revision plans to introduce new operations that will not follow it, as explained below.

For example, qNaN $\times$ 8, sNaN + 5, and $\sqrt{-2}$ all give qNaN.

An issue with the various rules on NaNs is that from a signaling NaN, one may get a non-NaN result. Thus users who wish to detect the use of uninitialized variables via signaling NaNs should test the invalid operation exception instead of the result. Moreover, these rules make the current operations that return the minimum or maximum of two data[10] non-associative, which is an issue in some contexts, such as parallel processing; for instance:

- $\mathrm{minNum}(1, \mathrm{minNum}(1, \mathrm{sNaN})) \rightarrow \mathrm{minNum}(1, \mathrm{qNaN}) \rightarrow 1;$
- $\mathrm{minNum}(\mathrm{minNum}(1, 1), \mathrm{sNaN}) \rightarrow \mathrm{minNum}(1, \mathrm{sNaN}) \rightarrow \mathrm{qNaN}.$

The next revision plans to replace these operations by ones that will be associative, thus with specific rules.[11]

We have seen in Section 3.1.1 that in the binary interchange formats, the least significant $p - 2$ bits of a NaN are not defined, and in the decimal interchange formats, the trailing significand bits of a NaN are not defined. These bits can be used for encoding the *payload* of the NaN, i.e., some information that can be transmitted through the arithmetic operation for diagnostic purposes. To preserve this diagnostic information, for an operation with quiet NaN inputs and a quiet NaN result, the returned result should be one of these input NaNs.

3.1.7.2 Arithmetic of infinities and zeros

The arithmetic of infinities and zeros follows the intuitive rules. For instance, $-1/(-0) = +\infty$, $-5/(+\infty) = -0$, $\sqrt{+\infty} = +\infty$ (the only somewhat counter intuitive property is $\sqrt{-0} = -0$). This very frequently allows one to get sensible results even when an underflow or an overflow has occurred. And yet, one should be cautious. Consider for instance, assuming round-to-nearest (with any choice in case of a tie), the computation of

$$f(x) = \frac{x}{\sqrt{1 + x^2}},$$

for $\sqrt{\Omega} < x \le \Omega$, where Ω is the largest finite floating-point number. The computation of x^2 will return an infinite result; hence, the computed value of $\sqrt{1 + x^2}$ will be $+\infty$. Since x is finite, by dividing it by an infinite value we

[10]These operations regard qNaN as missing data, but not sNaN.

[11]This means that in the case where a NaN is not regarded as missing data, qNaN must propagate even when the partial function is constant. For instance, the minimum function on (qNaN,$-\infty$) will return qNaN instead of $-\infty$.

will get $+0$. Therefore, the computed value of $f(x)$, for x large enough, will be $+0$, whereas the exact value of $f(x)$ is extremely close to 1. This shows, for critical applications, the need either to prove in advance that an overflow/underflow cannot happen, or to check, by testing the appropriate flags, that an overflow/underflow has not happened.

3.1.8 Recommended functions

The standard recommends (but does not require) that the following functions should be correctly rounded: e^x, $e^x - 1$, 2^x, $2^x - 1$, 10^x, $10^x - 1$, $\ln(x)$, $\log_2(x)$, $\log_{10}(x)$, $\ln(1+x)$, $\log_2(1+x)$, $\log_{10}(1+x)$, $\sqrt{x^2+y^2}$, $1/\sqrt{x}$, $(1+x)^n$, x^n, $x^{1/n}$ (n is an integer), $\sin(\pi x)$, $\cos(\pi x)$, $\arctan(x)/\pi$, $\arctan(y/x)/\pi$, $\sin(x)$, $\cos(x)$, $\tan(x)$, $\arcsin(x)$, $\arccos(x)$, $\arctan(x)$, $\arctan(y/x)$, $\sinh(x)$, $\cosh(x)$, $\tanh(x)$, $\sinh^{-1}(x)$, $\cosh^{-1}(x)$, $\tanh^{-1}(x)$ (note that the power function x^y is not in this list). This was not the case with the previous version (754-1985) of the standard.

See Section 10.5 for an introduction to the various issues linked with the correct rounding of transcendental functions.

3.2 On the Possible Hidden Use of a Higher Internal Precision

The processor being used may offer an internal precision that is wider than the precision of the variables of a program (a typical example is the double-extended format available on Intel platforms when we use the x87 instruction set, when the variables of the program are binary32 or binary64 floating-point numbers). This may sometimes have strange side effects, as we will see in this section.

Consider the C program (Program 3.1).

```
#include <stdio.h>

int main(void)
{
  double a = 1848874847.0;
  double b = 19954562207.0;
  double c;
  c = a * b;
  printf("c = %20.19e\n", c);
  return 0;
}
```

Program 3.1: A C program that might induce a double rounding.

Table 3.17 gives some results returned by this program, depending on the compilation options. In order to really test the arithmetic of the machine, it is important that the compiler does not optimize the multiplication by performing it at compile time (one controls even less what occurs at compile time); by default, GCC does not do such an optimization. Note that the double-precision number closest to the exact product is `3.6893488147419111424e+19`.

Switches on the GCC command line	Output
no switch (default)	c = 3.6893488147419111424e+19
-mfpmath=387	c = 3.6893488147419103232e+19

Table 3.17: Results returned by Program 3.1 on a Linux/Debian 64-bit Intel Xeon CPU platform, with GCC 7.2.0, depending on the compilation options. On a 64-bit platform, the default is to use the SSE registers.

What happened? The exact value of `a*b` is 36893488147419107329, whose binary representation is

$$\overbrace{\underbrace{10000000000000000000000000000000000000000000000000000}_{\text{53 bits}}10000000000}^{\text{64 bits}}01$$

On the processor used, with the `-mfpmath=387` switch, the product is first rounded to the precision of the `x87` registers (namely, "double-extended" precision), which gives (in binary)

$$\overbrace{\underbrace{10000000000000000000000000000000000000000000000000000}_{\text{53 bits}}10000000000}^{\text{64 bits}}\times 4$$

Then, that intermediate value is rounded to the binary64 destination format, which gives (using the round-to-nearest-even rounding mode)

$$\underbrace{10000000000000000000000000000000000000000000000000000}_{\text{53 bits}}\times 2^{13}$$

$$= 36893488147419103232_{10},$$

whereas the product `a*b` correctly rounded to the nearest binary64 number is

$$\underbrace{10000000000000000000000000000000000000000000000000001}_{\text{53 bits}}\times 2^{13}$$

$$= 36893488147419111424_{10}.$$

In general, by default,[12] the product is directly stored in the 64-bit Streaming SIMD Extension (SSE) registers. In that case, it is directly rounded to binary64, so that we get the expected result.

The problem we have faced here is called "double rounding." In this example, it appears during a multiplication, but it may also appear during another arithmetic operation. Another example (still with binary64 input values) is the addition of

$$9223372036854775808.0 = 2^{63}$$

and

$$1024.25.$$

Such examples are not so rare that they can be neglected. Assuming binary64 variables and an Intel "double-extended" internal format, if the chosen compilation switches (or the default mode) do not prevent the problem from occurring, the double rounding problem occurs when the binary expansion of the exact result of some operation is of the form

$$2^k \times \overbrace{1.xxxxx \cdots xx0}^{\text{53 bits}} \; \overbrace{10000000000}^{\text{11 bits}} \; 0 \; \overbrace{xxxxxxxxxxxxxxxxxxxxxx \cdots}^{\text{at least one 1 somewhere}}$$

or

$$2^k \times \overbrace{1.xxxxx \cdots xx1}^{\text{53 bits}} \; \overbrace{01111111111}^{\text{11 bits}} \; 1 \; \overbrace{xxxxxxxxxxxxxxxxxxxxxx \cdots}^{\text{at least one 0 somewhere}}.$$

Assuming equal probabilities of occurrence for the zeros and ones in the binary expansion of the result of an arithmetic operation,[13] the probability of a double rounding is $2^{-12} = 1/4096$, which means that without care with the compilation options, double roundings will occur in any computation of significant size. We must emphasize that this might be a problem with certain very specific algorithms (such as those presented in Chapter 4, see [409] for a discussion on that topic), but with most calculations, it will be unnoticed.

The possible, sometimes hidden, use of a larger internal precision may also lead to a management of overflow and underflow that is sometimes difficult to predict. Consider Program 3.2, due to Monniaux [423].

Compiled with GCC 4.9.2 under Linux on an Intel(R) Xeon(R) CPU E5-2695 v2 processor, the program returns $+\infty$. If we add the `-mfpmath=387` command line option, we get `1e+308`. What happened? Although in binary64 arithmetic, with the round-to-nearest ties-to-even (i.e., the default) rounding function, the multiplication `v * v` should return $+\infty$, in the program compiled with the `-mfpmath=387` option, the implicit variable representing this

[12] But it is the default on recent systems only.

[13] Which is not very realistic but suffices to get a rough estimate of the frequency of occurrences of double roundings.

```
#include <stdio.h>
int main(void)
{
  double v = 1E308;
  double x = (v * v) / v;
  printf("%g\n",x);
  return 0;
}
```

Program 3.2: This example is due to David Monniaux [423]. Compiled with GCC under Linux, we get `1e+308` *with the command line option* `-mfpmath=387` *and* $+\infty$ *without it.*

product was actually stored in a "double-extended" precision register of the `x87` instruction set of the processor. And since the product `v * v` is much below the overflow threshold in double-extended precision, the stored value was not $+\infty$, but the double-extended number closest to the exact product.

It is important to note that, in this case, the result obtained is *very accurate*, which is not so surprising: in most cases, using a larger internal precision for intermediate calculations leads to better calculations. What matters then is not to forbid the behavior, but to allow programmers to decide if they want all intermediate calculations to be performed in the format of the operands or in a format they clearly specify (which enhances portability and provability), or if they prefer these intermediate calculations to be performed in a wider format whenever available (typically, the largest format available in hardware, which in general improves the accuracy of the results). A tradeoff is to be found between portability, accuracy, and (frequently) speed. Choosing which among these criteria is the most important should be the programmer's task, not the compiler's. This is a reason for the introduction of the preferred width attributes in IEEE 754-2008 (see Section 3.1.2.3).

3.3 Revision of the IEEE 754-2008 Standard

At the time we are writing this book, the IEEE 754-2008 standard is under revision. The next version should be released in 2018. The goal of the revision committee is mainly to deal with the errata and improve clarity; hence the standard will be almost unchanged. Since the next version is not yet adopted, we cannot be definite about its contents. However, the main changes are expected to be:

- the replacement of the current Min/Max operations by new ones (see Section 3.1.7.1);
- recommended "augmented" addition and multiplication operations, which would more or less return the same results as the 2Sum and

Fast2Sum algorithms (described in Chapter 4): the sum or product x of two floating-point numbers would be expressed as a "double word," that is, a pair of floating-point numbers x_h and x_ℓ such that $x_h = \mathrm{RN}(x)$ and $x = x_h + x_\ell$. It is likely that in these operations, the RN function will be round to nearest ties-to-zero.

3.4 Floating-Point Hardware in Current Processors

Virtually all recent computers are able to support the IEEE 754 standard efficiently through a combination of hardware and software. This section reviews the hardware that can be found in mainstream systems. It should be understood that the line between hardware and software support is drawn by hardware vendors depending on the demand of the target market. To take just a few examples,

- IBM supports decimal floating-point in hardware [184, 548] (with the decimal encoding), while Intel provides well-tuned software implementations [117] (with the binary encoding). These encodings are presented in Section 3.1.1.2.
- Elementary functions are usually implemented in software, but some GPUs offer a limited set of hardware implementations [470].
- For the basic operations, the standard mandates rounding *to any supported format, from any combination of operand formats*. However, hardware usually only supports *homogeneous* operations, i.e., operations that have the same input and output format. The mixed-format operations are considered rare enough to be implemented in software only [400].
- One notable exception to the previous statement is the mixed-format FMA of the Kalray processors, where the addend and the result are binary64 while the two multiplicands are binary32 [83, 84]. Binary16/binary32 mixed-format operations are also appearing in GPUs. These are discussed in more detail in Section 7.8.2.

3.4.1 The common hardware denominator

Current processors for desktop computers offer hardware binary64 operators for floating-point addition, subtraction, and multiplication, and at least hardware assistance for division and square root. Peak performance is typically between 2 and 8 binary64 floating-point operations per clock cycle for $+$, $-$, and $\times$, with much slower division and square root [469, 271].

However, most processors go beyond this common denominator and offer larger precision and/or faster operators. The following sections detail these extensions.

3.4.2 Fused multiply-add

The *fused multiply-add* (FMA) instruction evaluates an expression of the form $a \times b + c$ with one rounding only, that is, if $\circ$ is the rounding function, the returned result is $\circ(a \times b + c)$ (see Section 2.4). The FMA has been the main floating-point operation in the most recent instruction sets (IBM POWER/PowerPC, HP/Intel IA-64, Kalray MPPA). It has also been added to older instruction sets:

- Intel IA-32 with AMD's SSE5 and Intel's AVX extensions, as detailed below;
- ARM since ARMv7;
- SPARC with HAL/Fujitsu SPARC64 VI[14];
- MIPS with the Loongson processor family.

These operations are all compatible with the FMA defined by IEEE 754-2008. As far as this operator is concerned, IEEE 754-2008 standardized already existing practice.

In terms of instruction set, the FMA comes in two variants:

- an FMA4 instruction specifies 4 registers a, b, c, and d, and stores in d the result $\circ(a \cdot b + c)$.
- an FMA3 instruction specifies only 3 registers: the destination d has to be one of the input registers, which is thus overwritten.

In each instruction set, there is actually a family of instructions that includes useful variations such as fused multiply-subtract.

FMA latency. In general, the FMA is pipelined and has a latency slightly larger than that of a floating-point addition or multiplication alone. For illustration, the FMA latency was 4 cycles in Itanium2 and 7 cycles on Power6. In a recent ARM implementation [399], the 7-cycle FMA is replaced with an 8-cycle one, built by connecting an adder and a multiplier (suitably modified). This enables additions and multiplications in 4 cycles instead of 7.

The chaotic introduction of the FMA in the IA-32-compatible processors. In 2007, AMD was the first to announce FMA support with its SSE5 extension [2]. Intel followed in 2008 by announcing an FMA4. Later this year, it switched to FMA3 with a different extension: AVX. Meanwhile, AMD

[14]Warning! The instructions called `FMADDs` and so on from SPARC64 V, which share the same name and the same encoding with SPARC64 VI, are *not* real FMA instructions as they perform two roundings. [208, page 56].

switched to an FMA4 compatible with Intel's first announce, and its first Bulldozer processors in 2005 supported this FMA4. Later processors introduced FMA3 instructions compatible with Intel's. Intel's Haswell was its first processor supporting an FMA3.

As of 2017, the situation has settled, and processors from both vendors support the same FMA3 instructions in the AVX instruction set extension.

3.4.3 Extended precision and 128-bit formats

As we have already seen in Sections 3.1.1 and 3.2, the legacy x87 instructions of the IA-32 instruction set can operate on a *double-extended* precision format with 64 bits of significand and 15 bits of exponent. The corresponding floating-point operators can be instructed to round to single, double, or double-extended precision.

The IA-64 instruction set also defines several double-extended formats, including a 80-bit format compatible with IA-32 and a 82-bit format with a 64-bit significand and a 17-bit exponent. The two additional exponent bits are designed to avoid intermediate "spurious" overflows in certain computations on 80-bit operands (a typical example is the evaluation of $\sqrt{a^2 + b^2}$).

Some instruction sets (SPARC, POWER, z/Architecture) have instructions operating on binary128 data. As far as we know, however, the internal data-paths are still optimized for 64-bit operations, and these instructions require several passes over 64-bit operators.

3.4.4 Rounding and precision control

Traditionally, the rounding precision (e.g., binary32, binary64, double-extended if available) and the rounding direction attributes used to be specified via a global status/control register. In IA-32, this was called `FPSR` for *Floating-Point Status Register*, or `MXCSR` for SSE/AVX extensions. Such a global register defines the behavior of the floating-point instructions.

However, recent processors are designed to execute, at a given time, many floating-point instructions launched out of order from several hardware threads. Keeping track of changes to the control/status word would be very costly in this context. Therefore, the prevailing choice in the recent years has been to ensure that all the instructions in flight share the same value of the status/control word. To achieve this, any instruction that changes the value of the control word must first wait until all in-flight instructions have completed execution.

In other words, before launching any new floating-point instruction with a new value of the control register, all current floating-point instructions have to terminate with the previous value. This can stall the processor for tens of cycles.

Unfortunately, some applications, such as interval arithmetic (see [428] and also Section 12.3.2), need frequent rounding direction changes. This performance issue could not be anticipated in 1985, when processor architectures were not yet pipelined. It also affects most processor instruction sets designed in the 1980s and 1990s.

To address this performance issue, more recent instruction sets make it possible to change the rounding direction attribute on a per-instruction basis without any performance penalty. Technically, the rounding direction attribute is defined in the instruction word, not in a global control register. This feature was for instance introduced in the HP/Intel IA-64 [118] instruction set and the Sun Microsystems' VIS extension to the SPARC instructions set [579]. It was introduced in the IA-32 instruction set by the AVX-512 extension[15] [272, Sec. 15.6.4]. Note that in all these cases, it is still possible to read the rounding precision and rounding direction from a global control/status register. In highly parallel processors such as Graphics Processing Units, rounding is also typically defined on a per-instruction basis.

The rounding direction specification in the IEEE 754-1985 standard (and hence in the language standards that were later on adapted to implement it) reflected the notion of a global status word. This meant in practice that per-instruction rounding specification could not be accessed from current high-level languages in a standard, portable way.

The IEEE 754-2008 standard corrected this, but the change will take time to percolate in programming languages and compilers. This issue will be addressed in more detail in Chapter 6, Languages and Compilers.

3.4.5 SIMD instructions

Most recent instruction sets also offer single instruction, multiple data (SIMD) instructions. An SIMD instruction applies the same operation to a vector of data, producing a vector of results. These vectors are kept in wide registers, for instance 128- to 512-bit registers in the IA-32 instruction set extensions, and up to 2048 bits for the ARM Scalable Vector Extension (SVE).

Such wide registers can store vectors of 8-bit, 16-bit, 32-bit, or 64-bit integers. SIMD instructions operating on vectors of small integers are often referred to as *multimedia instructions*. Indeed, in image processing, the color of a pixel may be defined by three 8-bit integers giving the intensity of the red, green, and blue components; in audio processing, sound samples are commonly digitized on 16 bits.

A wide register can also be considered as a vector of 16-bit, 32-bit, or 64-bit floating-point numbers.

[15] These instructions also suppress ("silent") exceptions, which induce similar problems in case of out-of-order execution (although their correct management in hardware must be implemented anyway).

Examples of vector instruction sets include AltiVec for the POWER/PowerPC family, and for the IA-32 instruction set, 3DNow!/MMX (64-bit vector), then SSE to SSE5 (128-bit vector), then AVX/AVX2 (256-bit vector) and AVX-512 (512-bit vector).

Each of these extensions comes with too many new instructions to be detailed here: not only arithmetic operations, but also data movement inside a vector, and complex operations such as scalar products, or sums of absolute values of differences.

At the time of writing this book, all new IA-32-compatible processors for workstation and personal computers implement the AVX extension (some low-power chips such as Intel's Atom are still limited to SSE3). AVX defines sixteen 256-bit registers, each of which can be considered either as a vector of eight binary32 numbers, or as a vector of four binary64 numbers. AVX instructions include the FMA and are fully IEEE 754-2008 compliant.[16]

3.4.6 Binary16 (half-precision) support

The 16-bit formats were introduced for graphics and gaming; the binary16 format of IEEE 754-2008 actually standardized existing practice. Although it was not considered as a basic format by the IEEE 754-2008 standard, binary16 is increasingly being used for computing. This is true in graphics, but also in the field of machine learning. In particular, it has been shown that certain convolution neural networks (CNN) can work with precisions as low as 8 bits. This justifies the use of the binary16 format, which provides a 11-bit significand.

This is leading to extensive binary16 arithmetic support in newer instruction sets such as ARMv8, Kalray Coolidge, and NVIDIA Volta.

3.4.7 Decimal arithmetic

At the time of writing this book, only high-end processors from IBM (POWER and zSeries) include hardware decimal support, using the binary encoding.

The latency of decimal operations is much larger than that of binary operations. Moreover, it depends on the operand values [184, 548], as illustrated by Table 3.18. This is due to several factors. The encoding itself is more complex (see Section 3.1.1.2). Since numbers are not necessarily normalized, significand alignment and rounding rules are also much more complex than in the binary case (see for instance the preferred exponent rules Section 3.1.3.3). Finally, the core of decimal units in current IBM processors is a pipelined 36-digit adder [548], and multiplication and division must iterate over it.

[16]The performance of AVX instructions can be degraded on some computations involving subnormals.

Cycles	decimal64 operands	decimal128 operands
addition/subtraction	12 to 28	16 to 31
multiplication	16 to 55	17 to 104
division	16 to 119	17 to 193

Table 3.18: Execution times in cycles of decimal operations on the IBM z10, from [548].

3.4.8 The legacy x87 processor

The Intel 8087 co-processor was a remarkable achievement when it was conceived. More than thirty years later, the floating-point instructions it defined are still available in IA-32. However, they are showing their age.

- There are only 8 floating-point registers, and their organization as a stack leads to data movement inefficiencies.
- The hidden use of extended precision entails a risk of double rounding (see Section 3.2).
- The dynamic rounding precision can introduce bugs in modern software, which is almost always made up of several components (dynamic libraries, plug-ins). For instance, the following bug in Mozilla's Javascript engine was discovered in 2006: if the rounding precision was reduced to single precision by a plug-in, then the `js_dtoa` function (double-to-string conversion) could overwrite memory, making the application behave erratically, e.g., crash. The cause was the loop exit condition being always false due to an unexpected floating-point error.[17]
- The x87 FPSR register defines the rounding precision (the significand size) but not the exponent size, which is always 15 bits. Even when instructed to round to single precision, the floating-point unit will signal overflows or underflows only for numbers out of the *double-extended* exponent range. True binary32 or binary64 overflow/underflow detection is performed only when writing the content of a floating-point register to memory. This two-step overflow/underflow detection can lead to subtle software problems, just like double rounding. It may be avoided only by writing all the results to memory, unless the compiler can prove in advance that there will be no overflows.

The SSE and AVX instructions were designed more recently. As they do not offer extended precision, they may result in less accurate results than the legacy x87 instructions. However, in addition to their obvious performance

[17] CVE-2006-6499 / https://bugzilla.mozilla.org/show_bug.cgi?id=358569.

advantage due to SIMD execution, they are fully IEEE 754-1985 (SSE) and IEEE 754-2008 (AVX) compliant. They permit better reproducibility (thanks to the static rounding precision) and portability with other platforms.

All these reasons push towards deprecating the use of the venerable x87 instructions. For instance, GCC uses SSE2 instructions by default on all 64-bit variants of GNU/Linux (since 64-bit capable processors all offer the SSE2 extension). The legacy x87 unit is still available to provide higher precision for platform-specific kernels that may require it, for instance some elementary function implementations.

3.5 Floating-Point Hardware in Recent Graphics Processing Units

Graphics processing units (GPUs), initially highly specialized for integer computations, quickly evolved towards more and more programmability and increasingly powerful arithmetic capabilities.

Binary floating-point units appeared in 2002-2003 in the GPUs of the two main vendors, ATI (with a 24-bit format in the R300 series) and NVIDIA (with a 32-bit format in the NV30 series). In both implementations, addition and multiplication were incorrectly rounded: according to a study by Collange et al. [109], instead of rounding the exact sum or product, these implementations typically rounded a $p+2$-bit result to the output precision of p bits.

Still, these units fueled interest in GPUs for general-purpose computing (GPGPU), as the theoretical floating-point performance of a GPU is up to two orders of magnitude times that of a conventional processor (at least in binary32 arithmetic). At the same time, programmability was also improved, notably to follow the evolution to version 10 of Microsoft's DirectX application programming interface. Specific development environments also appeared: first NVIDIA's C-based CUDA, soon followed by the Khronos Group's OpenCL.

Between 2007 and 2009, both ATI (now AMD) and NVIDIA introduced new GPU architectures with, among other things, improved floating-point support.

Currently, most GPUs support the binary32 and binary64 formats, with correct rounding in the four legacy rounding modes for basic operations and FMA, with subnormals [631]. It is worth mentioning that they also include hardware acceleration of some elementary functions [470]. The latest notable evolution of floating-point in GPUs has been the introduction of half precision (binary16) arithmetic to accelerate machine learning applications. Subnormals are fully supported in binary16.

The remaining differences with mainstream microprocessors, in terms of floating-point arithmetic, are mostly due to the highly parallel execution model of GPUs.

- Current GPUs do not raise floating-point exceptions, and there are no internal status flags to check for them.

- The rounding mode is part of the instruction opcode (thus statically determined at compile time) rather than stored in a global status register.

3.6 IEEE Support in Programming Languages

The IEEE 754-1985 standard was targeted mainly at processor vendors and did not focus on programming languages. In particular, it did not define bindings (i.e., how the IEEE 754 standard is to be implemented in the language), such as the mapping between native types of the language and the formats of IEEE 754 and the mapping between operators/functions of the language and the operations defined by IEEE 754. The IEEE 754-1985 standard did not even deal with what a language should specify or what a compiler is allowed to do. This has led to many misinterpretations, with users often thinking that the processor will do exactly what they have written in the programming language. Chapter 6 will survey in more detail floating-point issues in mainstream programming languages.

For instance, it is commonly believed that the `double` type of the ISO C language must correspond everywhere to the binary64 binary format of the IEEE 754 standard, but this property is "implementation-defined," and behaving differently is not a bug. Indeed the *destination* (as defined by the IEEE 754 standard) does not necessarily correspond to the C floating-point type associated with the value. This is the reason why an implementation using double-extended is valid.

The consequences are that one can get different results on different platforms. But even when dealing with a single platform, one can also get unintuitive results, as shown in Goldberg's article with the appendix *Differences Among IEEE 754 Implementations* [214] or in Chapter 6 of this book.

The bottom line is that the reader should be aware that a language will not necessarily follow standards as he or she might expect. Implementing the algorithms given in this book may require special care in some environments (languages, compilers, platforms, and so on). This book (in Chapter 6) will give some examples, but with no claim of exhaustiveness.

The IEEE 754-2008 standard clearly improves the situation, mainly in its clauses 10 (*Expression evaluation*) and 11 (*Reproducible floating-point results*) – see Section 3.1.2.5. For instance, it deals with the double-rounding problem (observed on x87, described in Section 3.4.8): "Language standards should disallow, or provide warnings for, mixed-format operations that would cause implicit conversion that might change operand values." However, it may take time until these improvements percolate from the IEEE 754-2008 standard into languages and their compilers.

3.7 Checking the Environment

Checking a floating-point environment (for instance, to make sure that a compiler optimization option is compliant with one of the IEEE standards) may be important for critical applications. Circuit or software manufacturers frequently use formal proofs to ensure that their arithmetic algorithms are correct [426, 534, 535, 536, 241, 242, 117, 6]. Also, when the algorithms used by some environment are known, it is possible to design test vectors that allow one to explore every possible branching. Typical examples are methods for making sure that every element of the table of a digit-recurrence division or square root algorithm [186] is checked. For some functions, one can find "hardest-to-round" values that may constitute good input values for tests. This can be done using Hensel lifting for multiplication, division and square root [484], or using some techniques briefly presented in Chapter 10 for the transcendental functions.

Checking the environment is more difficult for the end user, who generally does not have any access to the algorithms that have been used. When we check some environment as a "black box" (that is, without knowing the code, or the algorithms used in the circuits) there is no way of being absolutely sure that the environment will *always* follow the standards. Just imagine a buggy multiplier that always returns the right result but for one pair of input operands. The only way of detecting this would be to check all possible inputs, which would be extremely expensive in the binary32 format, and totally impossible in the binary64, decimal64, binary128 or decimal128 formats. This is not pure speculation: in binary32/single precision arithmetic, the divider of the first version of the Pentium circuit would produce an incorrect quotient with probability around 2.5×10^{-11} [108, 437, 183], assuming random inputs.

Since the early 1980s, various programs have been designed for determining the basic parameters of a floating-point environment and assessing its quality. We present some of them below. Most of these programs are merely of historical importance: thanks to the standardization of floating-point arithmetic, the parameters of the various widely available floating-point environments are not so different.

3.7.1 MACHAR

MACHAR was a program, written in FORTRAN by W. Cody [106], whose purpose was to determine the main parameters of a floating-point format (radix, "machine epsilon," etc.). This was done using algorithms similar to the one we give in Section 4.1.1 for finding the radix β of the system. Today, it is interesting for historical purposes only.

In their book on elementary functions, Cody and Waite [107] also gave methods for estimating the quality of an elementary function library. Their methods were based on mathematical identities such as

$$\sin(3x) = 3\sin(x) - 4\sin^3(x). \tag{3.2}$$

These methods were useful at the time they were published. And yet, they can no longer be used with current libraries. Recent libraries are either correctly rounded or have a maximal error very close to $\frac{1}{2}$ ulp. Hence, *they are far more accurate than the methods that are supposed to check them.*

3.7.2 Paranoia

Paranoia [329] is a program originally written in Basic by Kahan, and translated to Pascal by B.A. Wichmann and to C by Sumner and Gay in the 1980s, to check the behavior of floating-point systems. It finds the main properties of a floating-point system (such as its precision and its exponent range), and checks if underflow is gradual, if the arithmetic operations are properly implemented, etc. It can be obtained at http://www.netlib.org/paranoia/.

Today, Paranoia is essentially of historical importance. It can be useful as a debugging tool for someone who develops his or her own floating-point environment.

3.7.3 UCBTest

UCBTest can be obtained at http://www.netlib.org/fp/ucbtest.tgz. It is a collection of programs whose purpose is to test certain difficult cases of the IEEE floating-point arithmetic. Paranoia is included in UCBTest. The "difficult cases" for multiplication, division, and square root (i.e., almost hardest-to-round cases: input values for which the result of the operation is very near a breakpoint of the rounding mode) are constructed using algorithms designed by Kahan, such as those presented in [484].

3.7.4 TestFloat

J. Hauser designed an excellent software implementation of the IEEE 754 floating-point arithmetic. The package is named SoftFloat and can be downloaded at http://www.jhauser.us/arithmetic/SoftFloat.html (at the time of writing these lines, the most recent version was released in August 2017). He also designed a program, TestFloat, aimed at testing whether a system conforms to IEEE 754. TestFloat compares results returned by the system to results returned by SoftFloat. It is accessible from the SoftFloat page.

3.7.5 Miscellaneous

SRTEST is a FORTRAN program written by Kahan for checking implementation of SRT [186, 187] division algorithms. It can be accessed on Kahan's web page, at https://people.eecs.berkeley.edu/~wkahan/srtest/. Some useful software, written by Beebe, can be found at http://www.math.utah.edu/~beebe/software/ieee/. MPCHECK is a program written by Revol, Pélissier, and Zimmermann. It checks mathematical function libraries (for correct rounding, monotonicity, symmetry, and output range). It can be downloaded at https://members.loria.fr/PZimmermann/mpcheck/.

Part II

Cleverly Using Floating-Point Arithmetic

Chapter 4

Basic Properties and Algorithms

IN THIS CHAPTER, we present some short yet useful algorithms and some basic properties that can be derived from specifications of floating-point arithmetic systems, such as the ones given in the successive IEEE 754 standards. Thanks to these standards, we now have an accurate definition of floating-point formats and operations. The behavior of a sequence of operations becomes at least partially[1] predictable (see Chapter 6 for more details on this). We therefore can build algorithms and proofs that refer to these specifications.

This also allows for the use of formal proofs to verify pieces of mathematical software. More information can be found in Chapter 13.

4.1 Testing the Computational Environment

4.1.1 Computing the radix

The various parameters (radix, significand and exponent widths, rounding modes, etc.) of the floating-point arithmetic used in a computing system may strongly influence the result of a numerical program. Indeed, very simple and short programs that only use floating-point operations can expose to view these parameters. An amusing example of this is the C program (Listing 4.1), derived from an idea of Malcolm [212, 403], that returns the radix β of the floating-point system. It works if the active rounding is one of the rounding direction attributes of IEEE 754-2008 [267] (or one of the four rounding modes of IEEE 754-1985). It is important to make sure that a zealous compiler

[1] In some cases, intermediate calculations may be performed in a wider internal format. Some examples are given in Section 3.2.

J.-M. Muller et al., *Handbook of Floating-Point Arithmetic*,
https://doi.org/10.1007/978-3-319-76526-6_4

does not try to "simplify" expressions such as $(A + 1.0) - A$. See Chapter 6 for more information on how languages and compilers handle floating-point arithmetic.

C listing 4.1 Malcolm's algorithm (Algorithm 4.1, see below), written in C.

```
#include <stdio.h>
#include <math.h>

#pragma STDC FP_CONTRACT OFF
/*  contraction of a FP arithmetic expression is forbidden */

int main (void)
{
  double A, B;

  A = 1.0;
  while ((A + 1.0) - A == 1.0)
    A *= 2.0;
  B = 1.0;
  while ((A + B) - A != B)
    B += 1.0;
  printf ("Radix B = %g\n", B);
  return 0;
}
```

Let us describe the corresponding algorithm more precisely. Let $\circ$ be the rounding function. The algorithm is as follows.

Algorithm 4.1 Computing the radix of a floating-point system.

$A \leftarrow 1.0$
$B \leftarrow 1.0$
while $\circ(\circ(A + 1.0) - A) = 1.0$ **do**
  $A \leftarrow \circ(2 \times A)$
end while
while $\circ(\circ(A + B) - A) \neq B$ **do**
  $B \leftarrow \circ(B + 1.0)$
end while
return B

Incidentally, this example shows that analyzing algorithms sometimes depends on the whole specification of the arithmetic operations, and especially the fact that they are correctly rounded:

- If one assumes that the operations are exact, then one erroneously concludes that the first loop never ends (or ends with an error due to an overflow on variable A).

- If one tries to analyze this algorithm just by assuming that $\circ(x+y)$ is $(x+y)(1+\epsilon)$ where $|\epsilon|$ is bounded by some tiny value, it is impossible to prove anything. For instance, $\circ(\circ(A+1.0)-A)$ is just 1 plus some "noise."

And yet, assuming correctly rounded operations, it is easy to show that the final value of B is the radix of the floating-point system being used, as we show now.

Proof. Define A_i as the value of A after the i-th iteration of the loop:

$$\textbf{while } \circ(\circ(A+1.0)-A)=1.0.$$

Let β be the radix of the floating-point system and p its precision. One easily shows by induction that if $2^i \le \beta^p - 1$, then A_i equals 2^i exactly. In such a case, $A_i + 1 \le \beta^p$, which implies that $\circ(A_i+1.0) = A_i + 1$. Therefore, one deduces that $\circ(\circ(A_i+1.0)-A_i) = \circ((A_i+1)-A_i) = 1$. Hence, while $2^i \le \beta^p - 1$, we stay in the first loop.

Now, consider the first iteration j, such that $2^j \ge \beta^p$. We have $A_j = \circ(2A_{j-1}) = \circ(2 \times 2^{j-1}) = \circ(2^j)$. Since $\beta \ge 2$, we deduce

$$\beta^p \le A_j < \beta^{p+1}.$$

This implies that the floating-point successor of A_j is $A_j + \beta$. Therefore, depending on the rounding mode, $\circ(A_j + 1.0)$ is either A_j or $A_j + \beta$, which implies that $\circ(\circ(A_j+1.0)-A_j)$ is either 0 or β. In any case, this value is different from 1.0, so we exit the first loop.

So we conclude that, at the end of the first **while** loop, variable A satisfies $\beta^p < A < \beta^{p+1}$.

Now, let us consider the second **while** loop. We have seen that the floating-point successor of A is $A+\beta$. Therefore, while $B < \beta$, $\circ(A+B)$ is either A or $A+\beta$, which implies that $\circ(\circ(A+B)-A)$ is either 0 or β. In any case, this value is different from B, which implies that we stay in the loop.

Now, as soon as $B = \beta$, $\circ(A+B)$ is exactly equal to $A+B$; hence, $\circ(\circ(A+B)-A) = B$. We therefore exit the loop when $B = \beta$. □

4.1.2 Computing the precision

Algorithm 4.2, also introduced by Malcolm [403], is very similar to Algorithm 4.1. It computes the precision p of the floating-point system being used.

Algorithm 4.2 Computing the precision of a floating-point system. It requires the knowledge of the radix of the system, and that radix can be given by Algorithm 4.1.

```
Require: B is the radix of the FP system
  i ← 0
  A ← 1.0
  while ∘(∘(A + 1.0) − A) = 1.0 do
    A ← ∘(B × A)
    i ← i + 1
  end while
  return i
```

The proof is very similar to the proof of Algorithm 4.1, so we omit it.

Similar—yet more sophisticated—algorithms are used in inquiry programs such as Paranoia [329], which provide a means for examining your computational environment (see Section 3.7).

4.2 Exact Operations

Although most floating-point operations involve some sort of rounding, there are some cases where a single operation will be *exact*, i.e., without rounding error. Knowing these cases allows an experienced programmer to use them in critical algorithms. Typical examples of such algorithms are elementary function programs [142, 129]. Many other examples will follow throughout this book.

What are exact operations? For many of the possible floating-point operand combinations, mathematical operations do not yield a result representable as a floating-point number. In this case, the corresponding floating-point operations, which always return floating-point numbers, yield an output that is different (rounded) from that of their mathematical counterpart. But there are also many situations where the output of a mathematical operation over floating-point numbers is a floating-point number too. In this case the corresponding floating-point operation, whatever the rounding mode is, must yield this very floating-point number. What we call exact operations are restrictions (generally assumptions on the input operands) to the classic floating-point operations such that one can prove that they yield an exact result (i.e., identical to the one of the mathematical operation, unaffected by the application of the final rounding function).

4.2.1 Exact addition

An important result, frequently used when designing or analyzing algorithms, is Sterbenz's lemma.

Lemma 4.1 (Sterbenz [572]). *In a radix-β floating-point system with subnormal numbers available, if x and y are nonnegative finite floating-point numbers such that*

$$\frac{y}{2} \le x \le 2y,$$

then $x - y$ is exactly representable.

Sterbenz's lemma implies that, with any of the rounding functions presented in Section 2.2, if x and y satisfy the preceding conditions, then when computing $x - y$ in floating-point arithmetic, the obtained result is exact.

Proof. By assumption $x, y \ge 0$ and, up to exchanging x and y, we can assume that $y \le x \le 2y$. Let M_x and M_y be the integral significands of x and y, and let e_x and e_y be their exponents. (If x and y have several floating-point representations, we choose the ones with the smallest exponents.) We have

$$x = M_x \times \beta^{e_x - p + 1}$$

and

$$y = M_y \times \beta^{e_y - p + 1},$$

where

$$\begin{cases} e_{\min} \le e_x \le e_{\max}, \\ e_{\min} \le e_y \le e_{\max}, \\ 0 \le M_x \le \beta^p - 1, \\ 0 \le M_y \le \beta^p - 1. \end{cases}$$

From $0 \le y \le x$, we easily deduce that $e_y \le e_x$. Define $\delta = e_x - e_y$. We get

$$x - y = \left(M_x \beta^\delta - M_y\right) \times \beta^{e_y - p + 1}.$$

Define $M = M_x \beta^\delta - M_y$. We have

- $x \ge y$ implies $M \ge 0$;
- $x \le 2y$ implies $x - y \le y$, hence $M\beta^{e_y-p+1} \le M_y \beta^{e_y-p+1}$ and then

$$M \le M_y \le \beta^p - 1.$$

Therefore, $x - y$ is equal to $M \times \beta^{e-p+1}$ with $e_{\min} \le e \le e_{\max}$ and $|M| \le \beta^p - 1$. This shows that $x - y$ is a floating-point number, which implies that it is exactly computed. □

It is important to note that, in our proof, the only thing we have shown is that $x - y$ is representable with an integral significand M whose absolute value is less than or equal to $\beta^p - 1$. We have not shown (and it is *not* possible to show) that it can be represented with an integral significand of absolute value larger than or equal to β^{p-1}. Indeed, $|x-y|$ can be less than $\beta^{e_{\min}}$. In such a case, the availability of subnormal numbers is required for Sterbenz's lemma to be applicable (it suffices to consider the example given by Figure 2.1).

A slightly more general result is that if the exponent of $x-y$ is less than or equal to the minimum of e_x and e_y, then the subtraction is exactly performed (see for instance [199]). Hauser's theorem below (Theorem 4.2) is a particular case of that result.

Sterbenz's lemma might seem strange to those who remember their early lectures on numerical analysis: It is a common knowledge that subtracting numbers that are very near may lead to very inaccurate results. This kind of numerical error is called a *cancellation*, or a *catastrophic cancellation* when almost all digits of the result are lost. There is no contradiction: The subtraction of two floating-point numbers that are very near does not introduce any error in itself (since it is an exact operation), yet it *amplifies* a *pre-existing error*. Consider the following example in IEEE 754-2008 binary32 (a.k.a. single precision arithmetic) and round-to-nearest mode:

- $A = 10000$ and $B = 9999.5$ (both are exactly representable);
- $C = \mathrm{RN}(1/10) = 13421773/134217728$;
- $A' = \mathrm{RN}(A + C) = 5120051/512$;
- $\Delta = \mathrm{RN}(A' - B) = 307/512 = 0.599609375$.

Sterbenz's Lemma implies that Δ is exactly equal to $A' - B$. And yet, the computation $A' = \mathrm{RN}(A + C)$ introduced some error: A' is slightly different from $A+C$. This suffices to make Δ a rather bad approximation to $(A+C)-B$, since $(A + C) - B \approx 0.6000000015$.

In this example, the subtraction $A' - B$ was errorless, but it amplified the error introduced by the computation of A'.

Hauser [249] gives another example of exact additions. It shows, incidentally, that when a gradual underflow occurs (that is, the absolute value of the obtained result is less than $\beta^{e_{\min}}$ but larger than or equal to the smallest subnormal number $\alpha = \beta^{e_{\min}-p+1}$), this does not necessarily mean an inaccurate result.

Theorem 4.2 (Hauser). *If x and y are radix-β floating-point numbers, and if the number $\mathrm{RN}(x + y)$ is subnormal, then $\mathrm{RN}(x + y) = x + y$ exactly.*

Proof. It suffices to note that x and y (as all floating-point numbers) are integral multiples of the smallest nonzero floating-point number $\alpha = \beta^{e_{\min}-p+1}$. Hence, $x + y$ is an integral multiple of α. If it is a subnormal number, then it is less than $\beta^{e_{\min}}$. This implies that it is exactly representable. $\square$

4.2.2 Exact multiplications and divisions

In some cases, multiplications and divisions are exactly performed. A straightforward example is multiplication or division by a power of the radix: As long as there is no overflow or underflow,[2] the result is exactly representable.

Another example is multiplication of numbers with known zero digits at the lower-order part of the significand. For instance, assume that x and y are floating-point numbers whose significands have the form

$$\underbrace{x_0.x_1x_2x_3\cdots x_{k_x-1}}_{k_x \text{ digits}}\ \underbrace{000\cdots 0}_{p-k_x \text{ zeros}}$$

for x, and

$$\underbrace{y_0.y_1y_2y_3\cdots y_{k_y-1}}_{k_y \text{ digits}}\ \underbrace{000\cdots 0}_{p-k_y \text{ zeros}}$$

for y. If $k_x + k_y \leq p$ then the product of the significands of x and y fits in p radix-β digits. This implies that the product xy is exactly computed if neither overflow nor underflow occurs. This property is at the heart of Dekker's multiplication algorithm (see Section 4.4.2.2). It is also very useful for reducing the range of inputs when evaluating elementary functions (see Section 10.2).

4.3 Accurate Computation of the Sum of Two Numbers

Let a and b be radix-β precision-p floating-point numbers. Let s be $\mathrm{RN}(a+b)$, i.e., $a+b$ correctly rounded to the nearest precision-p floating-point number, with any choice here in case of a tie. It can easily be shown that, if the computation of s does not overflow, then the error of the floating-point addition of a and b, namely $t = (a+b) - s$, is exactly representable in radix β with p digits. Note that this property can be false with other rounding functions. For instance, in a radix-2 and precision-p arithmetic, assuming rounding toward $-\infty$, if $a - 1$ and $b - -2^{-3p}$, then

$$\begin{aligned} s &= \mathrm{RD}(a+b) = 0.\underbrace{111111\cdots 11}_{p} \\ &= 1 - 2^{-p}, \end{aligned}$$

and

$$t - s = \underbrace{1.1111111111\cdots 11}_{2p} \times 2^{-p-1},$$

[2]Beware! We remind the reader that by "no underflow" we mean that the absolute value of the result (before or after rounding, this depends on the definition) is not less than the smallest *normal* number $\beta^{e_{\min}}$. When subnormal numbers are available, as requested by the IEEE 754 standards, it is possible to represent smaller nonzero numbers, but with a precision that does not always suffice to represent the product exactly.

which cannot be exactly represented with precision p (it would require precision $2p$).

In the following sections, we present two algorithms for computing t. They are useful for computing the sum of many numbers very accurately (see Chapter 5). They are the basic building blocks of *double-word* arithmetic (see Chapter 14). They also are of interest for a very careful implementation of mathematical functions [129].

4.3.1 The Fast2Sum algorithm

The Fast2Sum algorithm was introduced by Dekker [158] in 1971, but the three operations of this algorithm already appeared in 1965 as a part of a summation algorithm, called "Compensated sum method," due to Kahan [313] (Algorithm 5.6). The name "Fast-Two-Sum" seems to have been coined by Shewchuk [555]. Note that the algorithm may return useful results even if the conditions of Theorem 4.3 are not satisfied (for instance in compensated summation algorithms).

Theorem 4.3 (Fast2Sum algorithm [158], and Theorem C of [342], page 236). *Assume the floating-point system being used has radix $\beta \leq 3$, has subnormal numbers available, and provides correct rounding with rounding to nearest.*

Let a and b be floating-point numbers such that $\mathrm{RN}(a+b)$ is a finite floating-point number, and assume that the exponent of a is larger than or equal to that of b (this condition might be difficult to check, but of course, if $|a| \geq |b|$, it will be satisfied). Algorithm 4.3 computes two floating-point numbers s and t that satisfy the following:

- *$s + t = a + b$ exactly;*
- *s is the floating-point number that is closest to $a + b$.*

Algorithm 4.3 The Fast2Sum algorithm [158].

input a, b
$s \leftarrow \mathrm{RN}(a+b)$
$z \leftarrow \mathrm{RN}(s-a)$
$t \leftarrow \mathrm{RN}(b-z)$
return (s, t)

(We remind the reader that $\mathrm{RN}(x)$ means "x rounded to nearest.")

Note that if a wider internal format is available (one more digit of precision is enough), and if the computation of z is carried on using that wider format, then the condition $\beta \leq 3$ is no longer necessary [158]. This may be useful when working with decimal arithmetic. The Fast2Sum algorithm is simpler when written in C, since all rounding functions are implicit, as one can see in Listing 4.2 (yet, it requires round-to-nearest mode, which is the default rounding mode).

In that program it is assumed that all the variables are declared as elements of the same floating-point format, say all `float` or all `double`, and the system is set up to ensure that all the computations are done in this format. A compiler that is compliant with the C11 standard including Annex F (IEEE 754 support) will not attempt to simplify these operations.

C listing 4.2 Fast2Sum.

```
/* fast2Sum.c */

#include <stdio.h>
#include <stdlib.h>

void fast2Sum(double a, double b, double *s, double *t)
{
  double z;

  *s =  a + b;
   z = *s - a;
  *t =  b - z;
}

int main(int argc, char **argv)
{
  double a;
  double b;
  double s;
  double t;

  /* The inputs are read from the command line. */
  a = strtod(argv[1], NULL);
  b = strtod(argv[2], NULL);

  printf("a = %1.16g\n", a);
  printf("b = %1.16g\n", b);

  fast2Sum(a, b, &s, &t);

  printf("s = %1.16g\n", s);
  printf("t = %1.16g\n", t);

  return 0;
}
```

Let us now give Dekker's proof for this algorithm.

Proof. Without loss of generality, we assume $a > 0$. Let e_a, e_b, and e_s be the exponents of a, b, and s. Let M_a, M_b, and M_s be their integral significands, and let p be the precision of the floating-point format being used. We recall that integral significands have absolute values less than or equal to $\beta^p - 1$.

First, let us show that $s - a$ is exactly representable. Note that, since $e_b \le e_a$, the number s can be represented with an exponent less than or equal to $e_a + 1$. This comes from

$$|a + b| \le 2(\beta^p - 1)\beta^{e_a-p+1} \le (\beta^p - 1)\beta^{e_a-p+2}.$$

1. **If** $e_s = e_a + 1$.

   Define $\delta = e_a - e_b$. We have

   $$M_s = \left\lceil \frac{M_a}{\beta} + \frac{M_b}{\beta^{\delta+1}} \right\rfloor,$$

   where $\lceil u \rfloor$ is the integer that is nearest to u (when u is an odd multiple of $1/2$, there are two integers that are nearest to u, and we choose the one that is even).

   Define $\mu = \beta M_s - M_a$. We easily find

   $$\frac{M_b}{\beta^\delta} - \frac{\beta}{2} \le \mu \le \frac{M_b}{\beta^\delta} + \frac{\beta}{2}.$$

   Since μ is an integer and $\beta \le 3$, this gives

   $$|\mu| \le |M_b| + 1.$$

   Therefore, since $|M_b| \le \beta^p - 1$, either $|\mu| \le \beta^p - 1$ or $|\mu| = \beta^p$. In both cases, since $s - a$ is $\mu\beta^{e_a-p+1}$, it is exactly representable.[3]

2. **If** $e_s \le e_a$.

   Define $\delta_1 = e_a - e_b$. We have

   $$a + b = \left(\beta^{\delta_1} M_a + M_b\right) \beta^{e_b-p+1}.$$

   If $e_s \le e_b$ then $s = a + b$, since $a + b$ is a multiple of β^{e_b-p+1}, and s is obtained by rounding $a+b$ to the nearest multiple of $\beta^{e_s-p+1} \le \beta^{e_b-p+1}$. This implies that $s-a = b$ is exactly representable. If $e_s > e_b$, then define $\delta_2 = e_s - e_b$. We have

   $$s = \left\lceil \beta^{\delta_1-\delta_2} M_a + \beta^{-\delta_2} M_b \right\rfloor \beta^{e_s-p+1},$$

   which implies

   $$\left(\beta^{-\delta_2} M_b - \frac{1}{2}\right) \beta^{e_s-p+1} \le s - a \le \left(\beta^{-\delta_2} M_b + \frac{1}{2}\right) \beta^{e_s-p+1}.$$

[3] When $|\mu| \le \beta^p - 1$, the difference $s - a$ is representable with exponent e_a, but not necessarily in normal form. This is why the availability of subnormal numbers is necessary.

Hence,

$$|s-a| \leq \left(\beta^{-\delta_2}|M_b| + \frac{1}{2}\right)\beta^{e_s-p+1},$$

and $s-a$ is a multiple of β^{e_s-p+1}, which gives $s-a = K\beta^{e_s-p+1}$, with

$$|K| \leq \beta^{-\delta_2}|M_b| + \frac{1}{2} \leq \beta^p - 1,$$

which implies that $s-a$ is exactly representable.

Therefore, in all cases, $z = \mathrm{RN}(s-a) = s-a$ exactly.

Second, let us show that $b-z$ is exactly representable. From $e_a \geq e_b$, we deduce that a and b are both multiples of β^{e_b-p+1}. This implies that s (obtained by rounding $a+b$), $s-a$, $z = s-a$, and $b-z$, are multiples of β^{e_b-p+1}. Moreover,

$$|b-z| \leq |b|. \tag{4.1}$$

This comes from $|b-z| = |a+b-s|$: If $|a+b-s|$ was larger than $|b| = |a+b-a|$, then a would be a better floating-point approximation to $a+b$ than s.

From (4.1) and the fact that $b-z$ is a multiple of β^{e_b-p+1}, we deduce that $b-z$ is exactly representable, which implies $t = b-z$.

Now, the theorem is easily obtained. From $t = b-z$ we deduce

$$t = b-(s-a) = (a+b)-s.$$

□

When designing algorithms for "nearly atomic" operations it is important to check if "spurious" underflows or overflows may occur. A spurious underflow (resp. overflow) occurs when an intermediate operation underflows (resp. overflows) whereas the exact result lies in the normal range. A typical example is the "naive" calculation of $\sqrt{a^2+b^2}$ with $a = 1$ and b slightly above $\beta^{e_{\max}/2+1}$: The computation of b^2 overflows, and yet the exact value of $\sqrt{a^2+b^2}$ is very close to b (i.e., far from the overflow threshold). No "spurious" overflow can occur in Fast2Sum: in radix-2 arithmetic, if the computation of $s = \mathrm{RN}(a+b)$ does not overflow, then there cannot be an overflow at lines 2 and 3 of Algorithm 4.3 [49]. Also, Hauser's theorem (Theorem 4.2) implies that underflows are harmless in Fast2Sum.

4.3.2 The 2Sum algorithm

The Fast2Sum algorithm, presented in the previous section, requires a preliminary knowledge of the orders of magnitude of the two operands (since we must know which of them has the largest exponent).[4]

[4]Do not forget that $|a| \geq |b|$ implies that the exponent of a is larger than or equal to that of b. Hence, it suffices to compare the two variables.

The following 2Sum algorithm (Algorithm 4.4), due to Knuth [342] and Møller [422], requires 6 floating-point operations instead of 3 for Fast2Sum, but does not require a preliminary comparison of a and b. On some processors the penalty due to a wrong branch prediction when comparing a and b costs much more than 3 additional floating-point operations. Also, unless a wider format is available, Fast2Sum does not work in radices greater than 3, whereas 2Sum works in any radix. This is of interest when using a radix-10 system.

The name "TwoSum" seems to have been coined by Shewchuk [555].

Algorithm 4.4 The 2Sum algorithm.

input a, b
$s \leftarrow \mathrm{RN}(a+b)$
$a' \leftarrow \mathrm{RN}(s-b)$
$b' \leftarrow \mathrm{RN}(s-a')$
$\delta_a \leftarrow \mathrm{RN}(a-a')$
$\delta_b \leftarrow \mathrm{RN}(b-b')$
$t \leftarrow \mathrm{RN}(\delta_a+\delta_b)$
return (s, t)

Knuth shows that if a and b are normal floating-point numbers, then for any radix β, provided that no underflow or overflow occurs, $a + b = s + t$. Boldo et al. [48] show that in radix 2, underflow does not hinder the result (and yet, obviously, overflow does). Formal proofs of 2Sum, Fast2Sum, and many other useful algorithms can be found in the Flocq library.[5]

In a way, 2Sum is optimal in terms of number of floating-point operations. More precisely, Kornerup et al. give the following definition [347]:

Definition 4.1 (RN-addition algorithm without branching). *We call* RN-addition algorithm without branching *an algorithm*

- *without comparisons, or conditional expressions, or min/max instructions;*
- *only based on floating-point additions or subtractions in round-to-nearest mode: At step i the algorithm computes* $\mathrm{RN}(a+b)$ *or* $\mathrm{RN}(a-b)$, *where a and b are either one of the input values or a previously computed value.*

For instance, 2Sum is an RN-addition algorithm without branching. It requires 6 floating-point operations. Only counting the operations just gives a rough estimate on the performance of an algorithm. Indeed, on modern architectures, pipelined arithmetic operators and the availability of several floating-point units (FPUs) make it possible to perform some operations in parallel, provided they are independent. Hence, the depth of the dependency

[5]http://flocq.gforge.inria.fr/.

graph of the instructions of the algorithm is an important criterion. In the case of the 2Sum algorithm, only two operations can be performed in parallel:

$$\delta_b = \mathrm{RN}(b - b')$$

and

$$\delta_a = \mathrm{RN}(a - a');$$

hence, we will say that the depth of the 2Sum algorithm is 5. Kornerup et al. prove the following two theorems [347]:

Theorem 4.4. *In any correctly rounded binary floating-point arithmetic of precision $p \geq 2$, an RN-addition algorithm without branching that computes the same results as 2Sum requires at least 6 arithmetic operations.*

Theorem 4.5. *In any correctly rounded binary floating-point arithmetic of precision $p \geq 2$, an RN-addition algorithm without branching that computes the same results as 2Sum has depth at least 5.*

These theorems show that, among the RN-addition algorithms without branching, if we do not have any information on the ordering of $|a|$ and $|b|$, 2Sum is optimal both in terms of number of arithmetic operations and in terms of depth. The proof was obtained by enumerating all possible RN-addition algorithms without branching that use 5 additions or less, or that have depth 4 or less. Each of these algorithms was run with a few well-chosen floating-point entries. None of the enumerated algorithms gave the same results as 2Sum for all chosen entries.

In algorithm 2Sum, a spurious overflow may happen, but in extremely rare cases only. More precisely, in radix-2, precision-p floating-point arithmetic, if the first input value a of Algorithm 4.4 satisfies $|a| < \Omega$ and if there is no overflow at line (2) of the algorithm, then there will be no overflow at lines (3) to (7) [49].

It may be possible that in a future release of the IEEE 754 standard, the 2Sum operation become specified, so that future processor instructions may directly deliver s and t from a and b without having to use Algorithm 4.4. See Section 3.3.

4.3.3 If we do not use rounding to nearest

The Fast2Sum and 2Sum algorithms rely on rounding to nearest. The example given at the beginning of Section 4.3 shows that if s is computed as $\mathrm{RD}(a+b)$ or $\mathrm{RU}(a+b)$, then $s - (a+b)$ may not be a floating-point number, hence the algorithms would fail to return an approximate sum and the error term.

Nonetheless, in his Ph.D. dissertation [496], Priest gives a longer algorithm that only requires faithful arithmetic (see definition in [158]). From two floating-point numbers a and b, it deduces two other floating-point numbers c and d such that

- $c + d = a + b$, and
- either $c = d = 0$, or $|d| < \mathrm{ulp}(c)$.

In particular, Algorithm 4.5 works if $\circ$ is any of the rounding functions presented in Section 2.2, provided neither underflow nor overflow occurs.

Algorithm 4.5 Priest's Sum and Roundoff error algorithm. It only requires faithful arithmetic.

if $|a| < |b|$ **then**
  $\mathrm{swap}(a, b)$
end if
$c \leftarrow \circ(a + b)$
$e \leftarrow \circ(c - a)$
$g \leftarrow \circ(c - e)$
$h \leftarrow \circ(g - a)$
$f \leftarrow \circ(b - h)$
$d \leftarrow \circ(f - e)$
if $\circ(d + e) \neq f$ **then**
  $c \leftarrow a$
  $d \leftarrow b$
end if
return (c, d)

Also, in radix-2 floating-point arithmetic with rounding functions different from round-to-nearest, even if the error $s - (a + b)$ is not a floating-point number, the returned result is meaningful and can be useful, for instance for designing compensated algorithms, detailed in Section 5.3.2. More precisely, Boldo, Graillat, and Muller [49] consider the following two algorithms:

Algorithm 4.6 Fast2Sum with faithful roundings [49]: $\circ_1, \circ_2, \circ_3$ are rounding functions taken in $\{\mathrm{RD}, \mathrm{RU}, \mathrm{RZ}, \mathrm{RN}\}$.

$s \leftarrow \circ_1(a + b)$
$z \leftarrow \circ_2(s - a)$
$t \leftarrow \circ_3(b - z)$
return (s, t)

Algorithm 4.7 2Sum with faithful roundings [49]: $\circ_i$, for $i = 1, \ldots, 6$, are rounding functions taken in $\{\text{RD}, \text{RU}, \text{RZ}, \text{RN}\}$.

$s \leftarrow \circ_1(a+b)$
$a' \leftarrow \circ_2(s-b)$
$b' \leftarrow \circ_3(s-a')$
$\delta_a \leftarrow \circ_4(a-a')$
$\delta_b \leftarrow \circ_5(b-b')$
$t \leftarrow \circ_6(\delta_a+\delta_b)$
return (s,t)

They show the following results:

Theorem 4.6. *In radix-2 floating-point arithmetic, if no overflow occurs and $e_a \geq e_b$, then the values s and t returned by Algorithm 4.6 satisfy*

$$t = \circ_3((a+b)-s),$$

i.e., t is a faithful rounding of the error of the FP sum $\circ_1(a+b)$.

Theorem 4.7. *In radix-2, precision-p floating-point arithmetic, if $p \geq 4$ and no overflow occurs, then the values s and t returned by Algorithm 4.7 satisfy*

$$t = (a+b) - s + \alpha,$$

with $|\alpha| < 2^{-p+1} \cdot \text{ulp}(a+b) \leq 2^{-p+1} \cdot \text{ulp}(s)$. Furthermore, if the floating-point exponents e_s and e_b of s and b satisfy $e_s - e_b \leq p-1$, then t is a faithful rounding of $(a+b)-s$.

4.4 Accurate Computation of the Product of Two Numbers

In the previous section, we have seen that, under some conditions, the error of a floating-point addition is a floating-point number that can be computed using a few operations. The same holds for floating-point multiplication:

- If x_1 and x_2 are radix-β precision-p floating-point numbers, whose exponents e_{x_1} and e_{x_2} satisfy

$$e_{x_1} + e_{x_2} \geq e_{\min} + p - 1 \tag{4.2}$$

  (where $e_{\min}$ is the minimum exponent of the system being considered), and

- if r is $\circ(x_1 x_2)$, where $\circ$ is a rounding function picked from $\{\text{RD}, \text{RU}, \text{RZ}\}$, $\{\text{RN}\}$, then $t = x_1 x_2 - r$ is a radix-β precision-p floating-point number.

Condition $e_{x_1} + e_{x_2} \geq e_{\min} + p - 1$ cannot be avoided: if it is not satisfied, the product may not be representable as the exact sum of two floating-point numbers ($|t|$ would be below the underflow threshold). Consider for instance the following example, in the decimal64 format of the IEEE 754-2008 standard ($\beta = 10, p = 16, e_{\min} = -383$).

- $x_1 = 3.141592653589793 \times 10^{-190}$;
- $x_2 = 2.718281828459045 \times 10^{-190}$;
- the floating-point number closest to $x_1 \cdot x_2$ is $r = 8.539734222673566 \times 10^{-380}$;
- the floating-point number closest to $x_1 \cdot x_2 - r$ is subnormal. Its value is $-0.000000000000322 \times 10^{-383}$, which differs from the exact value of $x_1 \cdot x_2 - r$, namely $-0.322151269472315 \times 10^{-395}$.

Actually, computing t is very easy if a fused multiply-add (FMA) operator is available, and rather complex without an FMA. Let us first present the easy solution.

4.4.1 The 2MultFMA Algorithm

If an FMA instruction is available, Algorithm 4.8 allows one to deduce, from two floating-point numbers x_1 and x_2, two other floating-point numbers r_1 and r_2 such that, if (4.2) holds,

$$r_1 + r_2 = x_1 \cdot x_2$$

exactly, and

$$r_1 = \mathrm{RN}(x_1 \cdot x_2).$$

That is, r_2 is the error of the floating-point multiplication of x_1 by x_2.

That algorithm only requires two consecutive operations, and works for any radix and precision. Although we present it for round-to-nearest mode, it works as well for the other rounding modes.

Algorithm 4.8 2MultFMA(x_1, x_2).

Ensure: $r_1 + r_2 = x_1 x_2$
  $r_1 \leftarrow \mathrm{RN}(x_1 \cdot x_2)$
  $r_2 \leftarrow \mathrm{RN}(x_1 \cdot x_2 - r_1)$
  return (r_1, r_2)

Algorithm 4.8 is very useful for implementing 2D rotations, complex multiplication and division (see Chapter 11), and double-word arithmetic (see Chapter 14). Listing 6.1 shows how to write this algorithm (and other ones for which we explicitly want an FMA instruction to be used) in standard C.

4.4.2 If no FMA instruction is available: Veltkamp splitting and Dekker product

Without an FMA, the best-known algorithm for computing t is Dekker's algorithm [158]. It requires 17 floating-point operations. Roughly speaking, it consists in first splitting each of the operands x and y into two floating-point numbers, the significand of each of them being representable with $\lfloor p/2 \rfloor$ or $\lceil p/2 \rceil$ digits only. The underlying idea is that (using a property given in Section 4.2.2) the pairwise products of these values should be exactly representable, which is not always possible if p is odd, since the product of two $\lceil p/2 \rceil$-digit numbers does not necessarily fit in p digits.[6] Then these pairwise products are added.

Let us now present that algorithm with more detail. We first show how to perform the splitting, using floating-point addition and multiplication only, by means of an algorithm due to Veltkamp [616, 617].

4.4.2.1 Veltkamp splitting

Before examining how we can compute exact products without an FMA, we need to see how we can "split" a precision-p radix-β floating-point number x into two floating-point numbers x_h and x_ℓ such that, for a given $s < p$, the significand of x_h fits in $p - s$ digits, the significand of x_ℓ fits in s digits, and $x = x_h + x_\ell$ exactly. Algorithms that use such a splitting may also require that x_h be sufficiently close to x, and, in an equivalent way, that $|x_\ell|$ be sufficiently small.

This is done using Veltkamp's algorithm (Algorithm 4.9). It uses a floating-point constant C equal to $\beta^s + 1$.

Algorithm 4.9 Split(x, s): Veltkamp's algorithm.

Require: $C = \beta^s + 1$
Require: $2 \le s \le p - 2$
  $\gamma \leftarrow \mathrm{RN}(C \cdot x)$
  $\delta \leftarrow \mathrm{RN}(x - \gamma)$
  $x_h \leftarrow \mathrm{RN}(\gamma + \delta)$
  $x_\ell \leftarrow \mathrm{RN}(x - x_h)$
  return (x_h, x_ℓ)

[6]In radix 2, we will use the fact that a $2g+1$-bit number can be split into two g-bit numbers. This explains why Dekker's algorithm works if the precision is even or if the radix is 2.

Dekker [158] proves this algorithm in radix 2, with the implicit assumption that neither overflow nor underflow occurs. Boldo [42] shows that for any radix β and any precision p, provided that $C \cdot x$ does not overflow, the algorithm works. More precisely:

- if $C \cdot x$ does not overflow, no other operation will overflow;
- there is no underflow problem: If x_ℓ is subnormal, the result still holds.

x_h is x rounded to the nearest precision-$(p - s)$ number (halfway values may be rounded in either direction).

Another property of Veltkamp's splitting that will be important for analyzing Dekker's multiplication algorithm is the following: If $\beta = 2$, the significand of x_ℓ actually fits in $s - 1$ bits.

Before giving a proof of Veltkamp's splitting algorithm, let us give an example.

Example 4.1 (Veltkamp's splitting). *Assume a radix-10 system, with precision 8. We want to split the significands of the floating-point numbers into parts of equal width; that is, we choose $s = 4$. This gives $C = 10001$. Assume $x = 1.2345678$. We successively find:*

- $C \cdot x = 12346.9125678$, *therefore*
$$\gamma = \mathrm{RN}(C \cdot x) = 12346.913;$$
- $x - \gamma = -12345.6784322$, *therefore*
$$\delta = \mathrm{RN}(x - \gamma) = -12345.678;$$
- $\gamma + \delta = 1.235$, *therefore*
$$x_h = \mathrm{RN}(\gamma + \delta) = 1.235;$$
- $x - x_h = -0.0004322$, *therefore*
$$x_\ell = \mathrm{RN}(x - x_h) = -0.0004322.$$

One can easily check that 1.2345678 equals $1.235 - 0.0004322$. In this example, the last two arithmetic operations are exact; that is, $x_h = \gamma + \delta$ and $x_\ell = x - x_h$ (no rounding error occurs). We will see in the proof that this always holds.

Now, let us give a proof of Algorithm 4.9. For simplicity, we assume that the radix is 2, that $s \geq 2$, and that neither underflow nor overflow occurs. For a more general and very rigorous proof, see the remarkable paper by Boldo [42].

Proof. If we multiply x by 2^k, then all the variables in the algorithm are scaled by a factor 2^k. Hence we can assume that $1 \le x < 2$ without loss of generality. Furthermore, since the behavior of the algorithm is fairly obvious when $x = 1$, we assume $1 < x < 2$, which gives (since x is a precision-p floating-point number)

$$1 + 2^{-p+1} \le x \le 2 - 2^{-p+1}.$$

We now consider the four successive operations in Algorithm 4.9.

Computation of γ. $Cx = (2^s + 1)x$ implies $2^s + 1 \le Cx < 2^{s+2}$. Therefore,

$$2^{s-p+1} \le \text{ulp}(Cx) \le 2^{s-p+2}.$$

This gives

$$\begin{cases} \gamma = (2^s + 1)x + \epsilon_1, \text{ with } |\epsilon_1| \le 2^{s-p+1}, \\ \gamma \text{ is a multiple of } 2^{s-p+1}. \end{cases}$$

Computation of δ. We have $x - \gamma = -2^s x - \epsilon_1$. From this we deduce

$$|x - \gamma| \le 2^s(2 - 2^{-p+1}) + 2^{s-p+1} = 2^{s+1}.$$

This implies that $\delta = \text{RN}(x - \gamma) = x - \gamma + \epsilon_2 = -2^s x - \epsilon_1 + \epsilon_2$, with $|\epsilon_2| \le 2^{s-p}$.

Also, $|x - \gamma| \ge 2^s(1 + 2^{-p+1}) - 2^{s-p+1} \ge 2^s$, which implies that δ is a multiple of 2^{s-p+1}.

Computation of x_h. Now, $-\delta = 2^s x + \epsilon_1 - \epsilon_2$ and $\gamma = 2^s x + x + \epsilon_1$ are quite close together. When $s \ge 2$, they are within a factor of 2 from each other. So Lemma 4.1 (Sterbenz's lemma) can be applied to deduce that $\gamma + \delta$ is computed exactly. Therefore,

$$x_h = \gamma + \delta = x + \epsilon_2.$$

Also, x_h is a multiple of 2^{s-p+1} (since $x_h = \gamma + \delta$, and γ and δ are multiples of 2^{s-p+1}). From $x < 2$ and $x_h = x - \epsilon_2$, one deduces $x_h < 2 + 2^{s-p}$, but the only multiple of 2^{s-p+1} between 2 and $2 + 2^{s-p}$ is 2, so $x_h \le 2$.

Computation of x_ℓ. Since $x_h = x + \epsilon_2$ and x are very close, we can use Lemma 4.1 again to show that $x - x_h$ is computed exactly. Therefore,

$$x_\ell = x - x_h = \epsilon_2.$$

As a consequence, $|x_\ell| = |\epsilon_2|$ is less than or equal to 2^{s-p}. Moreover, x_ℓ is a multiple of 2^{-p+1}, since x and x_h are multiples of 2^{-p+1}.

Thus, we have written x as the sum of two floating-point numbers x_h and x_ℓ. Moreover,

- $x_h \leq 2$ and x_h is a multiple of 2^{s-p+1} imply that x_h fits in $p-s$ bits;
- $|x_\ell| \leq 2^{s-p}$ and x_ℓ is a multiple of 2^{-p+1} imply that x_ℓ fits in $s-1$ bits.

□

4.4.2.2 Dekker's multiplication algorithm

Algorithm 4.10 was devised by Dekker, who presented it and proved it in radix 2, yet seemed to assume that it would work as well in higher radices. Later on, it was analyzed by Linnainmaa [391], who found the necessary condition in radices different from 2 (an even precision), and by Boldo [42], who examined the difficult problem of possible overflow or underflow in the intermediate operations, and gave a formal proof in Coq.

Here, we present the algorithm using Boldo's notation. We assume a floating-point arithmetic with radix β, subnormal numbers, and precision p. From two finite floating-point numbers x and y, the algorithm returns two floating-point numbers r_1 and r_2 such that $xy = r_1 + r_2$ exactly, under conditions that will be made explicit below.

Algorithm 4.10 Dekker's product.

Require: $s = \lceil p/2 \rceil$
Ensure: $r_1 + r_2 = x \cdot y$
  $(x_h, x_\ell) \leftarrow \text{Split}(x, s)$
  $(y_h, y_\ell) \leftarrow \text{Split}(y, s)$
  $r_1 \leftarrow \text{RN}(x \cdot y)$
  $t_1 \leftarrow \text{RN}(-r_1 + \text{RN}(x_h \cdot y_h))$
  $t_2 \leftarrow \text{RN}(t_1 + \text{RN}(x_h \cdot y_\ell))$
  $t_3 \leftarrow \text{RN}(t_2 + \text{RN}(x_\ell \cdot y_h))$
  $r_2 \leftarrow \text{RN}(t_3 + \text{RN}(x_\ell \cdot y_\ell))$
  return (r_1, r_2)

Listing 4.3 presents the same algorithm, written in C.

Here is Boldo's version [42] of the theorem that gives the conditions under which Dekker's multiplication algorithm returns a correct result.

Theorem 4.8. *Assume the minimal exponent $e_{\min}$ and the precision p satisfy*[7] *$p \geq 3$ and $\beta^{e_{\min}-p+1} \leq 1$. Let e_x and e_y be the exponents of the floating-point numbers x and y. If $\beta = 2$ or p is even, and if there is no overflow in the splitting of x and y or in the computation of $r_1 = \text{RN}(x \cdot y)$ and $\text{RN}(x_h \cdot y_h)$, then the floating-point numbers r_1 and r_2 returned by Algorithm 4.10 satisfy:*

[7]These assumptions hold on any "reasonable" floating-point system.

1. *If* $e_x + e_y \geq e_{\min} + p - 1$, *then* $xy = r_1 + r_2$ *exactly;*

2. *in any case,*

$$|xy - (r_1 + r_2)| \leq \frac{7}{2}\beta^{e_{\min}-p+1}.$$

As pointed out by Boldo, the "7/2" in Theorem 4.8 could probably be sharpened. In particular, if the radix is 2, that coefficient can be reduced to 3.

Condition "$e_x + e_y \geq e_{\min} + p - 1$" on the exponents is a necessary and sufficient condition for the error term of the product to be always representable (see [46]), whatever the significands of x and y might be. Condition "$\beta = 2$ or p is even" might seem strange at first glance, but is easily understood from the following remarks:

- $x_h \cdot y_h$ is exactly representable with a $2 \times \lfloor p/2 \rfloor$-digit significand, hence, it is representable in precision p:

$$\mathrm{RN}(x_h \cdot y_h) = x_h \cdot y_h;$$

- $x_h \cdot y_\ell$ and $x_\ell \cdot y_h$ are exactly representable with a $\lfloor p/2 \rfloor + \lceil p/2 \rceil = p$-digit significand;

- and yet, in many cases, $x_\ell \cdot y_\ell$ will be a $2 \times \lceil p/2 \rceil$-digit number.

Therefore, if the precision p is even, then $2 \times \lceil p/2 \rceil = p$, so $x_\ell \cdot y_\ell$ is exactly representable. And if $\beta = 2$, then we know (see Section 4.4.2.1) that, even if p is odd, x_ℓ and y_ℓ actually fit in $\lceil p/2 \rceil - 1$ bits, so their product fits in p bits.

For instance, with the decimal interchange formats specified by the IEEE 754-2008 standard (see Chapter 3), Algorithm 4.10 will not always work in the decimal32 format ($p = 7$), and yet it can be used safely in the decimal64 ($p = 16$) and decimal128 ($p = 34$) formats.

Conditions on the absence of overflow for $r_1 = \mathrm{RN}(x \cdot y)$ and $\mathrm{RN}(x_h \cdot y_h)$ might seem redundant: Since these values are very close, in general they will overflow simultaneously. And yet, it is possible to construct tricky cases where one of these computations will overflow, and the other will not.[8] Condition $\beta^{e_{\min}-p+1} \leq 1$ is always satisfied in practice.

Proof. For a full proof, see [42]. Here, we give a simplified proof, assuming radix 2, and assuming that no underflow/overflow occurs.

First, since the splittings have been performed to make sure that $x_h y_h$, $x_\ell y_h$, $x_h y_\ell$, and $x_\ell y_\ell$ should be exactly representable, we have

$$\mathrm{RN}(x_h y_h) = x_h y_h, \mathrm{RN}(x_\ell y_h) = x_\ell y_h, \mathrm{RN}(x_h y_\ell) = x_h y_\ell, \text{ and } \mathrm{RN}(x_\ell y_\ell) = x_\ell y_\ell.$$

[8]For example, in the IEEE 754 binary64 format, with $x = (2^{53} - 1) \cdot 2^{940}$ and $y = 2^{31}$, we obtain $x_h = 2^{993}$ and $y_h = y$. The floating-point multiplication $\mathrm{RN}(x_h y_h)$ overflows, whereas $\mathrm{RN}(xy) = \Omega = (2^{53} - 1) \cdot 2^{971}$.

C listing 4.3 Dekker's product.

```
/* Dekker (exact) double multiplication */

#include <stdlib.h>
#include <stdio.h>
#define SHIFT_POW 27      /* ceil(53/2) for binary64. */
void dekkerMult(double a, double b, double *p, double *t);
void veltkampSplit(double x, int sp, double *x_high, double *x_low);
int main(int argc, char **argv)
{
  double x;
  double y;
  double r_1;
  double r_2;
  x = strtod(argv[1], NULL);
  y = strtod(argv[2], NULL);
  printf("x          = %1.16a\n", x);
  printf("y          = %1.16a\n", y);
  dekkerMult(x, y, &r_1, &r_2);
  printf("r_1        = %1.16a\n", r_1);
  printf("r_2        = %1.16a\n", r_2);
  return 0;
}

void dekkerMult(double x, double y, double *r_1, double *r_2)
{
  double x_high, x_low;
  double y_high, y_low;
  double t_1;
  double t_2;
  double t_3;
  veltkampSplit(x, SHIFT_POW, &x_high, &x_low);
  veltkampSplit(y, SHIFT_POW, &y_high, &y_low);
  printf("x_high    = %1.16a\n", x_high);
  printf("x_low     = %1.16a\n", x_low);
  printf("y_high    = %1.16a\n", y_high);
  printf("y_low     = %1.16a\n", y_low);
  *r_1 = x * y;
  t_1  = -*r_1 + x_high * y_high  ;
  t_2  =   t_1 + x_high * y_low;
  t_3  =   t_2 + x_low  * y_high;
  *r_2 =   t_3 + x_low  * y_low;
}

void veltkampSplit(double x, int sp, double *x_high, double *x_low)
{
  unsigned long C = (1UL << sp) + 1;
  double gamma = C * x;
  double delta = x - gamma;
  *x_high      = gamma + delta;
  *x_low       = x - x_high;
}
```

Without loss of generality, we can assume $1 \le x < 2$ and $1 \le y < 2$. We have already seen that $x_h \le 2$, $y_h \le 2$, and that $|x - x_h| \le 2^{s-p}$ and $|y - y_h| \le 2^{s-p}$. From

$$(xy - x_h y_h) = (x - x_h)y + (y - y_h)x_h,$$

we deduce

$$|xy - x_h y_h| \le 2^{s-p+2}.$$

Since we also have

$$|xy - r_1| \le \frac{1}{2}\,\mathrm{ulp}(xy) \le 2^{-p+1},$$

we get

$$|r_1 - x_h y_h| \le 2^{-p+1} + 2^{s-p+2}.$$

This shows that r_1 and $\mathrm{RN}(x_h y_h) = x_h y_h$ are very close, so Sterbenz's lemma (Lemma 4.1) can be applied to their subtraction. As a consequence,

$$t_1 = -r_1 + x_h y_h.$$

Now, $xy - x_h y_h = x_h y_\ell + x_\ell y_h + x_\ell y_\ell$, so

$$\begin{aligned} |t_1 + x_h y_\ell| &= |-r_1 + x_h y_h + x_h y_\ell| \\ &= |-r_1 + xy + (x_h y_h + x_h y_\ell - xy)| \\ &\le |-r_1 + xy| + |x_\ell(y_h + y_\ell)| \\ &\le 2^{-p+1} + 2^{s-p+1} < 2^{s-p+2}. \end{aligned}$$

Since x_h is a multiple of 2^{s-p+1} and y_ℓ is a multiple of 2^{-p+1}, $x_h y_\ell$ is a multiple of 2^{s-2p+2}. This implies that $t_1 + x_h y_\ell$ is a multiple of 2^{s-2p+2}. This and $|t_1 + x_h y_\ell| < 2^{s-p+2}$ imply that $t_1 + x_h y_\ell$ is exactly representable. Therefore,

$$t_2 = -r_1 + x_h y_h + x_h y_\ell.$$

Now, $t_2 + x_\ell y_h = (-r_1 + xy) + x_\ell y_\ell$; therefore,

$$\begin{aligned} |t_2 + x_\ell y_h| &\le |-r_1 + xy| + |x_\ell y_\ell| \\ &\le 2^{-p+1} + 2^{2s-2p}. \end{aligned}$$

From $s = \lceil p/2 \rceil$, we deduce $2s - 2p = -p$ or $-p + 1$; therefore,

$$|t_2 + x_\ell y_h| \le 2^{-p+2}.$$

This and the fact that $t_2 + x_\ell y_h$ is a multiple of 2^{s-2p+2} imply that $t_2 + x_\ell y_h$ is exactly representable; therefore,

$$t_3 = t_2 + x_\ell y_h = -r_1 + x_h y_h + x_h y_\ell + x_\ell y_h.$$

Finally, $|t_3 + x_\ell y_\ell| = |-r_1 + xy| \leq 2^{-p+1}$, and it is a multiple of 2^{-2p+2}, thus $t_3 + x_\ell y_\ell$ is exactly computed. Therefore,

$$r_2 = xy - r_1.$$

□

Dekker's multiplication algorithm requires 17 floating-point operations: 7 multiplications, and 10 additions/subtractions. This may seem a lot, compared to the 6 floating-point additions/subtractions required by the 2Sum algorithm (Algorithm 4.4). Yet an actual implementation of Dekker's algorithm will not be 17/6 times slower than an actual implementation of 2Sum. Indeed, since many operations in Dekker's algorithm are independent, they can be performed in parallel or pipelined if the underlying architecture allows for it. In the summary given in Figure 4.1, all the operations on a same line can be performed in parallel.

$\gamma_x \leftarrow \mathrm{RN}(Cx)$	$\gamma_y \leftarrow \mathrm{RN}(Cy)$	$r_1 \leftarrow \mathrm{RN}(xy)$	
$\delta_x \leftarrow \mathrm{RN}(x - \gamma_x)$	$\delta_y \leftarrow \mathrm{RN}(y - \gamma_y)$		
$x_h \leftarrow \mathrm{RN}(\gamma_x + \delta_x)$	$y_h \leftarrow \mathrm{RN}(\gamma_y + \delta_y)$		
$x_\ell \leftarrow \mathrm{RN}(x - x_h)$	$y_\ell \leftarrow \mathrm{RN}(y - y_h)$	$\alpha_{11} \leftarrow \mathrm{RN}(x_h y_h)$	
$t_1 \leftarrow \mathrm{RN}(-r_1 + \alpha_{11})$	$\alpha_{12} \leftarrow \mathrm{RN}(x_h y_\ell)$	$\alpha_{21} \leftarrow \mathrm{RN}(x_\ell y_h)$	$\alpha_{22} \leftarrow \mathrm{RN}(x_\ell y_\ell)$
$t_2 \leftarrow \mathrm{RN}(t_1 + \alpha_{12})$			
$t_3 \leftarrow \mathrm{RN}(t_2 + \alpha_{21})$			
$r_2 \leftarrow \mathrm{RN}(t_3 + \alpha_{22})$			

Figure 4.1: Summary of the various floating-point operations involved in the Dekker product of x and y. The operations on a same line can be performed in parallel.

4.5 Computation of Residuals of Division and Square Root with an FMA

As we will see, the availability of an FMA instruction simplifies the implementation of correctly rounded division and square root. The following two theorems have been known for a long time (see for instance [41]). We prefer here to give the presentation by Boldo and Daumas [46], since it fully takes into account the possibilities of underflow.

In these theorems, assuming a radix-β, precision-p, floating-point system with minimal exponent $e_{\min}$, we will call a *representable pair* for a floating-point number x a pair (M, e) of integers[9] such that $x = M \cdot \beta^{e-p+1}$, $|M| \leq \beta^p - 1$, and $e_{\min} \leq e$ (such pairs are "floating-point representations" that are not necessarily normal, without an upper bound on the exponents).

Theorem 4.9 (Exact residual for division [46]). *Let x and y be floating-point numbers in the considered format. Let q be $\circ(x/y)$, where $\circ$ is round-to-nearest, or a directed rounding function (see Section 2.2). If q is neither an infinity nor a Not a Number (NaN) datum, then*

$$x - qy$$

is a floating-point number if and only if there exist two representable pairs (M_y, e_y) and (M_q, e_q) that represent y and q such that

- $e_y + e_q \geq e_{\min} + p - 1$ *and*
- $q \neq \alpha$ *or* $\alpha/2 \leq |x/y|$,

where $\alpha = \beta^{e_{\min}-p+1}$ is the smallest positive subnormal number.

Theorem 4.10 (Exact residual for square root [46]). *Let x be a floating-point number in the considered format. Let σ be $\sqrt{x}$ rounded to a nearest floating-point value. If σ is neither an infinity nor a NaN, then*

$$x - \sigma^2$$

is representable if and only if there exists a representable pair (M_σ, e_σ) that represents σ such that

$$2e_\sigma \geq e_{\min} + p - 1.$$

See [46] for the proofs of these theorems. Consider the following example, which illustrates that if the conditions of Theorem 4.9 are not satisfied, then $x - qy$ is not exactly representable.

Example 4.2 (A case where $x - qy$ is not exactly representable). *Assume $\beta = 2$, $p = 24$, and $e_{\min} = 1 - e_{\max} = -126$ (single-precision format of IEEE 754-1985, binary32 format of IEEE 754-2008). Let x and y be floating-point numbers defined as*

$$x = 2^{-104} + 2^{-105} = (1.1\overbrace{0000000000000000000000}^{22 \text{ zeros}})_2 \times 2^{-104}$$

and

$$y = 2^{-21} + 2^{-44} = (1.\overbrace{0000000000000000000000}^{22 \text{ zeros}}1)_2 \times 2^{-21}.$$

The floating-point number that is nearest to x/y is

$$q = (1.0\underbrace{1111111111111111111111}_{22 \text{ ones}})_2 \times 2^{-83},$$

[9]Caution: M is not necessarily the integral significand of x.

and the exact value of $x - qy$ is

$$x - qy = -(1.\underbrace{111111111111111111111}_{21 \text{ ones}})_2 \times 2^{-129},$$

which is not exactly representable. The (subnormal) floating-point number obtained by rounding $x - qy$ to nearest even is -2^{-128}.

In the example, we have $e_y + e_q = -104$ and $e_{\min} + p - 1 = -103$, so the condition "$e_y + e_q \geq e_{\min} + p - 1$" of Theorem 4.9 is not satisfied.

An important consequence of Theorem 4.9 is the following result, which will make it possible to perform correctly rounded divisions using Newton–Raphson iterations, provided that an FMA instruction is available (see Section 4.7).

Corollary 4.11 (Computation of division residuals using an FMA). *Assume x and y are precision-p, radix-β, floating-point numbers, with $y \neq 0$ and $|x/y|$ below the overflow threshold. If q is defined as*

- *x/y if it is exactly representable;*
- *one of the two floating-point numbers that surround x/y otherwise;*[10]

then

$$x - qy$$

is exactly computed using one FMA instruction, with any rounding mode, provided that

$$\begin{gathered} e_y + e_q \geq e_{\min} + p - 1, \\ \textit{and} \\ q \neq \alpha \text{ or } |x/y| \geq \tfrac{\alpha}{2}, \end{gathered} \tag{4.3}$$

where e_y and e_q are the exponents of y and q. In the frequent case $1 \leq x < \beta$ and $1 \leq y < \beta$ (straightforward separate handling of the exponents in the division algorithm), condition (4.3) is satisfied as soon as

$$e_{\min} \leq -p,$$

which holds in all usual formats.

Similarly, from Theorem 4.10, we deduce the following result.

[10]Or q is the largest finite floating-point number Ω, in the case where x/y is between that number and the overflow threshold (the same thing applies on the negative side).

Corollary 4.12 (Computation of square root residuals using an FMA). *Assume x is a precision-p, radix-β, positive floating-point number. If σ is $\sqrt{x}$ rounded to a nearest floating-point number, then*

$$x - \sigma^2$$

is exactly computed using one FMA instruction, with any rounding mode, provided that

$$2e_\sigma \geq e_{\min} + p - 1, \tag{4.4}$$

where e_σ is the exponent of σ. In the frequent case $1 \leq x < \beta^2$ (straightforward separate handling of the exponent in the square root algorithm), condition (4.4) is satisfied as soon as

$$e_{\min} \leq 1 - p,$$

which holds in all usual formats.

Corollary 4.12 is much weaker than Corollary 4.11: the "correcting term" $x - \sigma^2$ may not be exactly representable when σ is not a floating-point number nearest to $\sqrt{x}$, even if σ is one of the two floating-point numbers that surround $\sqrt{x}$. An example, in radix-2, precision-5 arithmetic is $x = 11110_2$ and $\sigma = 101.10_2$.

4.6 Another splitting technique: splitting around a power of 2

In this section, we assume a radix-2, precision-p floating-point arithmetic. We have presented in Section 4.4.2.1 a technique, Veltkamp splitting, that allows one to express a floating-point number as the sum of a $(p - s)$-digit number and an s-digit number. Several summation algorithms require a different kind of splitting: We need to express a floating-point number as the sum of a number that is multiple of some 2^h (where $h \in \mathbb{Z}$) and a number less than or equal to 2^h. Let us present an algorithm, introduced by Rump et al. under the name "ExtractScalar" [531].

Given a constant $h \in \mathbb{Z}$ and an input floating-point number x of absolute value less than or equal to 2^{h+p-1}, we wish to compute two floating-point numbers x_h and x_ℓ that satisfy:

$$\begin{cases} x_h \text{ is a multiple of } 2^h, \\ |x_\ell| \leq 2^h, \\ x = x_h + x_\ell. \end{cases} \tag{4.5}$$

This can be done as follows (we assume $p + h \leq e_{\max}$ so no overflow occurs).

Algorithm 4.11 ExtractScalar(x, h).

Require: $\sigma = 2^{p+h}$ (precomputed constant)
Ensure: (x_h, x_ℓ) satisfies (4.5)
  $y \leftarrow \mathrm{RN}(\sigma + x)$
  $x_h \leftarrow \mathrm{RN}(y - \sigma)$
  $x_\ell \leftarrow \mathrm{RN}(x - x_h)$
  return (x_h, x_ℓ)

Proof. First, Algorithm 4.11 is nothing but Fast2Sum (Algorithm 4.3) called with input values σ and x. From the proof of Algorithm 4.3, we deduce

- $x_h = y - \sigma$, and
- $x_\ell = x - x_h$.

Now, $|x| \le 2^{h+p-1}$ implies $-\frac{1}{2}\sigma \le x \le \frac{1}{2}\sigma$, so $\frac{1}{2}\sigma \le x + \sigma \le \frac{3}{2}\sigma$. Therefore,

- $y = \mathrm{RN}(x + \sigma)$ is a multiple of $\mathrm{ulp}(\frac{1}{2}\sigma) = 2^{-p}\sigma = 2^h$, and
- $|y - (x + \sigma)| \le 2^h$.

Therefore, since σ is a multiple of 2^h, $x_h = y - \sigma$ is a multiple of 2^h. We also have

$$|x_\ell| = |x - x_h| = |x - (y - \sigma)| = |(x + \sigma) - y| \le 2^h.$$

□

4.7 Newton–Raphson-Based Division with an FMA

Before using some of the results presented in the Section 4.5 to build algorithms for correctly rounded division, let us recall some variants of the Newton–Raphson iteration for reciprocation and division.

4.7.1 Variants of the Newton–Raphson iteration

Assume we wish to compute an approximation to b/a in a *binary*[11] floating-point arithmetic of precision p. We will first present some classic iterations, all derived from the Newton–Raphson root-finding iteration, that are used in several division algorithms [405, 120, 406, 118]. Some of these iterations make it possible to directly compute b/a, yet most algorithms first compute $1/a$: A multiplication by b, possibly followed by a correcting step, is necessary.

For simplicity, we assume that a and b satisfy

$$1 \le a, b < 2,$$

[11] Part of what we are going to explain does not generalize to decimal arithmetic.

i.e., a and b are significands of floating-point numbers. Caution: reducing the "general" case where a and b are arbitrary floating-point number to the case where they are between 1 and 2 is not as simple as it seems: a problem of *double rounding* may occur if the result lies in the subnormal range. We address this problem in Section 4.7.3.

The Newton–Raphson iteration is a well-known and useful technique for finding roots of functions. It was introduced by Newton around 1669 [458] to solve polynomial equations (without explicit use of the derivative). Raphson also used a similar technique. The method was generalized by Simpson, with an explicit use of the derivative, in 1740 [558]. Ypma's paper *Historical Development of the Newton-Raphson Method* [639] is a fascinating presentation of the history of this method.

For finding roots of function f, the method is based on the iteration

$$x_{n+1} = x_n - \frac{f(x_n)}{f'(x_n)}. \tag{4.6}$$

If x_0 is close enough to a root α of f, if f has a second derivative, and if $f'(\alpha) \neq 0$, then the iteration (4.6) converges *quadratically* to α. By "quadratic convergence" we mean that the distance between x_{n+1} and α is bounded by a value proportional to the square of the distance between x_n and α. If $\alpha \neq 0$, this implies that the number of common digits between x_n and α roughly doubles at each iteration.

In order to compute $1/a$, we look for the root of function $f(x) = 1/x - a$. Equation (4.6) becomes

$$x_{n+1} = x_n(2 - ax_n). \tag{4.7}$$

This iteration converges to $1/a$ for any $x_0 \in (0, 2/a)$. This is easy to see in Figure 4.2. And yet, of course, fast convergence requires a value of x_0 close to $1/a$.

In the case of iteration (4.7), we easily get

$$x_{n+1} - \frac{1}{a} = -a\left(x_n - \frac{1}{a}\right)^2, \tag{4.8}$$

which illustrates the quadratic convergence.

Caution: for the moment, we only deal with mathematical, exact iterations. When rounding errors are taken into account, the formulas become more complicated. Table 4.1 gives the first values x_n in the case $a = 1.5$ and $x_0 = 1$.

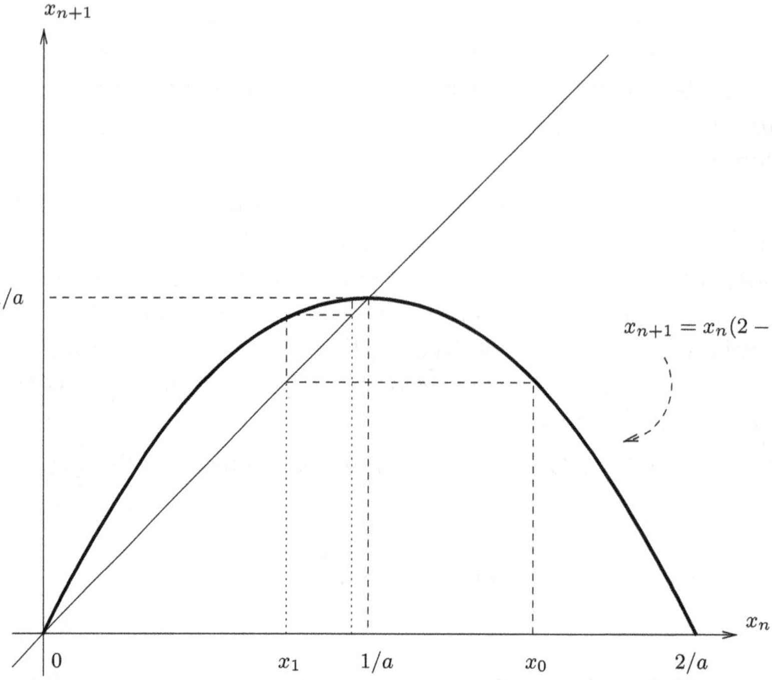

Figure 4.2: Iteration (4.7). We easily see that it converges to $1/a$ for $0 < x_0 < 2/a$.

n	x_n
1	0.5
2	0.625
3	0.6640625
4	0.666656494140625
5	0.666666666511446237564086914062 5
6	0.666666666666666666630526594250483185533084906637668 6 $\cdots$
7	0.6666666666666666666666666666666666666647075094152961 $\cdots$

Table 4.1: First values x_n given by iteration (4.7), in the case $a = 1.5$, $x_0 = 1$. The quadratic convergence is displayed by the fact that the number of common digits between x_n and the limit roughly doubles at each iteration.

Iteration (4.7) has a drawback: As the three floating-point operations (two if an FMA instruction is available) it requires are dependent, it leaves no room for parallelism. This can be a significant performance penalty, for instance, when the floating-point operations can be scheduled in a deep pipeline. And yet, this iteration has the great advantage of being "self-correcting": Small errors (e.g., rounding errors) when computing x_n do not change the value of the limit.

Although the iterations would converge with $x_0 = 1$ (they converge for $0 < x_0 < 2/a$, and we have assumed $1 \leq a < 2$), in practice, we drastically reduce their number by starting with a value x_0 close to $1/a$, obtained either by looking up an approximation to $1/a$ in a table[12]

addressed by a few most significant bits of a, or by using a polynomial approximation of very small degree [189, 492] to the reciprocal function. Many papers address the problem of cleverly designing a table that returns a convenient value x_0 [482, 130, 131, 132, 574, 156, 350]—see Section 8.2.8 for a review.

Now, by defining $\epsilon_n = 1 - ax_n$, one obtains the following iteration:

$$\begin{cases} \epsilon_n &= 1 - ax_n \\ x_{n+1} &= x_n + x_n\epsilon_n. \end{cases} \tag{4.9}$$

This iteration was implemented on the Intel Itanium processor [120, 406, 118]:

- it still has the "self-correcting" property;
- it is as sequential as iteration (4.7), since there is a dependency between x_{n+1} and ϵ_n;
- however, it has a nice property; under the conditions of Corollary 4.11 (that is, roughly speaking, if x_n is within one ulp from $1/a$, the "residual" $\epsilon_n = 1 - ax_n$ will be exactly computed with an FMA. As we will see later on, this is a key feature that allows for correctly rounded division.

Another *apparently different* way of devising fast division algorithms is to use the power series

$$\frac{1}{1-\epsilon} = 1 + \epsilon + \epsilon^2 + \epsilon^3 + \cdots \tag{4.10}$$

with $\epsilon = 1 - a$. Using (4.10) and the factorization

$$1 + \epsilon + \epsilon^2 + \epsilon^3 + \cdots = (1+\epsilon)(1+\epsilon^2)(1+\epsilon^4)(1+\epsilon^8)\cdots,$$

one gets a "new" fast iteration. By denoting

$$\epsilon_n = \epsilon^{2^n},$$

we get $\epsilon_{n+1} = \epsilon_n^2$, and

$$\frac{1}{1-\epsilon} = (1+\epsilon_0)(1+\epsilon_1)(1+\epsilon_2)(1+\epsilon_3)\cdots,$$

[12] For instance, the `frcpa` instruction of the IA-64 instruction set returns approximations to reciprocals with relative error less than or equal to $2^{-8.886}$. Such tables are easily implemented using the bipartite method, See [131].

denoting $x_n = (1+\epsilon_0)(1+\epsilon_1)\cdots(1+\epsilon_n)$, we get the following iteration:

$$\begin{cases} x_{n+1} &= x_n + x_n\epsilon_n \\ \epsilon_{n+1} &= \epsilon_n^2. \end{cases} \tag{4.11}$$

As for the Newton–Raphson iteration, one can significantly accelerate the convergence by starting from a value x_0 close to $1/a$, obtained from a table. It suffices then to choose $\epsilon_0 = 1 - ax_0$. Now, it is important to note that the variables ϵ_n of (4.11) and (4.9) are the same: from $\epsilon_n = 1 - ax_n$ in (4.9), one deduces

$$\begin{aligned} \epsilon_n^2 &= 1 - 2ax_n + a^2x_n^2 \\ &= 1 - ax_n(2 - ax_n) \\ &= 1 - ax_{n+1} \\ &= \epsilon_{n+1}. \end{aligned}$$

Hence, from a *mathematical* point of view, iterations (4.9) and (4.11) are equivalent: We have found again the same iteration through a totally different method. However, from a *computational* point of view, they are quite different:

- The computations of x_{n+1} and ϵ_{n+1} in (4.11) can be done in parallel, which is a significant improvement in terms of performance on most platforms;
- however, variable a no longer appears in (4.11). As a consequence, rounding errors during the computations will progressively blur the information about the input operand—this iteration is *not* self-correcting.

Of course, one can mix these various iterations. For instance, we may choose to compute ϵ_n as $1 - ax_n$ during the last iterations, because accuracy becomes crucial, whereas it may be preferable to compute it as ϵ_{n-1}^2 during the first iterations, because this can be done in parallel with the computation of x_n.

Another feature of iteration (4.11) is that one can directly compute b/a instead of first computing $1/a$ and then multiplying by b, *but this is not always efficient (see below)*. This is done by defining a new variable,

$$y_n = bx_n,$$

which gives

$$\begin{cases} y_{n+1} &= y_n + y_n\epsilon_n \\ \epsilon_{n+1} &= \epsilon_n^2. \end{cases} \tag{4.12}$$

The difficult point, however, is to get a sensible starting value y_0:

- Either we start the iterations with $y_0 = b$ (which corresponds to $x_0 = 1$) and $\epsilon_0 = 1 - a$, and in such a case, we may need many iterations if a is not very close to 1 (i.e., if x_0 is far from $1/a$);

- or we try to start the iterations with the (hidden) variable x_0 equal to a close approximation a^* to $1/a$, and $\epsilon_0 = 1 - aa^*$. In such a case, we must have $y_0 = ba^*$.

Now, one can design another iteration by defining

$$\left\{ \begin{array}{lcl} r_n & = & 1 - \epsilon_n \\ K_{n+1} & = & 1 + \epsilon_n, \end{array} \right. \tag{4.13}$$

which leads to the well-known *Goldschmidt iteration* [217]

$$\left\{ \begin{array}{lcl} y_{n+1} & = & K_{n+1} y_n \\ r_{n+1} & = & K_{n+1} r_n \\ K_{n+1} & = & 2 - r_n. \end{array} \right. \tag{4.14}$$

This iteration, still mathematically equivalent to the previous ones, also has interesting properties from a computational point of view. The computations of y_{n+1} and r_{n+1} are independent and can hence be performed in parallel, and the computation of K_{n+1} is very simple (by two's complementing r_n). In case of a hardware implementation, since both multiplications that appear in the iteration are by the same value K_{n+1}, some optimizations—such as a common Booth recoding [57, 187]—are possible. Unfortunately, the iteration is not self-correcting: After the first rounding error, exact information on a is lost forever.

Goldschmidt iteration was used in the IBM System/360 model 91 [15], and in the AMD K7 processor. A careful error analysis of Goldschmidt's division algorithm can be found in [197].

From (4.12) or (4.9), by defining

$$\left\{ \begin{array}{lcl} y_n & = & b x_n \\ \delta_n & = & b \epsilon_n, \end{array} \right. \tag{4.15}$$

one gets

$$\left\{ \begin{array}{lcl} \delta_n & = & b - a y_n \\ y_{n+1} & = & y_n + \delta_n x_n. \end{array} \right. \tag{4.16}$$

This last iteration was used in Intel and HP's algorithms for the Itanium. Once a correctly rounded approximation x_n to $1/a$ is obtained from (4.9)—we will see later how it can be correctly rounded—one computes a first approximation y_n to b/a by multiplying x_n by b. Then, this approximation is improved by applying iteration (4.16).

4.7.2 Using the Newton–Raphson iteration for correctly rounded division with an FMA

In this section, we assume that we wish to compute $\circ(b/a)$, where a and b are binary floating-point numbers of the same format, and $\circ$ is RN (round to nearest ties-to-even), RD, RU, or RZ. We will use some of the iterations presented in the previous section. Since the radix of the floating-point system is 2, we do not have to worry about how values halfway between two consecutive floating-point numbers are handled in the round-to-nearest mode. This is due to the following result.

Lemma 4.13 (Size of quotients in prime radices, adapted from [406]). *Assume that the radix β of the floating-point arithmetic is a prime number. Let $q = b/a$, where a and b are two floating-point numbers of precision p:*

- *Either q cannot be exactly represented with a finite number of radix-β digits;*
- *or q is a floating-point number of precision p (assuming unbounded exponent range).*

Proof. Assume that q is representable with a finite number of radix-β digits, but not with p digits or less. This means that there exist integers Q and e_q such that

$$q = Q \cdot \beta^{e_q - p + 1},$$

where $Q > \beta^p$, and Q is not a multiple of β.

Let A and B be the integral significands of a and b, and let e_a and e_b be their exponents. We have

$$\frac{B \cdot \beta^{e_b - p + 1}}{A \cdot \beta^{e_a - p + 1}} = Q \cdot \beta^{e_q - p + 1}.$$

Therefore, there exists an integer e, $e \geq 0$, such that

$$B = AQ \cdot \beta^e$$

or

$$B \cdot \beta^e = AQ.$$

This and the primality of β imply that B is a multiple of Q, which implies $B > \beta^p$. This is not possible since b is a precision-p floating-point number. □

In Lemma 4.13, the fact that the radix should be a prime number is necessary. For instance, in radix 10 with $p = 4$, 2.005 and 2.000 are floating-point numbers, and their quotient 1.0025 has a finite representation, but cannot be represented with precision 4. A consequence of Lemma 4.13 is that, in radix 2, a quotient is never exactly halfway between two consecutive floating-point numbers. Note that in prime radices greater than 2, Lemma 4.13 does not imply that quotients exactly halfway between two consecutive floating-point numbers do not occur.[13]

---

[13]When the radix is an odd number, values exactly halfway between two consecutive floating-point numbers are represented with infinitely many digits.

An interesting consequence of Lemma 4.13, since there is a finite number of quotients of floating-point numbers, is that there exists an *exclusion zone* around middles of consecutive floating-point numbers, where we cannot find quotients. Several, slightly different, "exclusion theorems" give lower bounds on the sizes of such exclusion zones. Here is one of them:

Lemma 4.14 (Exclusion lemma [118]). *Assume radix 2 and precision p, and let a and b be floating-point numbers, with $1 \le a, b < 2$. If c is either a floating-point number or the exact midpoint between two consecutive normal floating-point numbers, then either $b/a = c$ (which cannot happen if c is a midpoint), or*

$$\left|\frac{b}{a} - c\right| > 2^{-2p-2}\frac{b}{a}.$$

Another one is the following.

Lemma 4.15 (A slightly different exclusion lemma). *Assume radix 2 and precision p, and let a and b be floating-point numbers, with $1 \le a, b < 2$. If c is the exact midpoint between two consecutive normal floating-point numbers, then we have*

$$\left|\frac{b}{a} - c\right| > 2^{-p-1}\,\mathrm{ulp}\left(\frac{b}{a}\right).$$

Let us give a simple proof for Lemma 4.15.

Proof. Let A and B be the integral significands of a and b. We have

$$a = A \cdot 2^{-p+1}$$

and

$$b = B \cdot 2^{-p+1}.$$

Define

$$\delta = \begin{cases} 0 & \text{if } B \ge A \\ 1 & \text{otherwise.} \end{cases}$$

We have

$$\mathrm{ulp}\left(\frac{b}{a}\right) = 2^{-p+1-\delta}.$$

Also, the middle c of two consecutive normal floating-point numbers around b/a is of the form

$$c = (2C+1) \cdot 2^{-p-\delta},$$

where C is an integer, $2^{p-1} \le C \le 2^p - 1$.

Therefore,

$$\begin{aligned}
\frac{b}{a} - c &= \frac{B}{A} - (2C+1)\cdot 2^{-p-\delta} \\
&= \frac{1}{2}\,\mathrm{ulp}\left(\frac{b}{a}\right)\cdot\left(\frac{B}{A}\cdot 2^{p+\delta} - (2C+1)\right) \\
&= \frac{1}{2A}\,\mathrm{ulp}\left(\frac{b}{a}\right)\cdot\left(B\cdot 2^{p+\delta} - (2C+1)A\right).
\end{aligned}$$

Since $1/A > 2^{-p}$, and since $B \cdot 2^{p+\delta} - (2C+1)A$ is a nonzero integer, we deduce

$$\left|\frac{b}{a} - c\right| > 2^{-p-1}\,\mathrm{ulp}\left(\frac{b}{a}\right).$$

□

When the processor being used has an internal precision that is significantly wider than the "target" precision, a careful implementation of iterations such as (4.9) and (4.16) will allow one to obtain approximations to b/a accurate enough so that Lemma 4.14 or a variant can be applied to show that, once rounded to the target format, the obtained result will be the correctly rounded (to the nearest) quotient. The main difficulty is when we want to compute quotients of floating-point numbers in a precision that is the widest available. In such a case, the following result is extremely useful.

Theorem 4.16 (Markstein [406]). *Assume a precision-p binary floating-point arithmetic, and let a and b be floating-point numbers, with $1 \le a, b < 2$. If*

- *q is a faithful approximation to b/a, and*
- *y approximates $1/a$ with a relative error less than 2^{-p}, and*
- *the calculations*
$$r = \circ(b - aq), \qquad q' = \circ(q + ry)$$
are performed using a given rounding function $\circ$, taken among round to nearest even, round toward zero, round toward $-\infty$, round toward $+\infty$,

then q' is exactly $\circ(b/a)$ (that is, b/a rounded according to the same rounding function $\circ$).

Markstein also shows that the last FMA raises the inexact exception if and only if the quotient was inexact. If y approximates $1/a$ with an error less than or equal to $\frac{1}{2}\,\mathrm{ulp}(1/a)$, (that is, if y is $1/a$ rounded to nearest), then it approximates $1/a$ with a relative error less than 2^{-p}, so Theorem 4.16 can be applied.

Proof. Let us prove Theorem 4.16 in the case of round-to-nearest even mode (the other cases are quite similar). First, note that from Theorem 4.9, r is computed exactly. We will use the fact that

$$\frac{a}{2^p - 1} \le \mathrm{ulp}(a) \le \frac{a}{2^{p-1}}. \tag{4.17}$$

Also, if $a = 1$ then $q = b$, $r = 0$, and $q' = b = b/a$ rightfully, so in the following we assume $a > 1$. Figure 4.3 illustrates the various cases that may occur, if we wish to return b/a rounded to nearest:

- if $q - \frac{1}{2}\,\mathrm{ulp}\left(\frac{b}{a}\right) < \frac{b}{a} < q + \frac{1}{2}\,\mathrm{ulp}\left(\frac{b}{a}\right)$, we must return q;
- if $\frac{b}{a} > q + \frac{1}{2}\,\mathrm{ulp}\left(\frac{b}{a}\right)$, we must return $q + \mathrm{ulp}\left(\frac{b}{a}\right)$;
- if $\frac{b}{a} < q - \frac{1}{2}\,\mathrm{ulp}\left(\frac{b}{a}\right)$, we must return $q - \mathrm{ulp}\left(\frac{b}{a}\right)$.

Note that the case $\frac{b}{a} = q \pm \frac{1}{2}\,\mathrm{ulp}\left(\frac{b}{a}\right)$ cannot occur from Lemma 4.13. The returned value q' is obtained by adding a correcting term ry to q, and rounding the obtained result to nearest. One can easily find that to make sure that $q' = \mathrm{RN}(b/a)$:

- If $q - \frac{1}{2}\,\mathrm{ulp}\left(\frac{b}{a}\right) < \frac{b}{a} < q + \frac{1}{2}\,\mathrm{ulp}\left(\frac{b}{a}\right)$, it suffices that $|ry| < \frac{1}{2}\,\mathrm{ulp}\left(\frac{b}{a}\right)$;
- if $\frac{b}{a} > q + \frac{1}{2}\,\mathrm{ulp}\left(\frac{b}{a}\right)$, it is necessary and sufficient that ry be larger than $\frac{1}{2}\,\mathrm{ulp}\left(\frac{b}{a}\right)$ and less than $\frac{3}{2}\,\mathrm{ulp}\left(\frac{b}{a}\right)$;
- if $\frac{b}{a} < q - \frac{1}{2}\,\mathrm{ulp}\left(\frac{b}{a}\right)$, it is necessary and sufficient that ry be larger than $-\frac{3}{2}\,\mathrm{ulp}\left(\frac{b}{a}\right)$ and less than $-\frac{1}{2}\,\mathrm{ulp}\left(\frac{b}{a}\right)$.

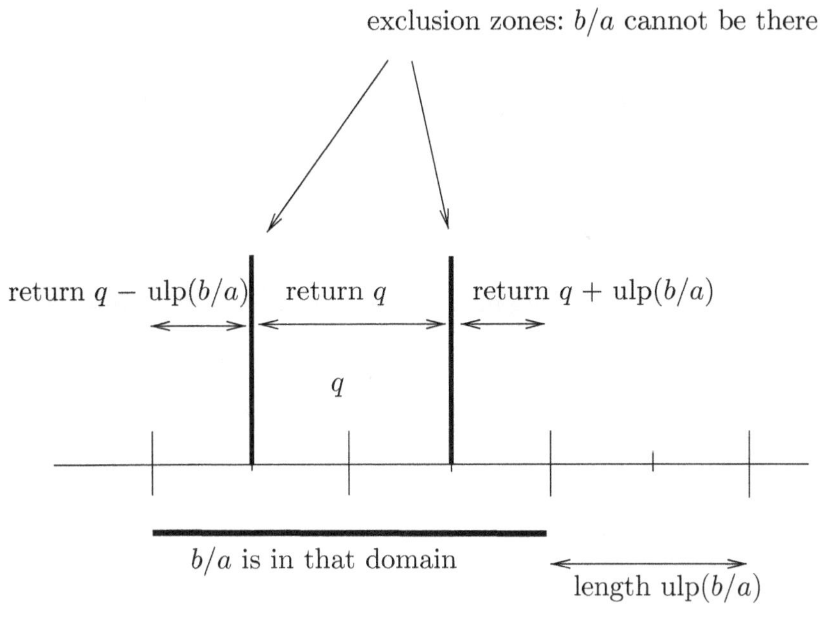

Figure 4.3: The various values that should be returned in round-to-nearest mode, assuming q is within one $\mathrm{ulp}(b/a)$ *from b/a.*

Since y approximates $1/a$ with relative error less than 2^{-p}, we have

$$\frac{1}{a} - \frac{1}{2^p a} < y < \frac{1}{a} + \frac{1}{2^p a}. \tag{4.18}$$

1. If $q - \frac{1}{2}\,\mathrm{ulp}\left(\frac{b}{a}\right) < \frac{b}{a} < q + \frac{1}{2}\,\mathrm{ulp}\left(\frac{b}{a}\right)$. Assume $\frac{b}{a} - q \geq 0$ (the other case is symmetrical). We have

$$0 \leq b - aq < \frac{a}{2}\,\mathrm{ulp}\left(\frac{b}{a}\right);$$

therefore, since $b - aq$ and $\frac{a}{2}\,\mathrm{ulp}\left(\frac{b}{a}\right)$ are floating-point numbers,

$$r = b - aq \leq \frac{a - \mathrm{ulp}(a)}{2}\,\mathrm{ulp}\left(\frac{b}{a}\right).$$

Using (4.17), we get

$$0 \leq r \leq \left(\frac{a}{2} - \frac{a}{2^{p+1} - 2}\right)\mathrm{ulp}\left(\frac{b}{a}\right).$$

This, along with (4.18), gives an upper bound on $|ry|$:

$$|ry| < \left(\frac{a}{2} - \frac{a}{2^{p+1} - 2}\right)\left(\frac{1}{a} + \frac{1}{2^p a}\right)\mathrm{ulp}\left(\frac{b}{a}\right),$$

from which we deduce

$$|ry| < \left(\frac{1}{2} - \frac{1}{2^{p-1}(2^{p+1} - 2)}\right)\mathrm{ulp}\left(\frac{b}{a}\right) < \frac{1}{2}\,\mathrm{ulp}\left(\frac{b}{a}\right).$$

Therefore, $\mathrm{RN}(q + ry) = q$, which is what we wanted to show.

2. If $\frac{b}{a} > q + \frac{1}{2}\,\mathrm{ulp}\left(\frac{b}{a}\right)$ (the case $\frac{b}{a} < q - \frac{1}{2}\,\mathrm{ulp}\left(\frac{b}{a}\right)$ is symmetrical). We have

$$\frac{a}{2}\,\mathrm{ulp}\left(\frac{b}{a}\right) < b - aq < a\,\mathrm{ulp}\left(\frac{b}{a}\right);$$

therefore, since $b - aq$, $\frac{a}{2}\,\mathrm{ulp}\left(\frac{b}{a}\right)$ and $a\,\mathrm{ulp}\left(\frac{b}{a}\right)$ are floating-point numbers,

$$\left(\frac{a + \mathrm{ulp}(a)}{2}\right)\mathrm{ulp}\left(\frac{b}{a}\right) \leq r = b - aq \leq (a - \mathrm{ulp}(a))\,\mathrm{ulp}\left(\frac{b}{a}\right).$$

Using (4.17), we get

$$\left(\frac{a + \frac{a}{2^p - 1}}{2}\right)\mathrm{ulp}\left(\frac{b}{a}\right) \leq r \leq \left(a - \frac{a}{2^p - 1}\right)\mathrm{ulp}\left(\frac{b}{a}\right).$$

This, along with (4.18), gives

$$\begin{aligned}
&\left(\frac{a + \frac{a}{2^p - 1}}{2}\right)\left(\frac{1}{a} - \frac{1}{2^p a}\right)\mathrm{ulp}\left(\frac{b}{a}\right)\\
&< ry\\
&< \left(a - \frac{a}{2^p - 1}\right)\left(\frac{1}{a} + \frac{1}{2^p a}\right)\mathrm{ulp}\left(\frac{b}{a}\right),
\end{aligned}$$

from which we deduce

$$\frac{1}{2}\,\text{ulp}\left(\frac{b}{a}\right) < ry < \left(1 - \frac{2^{1-p}}{2^p-1}\right)\text{ulp}\left(\frac{b}{a}\right) < \text{ulp}\left(\frac{b}{a}\right).$$

Therefore, $\text{RN}(q + ry) = q + \text{ulp}\left(\frac{b}{a}\right)$, which is what we wanted to show.

□

Hence, a careful use of Theorem 4.16 makes it possible to get a correctly rounded quotient b/a, once we have computed a very accurate approximation to the reciprocal $1/a$. Let us therefore focus on the computation of reciprocals. More precisely, we wish to always get $\text{RN}(1/a)$. The central result, due to Markstein, is the following one.

Theorem 4.17 (Markstein [406]). *In precision-p binary floating-point arithmetic, if y is an approximation to $1/a$ with an error less than* $\text{ulp}(1/a)$*, and the calculations*

$$r = \text{RN}(1 - ay), \; y' = \text{RN}(y + ry)$$

are performed using round-to-nearest-even mode, then y' is exactly $1/a$ rounded to nearest even, provided that the integral significand of a, namely $A = a/\,\text{ulp}(a)$, is different from $2^p - 1 = 11111 \cdots 11_2$.

A division algorithm can therefore be built as follows.

- First, an accurate enough approximation to $1/a$ is computed. In general, it is wise to use iteration (4.11) for the first steps because it is faster, and iteration (4.9) for the last steps because it is more accurate (both in round-to-nearest mode). A *very careful error analysis* must be performed to make sure that, after these iterations, we get an approximation to $1/a$ that is within 1 ulp from $1/a$. That error analysis depends on the accuracy of the table (or approximation of any kind, e.g., by a polynomial of small degree) that gives x_0, on the precision of the input and output values, on the available internal precision, etc.). A way to perform (and to automate) this error analysis is presented by Panhaleux [481].

- Then, Theorem 4.17 is applied, to get $\text{RN}(1/a)$ (except, possibly in the case where the integral significand of a is $2^p - 1 = 11111 \cdots 11_2$. This case is easily handled separately [406]).

- A first approximation to the quotient is computed by multiplying the previously obtained value $\text{RN}(1/a)$ by b.

- This approximation to the quotient is refined, in round-to-nearest mode, using iteration (4.16). Again, a careful error analysis is required to make sure that we get an approximation to b/a that is within $\text{ulp}(b/a)$.

- Finally, Theorem 4.16 is applied to get b/a correctly rounded in the desired rounding mode.

Several variants of division algorithms (depending on the required and internal precisions, depending on whether we wish to optimize the throughput or to minimize the latency) are given by Markstein in his excellent book [406].

4.7.3 Possible double roundings in division algorithms

We have examined how one can compute b/a, where a and b are floating-point numbers between 1 and 2. Let us now see how one can deal with arbitrary input values: We now assume that a and b are nonnegative floating-point numbers (separately handling the signs of the input operands is straightforward). Assume one wishes to obtain $\mathrm{RN}(b/a)$. To make sure that the conditions of Theorem 4.16 are satisfied one needs to *scale* the iterations. This can be done as follows. Let m_a and m_b be the significands of a and b, and let e_a and e_b be their exponents. We first perform the division m_b/m_a as explained above: Using iteration (4.9) and iteration (4.16) we obtain a "scaled approximate quotient" q^* that is a faithful approximation to m_b/m_a, and a "scaled approximate reciprocal" y^* that approximates $1/m_a$ with relative error less than 2^{-p}. We therefore can apply Theorem 4.16, i.e., we compute

$$\begin{array}{rcl} r & = & \mathrm{RN}\,(m_b - m_a q^*), \\ q' & = & \mathrm{RN}\,(q^* + r y^*), \end{array} \qquad (4.19)$$

Theorem 4.16 implies $q' = \mathrm{RN}(m_b/m_a)$. Note that q' is in the normal range.

- If $2^k q'$ is in the normal range too (i.e., if $2^{e_{\min}} \le |2^k q'| \le 2^{e_{\max}+1} - 2^{e_{\max}-p+1}$), then it is a floating-point number and $\mathrm{RN}(b/a) = 2^k q'$. We therefore return this value.

- Obviously, if $|2^k q'| > 2^{e_{\max}+1} - 2^{e_{\max}-p+1}$, then we return $+\infty$.

A problem may occur when $2^k q'$ falls in the subnormal range and is not a floating-point number. We cannot just round it because this new rounding may lead to the delivery of a wrong result. Consider the following example in binary32 arithmetic (precision $p = 24$, extremal exponents $e_{\min} = -126$ and $e_{\max} = 127$), and assume that RN is round-to-nearest-*ties-to-even*. Consider the two floating-point input values:

$$\begin{cases} b &= 1.01_2 \times 2^{-115} \\ &= 10485760_{10} \times 2^{-138}, \\ a &= 1.00000000000011001100111_2 \times 2^{23} \\ &= 8390247_{10}. \end{cases}$$

The number b/a is equal to

$$1.0011111111101111111111110100110011101000111101 0_2 \cdots \times 2^{-138},$$

so the correctly rounded, subnormal value that must be returned when computing b/a should be

$$\begin{aligned} \mathrm{RN}(b/a) &= 1.00111111111_2 \times 2^{-138} \\ &= 0.00000000000100111111111_2 \times 2^{-126}. \end{aligned}$$

The exact value of m_b/m_a is

$$1.0011111111101111111111110100110011101000111101 0_2 \cdots ,$$

so $q' = \mathrm{RN}(m_b/m_a) = 1.001111111111_2$. Once multiplied by $2^{e_b - e_a} = 2^{-138}$, this result would be equal to

$$1.001111111111_2 \times 2^{-138} = \mathrm{RN}(b/a) + 2^{-150},$$

which means, since it is in the subnormal range that, after rounding it to the nearest floating-point number, we would get

$$1.01_2 \times 2^{-138} \neq \mathrm{RN}(b/a).$$

This problem occurs each time $2^k q'$ is equal to a (subnormal) midpoint, and:

- Either $b/a > 2^k q'$ and $\mathrm{RN}(2^k q') < 2^k q'$;
- or $b/a < 2^k q'$ and $\mathrm{RN}(2^k q') > 2^k q'$.

If we are only given q' without any extra information, it is impossible to deduce whether the exact, infinitely precise, result is above or below the midpoint, so it is not possible to make sure that we return a correctly rounded value.

Fortunately, the number r computed during the "correction iteration" (4.19) gives enough information to obtain $\mathrm{RN}(b/a)$. More precisely, it is possible [440] to show that when $2^k q'$ is a midpoint:

1. If $r = 0$ then $q' = q^* = m_b/m_a$, hence $b/a = 2^k q'$ exactly. Therefore, we should return $\mathrm{RN}(2^k q')$;

2. if $q' \neq q^*$ and $r > 0$, then q' overestimates m_b/m_a. Therefore, we should return $2^k q'$ rounded down;

3. if $q' \neq q^*$ and $r < 0$, then q' underestimates m_b/m_a. Therefore, we should return $2^k q'$ rounded up;

4. if $q' = q^*$ and $r > 0$, then q' underestimates m_b/m_a. Therefore, we should return $2^k q'$ rounded up;

5. if $q' = q^*$ and $r < 0$, then q' overestimates m_b/m_a. Therefore, we should return $2^k q'$ rounded down.

See [440] for more details.

4.8 Newton–Raphson-Based Square Root with an FMA

The Newton–Raphson iteration can also be used to evaluate square roots. Again, the availability of an FMA instruction allows for rather easily obtained correctly rounded results. We will not present the methods in detail here (one can find them in [406, 120, 118]): we will focus on the most important results only.

4.8.1 The basic iterations

From the general Newton–Raphson iteration (4.6), one can derive two classes of algorithms for computing the square root of a positive real number a.

- If we look for the positive root of function $f(x) = x^2 - a$, we get

$$x_{n+1} = \frac{1}{2}\left(x_n + \frac{a}{x_n}\right). \tag{4.20}$$

  This "Newton–Raphson" square root iteration goes back to much before Newton's time. Al-Khwarizmi mentions this method in his arithmetic book [128]. Furthermore, it was already used by Heron of Alexandria (which explains why it is frequently named "the Heron iteration"), and seems to have been known by the Babylonians 2000 years before Heron [205]. One easily shows that if $x_0 > 0$, then x_n goes to $\sqrt{a}$. This iteration has a drawback: it requires a division at each step. Also, guaranteeing correct rounding does not seem to be a simple task.

- If we look for the positive root of function $f(x) = 1/x^2 - a$, we get

$$x_{n+1} = x_n(3 - ax_n^2)/2. \tag{4.21}$$

  This iteration converges to $1/\sqrt{a}$, provided that $x_0 \in (0, \sqrt{3}/\sqrt{a})$. To get a first approximation to $\sqrt{a}$ it suffices to multiply the obtained result by a. And yet, this does not always give a correctly rounded result; some refinement is necessary. To obtain fast, quadratic convergence, the first point x_0 must be a close approximation to $1/\sqrt{a}$, read from a table or obtained using a polynomial approximation of small degree. Iteration (4.21) still has a division, but that division (by 2) is very simple, especially in radix 2.

4.8.2 Using the Newton–Raphson iteration for correctly rounded square roots

From Equation (4.21), and by defining a "residual" ϵ_n as $1 - ax_n^2$, one gets

$$\begin{cases} \epsilon_n & = & 1 - ax_n^2 \\ x_{n+1} & = & x_n + \frac{1}{2}\epsilon_n x_n. \end{cases} \tag{4.22}$$

To decompose these operations in terms of FMA instructions, Markstein [406] defines new variables:

$$\left\{\begin{array}{rcl} r_n & = & \frac{1}{2}\epsilon_n \\ g_n & = & a x_n \\ h_n & = & \frac{1}{2}x_n. \end{array}\right. \tag{4.23}$$

From (4.22) and (4.23), one finds the following iteration:

$$\left\{\begin{array}{rcl} r_n & = & \frac{1}{2} - g_n h_n \\ g_{n+1} & = & g_n + g_n r_n \\ h_{n+1} & = & h_n + h_n r_n. \end{array}\right. \tag{4.24}$$

Variable h_n goes to $1/(2\sqrt{a})$, and variable g_n goes to $\sqrt{a}$. Iteration (4.24) is easily implemented with an FMA instruction. Some parallelism is possible since the computations of g_{n+1} and h_{n+1} can be performed simultaneously.

Exactly as explained for the division iteration, a very careful error analysis is needed, as well as a final refinement step. Here are some results that make it possible to build refinement techniques. See Markstein's book [406] for more details.

Theorem 4.18. *In any radix, the square root of a floating-point number cannot be the exact midpoint between two consecutive floating-point numbers.*

Proof. Assume that r is the exact middle of two consecutive radix-β, precision-p floating-point numbers, and assume that it is the square root of a floating-point number x. Note that r cannot be in the subnormal range. Without loss of generality we can assume that r has the form

$$r = (r_0.r_1r_2\cdots r_{p-1})_\beta + \frac{1}{2}\beta^{-p+1};$$

i.e., that $1 \le r < \beta$. Let $R = (r_0r_1r_2\cdots r_{p-1})_\beta$ be the integral significand of r. We have

$$2r\beta^{p-1} = 2R + 1;$$

i.e.,

$$4r^2\beta^{2p-2} = (2R+1)^2.$$

Since r^2 is a floating-point number between 1 and β^2, it is a multiple of β^{-p+1}, which implies that $r^2\beta^{2p-2}$ is an integer. Thus, $4r^2\beta^{2p-2}$ is a multiple of 4. This contradicts the fact that it is equal to the square of the odd number $2R+1$. $\square$

Theorem 4.19. *If the radix of the floating-point arithmetic is 2, then the square root reciprocal of a floating-point number cannot be the exact midpoint between two consecutive floating-point numbers.*

The proof is very similar to the proof of Theorem 4.18.

In nonbinary radices, Theorem 4.19 may not hold, even if the radix is prime. Consider for example a radix-3 arithmetic, with precision $p = 6$. The number

$$x = 324_{10} = 110000_3$$

is a floating-point number, and the reader can easily check that

$$\frac{1}{\sqrt{x}} = \frac{1}{3^8} \cdot \left(111111_3 + \frac{1}{2}\right),$$

which implies that $1/\sqrt{x}$ is the exact midpoint between two consecutive floating-point numbers.

Also, in radix 10 with precision 16 (which corresponds to the decimal64 format of the IEEE 754-2008 standard), the square-root reciprocal of

$$\underbrace{70.36874417766400}_{\text{16 digits}}$$

is

$$0.\underbrace{11920928955078125}_{\text{17 digits}},$$

which is the exact midpoint between two consecutive floating-point numbers.

The following *Tuckerman test* allows one to check if a floating-point number r is the correctly rounded-to-nearest square root of another floating-point number a. Markstein [406] proves the following theorem in prime radices, but it holds in any radix.

Theorem 4.20 (The Tuckerman test, adapted from Markstein's presentation [406]). *In radix β, if a and r are floating-point numbers, then r is $\sqrt{a}$ rounded to nearest if and only if*

$$r(r - \mathrm{ulp}(r^-)) < a \leq r(r + \mathrm{ulp}(r)) \tag{4.25}$$

where r^- is the floating-point predecessor of r.

Proof. Theorem 4.18 implies that $\sqrt{a}$ cannot be a "midpoint" (i.e., a value exactly halfway between two consecutive floating-point numbers). Therefore, we do not have to worry about tie-breaking rules. Also, if k is an integer such that $\beta^k r$ and $\beta^{2k} a$ do not overflow or underflow, then (4.25) is equivalent to

$$(\beta^k r)((\beta^k r) - \beta^k \,\mathrm{ulp}(r^-)) < \beta^{2k} a \leq (\beta^k r)((\beta^k r) + \beta^k \,\mathrm{ulp}(r));$$

which is equivalent, if we define $R = \beta^k r$ and $A = \beta^{2k} a$ to

$$R(R - \mathrm{ulp}(R^-)) < A \leq R(R + \mathrm{ulp}(R)).$$

Since $R = \mathrm{RN}(\sqrt{A})$ is straightforwardly equivalent to $r = \mathrm{RN}(\sqrt{a})$, we deduce that without loss of generality, we can assume that $1 \leq r < \beta$. Let us now consider two cases.

1. **If** $r = 1$. In this case, (4.25) becomes $1 - \beta^{-p} < a \le 1 + \beta^{-p+1}$; that is, since a is a floating-point number,

$$a \in \{1, 1 + \beta^{-p+1}\}. \tag{4.26}$$

   Since $\sqrt{a}$ cannot be a midpoint, $1 = \mathrm{RN}(\sqrt{a})$ is equivalent to

$$1 - \frac{1}{2}\beta^{-p} < \sqrt{a} < 1 + \frac{1}{2}\beta^{-p+1},$$

   which is equivalent to

$$1 - \beta^{-p} + \frac{1}{4}\beta^{-2p} < a < 1 + \beta^{-p+1} + \frac{1}{4}\beta^{-2p+2}. \tag{4.27}$$

   The only floating-point numbers lying in the real range defined by (4.27) are 1 and $1 + \beta^{-p+1}$, so (4.27) is equivalent to (4.26).

2. **If** $1 < r < \beta$. In this case $\mathrm{ulp}(r^-) = \mathrm{ulp}(r) = \beta^{-p+1}$ and (4.25) is thus equivalent to

$$r(r - \beta^{-p+1}) < a \le r(r + \beta^{-p+1}). \tag{4.28}$$

   Since $\sqrt{a}$ cannot be a midpoint, $r = \mathrm{RN}(\sqrt{a})$ is equivalent to

$$r - \frac{1}{2}\beta^{-p+1} < \sqrt{a} < r + \frac{1}{2}\beta^{-p+1}.$$

   By squaring all terms, this is equivalent to

$$r(r - \beta^{-p+1}) + \frac{1}{4}\beta^{-2p+2} < a < r(r + \beta^{-p+1}) + \frac{1}{4}\beta^{-2p+2}. \tag{4.29}$$

   Now, since r is a floating-point number between 1 and β, it is a multiple of β^{-p+1}. This implies that $r(r - \beta^{-p+1})$ and $r(r + \beta^{-p+1})$ are multiples of β^{-2p+2}.

   An immediate consequence is that there is no multiple of β^{-2p+2} between $r(r - \beta^{-p+1})$ and $r(r - \beta^{-p+1}) + \frac{1}{4}\beta^{-2p+2}$, or between $r(r + \beta^{-p+1})$ and $r(r + \beta^{-p+1}) + \frac{1}{4}\beta^{-2p+2}$.

   This implies that there is no *floating-point number* between $r(r - \beta^{-p+1})$ and $r(r - \beta^{-p+1}) + \frac{1}{4}\beta^{-2p+2}$, or between $r(r + \beta^{-p+1})$ and $r(r + \beta^{-p+1}) + \frac{1}{4}\beta^{-2p+2}$.

   As a consequence, since a is a floating-point number, (4.29) is equivalent to (4.28).

□

With an FMA instruction, and assuming that instructions for computing the floating-point predecessor and successor of r are available, the Tuckerman test is easily performed. For instance, to check whether

$$a \le r(r + \mathrm{ulp}(r)),$$

it suffices to compute

$$\delta = \mathrm{RN}(r \times \mathrm{successor}(r) - a)$$

using an FMA, and to test the sign of δ.

We hope that we have convinced the reader that, with some care, the Newton-Raphson iteration and some of its variants can be used for implementing correctly rounded division and square root. Efficient alternatives are high-radix digit-recurrence algorithms [186]. It is difficult to tell which family of algorithms will ultimately win.

4.9 Radix Conversion

4.9.1 Conditions on the formats

When the radix of the floating-point system is 2,[14] conversions from and to radix 10 must be provided, since humans read and write numbers in decimal. Early works on radix conversion were done by Goldberg [215] and by Matula [411]. Accurate algorithms for input and output radix conversion can be found in the literature [16, 86, 101, 102, 394, 517, 567], and are now implemented in most compilers. It is important to understand that radix conversion is not a fully innocuous operation.

First, some numbers that have a finite radix-10 representation do not have a finite binary one. A simple example is $0.1_{10} = 1/10$, whose binary representation is

$$0.0001100110011001100110011001100110011001100110011001100110\cdots.$$

Moreover, although all numbers with a finite radix-2 representation also have a finite decimal representation,[15] the number of digits of this decimal representation might sometimes be too large to be convenient. Consider for instance the following floating-point binary32 number:

$$0.11111111111111111111111_2 \times 2^{-126} = 2^{-126} - 2^{-149}.$$

[14] A very similar study can be done when it is a power of 2.

[15] A necessary and sufficient condition for all numbers representable in radix β with a finite number of digits to be representable in radix γ with a finite number of digits is that β should divide an integer power of γ.

Its exact decimal representation

$$1.17549421069244107548702944484928734882705242874589333385717453057158887047561890426550235133618116378784179687 5 \times 10^{-38}$$

is too large to be convenient for all applications. This is the worst case in binary32 arithmetic. If the two extremal exponents $e_{\min}$ and $e_{\max}$ of the binary format satisfy $e_{\min} \approx -e_{\max}$ (which holds for all usual formats) and if p_2 is its precision, then the largest width of a decimal significand we can obtain by exact conversion is[16]

$$-e_{\min} + p_2 + \left\lfloor (e_{\min} + 1)\log_{10}(2) - \log_{10}\left(1 - 2^{-p_2}\right) \right\rfloor \text{ digits.}$$

For instance, for the various basic binary formats of the IEEE 754-2008 standard (see Table 3.1), this gives 112 digits for binary32, 767 digits for binary64, and 11563 digits for binary128.

Hence, during radix conversions, numbers must be *rounded*. We assume here that we want to minimize the corresponding rounding errors (i.e., to round numbers to the nearest value in the target format whenever possible).

Other methods should be used when directed rounding modes are considered, since an experienced user will choose these rounding modes to get guaranteed lower or upper bounds on a numerical value. Therefore, it would be a waste to carefully design a numerical program in order to compute a binary value that is a definite lower bound to the exact result, only to eventually have this value rounded up during radix conversion.

A question that naturally arises is: For a given binary format, which decimal format is preferable if the user does not specify anything?

Assuming an internal binary format of precision p_2, the first idea that springs to mind is to have an input/output decimal format whose precision is the integer that is nearest to

$$p_2 \frac{\log(2)}{\log(10)}.$$

This would for instance give a decimal precision equal to 16 for the binary64 format ($p = 53$).

And yet, this is not the best idea, for the following reason. It is a common practice to write a floating-point value to a file, and to read it later, or (equivalently) to re-enter on the keyboard the result of a previous computation. One would like this operation (let us call it a "write-read cycle") to be

[16]This formula is valid for all *possible* values of p_2 and $e_{\min}$ (provided $e_{\min} \approx -e_{\max}$). And yet, for all *usual* formats, it can be simplified: A simple continued fraction argument (see Section A.1) shows that for $p_2 \geq 16$ and $e_{\min} \geq -28000$, it is equal to

$$-e_{\min} + p_2 + \lfloor (e_{\min} + 1)\log_{10}(2) \rfloor .$$

error-free: When converting a binary floating-point number x to the external decimal format, and back-converting the obtained result to binary, one would like to find x again, without any error. Of course, this is always possible by performing an "exact" conversion to decimal, using for the decimal representation of x a large number of digits, but, as we are about to see, such an exact conversion is not required. Furthermore, there is an important psychological point: As pointed out by Steele and White [567], if a system prints too many decimal digits, the excess digits will seem to reflect more information than the number actually contains.

Matula [411] shows the following result.

Theorem 4.21 (Base conversion). *Assume we convert a radix-β, precision-p floating-point number to the nearest number in a radix-γ, precision-q format, and then convert back the obtained value to the nearest number in the initial radix-β, precision-p format. If there are no positive integers i and j such that $\beta^i = \gamma^j$, then a necessary and sufficient condition for this operation to be the identity (provided no underflow/overflow occurs) is*

$$\gamma^{q-1} > \beta^p.$$

Let us explain Matula's result in the case of a write-read cycle (that is, the values β and γ of Theorem 4.21 are 2 and 10, respectively). Let p_2 be the precision of the "internal" binary format and p_{10} be the precision of the "external" radix-10 format. We assume in the following that the conversions are correctly rounded, in round-to-nearest mode. Hence, our problem is to find conditions on p_{10} to make sure that a write-read cycle is error-free.

Figure 4.4 shows that if p_{10} is not large enough, then after a write-read cycle we may end up with a binary number slightly different from the initial one.

In the neighborhood of the binary floating-point number x to be converted, let us call ϵ_2 the distance between two consecutive binary numbers (that is, $\epsilon_2 = \text{ulp}(x)$), and ϵ_{10} the distance between two consecutive decimal floating-point numbers of the "external" format.

- When x is converted to a decimal floating-point number x', since we assume round-to-nearest mode, this implies that $|x - x'|$ is less than or equal to $\epsilon_{10}/2$.

- When x' is back-converted to a binary floating-point number x'', this implies that $|x' - x''|$ is less than or equal to $\epsilon_2/2$.

To always have $x'' = x$ it suffices that

$$\epsilon_{10} < \epsilon_2. \tag{4.30}$$

Now, let us see what this constraint means in terms of p_2 and p_{10}. Consider numbers that are between two consecutive powers of 10, say, 10^r and

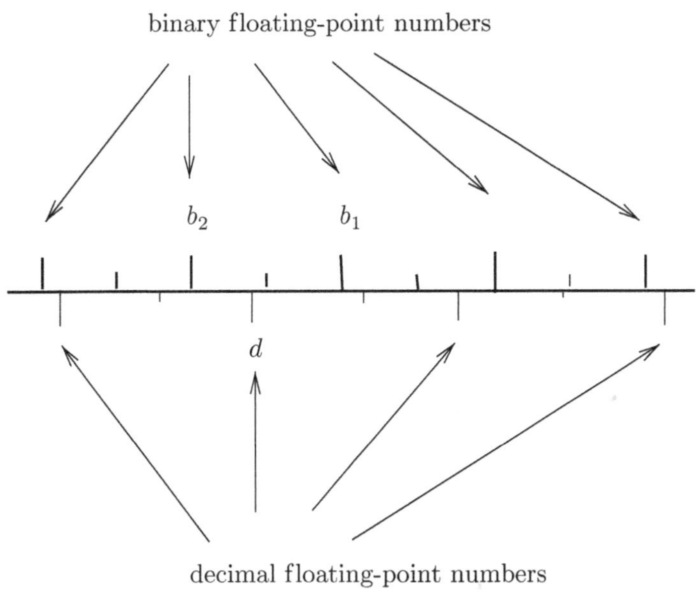

Figure 4.4: In this example, binary number b_1 is converted to the decimal number d, and d is converted back to the binary number b_2.

10^{r+1} (see Figure 4.5). In that domain,

$$\epsilon_{10} = 10^{r-p_{10}+1},$$

also, that domain contains at most four consecutive powers of 2, say, 2^q, 2^{q+1}, 2^{q+2}, and 2^{q+3}, so the binary ulp ϵ_2 varies from 2^{q-p_2} to 2^{q-p_2+4}. Therefore, condition (4.30) becomes

$$10^r \cdot 10^{-p_{10}+1} < 2^q \cdot 2^{-p_2}. \tag{4.31}$$

Now, since 2^q is larger than 10^r (yet, it can be quite close to 10^r), condition (4.31) will be satisfied if

$$2^{p_2} \leq 10^{p_{10}-1}. \tag{4.32}$$

Note that this condition is equivalent to $2^{p_2} < 10^{p_{10}-1}$, since $2^{p_2} = 10^{p_{10}-1}$ is impossible.

Therefore, the most convenient choice for p_{10} is the smallest integer for which (4.32) holds, namely,

$$p_{10} = 1 + \lceil p_2 \log_{10}(2) \rceil . \tag{4.33}$$

Table 4.2 gives such values p_{10} for various frequently used values of p_2.

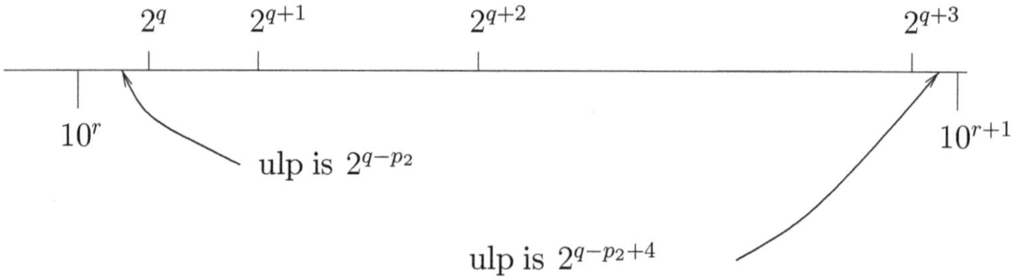

Figure 4.5: Various possible values of the (binary) ulp function between two consecutive powers of 10*. One can easily show that between two consecutive powers of* 10 *there are at most four consecutive powers of* 2*.*

p_2	24	53	64	113
p_{10}	9	17	21	36

Table 4.2: For various values of the precision p_2 of the internal binary format, minimal values of the external decimal precision p_{10} such that a write-read cycle is error-free, when the conversions are correctly rounded to nearest.

4.9.2 Conversion algorithms

4.9.2.1 Output conversion: from radix 2 to radix 10

The results given in the previous section correspond to *worst cases*. For many binary floating-point numbers x, the number of radix-10 digits that should be used to guarantee error-free write-read cycles will be less than what is given in Table 4.2. Consider for instance the floating-point number

$$x = 5033165 \times 2^{-24} = 0.300000011920928955078125.$$

It is exactly representable in the binary32 format of the IEEE 754-2008 standard, also known as single-precision ($p = 24$). Hence, we know from the study of the previous section and from Table 4.2 that if we convert x to the 9-digit decimal number $x^{(1)} = 0.300000012$, and convert back this decimal number to binary32 arithmetic, we will find x again. And yet, once converted to binary32 arithmetic, the 1-digit decimal number $x^{(2)} = 0.3$ also gives x. Hence, in this particular case, an error-free write-read cycle is possible with precision $p_{10} = 1$. One could object that $x^{(1)}$ is as legitimate as $x^{(2)}$ to "represent" x, but there is an important psychological aspect here, that should not be neglected. Someone typing in 0.3 on a keyboard (hence, getting x after conversion) will be disconcerted by seeing it displayed as 0.300000012.

This leads to a strategy suggested by Steele and White [566, 567]: When the output format is not specified, use for each binary floating-point number

x the *smallest* number of radix-10 significand digits that allows for an error-free write-read cycle. Steele and White designed an algorithm for that. Their algorithm was later improved by Burger and Dybvig [86] , and by Gay [211]. Gay's code is available for anyone to use, and is very robust.[17] Faster yet more complex algorithms have been introduced by Loitsch [394] and Andrysco et al. [16]. In the following, we present Burger and Dybvig's version of the conversion algorithm.

We assume that the internal floating-point system is binary[18] and of precision p. If x is a binary floating-point number, we denote by x^- and x^+ its floating-point predecessor and successor, respectively. In the following, we assume that the internal binary number to be converted is positive. The algorithm uses exact rational arithmetic. (Burger and Dybvig also give a more complex yet more efficient algorithm that only uses high-precision integer arithmetic and an efficient scale-factor estimator; see [86] for details.) The basic principle of Algorithm 4.12 for converting the binary number $x = X \times 2^{e-p+1}$ is quite simple:

- We scale x until it is between $1/10$ and 1, i.e., until it can be written $0.d_1d_2d_3d_4\cdots$ in decimal;
- the first digit d_1 is obtained by multiplying the scaled value by 10 and taking the integer part. The fractional part is used to compute the subsequent digits in a similar fashion.

4.9.2.2 Input conversion: from radix 10 to radix 2

The input conversion problem is very different from the previous one, primarily because the input decimal numbers may not have a predefined, bounded size. The number of input digits that need to be examined to decide which is the binary floating-point number nearest to the input decimal number may be arbitrarily large. Consider the following example. Assume that the internal format is the IEEE 754-2008 binary32 format (see Chapter 3), and that the rounding mode is round to nearest even. If the input number is

$$\begin{aligned} & 1.000000059604644775390625000000000000000000000000000000000000 \\ = \; & 1 + 2^{-24}, \end{aligned}$$

then the conversion algorithm should return 1, whereas if the input number is

$$\begin{aligned} & 1.000000059604644775390625000000000000000000000000000000000001 \\ = \; & 1 + 2^{-24} + 10^{-60}, \end{aligned}$$

[17] At the time of writing this book, it can be obtained at http://www.netlib.org/fp/ (file `dtoa.c`).

[18] The algorithm works for other radices. See [86] for details.

Algorithm 4.12 Conversion from radix 2 to radix 10 [86]. The input value is a precision-p binary floating-point number x, and the output value is a decimal number $V = 0.d_1d_2\cdots d_n \times 10^k$, where n is the smallest integer such that 1) $(x^- + x)/2 < V < (x + x^+)/2$, i.e., the floating-point number nearest to V is x, regardless of how the input rounding algorithm breaks ties (x^- is the floating-point predecessor of x, and x^+ is its floating-point successor); and 2) $|V - x| \leq 10^{k-n}/2$, i.e., V is correctly rounded in the precision-n output decimal format. Here $\{t\}$ denotes the fractional part of t. We assume that the calculations are done using exact arithmetic.

$\ell \leftarrow (x^- + x)/2$
$u \leftarrow (x + x^+)/2$
find the smallest k such that $u \leq 10^k$
$W \leftarrow x/10^{k-1}$
$d_1 \leftarrow \lfloor W \rfloor$
$W \leftarrow \{W\}$
$n \leftarrow 1$
while $0.d_1d_2d_3\cdots d_n \times 10^k \leq \ell$ and $\left(0.d_1d_2d_3\cdots d_n + \frac{1}{10^n}\right) \times 10^k \geq u$ **do**
  $n \leftarrow n + 1$
  $d_n \leftarrow \lfloor 10 \times W \rfloor$
  $W \leftarrow \{10 \times W\}$
end while
if $0.d_1d_2d_3\cdots d_n \times 10^k > \ell$ and $\left(0.d_1d_2d_3\cdots d_n + \frac{1}{10^n}\right) \times 10^k \geq u$ **then**
  return $0.d_1d_2d_3\cdots d_n \times 10^k$
else if $0.d_1d_2d_3\cdots d_n \times 10^k \leq \ell$ and $\left(0.d_1d_2d_3\cdots d_n + \frac{1}{10^n}\right) \times 10^k < u$ **then**
  return $\left(0.d_1d_2d_3\cdots d_n + \frac{1}{10^n}\right) \times 10^k$
else
  return the value closest to x among $0.d_1d_2d_3\cdots d_n \times 10^k$ and $\left(0.d_1d_2d_3\cdots d_n + \frac{1}{10^n}\right) \times 10^k$
end if

the conversion algorithm should return the floating-point successor of 1, namely $1 + 2^{-23}$.

The first efficient and accurate input conversion algorithms were introduced by Rump [517] and Clinger [101, 102]. Later on, Gay suggested improvements [211]. Let us describe Gay's version of Clinger's algorithm.

We assume that the floating-point number system being used is a binary system of precision p. When converting an input decimal number x to binary, the best result we can provide is a correctly rounded result: In this case, the obtained value v is $\circ(x)$, where $\circ$ is the chosen rounding function. Note that the requirements of the IEEE 754-1985 standard for floating-point arithmetic were not that strong.[19] Here, we assume that we want to correctly round

[19] In round-to-nearest modes, it required that the error introduced by the conversion should

to nearest ties-to-even (similar work can be done with the other rounding modes).

The input value d is an n-digit decimal number:

$$\begin{aligned} d &= 10^k \times [d_0.d_1d_2\cdots d_{n-1}]_{10} \\ &= \sum_{i=0}^{n-1} d_i 10^{k-i}. \end{aligned}$$

We want to return a precision-p binary floating-point number $b = \mathrm{RN}_{\mathrm{even}}(d)$. For simplicity, we assume that $d > 0$, and that neither underflow nor overflow will occur, that is:

$$2^{e_{\min}} \le d \le 2^{e_{\max}}\left(2 - 2^{1-p}\right),$$

where $e_{\min}$ and $e_{\max}$ are the extremal exponents of the binary floating-point format. Our problem consists in finding an exponent e and a significand $b_0.b_1b_2\cdots b_{p-1}$, with $b_0 \neq 0$, such that the number

$$\begin{aligned} b &= 2^e \times [b_0.b_1b_2\cdots b_{p-1}]_2 \\ &= \sum_{i=0}^{p-1} b_i 2^{e-i} \end{aligned}$$

satisfies

$$\begin{cases} \text{if } b = 2^e \text{ exactly} & \text{then } -2^{e-p-1} \le d - b \le 2^{e-p} \\ \text{otherwise} & |b-d| \le 2^{e-p}, \text{ and } |b-d| = 2^{e-p} \Rightarrow b_{p-1} = 0. \end{cases} \tag{4.34}$$

Figure 4.6 helps us to understand these conditions in the case $b = 2^e$.

First, note that some cases (in practice, those that occur most often!) are very easily handled. Denote

$$D = \frac{d}{10^{k-n+1}},$$

that is, D is the integer whose decimal representation is

$$d_0d_1d_2\cdots d_{n-1}.$$

1. If $10^n \le 2^p$ and $10^{|k-n+1|} \le 2^p$, then the integers D and $10^{|k-n+1|}$ are exactly representable in the binary floating-point system. In this case, it suffices to compute D exactly as[20]

$$(\cdots(((d_0 \times 10 + d_1) \times 10 + d_2) \times 10 + d_3)\cdots) \times 10 + d_{n-1},$$

be at most 0.97 ulps. The major reason for this somewhat weak requirement is that the conversion algorithms presented here were not known at the time that standard was designed.

[20]Another solution consists in using a precomputed table of powers of 10 in the binary format.

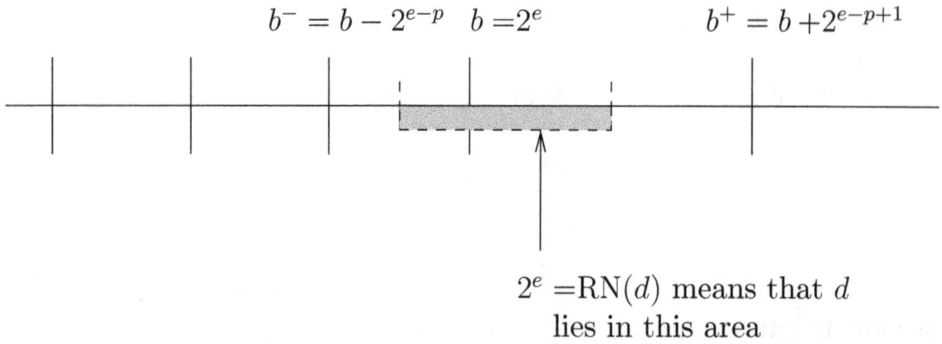

Figure 4.6: Illustration of the conditions (4.34) in the case $b = 2^e$.

and to compute $K = 10^{|k-n+1|}$ by iterative multiplications. We then get b by performing one floating-point multiplication or division:

$$b = \begin{cases} \mathrm{RN}(D \times K) & \text{if } k - n + 1 \geq 0 \\ \mathrm{RN}(D/K) & \text{otherwise.} \end{cases}$$

2. Even if the above conditions are not satisfied, if there exists an integer j, $1 \leq j < k$, such that the integers 10^{k-j} and $10^{n+j}\,[d_0.d_1d_2\cdots d_{n-1}]_{10}$ are less than or equal to 2^p, then 10^{k-j} and $10^{n+j}\,[d_0.d_1d_2\cdots d_{n-1}]_{10}$ are exactly representable and easily computed, and their floating-point product is b.

Now, if we are not in these "easy cases," we build a series of "guesses"

$$b^{(1)}, b^{(2)}, b^{(3)} \ldots$$

and stop at the first guess $b^{(m)}$ such that $b^{(m)} = b$. Let us show how these guesses are built, and how we can check if $b^{(j)} = b$.

The first guess is built using standard floating-point arithmetic in the target format.[21] One can find $b^{(1)}$ such that

$$\left| d - b^{(1)} \right| < c \cdot 2^{e_b - p + 1},$$

where c is a small constant and e_b is an integer (equal to the exponent of $b^{(1)}$). Let us give a possible solution to do that. We assume $d_0 \neq 0$. Let j be the smallest integer such that $10^j \geq 2^p$. Define

$$\begin{aligned} D^* &= d_0 d_1 \cdots d_{\min\{n-1,j\}}.d_{\min\{n-1,j\}+1} \cdots d_{n-1} \\ &= \sum_{m=0}^{n-1} d_m 10^{\min\{n-1,j\}-m}. \end{aligned}$$

[21] If a wider internal format is available, one can use it and possibly save one step.

Also define

$$\hat{D} = \lfloor D^* \rfloor = d_0 d_1 \cdots d_{\min\{n-1,j\}}.$$

If we compute in standard floating-point arithmetic an approximation to $\hat{D}$ using the sequence of operations

$$(\cdots((d_0 \times 10 + d_1) \times 10 + d_2)\cdots) \times 10 + d_{\min\{n-1,j\}},$$

then all operations except possibly the last multiplication and addition are performed exactly. A simple analysis shows that the computed result, say $\tilde{D}$, satisfies

$$\tilde{D} = \hat{D}(1 + \epsilon_1),$$

with $|\epsilon_1| \leq 2^{-p+1} + 2^{-2p}$. This gives

$$\tilde{D} = D^*(1 + \epsilon_1)(1 + \epsilon_2),$$

where $|\epsilon_2| \leq 10^{-j} \leq 2^{-p}$.[22] Now, we want to get a binary floating-point approximation to

$$d = D^* \times 10^{k-\min\{n-1,j\}}.$$

To approximate $K^* = 10^{k-\min\{n-1,j\}}$, several solutions are possible. We can assume that the best floating-point approximations to the powers of 10 are precomputed and stored in a table. An alternative solution [211] is to compute K^* on the fly (assuming we have stored the first powers of 10, and powers of the form $10^{(2^i)}$, to save time and accuracy). For simplicity, let us assume here that we get from a table the best floating-point approximation to K^*, i.e., that we get a binary floating-point number $\tilde{K}$ that satisfies

$$\tilde{K} = K^*(1 + \epsilon_3),$$

where $|\epsilon_3| \leq 2^{-p}$. We finally compute

$$b^{(1)} = \mathrm{RN}(\tilde{K}\tilde{D}) = \tilde{K}\tilde{D}(1 + \epsilon_4),$$

with $|\epsilon_4| \leq 2^{-p}$. Therefore, we get

$$\begin{aligned} b^{(1)} &= d \times (1 + \epsilon_1)(1 + \epsilon_2)(1 + \epsilon_3)(1 + \epsilon_4) \\ &= d \times (1 + \epsilon), \end{aligned}$$

with

$$|\epsilon| \leq 5 \cdot 2^{-p} + 10 \cdot 2^{-2p} + 10 \cdot 2^{-3p} + 5 \cdot 2^{-4p} + 2^{-5p},$$

which gives $|\epsilon| \leq 5.0000006 \times 2^{-p}$ as soon as $p \geq 24$.

[22] A straightforward analysis of the error induced by the truncation of the digit chain D^* would give $|\epsilon_2| \leq 10^{-\min\{n-1,j\}}$, but when $j \geq (n-1)$, $D^* = \hat{D}$ and there is no truncation error at all.

From this we deduce

$$|b^{(1)} - d| \leq 5.0000006 \times 2^{e_b-p+1}.$$

Once we have an approximation $b^{(j)}$ of exponent e_j, as said above, for $b^{(j)}$ to be equal to b, it is necessary that

$$|d - b^{(j)}| \leq 2^{e_j-p}. \tag{4.35}$$

Furthermore, if $b^{(j)} = 2^{e_j}$ exactly, it is also necessary that

$$d - b^{(j)} \geq -\frac{1}{2}2^{e_j-p}. \tag{4.36}$$

We will focus on condition (4.35) and show how Gay handles it. Define

$$M = \max\left\{1, 2^{p-e_j-1}\right\} \times \max\left\{1, 10^{n-k-1}\right\}.$$

Condition (4.35) is equivalent to

$$|2M(d - b^{(j)})| \leq M \times 2^{e_j-p+1}, \tag{4.37}$$

but since $2Md$, $2Mb^{(j)}$, and $M \times 2^{e_j-p+1}$ are integers, condition (4.37) can easily be checked using multiple-precision integer arithmetic. There are three cases to consider:

- If $|2M(d - b^{(j)})| < M \times 2^{e_j-p+1}$, then $b^{(j)}$ is equal to $\mathrm{RN}(d)$.
- If $|2M(d - b^{(j)})| = M \times 2^{e_j-p+1}$, then $\mathrm{RN}(d)$ is $b^{(j)}$ if the integral significand of $b^{(j)}$ is even (i.e., if the last bit of the significand of $b^{(j)}$ is a zero), and the floating-point number adjacent to $b^{(j)}$ in the direction of d otherwise.
- If $|2M(d - b^{(j)})| > M \times 2^{e_j-p+1}$, we must find a closer floating-point approximation, $b^{(j+1)}$, to d.

The approximation $b^{(j+1)}$ can be built as follows. Define

$$\delta^{(j)} = \frac{(d - b^{(j)})}{2^{e_j-p+1}}.$$

This value will be computed as the ratio of the multiple-precision integers used in the previous test, as

$$\delta^{(j)} = \frac{1}{2} \times \frac{2M \times (d - b^{(j)})}{M \times 2^{e_j-p+1}}.$$

We have $d = b^{(j)} + \delta^{(j)}2^{e_j-p+1}$, which means that $\delta^{(j)}$ is the number of ulps that should be added to $b^{(j)}$ to get d. In most cases, b will be obtained by adding to $b^{(j)}$, in floating-point arithmetic, a floating-point approximation to that correcting term $\delta^{(j)}$. Hence, once floating-point approximations to $M(d-b^{(j)})$ and $M \times 2^{e_j-p+1}$ are computed (from the integers computed for checking $b^{(j)}$), we compute $\tilde{\delta}_j$ as the quotient of these approximations, and we compute

$$b^{(j+1)} = \mathrm{RN}(b^{(j)} + \tilde{\delta}_j 2^{e_j-p+1}).$$

Some care is necessary to avoid loops (if $b^{(j+1)} = b^{(j)}$), see [211] for details on how to handle these cases. Gay [211] shows that the number of steps needed to have $b^{(m)} = b$ is at most 3. In most cases, b is $b^{(1)}$ or $b^{(2)}$. Indeed, the only cases for which $m = 3$ are those for which $|b^{(2)} - b| = 2^{e-p+1}$.

4.10 Conversion Between Integers and Floating-Point Numbers

Floating-point units may not allow for all possible conversions between floating-point numbers and integers to be performed in hardware. This section presents some algorithms for converting an integer into a floating-point number, as well as some algorithms for "truncating" a floating-point number to an integer. All these algorithms are part of the CompCert C compiler and have been verified using the Coq proof assistant [51] (see also Section 6.2.7). This makes them very reliable.

The types of interest are, on one side, the basic 32- and 64-bit binary floating-point formats denoted `f32` (binary32) and `f64` (binary64), and, on the other side, the integer types `s32`, `u32`, `s64`, and `u64` (signed or unsigned, 32 or 64 bits). We write *dst_src* for the conversion from type *src* to type *dst*. For example, `f64_u32` denotes the conversion from 32-bit unsigned integers to binary64 floating-point numbers. There are eight integer-to-floating-point conversions and eight floating-point-to-integer conversions.

In this section, the rounding function RN denotes a binary64 floating-point operation. Note that most floating-point operations in the algorithms presented below are exact, so the actual rounding mode does not matter. The last operation of a conversion, however, might not be exact, especially when rounding a wide integer to a narrow floating-point format; in this case, the rounding mode of that last operation matters and decides how the whole conversion is rounded.

4.10.1 From 32-bit integers to floating-point numbers

The `f64_u32` and `f64_s32` conversions can be performed as follows, using bit-level manipulations over the binary64 format, combined with a regular floating-point subtraction:

$$\begin{aligned}\texttt{f64\_u32}(n) &= \mathrm{RN}(\texttt{f64make}(\texttt{0x43300000}, n) - 2^{52}) && (4.38)\\ \texttt{f64\_s32}(n) &= \mathrm{RN}(\texttt{f64make}(\texttt{0x43300000}, n + 2^{31}) - (2^{52} + 2^{31})) && (4.39)\end{aligned}$$

We write f64make(h, ℓ), where h and ℓ are 32-bit integers, for the binary64 floating-point number whose in-memory representation is the 64-bit word obtained by concatenating h with ℓ (caution: The order of h and ℓ depends on the endianness). This f64make operation can easily be implemented by storing h and ℓ in two consecutive 32-bit memory words, then loading a binary64 floating-point number from the address of the first word. The reasons why these implementations produce correct results are that f64make(0x43300000, m) is equal to $2^{52} + m$ for $m \in [0, 2^{32})$, and that the floating-point subtraction is then exact.

If f64_s32 is provided in hardware (and thus faster than the software implementations above), its unsigned counterpart f64_u32 can be implemented from it by a case analysis that reduces the integer argument to the range $[0, 2^{31})$:

$$\begin{aligned} \texttt{f64\_u32}(n) = {} & \texttt{if } n < 2^{31} \\ & \texttt{then f64\_s32}(n) \\ & \texttt{else } \mathrm{RN}(\texttt{f64\_s32}(n - 2^{31}) + 2^{31}), \end{aligned} \tag{4.40}$$

Both the f64_s32 conversion and the floating-point addition in the else branch are exact, the latter because it is performed using binary64 arithmetic.

Conversions from 32-bit integers to binary32 are trivially implemented by first converting to binary64, then rounding to binary32:

$$\begin{aligned} \texttt{f32\_s32}(n) &= \texttt{f32\_f64(f64\_s32}(n)), \\ \texttt{f32\_u32}(n) &= \texttt{f32\_f64(f64\_u32}(n)). \end{aligned}$$

The inner conversion is exact and the outer f32_f64 conversion rounds the result according to the active rounding mode, as prescribed by the IEEE 754 standards and the ISO C standards, appendix F.

4.10.2 From 64-bit integers to floating-point numbers

Given conversions from 32-bit integers, one can convert a 64-bit integer by splitting it in two 32-bit halves, converting them, and combining the results. Writing $n = 2^{32} \cdot h + \ell$, where h and ℓ are 32-bit integers and ℓ is unsigned, we have

$$\texttt{f64\_s64}(n) = \mathrm{RN}(\mathrm{RN}(\texttt{f64\_s32}(h) \cdot 2^{32}) + \texttt{f64\_u32}(\ell)), \tag{4.41}$$

$$\texttt{f64\_u64}(n) = \mathrm{RN}(\mathrm{RN}(\texttt{f64\_u32}(h) \cdot 2^{32}) + \texttt{f64\_u32}(\ell)). \tag{4.42}$$

All operations are exact except the final floating-point addition, which performs the correct rounding. For the same reason, a fused multiply-add instruction can be used if available, without changing the result.

If f64_s32 and f64_u32 are not provided in hardware, one can combine Equations (4.38), (4.39), (4.41), and (4.42), which gives the following implementations after some simplifications. (See [51] for proof details.)

$$\begin{aligned}
\texttt{f64_u64}(n) &= \mathrm{RN}(\mathrm{RN}(\texttt{f64make}(\texttt{0x45300000}, h) - (2^{84} + 2^{52})) \\
&\qquad + \texttt{f64make}(\texttt{0x43300000}, \ell)), \\
\texttt{f64_s64}(n) &= \mathrm{RN}(\mathrm{RN}(\texttt{f64make}(\texttt{0x45300000}, h + 2^{31}) - (2^{84} + 2^{63} + 2^{52})) \\
&\qquad + \texttt{f64make}(\texttt{0x43300000}, \ell)).
\end{aligned}$$

If f64_s64 is provided in hardware (and thus faster than the software implementation above), its unsigned counterpart f64_u64 can be implemented as follows:

$$\begin{aligned}
\texttt{f64_u64}(n) = {} & \texttt{if } n < 2^{63} \\
& \texttt{then f64_s64}(n) \\
& \texttt{else } \mathrm{RN}(2 \cdot \texttt{f64_s64}((n \texttt{ >> } 1) \mid (n \,\&\, 1))).
\end{aligned} \tag{4.43}$$

If f64_s64 and f64_u64 are available, conversions from 64-bit integers to binary32 can be implemented as follows:

$$\begin{aligned}
\texttt{f32_u64}(n) &= \texttt{f32_f64}(\texttt{f64_u64}(\texttt{if } n < 2^{53} \texttt{ then } n \texttt{ else } n')), \\
\texttt{f32_s64}(n) &= \texttt{f32_f64}(\texttt{f64_s64}(\texttt{if } |n| < 2^{53} \texttt{ then } n \texttt{ else } n')), \\
\text{with} \quad n' &= (n \mid ((n \,\&\, \texttt{0x7FF}) + \texttt{0x7FF})) \,\&\, \texttt{\~{}0x7FF}.
\end{aligned}$$

Both here and in Equation (4.43), the bit-fiddling on n emulates a rounding to odd (see Section 5.3.4), thus ensuring that the final result is correctly rounded [52, 51].

4.10.3 From floating-point numbers to integers

Conversions from floating-point numbers to integers are more straightforward. The general specification, as given in the ISO C standards, is that they must round the given floating-point number f toward zero to obtain an integer. If the resulting integer falls outside the range of representable values for the target integer type (e.g., $[-2^{31}, 2^{31})$ for target type s32), or if the floating-point argument is infinity or NaN, the conversion has undefined behavior: It can produce an arbitrary integer result, but it can also abort the program.[23] The algorithms below assume that their floating-point input is finite and in the range of the target type.

If s32_f64 is available, its unsigned counterpart u32_f64 can be obtained by case analysis:

$$\begin{aligned}
\texttt{u32_f64}(f) = {} & \texttt{if } f < 2^{31} \\
& \texttt{then s32_f64}(f) \\
& \texttt{else s32_f64}(\mathrm{RN}(f - 2^{31})) + 2^{31}.
\end{aligned}$$

[23]The IEEE 754-2008 standard allows the conversion of out-of-range numbers, infinity or NaN. In that case, either there should be a dedicated signaling mechanism or the invalid operation exception should be signaled.

Since the conversion `u32_f64`(f) is defined only if $f \in [0, 2^{32})$, the floating-point subtraction $\mathrm{RN}(f - 2^{31})$ in the `else` branch is exact, and in either branch `s32_f64` is applied to an argument in the range $[0, 2^{31})$, where it is defined.

Assuming `s64_f64` is available, the same construction applies in the case of 64-bit unsigned integers:

$$\begin{aligned}\texttt{u64\_f64}(f) = \;&\texttt{if } f < 2^{63}\\ &\texttt{then s64\_f64}(f)\\ &\texttt{else s64\_f64}(\mathrm{RN}(f - 2^{63})) + 2^{63}.\end{aligned}$$

When there is no hardware support for converting floating-point numbers to integers, the code produced by the CompCert C compiler simply extracts the significand of the number and shifts it accordingly to the exponent of the number [51].

4.11 Multiplication by an Arbitrary-Precision Constant with an FMA

Many numerical algorithms require multiplications by constants that are not exactly representable in floating-point arithmetic. Typical examples of such constants are π, $1/\pi$, $\ln(2)$, e, as well as values of the form $\cos(k\pi/N)$ and $\sin(k\pi/N)$, which appear in fast Fourier transforms. A natural question that springs to mind is: *Can we, at low cost, perform these multiplications with correct rounding?*

Assume that C is an arbitrary-precision constant. We want to design an algorithm that always returns $\mathrm{RN}(Cx)$, for any input floating-point number x of a given format. We want the algorithm to be very simple (two consecutive operations only, without any test). We assume that the "target" format is a binary floating-point format of precision p. Two possible cases are of interest. In the first case, the intermediate calculations are performed in the target format. In the second case, the intermediate calculations are performed in a somewhat larger format. A typical example is when the target precision is the binary64 format of IEEE 754-2008, and the internal precision is that of the Intel double-extended precision format ($p = 64$). In this book, we will assume that the intermediate calculations are performed in the "target" format (see [74] for the general case).

The algorithm presented here was introduced by Brisebarre and Muller [74]. It requires that the two following floating-point numbers be precomputed:

$$\begin{cases} C_h &= \mathrm{RN}(C), \\ C_\ell &= \mathrm{RN}(C - C_h). \end{cases} \tag{4.44}$$

Algorithm 4.13 Multiplication by C with a multiplication and an FMA [74].

input x
$u_1 \leftarrow \text{RN}(C_\ell x)$
$u_2 \leftarrow \text{RN}(C_h x + u_1)$
return u_2

Beware: we do not claim that for *all* values of C, this algorithm will return $\text{RN}(Cx)$ for all x. Indeed, it is quite simple to construct counterexamples. What we claim is that,

- we have reasonably simple methods that make it possible, for a given value of C, to check if Algorithm 4.13 will return $\text{RN}(Cx)$ for all floating-point numbers x (unless underflow occurs), as we will explain now;
- in practice, for *most* usual values of C, Algorithm 4.13 returns $\text{RN}(Cx)$ for all floating-point numbers x.

Without the use of an FMA instruction, Algorithm 4.13 would fail to always return a correctly rounded result for all but a few simple values of C (e.g., powers of 2).

In this section, we will present a simple method that allows one to check, for a given constant C and a given format, whether Algorithm 4.13 returns $\text{RN}(Cx)$ for all x. That method is sometimes unable to give an answer. See [74] for a more sophisticated method that either certifies that Algorithm 4.13 always returns a correctly rounded result, or gives all the counterexamples. The method we are going to present is based on the continued fraction theory. Continued fractions are very useful in floating-point arithmetic (for instance, to get worst cases for range reduction of trigonometric functions, see Chapter 10 and [442]). We present some basic results on this theory in Appendix A.1.

Without loss of generality, we assume in the following that $1 < x < 2$, that $1 < C < 2$, that C is not exactly representable, and that $C - C_h$ is not a power of 2. Define $x_{\text{cut}} = 2/C$. We have to separately consider the cases $x < x_{\text{cut}}$ and $x > x_{\text{cut}}$, because the value of $\text{ulp}(Cx)$ is not the same for these two cases.

The middle of two consecutive floating-point numbers around Cx has the form

$$\frac{2A+1}{2^p} \quad \text{if } x < x_{\text{cut}},$$

and

$$\frac{2A+1}{2^{p-1}} \quad \text{if } x > x_{\text{cut}},$$

where A is an integer between 2^{p-1} and 2^p. Our problem comes down to checking if there can be such a midpoint between Cx and the value u_2 returned by Algorithm 4.13. If this is not the case, then, necessarily,

$$u_2 = \mathrm{RN}(Cx).$$

Hence, first, we must bound the distance between u_2 and Cx. One has the following property; see [74, Property 2]:

Property 4.22. *Define* $\epsilon_1 = |C - (C_h + C_\ell)|$.

- *If* $x < x_{cut} - 2^{-p+2}$, *then* $|u_2 - Cx| < \frac{1}{2}\,\mathrm{ulp}(u_2) + \eta$,
- *If* $x \geq x_{cut} + 2^{-p+2}$, *then* $|u_2 - Cx| < \frac{1}{2}\,\mathrm{ulp}(u_2) + \eta'$,

where

$$\begin{cases} \eta &= \dfrac{1}{2}\,\mathrm{ulp}(C_\ell x_{cut}) + \epsilon_1 x_{cut}, \\ \eta' &= \mathrm{ulp}(C_\ell) + 2\epsilon_1. \end{cases}$$

The algorithm must be checked separately with the four values of x satisfying $x - x_{\mathrm{cut}} \in [-2^{-p+2}; 2^{-p+2})$. Property 4.22 tells us that u_2 is within $\frac{1}{2}\,\mathrm{ulp}(u_2) + \eta$ or $\frac{1}{2}\,\mathrm{ulp}(u_2) + \eta'$ from Cx, where η and η' are very small. What does this mean?

- u_2 is within $\frac{1}{2}\,\mathrm{ulp}(u_2)$ from Cx. In such a case,[24] $u_2 = \mathrm{RN}(Cx)$, which is the desired result;
- or (see Figure 4.7), Cx is very close (within distance η or η') from a "midpoint," i.e., the exact middle of two consecutive floating-point numbers. Depending on whether $x < x_{\mathrm{cut}}$ or not, this means that Cx is very close to a number of the form $(2A+1)/2^p$ or $(2A+1)/2^{p-1}$, which, by defining

  $$X = 2^{p-1}x,$$

  means that $2C$ or C is very close to a rational number of the form

  $$\frac{2A+1}{X}.$$

Hence, our problem is reduced to examining the best possible rational approximations to C or $2C$, with denominator bounded by $2^p - 1$. This is a typical continued fraction problem. Using Theorem A.3 in the Appendix, one can prove the following result [74].

Theorem 4.23 (Conditions on C and p). *Assume* $1 < C < 2$. *Let* $x_{cut} = 2/C$ *and* $X_{cut} = \lfloor 2^{p-1} x_{cut} \rfloor$.

[24] Unless u_2 is a power of 2, but this case is easily handled separately.

- *If*
$$X = 2^{p-1}x \leq X_{cut}$$
and
$$\epsilon_1 x_{cut} + \frac{1}{2}\,\mathrm{ulp}(C_\ell x_{cut}) \leq \frac{1}{2^{p+1}X_{cut}}$$
(where ϵ_1 is defined as in Property 4.22), then Algorithm 4.13 will always return a correctly rounded result, except possibly if X is a multiple of the denominator of a convergent n/d of $2C$ for which
$$|2Cd - n| < \frac{2^p}{\lceil 2^{p-1}/d \rceil}\left(\epsilon_1 x_{cut} + \frac{1}{2}\,\mathrm{ulp}(C_\ell x_{cut})\right)$$
(in such a case, Cx can be within η of a midpoint);

- *If*
$$X = 2^{p-1}x > X_{cut}$$
and
$$2^{2p+1}\epsilon_1 + 2^{2p-1}\,\mathrm{ulp}(2C_\ell) \leq 1$$
then Algorithm 4.13 will always return a correctly rounded result, except possibly if X is a multiple of the denominator of a convergent n/d of C for which
$$|Cd - n| < \epsilon_1 d + \frac{2^{n-1}}{\lceil X_{cut}/d \rceil}\,\mathrm{ulp}(C_\ell)$$
(in such a case, Cx can be within η' of a midpoint).

Hence, to check whether Algorithm 4.13 will always return a correctly rounded result, it suffices to compute the first convergents of C and $2C$ (those of denominator less than 2^p).

Table 4.3 gives some results obtained using this method ("Method 2" in the table) and two other methods, presented in [74]. Method 3 is the most complex, but it always gives an answer (either it certifies, for a given C, that the algorithm will always return $\mathrm{RN}(Cx)$, or it returns all the counterexamples). From Table 4.3, one can for instance deduce the following result.

Theorem 4.24 (Correctly rounded multiplication by π [74]). *Algorithm 4.13 always returns a correctly rounded result in the binary64 format with $C = 2^j\pi$, where j is any integer, provided no under/overflow occurs.*

Hence, in this case, multiplying by π with correct rounding only requires two consecutive FMAs.

If a wider internal format is available, then it is possible to slightly modify Algorithm 4.13 to get an algorithm that works for more values of C. See [74] for more details.

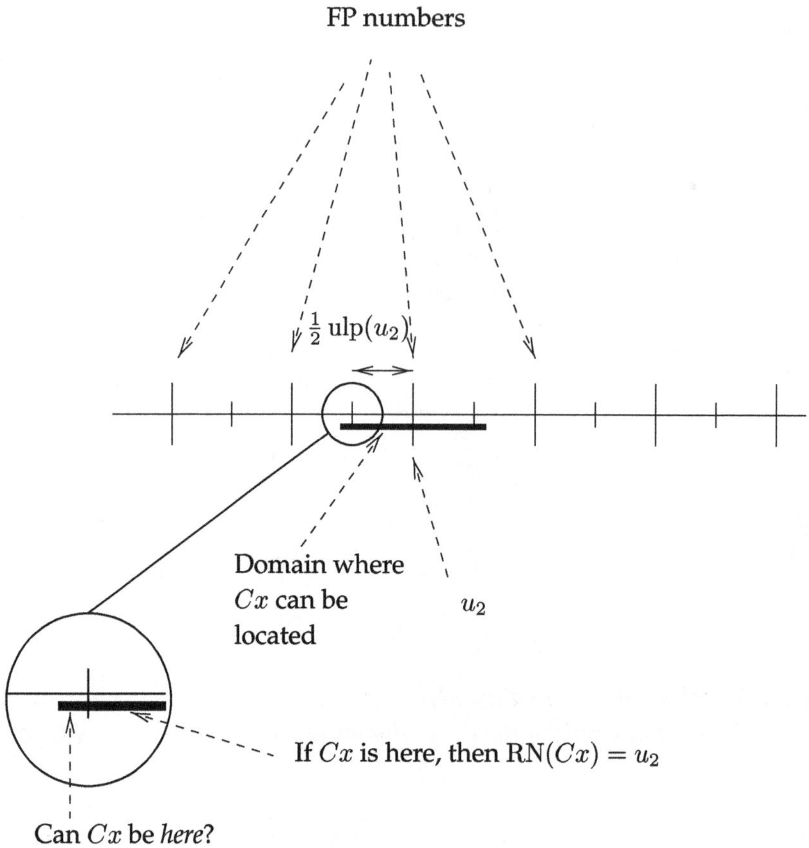

Figure 4.7: We know that Cx is within $\frac{1}{2}\,\mathrm{ulp}(u_2) + \eta$ (or η') from the floating-point (FP) number u_2, where η is less than 2^{-2p+1}. If we can show that Cx cannot be at a distance less than or equal to η (or η') from the midpoint of two consecutive FP numbers, then u_2 will be the FP number that is closest to Cx [74]. © IEEE, with permission.

4.12 Evaluation of the Error of an FMA

We have previously seen that, under some conditions, the error of a floating-point addition or multiplication is exactly representable using one floating-point number, and is readily computable using simple algorithms. When dealing with the FMA instruction, two natural questions arise:

- How many floating-point numbers does it take to exactly represent the error of an FMA operation?
- Can these numbers be easily calculated?

One easily sees that the error of an FMA is not always a floating-point number. Consider for instance, in radix-2, precision-p, rounded-to-nearest arith-

C	p	Method 1	Method 2	Method 3
π	8	Does not work for 226	Does not work for 226	OK unless $X =$ 226
π	24	?	?	OK
π	53	OK	?	OK
π	64	?	OK	OK
π	113	OK	OK	OK
$1/\pi$	24	?	?	OK
$1/\pi$	53	Does not work for 6081371451248382	?	OK unless $X =$ 6081371451248382
$1/\pi$	64	OK	OK	OK
$1/\pi$	113	?	?	OK
$\ln 2$	24	OK	OK	OK
$\ln 2$	53	OK	?	OK
$\ln 2$	64	OK	?	OK
$\ln 2$	113	OK	OK	OK
$\frac{1}{\ln 2}$	24	?	OK	OK
$\frac{1}{\ln 2}$	53	OK	OK	OK
$\frac{1}{\ln 2}$	64	?	?	OK
$\frac{1}{\ln 2}$	113	?	?	OK

Table 4.3: Some results obtained using the method presented here (Method 2), as well as Methods 1 and 3 of [74]. The results given for constant C hold for all values $2^{\pm j}C$. "OK" means "always works" and "?" means "the method is unable to conclude." "OK unless $X = n$" means that the algorithm returns a correctly rounded product for all values of X but n, and "Does not work for n means that we know at least one value for which the algorithm does not return a correctly rounded product [74], © IEEE, 2008, with permission.

metic, the numbers $a = x = 2^{p-1} + 1$, and $y = 2^{-p}$, and assume that we use an FMA instruction in order to compute $ax + y$. The returned result will be $\text{RN}(ax + y) = 2^{2p-2} + 2^p$, whereas the exact result is $2^{2p-2} + 2^p + 1 + 2^{-p}$. The error of the FMA operation is $1 + 2^{-p}$, which is not a floating-point number (it is a $(p+1)$-bit number).

Boldo and Muller [55] studied that problem, in the case of radix-2 arithmetic and assuming rounding to nearest. They showed that two floating-point numbers always suffice to exactly represent the error of an FMA operation, and they devised an algorithm for computing these two numbers. This is Algorithm 4.14, given below. It uses Algorithm 4.8 (2MultFMA), presented at the beginning of this chapter, as well as Algorithms 4.3 (Fast2Sum) and 4.4 (2Sum). This result generalizes to any arithmetic with an even radix.

The total number of floating-point operations it requires is 20. If one only wants the floating-point number nearest to the error of an FMA operation, one can use a simpler algorithm presented in [55].

Algorithm 4.14 ErrFma(a,x,y).

Require: a, x, y floating-point numbers
Ensure: $ax + y = r_1 + r_2 + r_3$
  $r_1 \leftarrow \mathrm{RN}(ax + y)$
  $(u_1, u_2) \leftarrow \mathrm{2MultFMA}(a, x)$
  $(\alpha_1, \alpha_2) \leftarrow \mathrm{2Sum}(y, u_2)$
  $(\beta_1, \beta_2) \leftarrow \mathrm{2Sum}(u_1, \alpha_1)$
  $\gamma \leftarrow \mathrm{RN}(\mathrm{RN}(\beta_1 - r_1) + \beta_2)$
  $(r_2, r_3) \leftarrow \mathrm{Fast2Sum}(\gamma, \alpha_2)$
  return (r_1, r_2, r_3)

One can show that if neither underflow nor overflow occurs, then the floating-point numbers r_1, r_2, and r_3 returned by Algorithm 4.14 satisfy:

- $ax + y = r_1 + r_2 + r_3$ exactly;
- $|r_2 + r_3| \leq \frac{1}{2}\,\mathrm{ulp}(r_1)$;
- $|r_3| \leq \frac{1}{2}\,\mathrm{ulp}(r_2)$.

Boldo wrote a formal proof in Coq of this result using Flocq [54]. Now, formal proof is a very mature domain that can help to certify nontrivial arithmetic algorithms. The interested reader will find useful information in Chapter 13.

Chapter 5

Enhanced Floating-Point Sums, Dot Products, and Polynomial Values

In this chapter, we focus on the computation of sums and dot products, and on the evaluation of polynomials in IEEE 754 floating-point arithmetic.[1] Such calculations arise in many fields of numerical computing. Computing sums is required, e.g., in numerical integration and the computation of means and variances. Dot products appear everywhere in numerical linear algebra. Polynomials are used to approximate many functions (see Chapter 10).

Many algorithms have been introduced for each of these tasks (some of them will be presented later on in this chapter), usually together with some backward, forward, or running/dynamic error analysis. See for example [258, Chapters 3, 4, 5] and [342, 166, 167, 161, 531, 396, 521, 361].

Our goal here is not to add to these algorithms but rather to observe how a floating-point arithmetic compliant with the IEEE 754-2008 standard can be used to provide validated *running error bounds* on and/or *improved accuracy* of the results computed by various algorithms. The consequence is enhanced implementations that need neither extended precision nor interval arithmetic but only the current working precision. In all that follows, we assume that the arithmetic is correctly rounded, and, more specifically, that it adheres to the standard mentioned above.

Providing a validated *running error bound* means being able to compute on the fly, during the calculation of a sum, a dot product, or a polynomial value, a floating-point number that is a mathematically true error bound on the result of that calculation, that can be computed using only standard

[1] Section 8.8 will survey how these tasks may be accelerated using specific hardware.

J.-M. Muller et al., *Handbook of Floating-Point Arithmetic*,
https://doi.org/10.1007/978-3-319-76526-6_5

floating-point operations (just like the error term of, say, $a + b$, detailed in Section 4.3), and that takes into account the intermediate quantities actually computed at each stage of the algorithm (and not just its input values). Such bounds follow from some basic properties of IEEE 754-2008 floating-point arithmetic, which we shall review first.

Providing *improved accuracy* means that we are able to return a useful result even if the problem is ill conditioned (e.g., when computing the sum $a_1 + a_2 + \cdots + a_n$, if $|\sum_{i=1}^{n} a_i|$ is very small in front of $\sum_{i=1}^{n} |a_i|$). More precisely, we wish to obtain results that are approximately as accurate as if the intermediate calculations were performed in, say, double-word or triple-word arithmetic (see Chapter 14), without having to pay the cost (in terms of computation time, of code size, and clarity) of such an arithmetic. To do that, we will frequently use *compensated algorithms*: In Chapter 4, we have studied some tricks (2Sum, Fast2Sum, 2MultFMA, Dekker product) that allow one to retrieve the error of a floating-point addition or multiplication.[2] It is therefore tempting to use these tricks to somehow *compensate* for the rounding errors that occur in a calculation. The first compensated algorithm was due to Kahan (see Section 5.3.2).

Although we will only focus here on sums, dot products, and polynomials, compensated algorithms can be built for other numerical problems. Some numerical algorithms that are simple enough (we need some kind of "linearity") can be transformed into compensated algorithms automatically. This is the underlying idea behind Langlois's CENA method [367]. Rump and Böhm also suggested a way of automatically improving some numerical calculations [526].

In this chapter we ignore the possibility of overflow (and, sometimes, of underflow as well), and we assume that all input values are exactly representable by floating-point numbers (which means that we do not include a possible preliminary rounding in the error analyses).

5.1 Preliminaries

We collect here some notation and basic facts needed later that follow from the definition of floating-point arithmetic given in Chapter 2. Most of these basic facts have already been mentioned in previous chapters, but we review them here to make this chapter (almost) self-contained. This section also includes some recent results, which are of independent interest, about the possibility of refining the most classical error bounds.

We assume that the radix of the floating-point arithmetic is β, that the precision is p, and we denote by $e_{\min}$ and $e_{\max}$ the extremal exponents. Notation $\circ$ will indicate a "generic" rounding function (i.e., $\circ =$ $\mathrm{RN}_{\mathrm{even}}, \mathrm{RN}_{\mathrm{away}}$, RU, RD, or RZ).

[2] Under some conditions. See Chapter 4 for more details.

Recall that the smallest positive subnormal number is

$$\alpha = \beta^{e_{\min}-p+1},$$

that the smallest positive normal number is

$$\beta^{e_{\min}},$$

and that the largest finite floating-point number is

$$\Omega = (\beta - \beta^{1-p}) \cdot \beta^{e_{\max}}.$$

Also (see Definition 2.3 in Chapter 2, page 27), we remind the reader that the *unit roundoff* $\mathbf{u}$ is

$$\mathbf{u} = \begin{cases} \frac{1}{2}\beta^{1-p} & \text{if } \circ = \text{RN} \\ \beta^{1-p} & \text{otherwise.} \end{cases}$$

5.1.1 Floating-point arithmetic models

Several models have been proposed in the literature to analyze numerical algorithms that use floating-point arithmetic [266, 76, 36, 81, 258]. A widely used property is the following. Let x and y be two floating-point numbers and let $\text{op} \in \{+, -, \times, /\}$. If $\beta^{e_{\min}} \le |x \,\text{op}\, y| \le \Omega$ then no underflow/overflow[3] occurs when computing $x \,\text{op}\, y$, and there exist real numbers ϵ_1 and ϵ_2 such that

$$\circ(x \,\text{op}\, y) = (x \,\text{op}\, y)(1+\epsilon_1) \tag{5.1a}$$

$$= (x \,\text{op}\, y)/(1+\epsilon_2), \qquad |\epsilon_1|, |\epsilon_2| \le \mathbf{u}. \tag{5.1b}$$

See Section 2.3.1, page 25, for an explanation. Identity (5.1a) corresponds to what is called the *standard model* of floating-point arithmetic (see [258, p. 40]).

Sometimes, (5.1a) and (5.1b) are also referred to as the first and second standard models, respectively (see for example [530]). In particular, the second standard model (5.1b) proves to be extremely useful for running error analysis and exact error-bound derivation, as we will see below.

The identities in (5.1) assume that no underflow occurs. If we want to take into account the possibility of underflow, we must note that:

- if underflow occurs during addition/subtraction, then *the computed result is the exact result* (this is Theorem 4.2 in Chapter 4, page 102, see also [249], [258, Problem 2.19]). Thus, (5.1) still holds (indeed, with $\epsilon_1 = \epsilon_2 = 0$) for $\text{op} \in \{+, -\}$ in the case of underflows;

[3] See the *note on underflow*, Section 2.1.3.

- however, if op $\in \{\times, /\}$ and underflow may occur, then the preceding model must be modified as follows: there exist some real numbers ϵ_1, ϵ_2, η_1, η_2 such that

$$\begin{aligned} \circ(x \operatorname{op} y) &= (x \operatorname{op} y)(1+\epsilon_1) + \eta_1 && \text{(5.2a)} \\ &= (x \operatorname{op} y)/(1+\epsilon_2) + \eta_2, && \text{(5.2b)} \end{aligned}$$

with

$$|\epsilon_1|, |\epsilon_2| \leq \mathbf{u}, \quad |\eta_1|, |\eta_2| \leq \alpha, \quad \epsilon_1\eta_1 = \epsilon_2\eta_2 = 0. \tag{5.2c}$$

From now on we will restrict our discussion to (5.1) for simplicity, thus assuming that no underflow occurs. More general results that do take possible underflows into account using (5.2) can be found in the literature (see for instance [396, 531, 522]).

5.1.2 Notation for error analysis and classical error estimates

Error analysis using the two standard models in (5.1) often makes repeated use of factors of the form $1+\epsilon_1$ or $1/(1+\epsilon_2)$, with $|\epsilon_1|, |\epsilon_2| \leq \mathbf{u}$ (see for instance the example of iterated products, given in Section 2.3.4, page 35). A concise way of handling such terms is through the θ_n and γ_n notation defined by Higham in [258, p. 63]:

Definition 5.1 (θ_n and γ_n). *For ϵ_i such that $|\epsilon_i| \leq \mathbf{u}$, $1 \leq i \leq n$, and assuming $n\mathbf{u} < 1$,*

$$\prod_{i=1}^{n}(1+\epsilon_i)^{\pm 1} = 1 + \theta_n,$$

where

$$|\theta_n| \leq \frac{n\mathbf{u}}{1-n\mathbf{u}} =: \gamma_n.$$

Note that if $n \ll 1/\mathbf{u}$, then $\gamma_n \approx n\mathbf{u}$. Such quantities enjoy many properties, among which (see [258, p. 67]):

$$\gamma_n \leq \gamma_{n+1}. \tag{5.3}$$

Let us now see the kind of error bounds that can be obtained by combining the standard model with the above θ_n and γ_n notation, focusing for instance on the following three classical algorithms (Algorithms 5.1 through 5.3)[4]:

[4]These bounds are given in [258, p. 82, p. 63, p. 95].

Algorithm 5.1 RecursiveSum.

```
r ← a_1
for i = 2 to n do
    r ← ∘(r + a_i)
end for
return r
```

Algorithm 5.1 is the straightforward algorithm for evaluating the sum

$$a_1 + a_2 + \cdots + a_n.$$

More sophisticated algorithms will be given in Section 5.3. Similarly, Algorithm 5.2 is the straightforward algorithm for evaluating the dot product

$$a_1 \cdot b_1 + a_2 \cdot b_2 + \cdots + a_n \cdot b_n.$$

Algorithm 5.2 Algorithm RecursiveDotProduct(a,b).

```
r ← ∘(a_1 × b_1)
for i = 2 to n do
    r ← ∘(r + ∘(a_i × b_i))
end for
return r
```

Algorithm 5.3 evaluates the polynomial $p(x) = a_n x^n + a_{n-1}x^{n-1} + \cdots + a_0$ using Horner's rule. See Section 5.5 for a "compensated" Horner algorithm.

Algorithm 5.3 Algorithm Horner(p,x).

```
r ← a_n
for i = n − 1 downto 0 do
    r ← ∘(∘(r × x) + a_i)
end for
return r
```

Let us first consider recursive summation (Algorithm 5.1). In that algorithm, the first value of variable r (after the first iteration of the **for** loop) is

$$(a_1 + a_2)(1 + \epsilon_1),$$

with $|\epsilon_1| \leq \mathbf{u}$. That is, that value of r can be rewritten as

$$(a_1 + a_2)(1 + \theta_1),$$

where the notation θ_1 is introduced in Definition 5.1. The second value of variable r (after the second iteration of the **for** loop) is

$$\begin{aligned} &((a_1 + a_2)(1 + \epsilon_1) + a_3)(1 + \epsilon_2) \quad \text{with} \quad |\epsilon_2| \leq \mathbf{u} \\ &= (a_1 + a_2)(1 + \theta_2) + a_3(1 + \theta_1), \end{aligned}$$

where now $\theta_1 = \epsilon_2$ and $\theta_2 = (1+\epsilon_1)(1+\epsilon_2)-1$. By a straightforward induction, the last value of r has the form

$$(a_1 + a_2)(1 + \theta_{n-1}) + a_3(1 + \theta_{n-2}) + a_4(1 + \theta_{n-3}) + \cdots + a_n(1 + \theta_1). \quad (5.4)$$

Using (5.3), we obtain the absolute forward error bound

$$\left|\mathrm{RecursiveSum}(a) - \sum_{i=1}^{n} a_i\right| \leq \gamma_{n-1} \sum_{i=1}^{n} |a_i|. \quad (5.5)$$

Note that if underflow has occurred during summation, then this forward error bound remains true, since the model (5.1a) we used to derive it still applies in such a case.

Similarly, it follows for Algorithm 5.2 that

$$\left|\mathrm{RecursiveDotProduct}(a, b) - \sum_{i=1}^{n} a_i \cdot b_i\right| \leq \gamma_n \sum_{i=1}^{n} |a_i \cdot b_i|, \quad (5.6)$$

and, for Algorithm 5.3, that

$$|\mathrm{Horner}(p, x) - p(x)| \leq \gamma_{2n} \sum_{i=0}^{n} |a_i| \cdot |x|^i. \quad (5.7)$$

Note that if a fused multiply-add (FMA, see Section 2.4, page 37) instruction is available, Algorithms 5.2 and 5.3 can be rewritten so that they become simpler, faster (the number of operations is halved) and, in general, slightly more accurate (for polynomial evaluation, γ_{2n} is replaced by γ_n so that the error bound is roughly halved).[5]

An important notion in numerical analysis is that of *backward error*, introduced by Wilkinson [634] and studied extensively in the books of Higham [258] and Corless and Fillion [115]. Assume we wish to compute $y = f(x)$. Instead of y we compute some value $\hat{y}$ (that we hope is close to y). In most cases, $\hat{y}$ is the exact value of f at some locally unique point $\hat{x}$ (that we hope is close to x).

- The *backward error* of this computation is

$$|x - \hat{x}|,$$

- and the *relative backward error* is

$$\left|\frac{x - \hat{x}}{x}\right|.$$

[5]Which, of course, does not imply that the error itself is halved.

When there might be some ambiguity, the usual absolute and relative errors

$$|y - \hat{y}|$$

and

$$\left|\frac{y - \hat{y}}{y}\right|$$

are called the *forward error* and *relative forward error*, respectively.

In Equations (5.5), (5.6), and (5.7), the values γ_{n-1}, γ_n, and γ_{2n} are upper bounds on the backward relative error of the computation. These equations show that recursive summation and dot product have a small backward error if $n\mathbf{u} \ll 1$, as well as Horner's algorithm if $2n\mathbf{u} \ll 1$ (which almost always holds in practice: for instance, in double-precision/binary64 arithmetic, nobody evaluates a polynomial of degree around 2^{50}).

And yet, the *forward* relative errors of these algorithms can be arbitrarily large if the *condition numbers*

$$C_{\text{summation}} = \frac{\sum_{i=1}^{n} |a_i|}{\left|\sum_{i=1}^{n} a_i\right|} \qquad \text{(summation)}$$

$$C_{\text{dot product}} = \frac{2 \cdot \sum_{i=1}^{n} |a_i \cdot b_i|}{\left|\sum_{i=1}^{n} a_i \cdot b_i\right|} \qquad \text{(dot product)}$$

$$C_{\text{Horner}} = \frac{\sum_{i=0}^{n} |a_i| \cdot |x|^i}{\left|\sum_{i=0}^{n} a_i \cdot x^i\right|} \qquad \text{(polynomial evaluation)}$$

are too large.

5.1.3 Some refined error estimates

The classical error bounds presented in the previous section in Equations (5.5), (5.6), and (5.7) all involve a so-called *constant factor* of the form

$$\gamma_n = \frac{n\mathbf{u}}{1 - n\mathbf{u}}$$

for some suitable value of n implicitly assumed to be such that

$$n\mathbf{u} < 1.$$

For given values of the unit roundoff $\mathbf{u}$ and of the dimension n, such a term γ_n is indeed a constant and does not depend on the input floating-point values of the summation, dot product, and polynomial evaluation algorithms.

If we fix only n and allow $\mathbf{u}$ to tend to zero (for example by letting the precision p increase for a given value of the radix β), then we have the following asymptotic expansion, which reveals the presence of terms at least quadratic in $\mathbf{u}$:

$$\begin{aligned} \gamma_n &= n\mathbf{u} + n^2\mathbf{u}^2 + n^3\mathbf{u}^3 + \cdots \\ &= n\mathbf{u} + \mathcal{O}(\mathbf{u}^2), \qquad \mathbf{u} \to 0. \end{aligned}$$

While using this γ_n notation is in most cases sufficient to analyze the stability and accuracy of many numerical algorithms (as illustrated well by [258]), it is natural to ask whether such constants can be replaced by better ones, which would be less restrictive and simpler (and, of course, at least as sharp). In other words, can the requirement that $n < \mathbf{u}^{-1}$ and the quadratic term $\mathcal{O}(\mathbf{u}^2)$ be removed? In this section we review some recent progress made towards answering this question.

Note first that, for small-enough values of n and rounding to nearest, removing the quadratic term in $\mathbf{u}$ is sometimes straightforward. Indeed, recalling from Chapter 2 that the relative error due to rounding to nearest is in fact bounded by $\mathbf{u}/(1+\mathbf{u})$ and not just $\mathbf{u} = \frac{1}{2}\beta^{1-p}$, we can first refine the first standard model (5.1a) as follows:

$$\mathrm{RN}(x \mathbin{\mathrm{op}} y) = (x \mathbin{\mathrm{op}} y)(1+\epsilon_1), \qquad |\epsilon_1| \le \frac{\mathbf{u}}{1+\mathbf{u}}. \tag{5.8}$$

Then, by applying this refined model, we deduce for example for recursive summation that the terms $\theta_{n-1}, \ldots, \theta_1$ appearing in (5.4) can be bounded as follows:

$$\left(1 - \frac{\mathbf{u}}{1+\mathbf{u}}\right)^k \le 1 + \theta_k \le \left(1 + \frac{\mathbf{u}}{1+\mathbf{u}}\right)^k, \qquad k = 1, \ldots, n-1.$$

In the above enclosure, it is easily checked that, since $0 \le \mathbf{u} < 1$, the lower bound satisfies

$$\left(1 - \frac{\mathbf{u}}{1+\mathbf{u}}\right)^k \ge (1-\mathbf{u})^k \ge 1 - k\mathbf{u} \qquad \text{for all } k \ge 1.$$

Concerning the upper bound, it is also easy to see that it satisfies

$$\left(1 + \frac{\mathbf{u}}{1+\mathbf{u}}\right)^k \le 1 + k\mathbf{u} \qquad \text{for } k \in \{1, 2, 3\}$$

and, for $k \geq 4$, that it is strictly larger than $1 + k\mathbf{u}$ and has the form $1 + k\mathbf{u} + \mathcal{O}(\mathbf{u}^2)$ as $\mathbf{u} \to 0$. Consequently, the constant γ_{n-1} in Equation (5.5) can be replaced by $(n-1)\mathbf{u}$ for all $n \leq 4$.

Similarly, for the recursive dot product and Horner's scheme, using (5.8) makes it possible to directly replace the constants γ_n and γ_{2n} in (5.6) and (5.7) by, respectively, $n\mathbf{u}$ when $n \leq 3$, and $2n\mathbf{u}$ when $n \leq 2$.

For arbitrary values of n or directed roundings, things are more complicated, but there are situations where a constant of the form γ_n can be replaced by $n\mathbf{u}$, and this without any restriction on n. A first result in this direction is Rump's bound for recursive summation with rounding to nearest, presented in Theorem 5.1 below (although this bound is given in [522] for radix 2, its generalization to any radix $\beta \geq 2$ follows using exactly the same arguments).

Theorem 5.1 (Rump [522]). *If* $\circ = \mathrm{RN}$ *and no overflow occurs, then recursive summation (Algorithm 5.1) satisfies the following forward relative error bound:*

$$\left| \mathrm{RecursiveSum}(a) - \sum_{i=1}^{n} a_i \right| \leq (n-1)\mathbf{u} \sum_{i=1}^{n} |a_i| \qquad \textit{for all } n. \tag{5.9}$$

Proof. We proceed by induction on n, the result being clear for $n = 1$. Let $n \geq 2$ and assume that the result is true for any dimension up to $n-1$. Also, for $1 \leq k \leq n$, let r_k denote the floating-point number returned by RecursiveSum on input $a_1, \ldots, a_k$, let $s_k = a_1 + \cdots + a_k$ (the exact kth partial sum), and let $\widetilde{s}_k = |a_1| + \cdots + |a_k|$. Then, defining

$$\delta = \mathrm{RN}(r_{n-1} + a_n) - (r_{n-1} + a_n)$$

and using the inductive assumption, we deduce that

$$\begin{aligned} |r_n - s_n| &= |\delta + (r_{n-1} + a_n) - (s_{n-1} + a_n)| \\ &\leq |\delta| + |r_{n-1} - s_{n-1}| \\ &\leq |\delta| + (n-2)\mathbf{u}\widetilde{s}_{n-1}. \end{aligned} \tag{5.10}$$

Now note that by definition of rounding to nearest, $|\delta|$ minimizes the distance $|f - (r_{n-1} + a_n)|$ over all floating-point numbers f. Hence, r_{n-1} being a floating-point number, we have in particular

$$|\delta| \leq |a_n|. \tag{5.11}$$

By using this inequality together with

$$|\delta| \leq \mathbf{u}|r_{n-1} + a_n|, \tag{5.12}$$

we can conclude as follows, with two separate sub-cases.

First, if $|a_n| \le \mathbf{u}\widetilde{s}_{n-1}$, then we deduce from (5.10) and (5.11) that

$$\begin{aligned}|r_n - s_n| &\le \mathbf{u}\widetilde{s}_{n-1} + (n-2)\mathbf{u}\widetilde{s}_{n-1}\\ &= (n-1)\mathbf{u}\widetilde{s}_{n-1}\\ &\le (n-1)\mathbf{u}\widetilde{s}_n.\end{aligned}$$

Assume now that $\mathbf{u}\widetilde{s}_{n-1} < |a_n|$. Using (5.10) and (5.12), we obtain

$$|r_n - s_n| \le \mathbf{u}|r_{n-1} + a_n| + (n-2)\mathbf{u}\widetilde{s}_{n-1},$$

and, using again the inductive assumption and the definition of $\widetilde{s}_n$, we can bound $|r_{n-1} + a_n|$ as follows:

$$\begin{aligned}|r_{n-1} + a_n| &\le |r_{n-1} - s_{n-1}| + |s_{n-1} + a_n|\\ &\le (n-2)\mathbf{u}\widetilde{s}_{n-1} + \widetilde{s}_n\\ &\le (n-2)|a_n| + \widetilde{s}_n.\end{aligned}$$

Consequently,

$$\begin{aligned}|r_n - s_n| &\le \mathbf{u}(n-2)|a_n| + \mathbf{u}\widetilde{s}_n + (n-2)\mathbf{u}\widetilde{s}_{n-1}\\ &= (n-1)\mathbf{u}\widetilde{s}_n.\end{aligned}$$

Hence, for $n \ge 2$, the desired bound $(n-1)\mathbf{u}\widetilde{s}_n$ has been obtained in both cases, and the conclusion follows by induction. □

Since its publication, Theorem 5.1 has been extended in several ways, which we briefly present now. The following results hold provided that no overflow occurs and, except for those related to summation, that underflow does not occur either.

Assume first that $\circ = \mathrm{RN}$, that is, rounding is to nearest, with any tie breaking rule. In [302] it is shown that the refined bound $(n-1)\mathbf{u}\sum_{i=1}^{n}|a_i|$ in fact holds not only for recursive summation, but for any summation scheme using $n-1$ floating-point additions, regardless of the order used to perform these additions. There, it is also proved that the classical constant γ_n associated with the recursive dot product in dimension n (see (5.6)) can be simplified into $n\mathbf{u}$ without any restriction on n:

$$\left|\mathrm{RecursiveDotProduct}(a,b) - \sum_{i=1}^{n} a_i \cdot b_i\right| \le n\mathbf{u}\sum_{i=1}^{n}|a_i \cdot b_i| \qquad \text{for all } n.$$

This bound applies more generally to any dot product obtained using n floating-point multiplications and $n-1$ floating-point additions, independently of the order in which the additions are performed. Some direct consequences are refined error bounds for the classical matrix multiplication and the evaluation of the Euclidean norm of a vector [302, 303]. Using more sophisticated inductions than the one appearing in the proof of Theorem 5.1, similar refinements can also be obtained for other linear algebra tasks, such as computing various triangular factors (L, U, Cholesky) of a given matrix and solving a triangular linear system by substitution; see [528]. For the evaluation of $p(x) = \sum_{i=0}^{n} a_i x^i$ using Horner's scheme, the constant γ_{2n} in (5.7) can also be replaced by $2n\mathbf{u}$, at least in the case where n is not larger than about $1/\sqrt{\mathbf{u}}$. More precisely, it is shown in [527] that

$$|\text{Horner}(p,x) - p(x)| \leq 2n\mathbf{u}\sum_{i=0}^{n} |a_i| \cdot |x|^i \qquad \text{if } n < \frac{1}{2}\left(\sqrt{\frac{\omega}{\beta}}\mathbf{u}^{-1/2} - 1\right),$$

where $\omega = 2$ if β is even (which always holds in practice: β is 2 or 10), and $\omega = 1$ otherwise. A similar result holds in the case of the product $x_1 x_2 \cdots x_n$ of n floating-point numbers, where the classical relative error bound γ_{n-1} can be replaced by $(n-1)\mathbf{u}$ if $n < \sqrt{\omega/\beta}\mathbf{u}^{-1/2} + 1$. This upper bound on n is not a mere "proof artifact": as noted in [527], requiring that n be at most of the order of $\mathbf{u}^{-1/2}$ is necessary in radix $\beta = 2$ and precision $p \geq 4$ (it is possible to construct counterexamples if the condition is not satisfied). For this case and the special one where $p(x) = x^n$ in radix two, see also [223].[6] The latter example thus indicates that, already for rounding to nearest, it is sometimes impossible to simultaneously replace γ_n by $n\mathbf{u}$ and remove the restriction on n.

Recently, Theorem 5.1 has also been extended to directed roundings, thus assuming $\mathbf{u} = \beta^{1-p}$ instead of $\mathbf{u} = \frac{1}{2}\beta^{1-p}$. In particular, for radix 2 and $\circ$ = RU (rounding toward $+\infty$), it is shown in [479] that if $n\mathbf{u} < 1/4$, then the bound in (5.9) holds for recursive summation and any other summation ordering. In [361], this result is generalized further to radix $\beta \geq 2$, faithful rounding (so that every floating-point addition is rounded in an arbitrary direction), and with the necessary condition that n satisfies $n \leq 1 + (\beta - 1)\mathbf{u}^{-1}$.

Finally, note that although the error bounds obtained by replacing γ_n by nu are clearly simpler and slightly sharper, they still need not be attained. For example, it was shown in [410] that if $(n-1)\mathbf{u} \leq 1/20$, then the term $(n-1)\mathbf{u}$ appearing in (5.9) can be decreased further to $(n-1)\mathbf{u}/(1+(n-1)\mathbf{u})$ and that the resulting bound is attained for $(a_1, a_2, \ldots, a_n) = (1, \mathbf{u}, \ldots, \mathbf{u})$ and $\circ$ = RN with the *tie-to-even* rule. More generally, it was proved in [363] that this bound holds for any summation ordering and under the weaker (yet necessary) restriction $(n-1)\mathbf{u} \leq (\beta - 1)/2$. We also refer to [363] for several applications (especially to sums of products, multiplication of a Vandermonde matrix by a vector, blocked summation, ...), where the resulting error bounds are similar to the one shown above for Horner's rule (that is, simpler and sharper for n not too large).

[6] We also refer to this paper for some interesting examples showing that in practice such bounds are reasonably tight.

5.1.4 Properties for deriving validated running error bounds

Together with the definition of $\mathbf{u}$, Equation (5.1) yields a number of properties that will prove useful for deriving validated error bounds.

Property 5.2. *Let x, y, z be nonnegative floating-point numbers. If underflow does not occur, then $xy + z \le \circ(\circ(x \times y) + z)(1 + \mathbf{u})^2$.*

Proof. Applying (5.1b) to xy gives $xy + z = \circ(x \times y)(1 + \epsilon) + z$, $|\epsilon| \le \mathbf{u}$. Since $xy + z$, $\circ(x \times y)$, and z are all nonnegative, we deduce that

$$\begin{aligned} xy + z &\le \circ(x \times y)|1 + \epsilon| + z \\ &\le \circ(x \times y)(1 + \mathbf{u}) + z \\ &\le (\circ(x \times y) + z)(1 + \mathbf{u}). \end{aligned}$$

Applying (5.1b) to the sum $\circ(x \times y) + z$ gives further

$$\begin{aligned} \circ(x \times y) + z &= \circ(\circ(x \times y) + z)(1 + \epsilon'), \quad |\epsilon'| \le \mathbf{u}, \\ &= \circ(\circ(x \times y) + z)|1 + \epsilon'| \\ &\le \circ(\circ(x \times y) + z)(1 + \mathbf{u}), \end{aligned}$$

which concludes the proof. □

Property 5.3. *If the radix β is even, $p \le -e_{\min}$ (these two conditions always hold in practice), and if k is a positive integer such that $k\mathbf{u} < 1$, then $1 - k\mathbf{u}$ is a normal floating-point number.*

Proof. Recall that u is equal to $\frac{1}{2}\beta^{1-p}$ in round-to-nearest, and to β^{1-p} in the other rounding modes. Since $p > 0$ both k and $\mathbf{u}^{-1}$ are positive integers. Thus, $k\mathbf{u} < 1$ implies $\mathbf{u} \le 1 - k\mathbf{u} < 1$. Since $p \le -e_{\min}$ and $\beta \ge 2$ is even, both possible values for $\mathbf{u}$ are at least $\beta^{e_{\min}}$. Consequently, $\beta^{e_{\min}} \le 1 - k\mathbf{u} < 1$.

Hence, there exist a real μ and an integer e such that

$$1 - k\mathbf{u} = \mu \cdot \beta^{e-p+1},$$

with $\beta^{p-1} \le \mu < \beta^p$ and $e_{\min} \le e < 0$. Writing $\mu = (\mathbf{u}^{-1} - k)\mathbf{u}\,\beta^{-e+p-1}$, it follows that μ is a positive integer as the product of the positive integers $\mathbf{u}^{-1} - k$ and $\mathbf{u}\,\beta^{-e+p-1} = \frac{1}{2}\beta^{-e}$ (with β even). □

Property 5.4 is a direct consequence of [472, Lemma 2.3] and Property 5.3.

Property 5.4 (Rump et al. [472]). *Let k be a positive integer such that $k\mathbf{u} < 1$ and let x be a floating-point number such that $\beta^{e_{\min}} \le |x| \le \Omega$. Then*

$$(1 + \mathbf{u})^{k-1}|x| \le \circ\left(\frac{|x|}{1 - k\mathbf{u}}\right).$$

5.2 Computing Validated Running Error Bounds

Equations (5.5), (5.6), and (5.7) give error bounds for three basic algorithms. These error bounds involve an exact quantity (such as in $\gamma_{2n} \sum_{i=0}^{n} |a_i| \cdot |x|^i$ for polynomial evaluation) that make them difficult to check using floating-point arithmetic only, and they also do not take into account the intermediate floating-point quantities computed by the algorithm. (Of course, the same holds for the refined error bounds seen in Section 5.1.3.) Hence, it is interesting to design algorithms that compute a validated error bound on the fly.

Ogita, Rump, and Oishi give a compensated algorithm for the dot product with running error bound in [471]. Below, Algorithm 5.4 is a modification of [258, Algorithm 5.1] which evaluates a polynomial using Horner's rule and provides a validated running error bound.

Algorithm 5.4 This algorithm computes a pair (r, b) of floating-point numbers such that $r = \text{Horner}(p, x)$ and $|r - p(x)| \leq b$, provided no underflow or overflow occurs.

$r \leftarrow a_n$
$s \leftarrow \circ(|a_n|/2)$
for $i = n-1$ downto 1 **do**
  $r \leftarrow \circ\left(\circ(r \times x) + a_i\right)$
  $s \leftarrow \circ\left(\circ(s \times |x|) + |r|\right)$
end for
$r \leftarrow \circ\left(\circ(r \times x) + a_0\right)$
$b \leftarrow \circ\left(2 \times \circ(s \times |x|) + |r|\right)$
$b \leftarrow \mathbf{u} \times \circ\left(b/(1-(2n-1)\mathbf{u})\right)$
return (r, b)

If an FMA instruction is available, then the core of Horner's loop in Algorithm 5.4 obviously becomes $r \leftarrow \circ(r \times x + a_i)$. This results in a faster and slightly better algorithm.

Following the analysis of Horner's rule given in [258, page 95], we arrive at the result below.

Theorem 5.5. *If no underflow or overflow occurs, then Algorithm 5.4 computes in* $4n+1$ *flops a pair of normal floating-point numbers* (r, b) *such that*

$$\left| r - \sum_{i=0}^{n} a_i x^i \right| \leq b.$$

Proof. Recalling that Horner's rule in degree n takes exactly $2n$ flops and ignoring operations such as absolute value and multiplication/division by 2, $\mathbf{u}$, or n, we deduce the flop count of $4n+1$ for Algorithm 5.4.

Now, for the error bound, following [258, page 95], let r_i be the value of r after the loop of index i, so that

$$r_i = \circ\big(\circ(r_{i+1} \cdot x) + a_i\big)$$

for $0 \le i < n$, and $r_n = a_n$. Using (5.1a) and (5.1b), we obtain

$$(1+\epsilon_i) r_i = r_{i+1} x (1+\delta_i) + a_i, \quad |\epsilon_i|, |\delta_i| \le \mathbf{u}.$$

For $0 \le i \le n$, define $q_i = \sum_{h=i}^{n} a_h x^{h-i}$ and $e_i = r_i - q_i$. We have $e_n = 0$ and, for $1 \le i < n$, using $q_i = q_{i+1}x + a_i$ allows one to deduce from the above equation that

$$e_i = x e_{i+1} + \delta_i x r_{i+1} - \epsilon_i r_i.$$

Taking absolute values, we get $|e_i| \le |x||e_{i+1}| + \mathbf{u}\big(|x||r_{i+1}| + |r_i|\big)$. Using $e_n = 0$ further leads to

$$\left| r - \sum_{i=0}^{n} a_i x^i \right| = |e_0| \le \mathbf{u} E_0,$$

with E_0 given by the following recurrence:

$$E_n = 0 \quad \text{and, for } n > i \ge 0, \quad E_i = \big(E_{i+1} + |r_{i+1}|\big)|x| + |r_i|.$$

Therefore, $E_0 = |r_n x^n| + 2\sum_{i=1}^{n-1} |r_i x^i| + |r_0|$. Since $r_n = a_n$ and $r_0 = r$, this can be rewritten as

$$E_0 = 2S(|x|) \cdot |x| + |r|,$$

where

$$S(x) = \frac{|a_n|}{2} x^{n-1} + \sum_{i=0}^{n-2} |r_{i+1}| x^i.$$

Since S is a polynomial of degree $n-1$ with nonnegative coefficients only, we deduce from Property 5.2 that $S(|x|) \le s \cdot (1+\mathbf{u})^{2n-2}$, where s is the floating-point number obtained at the end of Algorithm 5.4. The conclusion follows using Property 5.4 with $k = 2n-1$. □

Note that the validated running error bound b computed by Algorithm 5.4 is obtained at essentially the same cost as the running error estimate of [258, Algorithm 5.1]. The paper by Ogita, Rump, and Oishi [471] and Louvet's Ph.D. dissertation [396] are good references for other examples of validated running error bounds.

More recently, Rump introduced in [522] various tight and computable error bounds, expressed in terms of the unit in the first place (ufp) of the computed result. For applications to linear algebra and computational geometry, we refer further to [525] and [478], respectively.

5.3 Computing Sums More Accurately

As stated in the beginning of this chapter, many numerical problems require the computation of sums of many floating-point numbers. In [257] and later on in [258], Higham gives a survey on summation methods. Interesting information can also be found in [531]. Here, we will just briefly present the main results: the reader should consult these references for more details.

We will first deal with methods that generalize the straightforward RecursiveSum algorithm (Algorithm 5.1). After that, we will present some methods that use the Fast2Sum and 2Sum algorithms presented in Chapter 4, pages 104 and 108 (we remind the reader that these algorithms compute the error of a rounded-to-nearest floating-point addition. It is therefore tempting to use them to somehow compensate for the rounding errors).

In this section, we want to evaluate, as accurately as possible, the sum of n floating-point numbers, $x_1, x_2, \ldots, x_n$.

5.3.1 Reordering the operands, and a bit more

When considering the RecursiveSum algorithm (Algorithm 5.1), conventional methods for improving accuracy consist in preliminarily sorting the input values, so that

$$|x_1| \le |x_2| \le \cdots \le |x_n|$$

(increasing order), or even sometimes

$$|x_1| \ge |x_2| \ge \cdots \ge |x_n|$$

(decreasing order). Another common strategy (yet expensive in terms of comparisons), called *insertion* summation, consists in first sorting the x_i's by increasing order of magnitude, then computing $\circ(x_1 + x_2)$, and inserting that result in the list $x_3, x_4, \ldots, x_n$, so that the increasing order is kept, and so on. We stop when there remains only one element in the list: that element is the approximation to $\sum_{1 \le i \le n} x_i$.

To analyze a large class of similar algorithms, Higham defines in [258, page 81] a general algorithm expressed as Algorithm 5.5.

Algorithm 5.5 General form for a large class of addition algorithms (Higham, Algorithm 4.1 of [258]).

```
while S contains more than one element do
  remove two numbers x and y from S and add ∘(x + y) to S
end while
return the remaining element of S
```

Note that since the number of elements of S decreases by one at each iteration, this algorithm always performs $n - 1$ floating-point additions (the **while** loop can be replaced by a **for** loop).

If T_i is the result of the i-th addition of Algorithm 5.5, Higham shows that the final returned result, say $s = T_{n-1}$, satisfies

$$\left| s - \sum_{i=1}^{n} x_i \right| \leq \mathbf{u} \sum_{i=1}^{n-1} |T_i|, \tag{5.13}$$

where, as in the previous sections, $\mathbf{u}$ is the *unit roundoff* defined in Chapter 2, page 27 (Definition 2.3).

Hence, a good strategy is to minimize the terms $|T_i|$. This explains some properties:

- although quite costly, insertion summation is a rather accurate method (as pointed out by Higham, if all the x_i's have the same sign, this is the best method among those that are modeled by Algorithm 5.5);
- when all the x_i's have the same sign, ordering the input values in increasing order gives the smallest bound among the recursive summation methods.[7]

Also, when there is much cancellation in the computation (that is, when $|\sum_{i=1}^{n} x_i|$ is much less than $\sum_{i=1}^{n} |x_i|$), Higham suggests that recursive summation with the x_i sorted by decreasing order is likely to give better results than that using increasing ordering (an explanation of this apparently strange phenomenon is Sterbenz's lemma, Chapter 4, page 101: when x is very close to y, the subtraction $x - y$ is performed exactly). Table 5.1 presents an example of such a phenomenon.

5.3.2 Compensated sums

The algorithms presented in the previous section could be at least partly analyzed just by considering that, when we perform an addition $a + b$ of two floating-point numbers, the computed result is equal to

$$(a + b)(1 + \epsilon),$$

with $|\epsilon| \leq \mathbf{u}$. Now, we are going to consider algorithms that cannot be so simply analyzed. They use the fact that when the arithmetic operations are correctly rounded (to the nearest), floating-point addition has specific properties that allow for the use of tricks such as Fast2Sum (Algorithm 4.3, page 104).

In 1965 Kahan suggested the following *compensated summation* algorithm (Algorithm 5.6) for computing the sum of n floating-point numbers. Babuška [24] independently found the same algorithm.

[7]It gives the smallest *bound*, which does not mean that it will always give the smallest *error*.

Algorithm 5.6 Original version of Kahan's summation algorithm.

$s \leftarrow x_1$
$c \leftarrow 0$
for $i = 2$ to n **do**
  $y \leftarrow \circ(x_i - c)$
  $t \leftarrow \circ(s + y)$
  $c \leftarrow \circ(\circ(t - s) - y)$
  $s \leftarrow t$
end for
return s

Presented like this, Kahan's algorithm may seem very strange. But we note that if we assume that the rounding function is round-to-nearest the second and third lines of the **for** loop constitute the Fast2Sum algorithm (Algorithm 4.3, page 104), so that Kahan's algorithm can be rewritten as Algorithm 5.7.

Algorithm 5.7 Kahan's summation algorithm rewritten with a Fast2Sum.

$s \leftarrow x_1$
$c \leftarrow 0$
for $i = 2$ to n **do**
  $y \leftarrow \circ(x_i + c)$
  $(s, c) \leftarrow \mathrm{Fast2Sum}(s, y)$
end for
return s

Can we safely use the Fast2Sum algorithm? Note that the conditions of Theorem 4.3 (Chapter 4, page 104) are not necessarily fulfilled[8]:

- we are not sure that the exponent of s will always be larger than or equal to the exponent of y;
- furthermore, Algorithm 5.6 is supposed to be used with various rounding functions, not only round-to-nearest (however, Theorem 4.6 Page 111 shows that Fast2Sum returns a useful result with all rounding functions).

And yet, even if we cannot be certain (since the conditions of Theorem 4.3 may not hold) that after the line

$$(s, c) \leftarrow \mathrm{Fast2Sum}(s, y),$$

[8]Indeed, when Kahan introduced his summation algorithm, Theorem 4.3 of Chapter 4 was not known: Dekker's paper was published in 1971. Hence, it is fair to say that the Fast2Sum algorithm first appeared in Kahan's paper, even if it was without the conditions that must be fulfilled for it to always return the *exact* error of a floating-point addition.

the new value of s plus c will be *exactly* equal to the old value of s plus y, in practice, they will be quite close. Thus, *$-c$ is a good approximation to the rounding error committed when adding y to s*. The elegant idea behind Kahan's algorithm is therefore to subtract that approximation from the next term of the sum, in order to (at least partly) compensate for that rounding error.

Knuth and Kahan show that the final value s returned by Algorithm 5.6 satisfies

$$\left| s - \sum_{i=1}^{n} x_i \right| \le \left(2\mathbf{u} + \mathcal{O}(n\mathbf{u}^2)\right) \sum_{i=1}^{n} |x_i|. \tag{5.14}$$

This explains why, in general, Algorithm 5.6 will return a very accurate result.

And yet, if there is much cancellation in the computation (that is, if $|\sum_{i=1}^{n} x_i| \ll \sum_{i=1}^{n} |x_i|$), the relative error on the sum can be very large. Priest gives the following example [496]:

- assume binary, precision-p, rounded-to-nearest arithmetic, with $n = 6$, and
- set $x_1 = 2^{p+1}$, $x_2 = 2^{p+1} - 2$, and $x_3 = x_4 = x_5 = x_6 = -(2^p - 1)$.

The exact sum is 2, whereas the sum returned by Algorithm 5.6 is 3.

To deal with such difficulties, Priest [495, 496] comes up with another idea. He suggests to first sort the input numbers x_i in descending order of magnitude, then to perform the *doubly compensated summation algorithm* shown in Algorithm 5.8.

Algorithm 5.8 Priest's doubly compensated summation algorithm.

$s_1 \leftarrow x_1$
$c_1 \leftarrow 0$
for $i = 2$ to n **do**
  $y_i \leftarrow \circ(c_{i-1} + x_i)$
  $u_i \leftarrow \circ(x_i - \circ(y_i - c_{i-1}))$
  $t_i \leftarrow \circ(y_i + s_{i-1})$
  $v_i \leftarrow \circ(y_i - \circ(t_i - s_{i-1}))$
  $z_i \leftarrow \circ(u_i + v_i)$
  $s_i \leftarrow \circ(t_i + z_i)$
  $c_i \leftarrow \circ(z_i - \circ(s_i - t_i))$
end for

Again, the algorithm looks much less arcane if we rewrite it with Fast2Sums as shown in Algorithm 5.9.

Algorithm 5.9 Priest's doubly compensated summation algorithm, rewritten with Fast2Sums.

$s_1 \leftarrow x_1$
$c_1 \leftarrow 0$
for $i = 2$ to n **do**
  $(y_i, u_i) \leftarrow \mathrm{Fast2Sum}(c_{i-1}, x_i)$
  $(t_i, v_i) \leftarrow \mathrm{Fast2Sum}(s_{i-1}, y_i)$
  $z_i \leftarrow \circ(u_i + v_i)$
  $(s_i, c_i) \leftarrow \mathrm{Fast2Sum}(t_i, z_i)$
end for

Priest shows the following result.

Theorem 5.6 (Priest [496]). *In radix-β, precision-p arithmetic, assuming round-to-nearest,*[9] *if $|x_1| \geq |x_2| \geq \cdots \geq |x_n|$ and $n \leq \beta^{p-3}$, then the floating-point number s_n returned by the algorithm satisfies*

$$\left| s_n - \sum_{i=1}^{n} x_i \right| \leq 2\mathbf{u} \left| \sum_{i=1}^{n} x_i \right|.$$

As soon as the x_i's do not all have the same sign, this is a significantly better bound than the one given by formula (5.14). Indeed, if n is not huge ($n \leq \beta^{p-3}$), then the relative error is bounded by $2\mathbf{u}$ even for an arbitrarily large condition number. On the other hand, due to the need for preliminarily sorted input values, this algorithm will be significantly slower than Kahan's algorithm: one should therefore reserve Priest's algorithm for cases where we need very accurate results and we know that there will be some cancellation in the summation (i.e., $|\sum_{i=1}^{n} x_i| \ll \sum_{i=1}^{n} |x_i|$).

In Kahan's algorithm (Algorithm 5.7), in many practical cases c will have a much smaller magnitude than x_i, so that when adding them together, a large part of the information contained in variable c may be lost. Indeed, Priest's algorithm also compensates for the error of this addition, hence the name "doubly compensated summation."

To deal with that problem, Pichat [490] and Neumaier [454] independently found the same idea: at step i, the rounding error,[10] say e_i, is still computed due to the addition of x_i. However, instead of immediately subtracting e_i from the next operand, the terms e_k are added together, to get a correcting term e that will be added to s at the end of the computation. See Algorithm 5.10.

[9] Priest proves that result in a more general context, just assuming that the arithmetic is faithful and satisfies a few additional properties. See [496] for more details.

[10] It was then possible to evaluate that error exactly: Dekker's result was known when Pichat published her paper.

Algorithm 5.10 Pichat and Neumaier's summation algorithm [490, 454]. Note, since Fast2Sum is used, that the radix of the floating-point system must be at most 3 (which means, in practice, that this algorithm should be used in radix 2 only).

```
s ← x_1
e ← 0
for i = 2 to n do
   if |s| ≥ |x_i| then
      (s, e_i) ← Fast2Sum(s, x_i)
   else
      (s, e_i) ← Fast2Sum(x_i, s)
   end if
   e ← RN(e + e_i)
end for
return RN(s + e)
```

To avoid tests, the algorithm of Pichat and Neumaier can be rewritten using the 2Sum algorithm (it also has the advantage of working in any radix). This gives the *cascaded summation* algorithm of Rump, Ogita, and Oishi [471]. It always gives exactly the same result as Pichat and Neumaier's algorithm, but it does not need comparisons. See Algorithm 5.11.

Algorithm 5.11 By rewriting Pichat and Neumaier's summation algorithm with a 2Sum, we get the *cascaded summation* algorithm of Rump, Ogita, and Oishi [471].

```
s ← x_1
e ← 0
for i = 2 to n do
   (s, e_i) ← 2Sum(s, x_i)
   e ← RN(e + e_i)
end for
return RN(s + e)
```

Rump, Ogita, and Oishi show the following result.

Theorem 5.7 (Rump, Ogita, and Oishi [471]). *If Algorithm 5.11 is applied to floating-point numbers x_i, $1 \le i \le n$, and if $n\mathbf{u} < 1$, then, even in the presence of underflow, the final result σ returned by the algorithm satisfies*

$$\left| \sigma - \sum_{i=1}^{n} x_i \right| \le \mathbf{u} \left| \sum_{i=1}^{n} x_i \right| + \gamma_{n-1}^2 \sum_{i=1}^{n} |x_i|.$$

The same result also holds for Algorithm 5.10, since both algorithms output the very same value.

Rump, Ogita, and Oishi generalize their method, by reusing the same algorithm for summing the e_i, in a way very similar to what Pichat suggested in [490]. To present their *K-fold* algorithm, we first modify Algorithm 5.11 as shown in Algorithm 5.12.

Algorithm 5.12 VecSum(x) [471]. Here, p and x are vectors of floating-point numbers: $x = (x_1, x_2, \ldots, x_n)$ represents the numbers to be summed. If we compute $p = \text{Vecsum}(x)$, then p_n is the final value of variable s in Algorithm 5.11, and p_i (for $1 \leq i \leq n-1$) is variable e_i of Algorithm 5.11.

```
p ← x
for i = 2 to n do
  (p_i, p_{i-1}) ← 2Sum(p_i, p_{i-1})
end for
return p
```

Then, here is Rump, Ogita, and Oishi's K-fold summation algorithm (Algorithm 5.13).

Algorithm 5.13 K-fold algorithm [471]. It takes a vector $x = (x_1, x_2, \ldots, x_n)$ of floating-point numbers to be added and outputs a result whose accuracy is given by Theorem 5.8.

```
for k = 1 to K − 1 do
  x ← VecSum(x)
end for
c = x_1
for i = 2 to n do
  c ← c + x_i
end for
return c
```

Rump, Ogita, and Oishi prove the following result.

Theorem 5.8 (Rump, Ogita, and Oishi [471]). *If Algorithm 5.13 is applied to floating-point numbers x_i, $1 \leq i \leq n$, and if $4n\mathbf{u} < 1$, then, even in the presence of underflow, the final result σ returned by the algorithm satisfies*

$$\left|\sigma - \sum_{i=1}^{n} x_i\right| \leq (\mathbf{u} + \gamma_{n-1}^2)\left|\sum_{i=1}^{n} x_i\right| + \gamma_{2n-2}^K \sum_{i=1}^{n} |x_i|.$$

Theorem 5.8 shows that the K-fold algorithm is almost as accurate as a conventional summation in precision Kp followed by a final rounding to precision p. Klein [339] suggests very similar algorithms.

Tables 5.1 and 5.2 give examples of errors obtained using some of the summation algorithms presented in this section. They illustrate the fact that

Pichat and Neumaier's and Priest's algorithms give very accurate results. Although slightly less accurate, Kahan's compensated summation algorithm is still of much interest, since it is very simple and fast.

Method	Error in ulps
increasing order	18.90625
decreasing order	16.90625
compensated (Kahan)	6.90625
doubly compensated (Priest)	0.09375
Pichat and Neumaier; or Rump, Ogita, and Oishi	0.09375

Table 5.1: Errors of the various methods for $x_i = \mathrm{RN}(\cos(i))$, $1 \le i \le n$, with $n = 5000$ and binary32 arithmetic. Note that all the x_i are exactly representable. The methods of Priest; Pichat and Neumaier; and Rump, Ogita, and Oishi give the best possible result (that is, the exact sum rounded to the nearest binary32 number). The recursive summation method with decreasing ordering is slightly better than the same method with increasing order (which is not surprising: there is much cancellation in this sum), and Kahan's compensated summation method is significantly better than the recursive summation.

Incidentally, one could wonder whether it is possible to design a very simple summation algorithm that would always return correctly rounded sums. Kornerup, Lefèvre, Louvet, and Muller [347] have shown that, under simple conditions, an *RN-addition algorithm without branching* (that is, an algorithm that only uses rounded-to-nearest additions and subtractions, without any test; see Definition 4.1 in Chapter 4, page 108) cannot always return the correctly rounded-to-nearest sum of 3 or more floating-point numbers. This shows that an "ultimately accurate" floating-point summation algorithm cannot be very simple. And yet, if we accept tests and/or changes of rounding functions, getting the correctly rounded sum of several floating-point numbers is indeed possible, as we will see in Section 5.3.4 in the case of 3 numbers. Note that Rump, Ogita, and Oishi have designed an efficient algorithm, with tests, for the rounded-to-nearest sum of n numbers [532].

5.3.3 Summation algorithms that somehow imitate a fixed-point arithmetic

One can easily note that in binary, precision-p arithmetic, if n floating-point numbers are all multiple of the same power of 2, say 2^k, and less than or equal to $2^{k+w} - 2^k$, then as long as $n \cdot (2^w - 1) \le 2^p - 1$, their sum is exactly computed in floating-point arithmetic, whatever the order in which the terms are added: the additions are performed very much like in fixed-point arithmetic.

Method	Error in ulps
increasing order	6.86
decreasing order	738.9
compensated (Kahan)	0.137
doubly compensated (Priest)	0.137
Pichat and Neumaier; or Rump, Ogita, and Oishi	0.137

Table 5.2: Errors of the various methods for $x_i = \mathrm{RN}(1/i)$, $1 \le i \le n$, with $n = 10^5$ and binary32 arithmetic. Note that all the x_i are exactly representable. The methods of Kahan; Priest; Pichat and Neumaier; and Rump, Ogita, and Oishi give the best possible result (that is, the exact sum rounded to the nearest binary32 number). The recursive summation method with increasing ordering is much better than the same method with decreasing order, which is not surprising since all the x_i's have the same sign.

This leads to the following idea: for adding floating-point numbers, one can try to "split" them, so that the most significant parts resulting from the splitting satisfy the above-given conditions. We already know how such a splitting can be done: it suffices to use Algorithm ExtractScalar (Algorithm 4.11, presented in Chapter 4). Rump [521] attributes this idea to Zielke and Drygalla [644] in the special case of dot-product computation, and uses it for performing fast and very accurate summation. It was used later on in [165] to perform "reproducible summations" (see Section 5.3.3.2), and by Muller, Popescu, and Tang to multiply floating-point expansions (see Section 14.2.3.2).

Assume we are given n floating-point numbers $p_0, p_1, \ldots, p_{n-1}$ to be added.

Algorithm 5.14 ExtractVector(σ, p): computes an approximation τ to $\sum p_i$ and an array p_i^ℓ of floating-point numbers such that $\sum p_i = \tau + \sum p_i^\ell$ [521]. The number σ must be a large enough power of 2 (see Theorem 5.9).

$\tau \leftarrow 0$
for $i = 0$ to $n-1$ **do**
  $p_i^h \leftarrow \mathrm{RN}(\mathrm{RN}(\sigma + p_i) - \sigma)$
  $p_i^\ell \leftarrow \mathrm{RN}(p_i - p_i^h)$
  $\tau \leftarrow \mathrm{RN}(\tau + p_i^h)$
end for

Note that Algorithm 5.14 consists in applying Algorithm 4.11 to each term p_i, with the same constant σ (i.e., expressing each p_i as a sum $p_i^h + p_i^\ell$

of floating-point numbers such that p_i^h is a multiple of $\sigma \cdot 2^{-p}$ and $|p_i^\ell| \leq \sigma \cdot 2^{-p}$), and adding together the values p_i^h. The constraints in the conditions of Theorem 5.9 ensure that the sum of the p_i^h is computed exactly.

We have the following result.

Theorem 5.9 (Rump [521]). *Assume $\sigma = 2^k$ for some integer k, and that $n < 2^M$. If $\max |p_i| \leq 2^{-M}\sigma$, then*

$$\sum_{i=0}^{n-1} p_i = \tau + \sum_{i=0}^{n-1} p_i^\ell, \qquad \max |p_i^\ell| \leq 2^{-p}\sigma,$$

and τ is a multiple of $2^{-p}\sigma$.

Let us now see some possible applications of Algorithm 5.14.

5.3.3.1 Rump, Ogita, and Oishi's faithful summation

Assume we wish to compute $a_0 + a_1 + \cdots + a_{n-1}$. The idea behind the summation algorithm presented by Rump, Ogita, and Oishi in [531] is the following. A first call to Algorithm 5.14 gives an approximation τ_0 to the sum and a new vector $a_0^{(1)} + a_1^{(1)} + \cdots + a_{n-1}^{(1)}$ of floating-point numbers such that $\sum a_i = \tau_0 + \sum a_i^{(1)}$. We then call again Algorithm 5.14 with the vector $a_0^{(1)} + a_1^{(1)} + \cdots + a_{n-1}^{(1)}$, which gives an approximated sum τ_1 and a new vector $a_0^{(2)} + a_1^{(2)} + \cdots + a_{n-1}^{(2)}$, and so on (the stopping criterion for obtaining faithful rounding is explicited in [531]). Then (assuming we stop after having computed the terms $a_i^{(m)}$), the sum $\tau_0 + \tau_1 + \cdots + \tau_{m-1} + \sum a_i^{(m)}$ can be computed as follows:

- $s_m \approx \sum a_i^{(m)}$ is computed using the recursive summation algorithm;
- defining $\pi_1 = \tau_0$ and $e_1 = 0$, we compute the terms $\pi_2, \pi_3, \ldots, \pi_{m-1}$ and $e_2, e_3, \ldots, e_{m-1}$ defined as $(\pi_k, q_k) =$Fast2Sum(π_{k-1}, τ_k) and $e_k = \mathrm{RN}(e_{k-1} + q_k)$;
- we return $\mathrm{RN}(\pi_{m-1} + \mathrm{RN}(e_{m-1} + s_m))$.

5.3.3.2 Demmel, Ahrens, and Nguyen's reproducible summation

Demmel, Ahrens, and Nguyen [165] wish to provide an accurate summation algorithm that allow for reproducible computations (see Section 2.8). They do not want to perform multiple passes over the data (which would be necessary with the algorithm presented in the previous section, in order to find a suitable value of σ before using Algorithm 4.11). For that purpose, they break the whole exponent range into fixed *bins* of a given width. Each input number

a_i is then split into slices (using for instance the ExtractScalar algorithm—Algorithm 4.11, presented in Chapter 4) such that each slice fits into a bin. All slices in a given bin can be added exactly, whatever the order of summation. Only the bins corresponding to the largest exponent ranges are considered. See [165] for more details.

5.3.3.3 Towards accurate summation hardware?

To conclude this section, let us mention that Kulisch [353, 354, 355] advocated enhancing floating-point units with a very large register (or accumulator) that would allow sums of products to be performed exactly, without any rounding. This idea will be reviewed in Section 8.7.2.

5.3.4 On the sum of three floating-point numbers

Computing the correctly rounded sum of three numbers is sometimes useful. For instance, in the CRlibm[11] elementary function library, several calculations are done using a "triple-double" intermediate format (see Section 14.1, page 514), using functions due to Lauter [370]. To return a correctly rounded result in double-precision/binary64 arithmetic, one must convert a "triple-double" into a binary64 number: this reduces to computing the correctly rounded sum of three floating-point numbers.

For that purpose, Boldo and Melquiond [52] introduce a new rounding function, *round-to-odd*, $\circ_{\text{odd}}$, defined as follows:

- if x is a floating-point number, then $\circ_{\text{odd}}(x) = x$;
- otherwise, $\circ_{\text{odd}}(x)$ is the value among $\text{RD}(x)$ and $\text{RU}(x)$ whose integral significand is an odd integer.[12]

This rounding function is not implemented on current architectures, but that could easily be done. Interestingly enough, Boldo and Melquiond show that in radix-2 floating-point arithmetic, using only one rounded-to-odd addition (and a few rounded-to-nearest additions/subtractions), one can easily compute

$$\text{RN}(a + b + c),$$

where a, b, and c are floating-point numbers. Their algorithm is presented in Figure 5.1. They also explain how to emulate rounded-to-odd additions (with a method that requires testing). Listing 5.1 presents a C program that implements their method in the particular case where the input operands are ordered.

---

[11]CRlibm was developed by the Arénaire/AriC team of CNRS, INRIA, and ENS Lyon, France. It is available at https://gforge.inria.fr/scm/browser.php?group_id=5929&extra=crlibm.

[12]This means, in radix 2, that the least significant bit of the significand is a one.

C listing 5.1 Boldo and Melquiond's program [52] for computing the correctly rounded-to-nearest sum of three binary floating-point numbers x_h, x_m, and x_l, assuming $|x_h| \geq |x_m| \geq |x_l|$. The encoding of the binary64 floating-point numbers specified by the IEEE 754 standard is necessary here, as it ensures that `thdb.l++` is the floating-point successor of `thdb.l`.

```
double CorrectRoundedSum3(double xh, double xm, double xl) {
  double th, tl;
  // Dekker's error-free adder of two ordered numbers
  Fast2Sum(&th, &tl, xm, xl);
  // round th to odd if tl is not zero
  if (tl != 0.0) {
    db_number thdb;
    thdb.d = th;  // thdb.l is the binary representation of th
    // if the significand of th is odd, there is nothing to do
    if (!(thdb.l & 1)) {
      // choose the rounding direction
      // depending on the signs of th and tl
      if ((tl > 0.0) ^ (th < 0.0))
        thdb.l++;
      else
        thdb.l--;
      th = thdb.d;
    }
  }
  // final addition rounded to the nearest
  return xh + th;
}
```

Algorithm 5.15, introduced by Kornerup et al. [347], is another way of emulating the rounded-to-odd addition.

Algorithm 5.15 OddRoundSum(a,b): computes $\circ_{\text{odd}}(a+b)$ in radix-2 floating-point arithmetic.

$d \leftarrow \text{RD}(a+b)$
$u \leftarrow \text{RU}(a+b)$
$e' \leftarrow \text{RN}(d+u)$
$e \leftarrow e' \times 0.5$
$z' \leftarrow u - e$
$z \leftarrow z' + d$
return z

5.4 Compensated Dot Products

The dot product of two vectors $[a_i]_{1\le i\le n}$ and $[b_i]_{1\le i\le n}$ is $\sum_{1\le i\le n} a_i \cdot b_i$. When the condition number

$$C_{\text{dot product}} = \frac{2 \cdot \sum_{i=1}^{n} |a_i \cdot b_i|}{\left|\sum_{i=1}^{n} a_i \cdot b_i\right|}$$

is not too large, Algorithm 5.2 can be used safely. When this is not the case, one can design a compensated dot product algorithm.

Note that one can easily reduce the problem of computing the dot product of two vectors of dimension n to the problem of computing the sum of $2n$ floating-point numbers, since the Dekker product (Algorithm 4.10, page 116) and the 2MultFMA algorithm (Algorithm 4.8, page 112) make it possible (under some conditions) to deduce, from two floating-point numbers a and b, two floating-point numbers r_1 and r_2 such that

$$r_1 + r_2 = a \cdot b \qquad \text{and} \qquad |r_2| \le \frac{1}{2}\,\text{ulp}(r_1). \tag{5.15}$$

Hence, many methods presented in Section 5.3 can readily be adapted to the computation of dot products. And yet, by doing that, we do not use the fact that, in Equation (5.15), r_2 is very small in front of r_1: the sum we have to compute is not an arbitrary sum of $2n$ numbers, some are much smaller than others, and that property can be used to design faster algorithms.

Let us now give the compensated dot product algorithm introduced by Ogita, Rump, and Oishi [471]. See Algorithm 5.16. In that algorithm, 2Prod will denote either the 2MultFMA algorithm (if an FMA instruction is available), or the Dekker product. Remember that to be able to use the Dekker product we need the radix of the floating-point system to be equal to 2 or the precision p to be even. Remember also that Equation (5.15) holds if $e_a + e_b \ge e_{\min} + p - 1$, where e_a and e_b are the exponents of a and b, respectively (see Chapter 4 for more details).

Algorithm 5.16 Algorithm CompensatedDotProduct(a,b) [471] for the evaluating $a_1 \cdot b_1 + a_2 \cdot b_2 + \cdots + a_n \cdot b_n$.

$(s_1, c_1) \leftarrow \text{2Prod}(a_1, b_1)$
for $i = 2$ to n **do**
  $(p_i, \pi_i) \leftarrow \text{2Prod}(a_i, b_i)$
  $(s_i, \sigma_i) \leftarrow \text{2Sum}(p_i, s_{i-1})$
  $c_i \leftarrow \text{RN}(c_{i-1} + \text{RN}(\pi_i + \sigma_i))$
end for
return $\text{RN}(s_n + c_n)$

Many variants of the algorithm can be designed, possibly inspired from the variants of the compensated summation algorithm presented in Section 5.3. Ogita, Rump, and Oishi have shown the following theorem, which says (it suffices to compare to Equation (5.6)) that, in precision p, the result of Algorithm 5.16 is as accurate as the value computed by the straightforward algorithm (Algorithm 5.2) in precision $2p$ and rounded back to the working precision p.

Theorem 5.10 (Ogita, Rump, and Oishi [471]). *If no underflow or overflow occurs, in radix 2,*

$$\left|\mathrm{CompensatedDotProduct}(a,b) - \sum_{i=1}^{n} a_i \cdot b_i\right| \leq \mathbf{u}\left|\sum_{i=1}^{n} a_i \cdot b_i\right| + \gamma_n^2 \sum_{i=1}^{n} |a_i \cdot b_i|.$$

5.5 Compensated Polynomial Evaluation

Graillat, Langlois, and Louvet [222, 396] introduced a *compensated algorithm* for polynomial evaluation. Let us present it briefly. Assume we wish to compute

$$p(x) = a_n x^n + a_{n-1} x^{n-1} + \cdots + a_0,$$

where x and all the a_i are floating-point numbers. In the following, 2Prod will denote the 2MultFMA algorithm (Algorithm 4.8, page 112) if an FMA instruction is available, and the Dekker product (Algorithm 4.10, page 116) otherwise. Graillat, Langlois, and Louvet first define the following "error-free transformation" (see Algorithm 5.17).

Algorithm 5.17 The Graillat–Langlois–Louvet error-free transformation [222, 396]. Input: a degree-n polynomial $p(x) = a_n x^n + a_{n-1} x^{n-1} + \cdots + a_0$. Output: the same result r_0 as the conventional Horner's evaluation of p, and two degree-$(n-1)$ polynomials $\pi(x)$ and $\sigma(x)$, of degree-i coefficients π_i and σ_i, such that $p(x) = r_0 + \pi(x) + \sigma(x)$ exactly.

$r_n \leftarrow a_n$
for $i = n-1$ downto 0 **do**
  $(p_i, \pi_i) \leftarrow \mathrm{2Prod}(r_{i+1}, x)$
  $(r_i, \sigma_i) \leftarrow \mathrm{2Sum}(p_i, a_i)$
end for
return r_0, $(\pi_0, \pi_1, \ldots, \pi_{n-1})$, $(\sigma_0, \sigma_1, \ldots, \sigma_{n-1})$

Define Horner(p, x) as the result returned by Horner's rule (Algorithm 5.3) with polynomial p and input variable x. Also, for a polynomial $q = \sum_{i=0}^{n} q_i x^i$, define

$$\tilde{q}(x) = \sum_{i=0}^{n} |q_i| x^i.$$

We have the following result.

Theorem 5.11 (Langlois and Louvet [368]). *The values* r_0, $(\pi_0, \pi_1, \ldots, \pi_{n-1})$, *and* $(\sigma_0, \sigma_1, \ldots, \sigma_{n-1})$ *returned by Algorithm 5.17 satisfy*

- $r_0 = \text{Horner}(p, x)$;
- $p(x) = r_0 + \pi(x) + \sigma(x)$;
- $\widetilde{(\pi + \sigma)}(|x|) \leq \gamma_{2n}\, \tilde{p}(|x|)$.

Theorem 5.11 suggests to transform Algorithm 5.17 into a compensated polynomial evaluation algorithm as shown in Algorithm 5.18.

Algorithm 5.18 The Graillat–Langlois–Louvet compensated polynomial evaluation algorithm [222, 396]. Input: a degree-n polynomial $p(x) = a_n x^n + a_{n-1}x^{n-1} + \cdots + a_0$. Output: an approximation r to $p(x)$. In practice, the evaluation of $\text{Horner}(q, x)$ with $q(x) = \sum_{0 \leq i < n} q_i x^i$ would be done in the **for** loop: we wrote it as shown here for the sake of clarity.

$r_n \leftarrow a_n$
for $i = n-1$ downto 0 **do**
  $(p_i, \pi_i) \leftarrow \text{2Prod}(r_{i+1}, x)$
  $(r_i, \sigma_i) \leftarrow \text{2Sum}(p_i, a_i)$
  $q_i \leftarrow \text{RN}(\pi_i + \sigma_i)$
end for
$r \leftarrow \text{RN}(r_0 + \text{Horner}(q, x))$
return r

An immediate consequence of Theorem 5.11 is the following.

Theorem 5.12 (Langlois and Louvet [368]). *The value* r *returned by Algorithm 5.18 satisfies*

$$|r - p(x)| \leq \mathbf{u}|p(x)| + \gamma_{2n}^2\, \tilde{p}(|x|).$$

Note the similarity with Theorem 5.10. Just like for dot products, the above result says that, as soon as $\tilde{p}(|x|) = \sum_{i=0}^{n} |a_i| \cdot |x|^i$ is not huge in front of $|p(x)|$, Algorithm 5.18 will return a very accurate result (namely, for a working precision p, as accurate as if we had used Horner's rule in twice that working precision and then rounded back). If this is not the case, it is possible to define *K-fold compensated polynomial evaluation algorithms* by recursively using the same method for evaluating $\sigma(x)$ and $\pi(x)$. See Louvet's Ph.D. dissertation [396] for details.

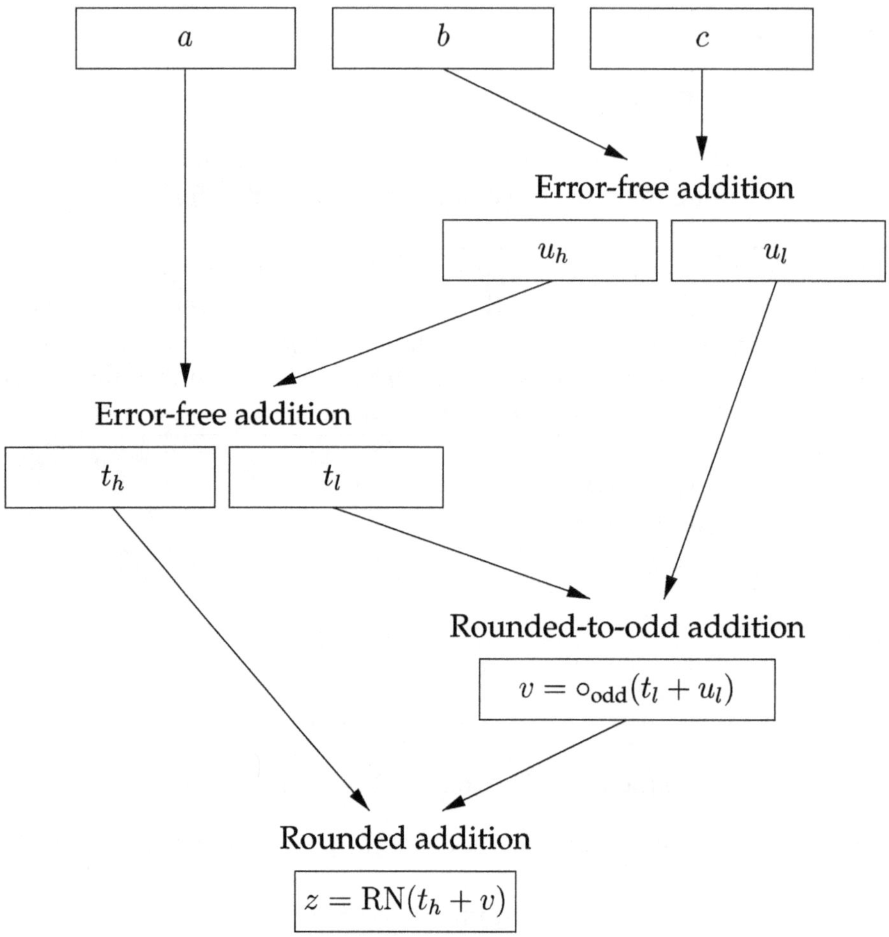

Figure 5.1: Boldo and Melquiond's algorithm [52] for computing $\text{RN}(a + b + c)$ *in radix-2 floating-point arithmetic. It requires a "rounded-to-odd" addition. The error-free additions are performed using the 2Sum algorithm (unless we know for some reason the ordering of the magnitude of the variables, in which case the Fast2Sum algorithm can be used). © IEEE, with permission.*

Chapter 6

Languages and Compilers

THE PREVIOUS CHAPTERS have given an overview of interesting properties and algorithms that can be built on an IEEE 754-compliant floating-point arithmetic. In this chapter, we discuss the practical issues encountered when trying to implement such algorithms in actual computers using actual programming languages. In particular, we discuss the relationship between standard compliance, portability, accuracy, and performance. This chapter is useful to programmers wishing to obtain a standard-compliant behavior from their programs, but it is also useful for understanding how performance may be improved by relaxing standard compliance and also what traps one may fall into.

6.1 A Play with Many Actors

Even with a computer supporting one of the IEEE 754 standards for floating-point arithmetic, it still requires some effort to take control of the details of the floating-point computations of one's program (for instance, to ensure portability).

Most programming languages will allow one to write an expression such as `a+b+c*d`, using variables `a`, `b`, `c`, and `d`, of some (possibly implicit) floating-point type. However, when the program is run, the sequence of floating-point operations that is actually executed may differ widely, depending on the language, but also on the environment used, including:

- the compiler, and the optimization options that were passed to it;
- the processor, which may or may not have a given floating-point capability;
- the operating system, which is responsible for making the processor's capabilities available for user programs;

J.-M. Muller et al., *Handbook of Floating-Point Arithmetic*,
https://doi.org/10.1007/978-3-319-76526-6_6

- the system libraries, for the mathematical functions (when they are not handled by the compiler).

Let us now review these elements to give a taste of what may happen. The following sections will then detail in more depth the specifics of some widely used programming environments.

6.1.1 Floating-point evaluation in programming languages

Consider the evaluation of `a+b+c+d`. It obviously involves three floating-point additions. A first question is: in which order should these three additions be performed? A second question is related to the data-types of the variables `a`, `b`, `c`, and `d`: if they differ, what is the type of the result? What is the type of the intermediate (implicit) results?

For each of these questions, the answer chosen by language designers may impact the accuracy, the predictability, and the performance of the program.

6.1.1.1 Expression evaluation order

In which order should these operations be executed? In other terms, what is the implicit parenthesizing used when evaluating `a+b+c+d`? Alternatives are, on this example, `((a+b)+c)+d`, `(a+b)+(c+d)`, `a+(b+(c+d))`, `a+((b+c)+d)`, `(a+(b+c))+d`. Related questions are: Should addition be considered associative? Is `(a+d)+(b+c)` also an option?

There is a tradeoff here. On one side, it should be clear to the reader, after having read the previous chapters, that floating-point addition is not associative. A language should allow one to express useful algorithms such as the 2Sum algorithm, which requires computing `(a+b)-b` without changing the order of evaluation or using a larger intermediate precision (see Section 4.3.2). On the other side, such situations are fairly rare and well identified. In more general floating-point codes, rewriting freedom allows for many optimizations.

- `(a+b)+(c+d)` exposes more parallelism for processors able to perform two additions in parallel.
- In general, in a pipelined processor, the fastest parenthesizing depends on the order of availability of the four variables, which itself depends on previous computations.
- If `a` and `d` are compile-time constants, computing the sum `a+d` at compile time will save one addition at execution time.
- The parenthesizing may impact register allocation, sub-expression sharing, etc.

6.1.1.2 Precision of intermediate computations

Many languages require the programmer to declare the variables of a given type, say binary32 (`float` in C) or binary64 (`double` in C). However, they usually do not require the precision to be declared for each *operation* of the code (assembly languages, of course, do). Deducing the precision of the operations from the precision of the data is not straightforward. Consider, for instance, the following situations.

- If `a`, `b`, `c`, and `d` are declared of binary32 type, and the hardware is able to compute on binary64 as fast as on binary32, shouldn't this "free" extra accuracy be used?

- In an assignment such as `r=a+b+c+d`, where `r` is declared as binary64 and the other variables are declared as binary32, should the computations be performed using binary32 addition or binary64 addition?

- If `a` and `d` are declared binary32, and `b` and `c` are declared binary64, what will be the precision of the operations? Note that this question makes sense only after a parenthesizing has been chosen.

These questions are not purely academic. For most applications, it makes perfect sense to use binary32 as a storage format, as it requires half the space of binary64 and holds more precision than most instruments can measure. Besides, this helps to lower the necessary data bandwidth between main memory and the processor (or its caches), which is nowadays a common limiting factor in HPC. Nevertheless it makes sense to use binary64 registers (as long as data are never flushed back to narrower memory locations) to carry out computations that may involve millions of iterations.

6.1.1.3 Antagonistic answers

The languages C and FORTRAN, probably the two languages that are most used in numerical computing, offer perfectly antagonistic answers to the previous questions.

- In C, an expression of successive add or multiply operators is interpreted with left-associativity,[1] i.e., `a+b+c+d` is syntactically equivalent to `((a+b)+c)+d`. Concerning the precision, each operation may well be performed in an internal precision wider than the precision of the type (we already discussed this issue of double rounding in Chapter 3 and a hint on the behavior can be provided by checking the value of the `FLT_EVAL_METHOD` macro, see Table 6.1); besides, the expression may also be contracted (see Section 6.2.3.2).

[1]This requirement was introduced in C89 and kept in the subsequent C99 and C11 standards. The original Kernighan and Ritchie C [333] allowed regrouping, *even with explicit parentheses* in expressions.

- In contrast, FORTRAN sets the precision but does not guarantee the parenthesizing, so the expression `a+b+c+d` may validly be evaluated as `(a+b)+(c+d)` or `(a+d)+(b+c)`.

Each language has a rationale for these choices, and we will explore it in the following sections.

Note that the IEEE 754-2008 standard attempts to address this problem; see Section 3.1.2. Also, we only discussed expression evaluation here, but similar issues arise for instance when one considers the active rounding mode.

6.1.2 Processors, compilers, and operating systems

The compiler is in charge of translating the program into a succession of elementary processor instructions. Modern compilers spend most of the compilation time in optimizations [434]. We have seen some optimizations related to floating-point evaluation order and precision, but there also exist optimizations that are more directly related to the processor's available hardware.

As an example, consider a processor which offers hardware implementations of the fused multiply-add (FMA, see Section 2.4). To comply with the IEEE 754 standard, a compiler should not generate FMA instructions for such processors when the user expresses a multiplication followed by an addition. Of course, the default behavior of most compilers will be to try to fuse additions and multiplications (i.e., to use FMA instructions), which usually improves both speed and accuracy. Caution: an expression such as `x*x + y*y` will no longer be symmetrical in `x` and `y` if one of the multiplications is fused into the addition.

If one wants portability between, for instance, a platform without FMA and one with FMA, one has to find special directives (such as C's `FP_CONTRACT` pragma) or compiler options (such as GCC's `-mno-fused-madd` with processors that support this option) that prevent fusing $\times$ and $+$.

There is a similar tradeoff between portability and improved accuracy on processors which offer hardware "double extended" arithmetic with 64-bit precision (see Section 3.4.3). Here the extended accuracy, although providing more accurate results in most cases, incurs additional risks, such as subtle bugs due to double rounding (see Section 3.2).

Again, much effort has been spent to address such issues in IEEE 754-2008. There was a clear consensus on the fact that programmers who want portability should be able to get it, while programmers who want performance also should be able to get it. The consensus was not so clear on what should be the default.

Finally, the *operating system* (kernel and libraries) is in charge of initializing the state of the processor prior to the program execution. For example, it will set the dynamic rounding precision on x87 hardware. Considering the previous tradeoff between accuracy and portability, different systems make different choices. For instance, the same conforming C program, compiled by the same compiler with the same options, may yield different results on the same hardware under OpenBSD and Linux. By default, OpenBSD chooses to enhance portability and configures the x86 traditional floating-point unit (FPU) to round to binary64. Linux, in contrast, favors better accuracy and configures the FPU to round to double-extended precision by default.

To summarize, the behavior of each of these actors may influence the others. A program may change the processor state because of an operating system call, for instance, to request rounding toward zero. An ill effect will often be that the mathematical library (a.k.a. `libm`, also a part of the operating system) no longer functions properly because it expects the processor to round to nearest.

6.1.3 Standardization processes

It is useful to have an idea of the process of elaborating a standard in order to appreciate what standard compliance means on a given system at a given time.

As far as floating-point is concerned, we have on the one hand the floating-point standard, and on the other hand the standards for the various languages that support floating-point. As the development of each of these standards is very complex and time-consuming, it is not possible to wait for the end of the standardization of floating-point itself before starting the work on language standards that depend on it. Hence standards are somehow developed in parallel, and floating-point is but an example of such parallel development.

6.1.3.1 Standard bodies

Floating-point standardization involves two different standard bodies: IEEE-SA and ISO/IEC (we count here ISO/IEC as a single standard organization because both bodies[2] join forces for the IT standards that are of interest here). In a nutshell, IEEE-SA is a "consensus building organization" among a wide range of stakeholders (individuals, companies, organizations…). It creates standards that have no more normative strength than that given by the trustworthiness of the committee and the willingness to comply of the community. In contrast, ISO and IEC standards are supposed to be normative and have more of an enforcement power.

In addition, the standardization of each language involves more bodies and committees. Standardization of the C language, for instance, takes place at ISO/IEC.

[2]The International Standard Organization and the International Electrotechnical Commission are two sister international organizations whose members are, respectively, 163 national standard bodies and 83 national committees.

6.1.3.2 Floating-point standards

The initial and actual work takes place in the IEEE. Two main standards have been produced so far in the floating-point realm: IEEE 754-1987 and IEEE 754-2008. These have been endorsed, with little modification, by ISO and IEC as ISO/IEC 60559:1989 and ISO/IEC 60559:2011. The difference in the dates of publication essentially reflects the complexity of the bureaucratic ratification process rather than a significant work to add extra value to what IEEE has already done.

6.1.3.3 C language standards

The standards elaboration for the C language takes place at ISO/IEC. Several versions have been published over time. They come with an official denomination (as ISO/IEC 9899:1999) and an informal name (as C99). At the time of writing the last official issue is ISO/IEC 9899:2011, also known as C11. As the effort is parallel to that of floating-point standardization, it takes some time to incorporate the advances (in both normative and recommendation form) from this effort. This is why, for instance and among other reasons, ISO/IEC 9899:2011 does not completely take into account ISO/IEC 60559:2011 (aka IEEE 754-2008). As a consequence, intermediate specifications are issued between main releases. An example are the five ISO/IEC TS 18661-[1:5] standards [279, 280, 281, 282, 283], published between 2014 and 2016, that eventually update the C language standard to ISO/IEC 60559:2011 level. The upcoming major C language standard release should incorporate these intermediate standards into the next "main" issue, currently dubbed C2X.

The same applies to C++, which is also standardized through ISO/IEC, with an extra twist: for some aspects, the C++ language standard refers to C. As for floating-point aspects, the current C++14 does not refer to the latest C11 version but to the previous C99. Again, the complexity and length of the standardization process explain this gap.

Knowing the standard denominations in ISO/IEC parlance is useful (even if not self-evident) since this is how they are referred to in subsequent standards. Eventually, the C and C++ languages standards introduce normative directives of their own, as explained above. Figure 6.1 displays the standardization timeline.

6.1.3.4 Implementations

For all this standardization effort to become a practical reality, it must be supported by compilers and their associated libraries (essentially "math" ones are of interest here). And yet, this is a different story. First, compiler development is not only (nor mainly) driven by standard compliance. Many other factors are taken into account, in particular performance of the generated

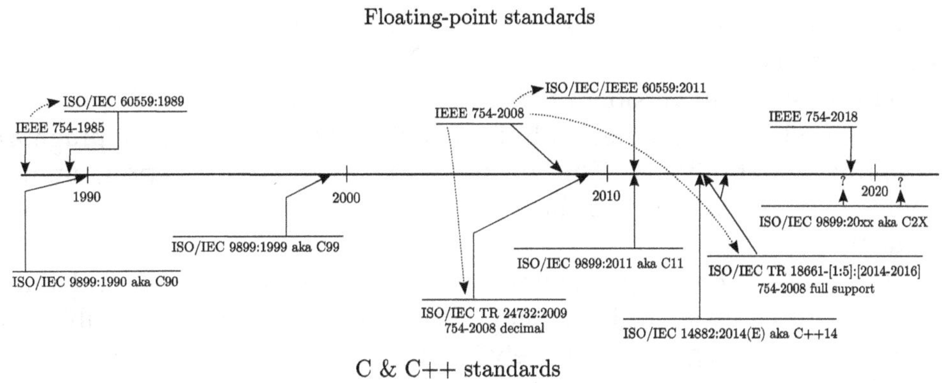

Figure 6.1: The tangled normalization timeline of floating-point (on top) and C language (at bottom). Implementations (materialized as compiler versions) do not perfectly fit in this timeline since a given release can incorporate a variable number of features picked from different standards. Besides, compiler options can also alter the conformance level.

code. Besides, compiler development teams have limited manpower. This, and the very difficulty of implementing some features, makes it difficult to comply both fully and quickly with standards. Compiler implementation will often concentrate on the most important or popular features and leave aside more seldom used ones. The priority can even go to the incorporation of features that are not standardized yet, because they are felt more useful to the programmers, or help them to take the best advantage of the hardware (especially when, as IBM and Intel, the compiler supplier is also a hardware vendor). Hence, standard compliance must be taken with a grain of salt. Another aspect of the problem is that compiler releases are much more frequent than standards updates. From a version to the next, the level of compliance varies. Every major compiler vendor publishes compatibility sheets that list, for a particular compiler version, which features of the standards are or are not implemented. But this is not the final word about compliance. What you read from the aforementioned sheets is that a particular compiler release has the capability to comply with some standard feature. However, this compliance may require some compiler options to be activated, or may be jeopardized by some other options (e.g., the `-ffast-math` option for GCC).

6.1.4 In the hands of the programmer

Standard compliance enhances portability, but usually degrades performance, and sometimes even accuracy. For this reason, the default behavior of a computing system will usually be a compromise between performance, accuracy, and portability. A notable exception is Java, which was designed for portability from the ground up. Section 6.5 will show how difficult fulfilling this ambition is.

However, recent versions of Java offer means to relax portability for performance under programmer control, while the C99 standard added pragmas to improve portability in C. This illustrates the consensus that a programmer should be given the ability to choose the behavior of the floating-point environment.

The important message in this chapter is that the floating-point behavior of a given program in a given computer is not arbitrary. It is usually well documented, although unfortunately in various places (language standards, compiler manuals, operating system specifications, web pages, and so on). It is thus possible for programmers to control to the last bit the behavior of every last floating-point operation of their programs.

Another important message is therefore "Read The Friendly Manual!" This includes the various standards mentioned in the remainder of this chapter, and also the compiler manual, paying attention to its version (there is no such thing as a "generic GCC manual") and the default options, which may change drastically from a release to another, or from one architecture to another (e.g., Intel 32-bit or 64-bit architecture).

Let us now consider some mainstream programming environments in more detail.

6.2 Floating Point in the C Language

The C language was designed to replace the assembly language for rewriting an operating system (UNIX). This explains why C is very close to the hardware. It also explains why it was not specifically designed for numeric and scientific computations, since OS internals mainly rely on integer arithmetic. As a consequence, important issues such as reproducibility of the results in floating-point arithmetic were initially not given all the necessary attention. Floating-point code could behave very differently from one platform to another, or even from one compiler to another on the same system. As time went by, C compilers were retrofitted with features that were common in languages like FORTRAN, which were designed from the start for numeric and scientific computing. It was only in the C99 standard and later that support for IEEE 754-1985 (mostly overlooked in the 1989/1990 C standard revision) was paid some attention.

The remainder of this section describes floating-point features of the C11 standard. Programmers should be aware that C11 compliance may not be the default for all compilers. However, there is almost always a compiler option to enable it.

Note that, at the time the C11 standard was prepared, IEEE 754-2008 (which the ISO standards refer to as IEC 60559:2011) had not yet been released. So C11 is only meant to support IEEE 754-1985 (referred to as IEC 60559:1989). Support for IEEE 754-2008 is planned for the C2X standard through the five parts of the ISO/IEC TS 18661 technical specification [279, 280, 281, 282, 283].

6.2.1 Standard C11 headers and IEEE 754-1985 support

Four headers, `<float.h>`, `<math.h>`, `<tgmath.h>`, and `<fenv.h>`, define the macros and functions that are necessary for dealing with floating-point numbers in C, to which we could add `<complex.h>` for complex numbers.

- The `<float.h>` header gathers the description of the characteristics of the floating-point environment. Indeed, the C standard requires very little.

  First, the radix for the standard floating-point types (`float`, `double`, and `long double`) is implementation defined, in other words not defined by the C standard. It can be obtained with the `FLT_RADIX` macro. In most cases, it is equal to 2. But particular platforms may choose another radix. For instance, radix 10 has been chosen by the TIGCC project for Texas Instruments calculators[3] (see Section 2.7 for a discussion on the choice of the radix). Interest in decimal floating-point arithmetic has increased even for desktop computers. Extending C to support decimal types is the purpose of TS 18661-2 [280].

  Other macros (such as `DBL_MAX`, `DBL_EPSILON`...) provide information on the range and precision of each standard floating-point type and on how rounding is performed.

- In `<math.h>`, one finds, apart from the expected mathematical functions (`sin`, `cos`...), most of the functions and predicates (`isnan`, `isunordered`...), that were recommended by the Appendix to the IEEE 754-1985/IEC 60559:1989 standard and that can be found nowadays in Section 5.7.2 of IEEE 754-2008. One also finds additional types and macros. Last but not least, one finds the `fma` and `fmaf` functions, which enable to access a hardware FMA when available.

- The `<tgmath.h>` header provides function-like type-generic macros. Depending on the type of their arguments, these macros call the corresponding function from `<math.h>` or `<complex.h>`. For instance, the `exp` macro calls the `expf` function when passed an argument of type `float`, while it calls the `cexpl` function when passed an argument of type `long double complex`. Arguments of integer type behave as if they were of type `double` for the purpose of selecting the target function.

[3]http://tigcc.ticalc.org/doc/float.html#FLT_RADIX.

- In <fenv.h>, one finds the necessary tools for manipulating the floating-point environment (e.g., changing the rounding mode, accessing the status flags...).

Most of the materials related to the optional support of IEEE 754 can be found in the normative Annex F of the C99 and C11 standards [273, 278]. Recent compilers (IBM's XL C/C++ 9.0 Linux PowerPC64, Sun's C 5.9 Linux i386, Intel's icc 10.1 Linux IA-64, for instance) and C libraries (since 1997, the GNU C library, a.k.a. glibc, for instance) define the `__STDC_IEC_559__` macro, which guarantees their conformance to Annex F, even though, in practice, the conformance is known to be incomplete. Moreover, Annex F allows the evaluation of operations in extended intermediate precision, as allowed by the IEEE 754-1985 standard, but in such a case, the IEEE 754-1985 standard mandates the implementation to allow the user to control the rounding precision, and the C11 standard does not provide a mean to do that.

6.2.2 Types

When `__STDC_IEC_559__` is defined, two of the basic binary formats (single-precision/binary32 and double-precision/binary64) are directly supported by the `float` and `double` types, respectively; but the operations may be rounded in a larger precision (see Section 6.2.3).

What `long double` means is a different story. In C11, whether `__STDC_IEC_559__` is defined or not, the `long double` type can be almost anything, provided that a `long double` has at least the precision and the range of a `double`. Note that these requirements are much weaker than those imposed on IEEE extended formats (see Table B.1, in Appendix B, for the IEEE 754-1985 requirements). The format of the `long double` type and the associated arithmetic depend on both the processor and the operating system. Sometimes it can also be changed by compiler options.

A program can obtain information on the arithmetic behind `long double` through the macros `LDBL_MANT_DIG`, `LDBL_MIN_EXP`, and `LDBL_MAX_EXP` (defined in <float.h>), which respectively provide the precision and the extremal exponents $e_{\min}$ and $e_{\max}$ (these 3 parameters are not well specified for formats like "double-double," as mentioned below); the `FLT_RADIX` macro, already mentioned, gives the radix β. Caution: The extremal exponents given by `LDBL_MIN_EXP` and `LDBL_MAX_EXP` do not correspond to our definition of the exponent (given in Section 2.1.1). There, we assumed significands of normal numbers in radix β to be between 1 and β, whereas these macros assume significands between $1/\beta$ and 1. Hence, having `LDBL_MAX_EXP` $= 1024$ corresponds, with our notation, to having $e_{\max} = 1023$.

For illustration, here are four arithmetics that have been found among various C implementations. The numbers in parentheses correspond to `LDBL_MANT_DIG`, `LDBL_MIN_EXP`, and `LDBL_MAX_EXP`, respectively.

Double precision (53 / −1021 / 1024): This arithmetic (the same as the one of the `double` type) has been found on ARM processors under Linux, and this is the choice originally made for the *ARM Developer Suite* [17, Section 3.3.2]. This choice could be surprising, as the floating-point accelerator (FPA) architecture (ARM's first floating-point implementation, now obsolete) supports extended precision [18, Section 2.9.2]. But ARM processors originally did not have floating-point hardware, some ARM processors still do not (thus floating-point arithmetic must be emulated in software), and the current floating-point architecture (VFP[4]) does not support extended precision [19, Section 2.6].

80-bit extended precision (64 / −16381 / 16384): This arithmetic has been found on x86, `x86_64`, and IA-64 architectures, because of hardware support.

Double-double arithmetic (106 / −968 / 1024): This arithmetic (also called double-word arithmetic) has been found on PowerPC, under both Darwin (Mac OS X) and Linux (with recent GCC/glibc), and comes from IBM's AIX operating system. A number is representable in this arithmetic if it can be written as the unevaluated sum of two double-precision/binary64 floating-point numbers, the first one being the sum rounded to nearest. That *very roughly* emulates a 106-bit precision (see Section 14.1). This is not a genuine floating-point arithmetic, but can almost be regarded as an extension of a floating-point arithmetic whose precision and exponent-range parameters are the numbers given in parentheses. This conforms to the C standard, which allows, in addition to normalized floating-point numbers, other kinds of floating-point data (such as the subnormal numbers, infinities, and NaNs of the IEEE 754 standard). The range of numbers that can be represented is roughly the same as the range of double-precision numbers. However, there are still open specification and/or implementation issues concerning the largest exponent[5] to regard this format as a valid `long double` type. The cause is that to be able to represent some numbers of exponent e, the first "double" component may need to have exponent $e+1$, but this is not possible when e is the maximum exponent of the `double` type. Moreover, some properties requiring strict floating-point arithmetic (such as Sterbenz's lemma: Lemma 4.1 in Chapter 4) will not always be true for the `long double` type, and corresponding floating-point algorithms may no longer work.

Quadruple precision (113 / −16381 / 16384): This arithmetic, now called binary128, has been found under HP-UX (HPPA and IA-64) and Solaris

[4]VFP stands for "Vector Floating-Point" but the vector mode was removed shortly after its introduction.

[5]See https://gcc.gnu.org/bugzilla/show_bug.cgi?id=61399 for instance.

(SPARC only [578]), and is implemented in software.[6] It is available in hardware with the IBM Power9 processor.

The arithmetic does not completely define the type. Indeed the encoding may depend on the platform (e.g., due to a different endianness, see Section 3.1.1.5). Even the type size (as per the `sizeof` operator) can vary, for alignment reasons (to provide better memory access speed). For instance, with GCC, the default width of a `long double` is 12 bytes long for the x86 (32-bit) application binary interface (ABI) and rises to 16 bytes long for the `x86_64` (64-bit) ABI. However, compiler options can alter this behavior. For instance, the `-m96bit-long-double` and `-m128bit-long-double` switches control the storage size of `long double` values. But note that this has no effect on the floating-point results: range and precision remain the same.

6.2.2.1 Infinities, NaNs, and signed zeros

Support for (signed or unsigned) infinities and signed zeros is optional. The `INFINITY` macro from `<math.h>` represents a positive or unsigned infinity, when available. However, a `HUGE_VAL` (for the `double` type) macro is always available, which typically is an infinity when supported (this is allowed by the C11 standard and required when its Annex F is in effect).

Support of Not a Number (NaN, also optional) is limited to the quiet flavor (qNaN). Signaling NaNs (sNaN) were not included in the C99 and C11 standards since its authors felt it entailed a lot of trouble for a limited usefulness, and that qNaNs were sufficient for the closure of the floating-point arithmetic algebraic structure.

6.2.3 Expression evaluation

Except for assignment and cast, the C11 standard states [278, Section 5.2.4.2.2]:

> the values yielded by operators with floating operands and values subject to the usual arithmetic conversions and of floating constants are evaluated to a format whose range and precision may be greater than required by the type.

While this will be a bonus in many situations, it may also break the assumptions underlying some algorithms presented in this book. It may also lead to the "double rounding" problem depicted in Section 3.2, which can occur even with a single operation, such as `a = b + c`. Splitting an expression by using temporary variables and only one operation per statement will force any intermediate result to be stored in the required precision; this workaround does

[6]The SPARC architecture has instructions for quadruple precision, but in practice, current processors generate traps to compute the results in software.

not avoid the "double rounding" problem, but one gets a faithful rounding (see Section 2.2), which may be sufficient for some algorithms.

The C11 standard provides a macro, FLT_EVAL_METHOD, whose value gives a clue about what is actually going on. Table 6.1 shows which intermediate type (both range and precision) is used for the evaluation of an operation of a given type.

FLT_EVAL_METHOD	float	double	long double
-1	"indeterminable" (inconsistent)		
0	float	double	long double
1	double	double	long double
2	long double	long double	long double

Table 6.1: FLT_EVAL_METHOD macro values.

FLT_EVAL_METHOD = 2 is typically used with x87 arithmetic (when the processor is configured to round in extended precision). Processors with static rounding format (range and precision) will generally use FLT_EVAL_METHOD = 0. In addition to these values, FLT_EVAL_METHOD = -1 can be used when the evaluation method is not determined (e.g., because of some optimizations that can change the results). Other negative values are reserved for implementation-defined methods.

Unfortunately, FLT_EVAL_METHOD is an information macro only. There is no standard way of changing the behavior it indicates. This does not mean that it cannot be changed, but it may require compiler switches (Section 3.2 contained some examples) or nonstandard features (e.g., operating system calls).

The header <math.h> defines two floating-point types float_t and double_t, which correspond to the intermediate types used for the evaluation of an operation on float and double values respectively, when FLT_EVAL_METHOD is 0, 1 or 2, as given by Table 6.1. For instance, when FLT_EVAL_METHOD = 2, they are both long double. These types can be used for assignment instead of the usual float and double types in order to avoid double rounding.

6.2.3.1 Operators and functions

The +, -, *, / operators and the sqrt, remainder, and fma functions provide the basic operations as expected in IEEE 754-2008. More functions and macros cover conversions, comparison, and floating-point environment manipulation as well as the set of recommended functions.

Note that the fma function can be used even if the FMA instruction is not directly supported by the processor. For instance, Listing 6.1 shows the implementation in C of Algorithm 4.8. This program can be compiled using `gcc` (tested with version 6.3) or `clang` (tested with version 4.0). The generated code can contain either

- the FMA instruction, such as `vfmsub132sd` on an Intel Core i7. Some compiler option may be needed to enable such an instruction, for instance `-mfma` or `-march=native` on x86;
- or a call to the libm implementation of the `fma` function, in which case the option `-lm` may be needed.

C listing 6.1 Expressing the 2MultFMA algorithm in C

```c
#include <stdio.h>
#include <stdlib.h>
#include <math.h>
#pragma STDC FP_CONTRACT OFF
int main (int argc, char **argv) {
  double a, b, p, r;
  a = atof(argv[1]);
  b = atof(argv[2]);

  /* A 2MultFMA */
  p = a*b;
  r = fma(a,b,-p);
  printf ("With explicit FMA: %.20g * %.20g = %.20g + %.20g\n", a,b,p,r);

  /* Contraction disabled by STDC FP_CONTRACT OFF */
  p = a*b;
  r = p-a*b;
  printf ("With r=p-a*b      : %.20g * %.20g = %.20g + %.20g\n", a,b,p,r);
}
```

Here is an example of execution of that program where we compute the square of the integer 123456789 (which happens to require exactly 54 bits):

```
$ ./a.out 123456789 123456789
With explicit FMA: 123456789 * 123456789  =  15241578750190520 + 1
With r=p-a*b     : 123456789 * 123456789  =  15241578750190520 + 0
```

Beware that some versions of the mathematical library that provides the `fma` function, including older versions of the GNU library used here, do not comply with the requirements of the C11 and IEEE 754-2008 standards: they typically implement it as a multiplication followed by an addition. Your mileage may vary: the program above may allow you to check this.

6.2.3.2 Contracted expressions

As stated by the C11 standard [278, Section 6.5], "a floating expression may be contracted, that is, evaluated as though it were a single operation, thereby omitting rounding errors implied by the source code and the expression evaluation method." This was meant to allow the use of mixed-format operations (with a single rounding, when supported by the processor) and hardware compound operators such as FMA. For instance, the default behavior of the main compilers (GCC, IBM's XL C/C++ 9.0 Linux PowerPC64 compiler, Intel's compiler icc, Microsoft Visual C/C++ compiler) is to contract `x * y + z` to `fma(x,y,z)` when a hardware FMA is available.

Most of the time, this is beneficial in terms of both performance and accuracy. However, it will break algorithms that rely on evaluation as prescribed by the code. For instance, `sqrt(a * a - b * b)` may be contracted as `sqrt(fma(a, a, - b * b))`, and if `a` and `b` are equal, the result can be nonzero because of the broken symmetry (see example below).

The `FP_CONTRACT` pragma (from `<math.h>`) gives some control on this issue to the programmer. When set to *on*, it allows contracting expressions. When set to *off*, it prevents it. As contracting expressions is potentially dangerous, a C implementation (compiler and associated libraries) must document the default state and the way in which expressions are contracted. Note that GCC does not currently support this pragma, so that so far, it assumes that this pragma is *off* when `-std=c99` or `-std=c11` is provided, in order to comply with these standards.[7]

Following the same example, Listing 6.2 computes the result of the expression `a >= b ? sqrt (a * a - b * b) : 0` in the particular case `a = b`. By default, contraction is disabled, but compiling this program with `-DFP_CONTRACT` sets the pragma to *on*, thus enabling contraction. For instance, icc 10.1 on IA-64 gives on the inputs 1, 1.1, and 1.2:

```
Test of a >= b ? sqrt (a * a - b * b) : 0 with FP_CONTRACT OFF
test(1) = 0
test(1.10000000000000000888) = 0
test(1.19999999999999999556) = 0

Test of a >= b ? sqrt (a * a - b * b) : 0 with FP_CONTRACT ON
test(1) = 0
test(1.10000000000000000888) = 2.98023223876953265625e-09
test(1.19999999999999999556) = nan
```

[7]Details on https://gcc.gnu.org/bugzilla/show_bug.cgi?id=37845, fixed for GCC 4.9.

C listing 6.2 Testing the effect of the contraction of a floating-point expression to FMA.

```
#include <stdio.h>
#include <stdlib.h>
#include <math.h>

#ifdef FP_CONTRACT
#undef FP_CONTRACT
#define FP_CONTRACT "ON"
#pragma STDC FP_CONTRACT ON
#else
#define FP_CONTRACT "OFF"
#pragma STDC FP_CONTRACT OFF
#endif

static double fct (double a, double b)
{
  return a >= b ? sqrt (a * a - b * b) : 0;
}

/* "volatile" and "+ 0.0" may be needed to avoid optimizations. */
static void test (volatile double x)
{
  printf ("test(%.20g) = %.20g\n", x, fct (x, x + 0.0));
}

int main (int argc, char **argv)
{
  int i;

  printf ("Test of a >= b ? sqrt (a * a - b * b) : 0 with FP_CONTRACT "
          FP_CONTRACT "\n");
  for (i = 1; i < argc; i++)
    test (atof (argv[i]));
  return 0;
}
```

6.2.3.3 Constant expressions, initialization, and exceptions

There are some issues specific to C regarding the relationship between translation (compilation) time and execution time, on the one hand, and exceptions, on the other hand. As stated by the C11 standard [278, Section 7.6.1],

> the FENV_ACCESS pragma provides a means to inform the implementation when a program might access the floating-point environment to test floating-point status flags or run under non-default floating-point control modes.

When the state of this pragma is *off*, the compiler is allowed to do some optimizations on code that would otherwise have side effects on the floating-point environment, such as constant folding, common subexpression elimination, code motion, and so on. Otherwise, if the state is *on*, such potential side effects impose an evaluation at execution time and a temporal ordering of floating-point expressions (unless the compiler can deduce that an optimization will not change the result, including exceptions). However, this does not affect initialization of objects with *static storage duration*, necessarily using constant expressions, which are evaluated at translation time and do not raise exceptions.

6.2.3.4 Special values of mathematical functions

In its conditionally normative Annex F, the C standard also specifies the values of the elementary functions for particular arguments. These definitions, which came from C99 (thus before IEEE 754-2008 came out with its own recommendations) and are explained in the C99 rationale [275], can cause some head scratching. Consider for instance, the following requirement for the `pow` function: $\texttt{pow}(-1, \pm\infty) = 1$. This may be a bit puzzling until one understands that any sufficiently large binary floating-point number is an even integer: hence, "by taking the limit," the prescribed value of 1. The result of $\texttt{pow}(x, \pm 0)$ is 1 for any x, including 0, because this result was found to be more useful in practice than a NaN.[8] Moreover, when a partial function is constant, such as $\texttt{pow}(x, \pm 0)$, it was chosen to return this constant even when the variable is a quiet NaN (let us recall that C11 does not specify the behavior of signaling NaNs): hence, $\texttt{pow}(\texttt{NAN}, \pm 0) = 1$.

Concerning the special cases, IEEE 754-2008 chose to remain compatible with C99, but added three variants of `pow`: powr (whose special cases are derived by considering that it is defined as $e^{y \log x}$) and pown/rootn (which deal with integral exponents only).

6.2.4 Code transformations

Many common transformations of a code into some *naively* equivalent code become impossible if one takes into account special values such as NaNs, signed zeros, and rounding modes. For instance, the expression `x + 0` is not equivalent to the expression `x` if `x` is `-0` and the rounding is to nearest.

Similarly, some transformations of code involving relational operators become impossible due to the possibility of unordered values (see Section 3.1.4). This is illustrated in Listing 6.3.

[8]The C11 standard still does not specify a power function with an integer type as the second argument.

C listing 6.3 Strangely behaving relational operators (excerpt from Annex F of the C11 standard). These two pieces of code may seem equivalent, but behave differently if a and b are unordered.

```
// calls g and raises "invalid" if a and b are unordered
if (a < b)
  f();
else
  g();

// calls f and raises "invalid" if a and b are unordered
if (a >= b)
  g();
else
  f();
```

As strict support for signaling NaNs is not mandated by C11 [278, Section F.9.2], C implementations that support them must do so with special care. For instance, the expressions `1 * x` and `x` are considered equivalent in C11, while they behave differently when `x` is an sNaN.

6.2.5 Enabling unsafe optimizations

6.2.5.1 Complex arithmetic in C11

The C11 standard defines another pragma allowing a more efficient implementation of some operations for multiplication, division, and absolute value of complex numbers, for which the usual, straightforward formulas can give incorrect results on infinities or spurious exceptions (overflow or underflow) on huge or tiny inputs (see Chapter 11). Programmers can set the `CX_LIMITED_RANGE` pragma to *on* if they guarantee that the following straightforward definitions are safe (e.g., do not cause overflow nor underflow with the expected input):

- $(x+iy) \times (u+iv) = (xu - yv) + i(yu + xv)$,
- $(x+iy)/(u+iv) = ((xu + yv) + i(yu - xv))/(u^2 + v^2)$,
- $|x+iy| = \sqrt{x^2 + y^2}$.

In this case the compiler can choose to use them instead of a slower code that would work on (almost) any input. The default state of this pragma is *off*, for safer computations.

6.2.5.2 Range reduction for trigonometric functions

For ARM processors using the ARM Developer Suite or the RealView tools, the default trigonometric range reduction is inaccurate for very large arguments. This is valid for most programs: if a floating-point number is so large that the value of its ulp is several times the period 2π, it usually makes little sense to compute its sine accurately. Conversely, if the input to the sine is bound by construction to reasonably small values, the default range reduction is perfectly accurate. The situation is comparable to using the previous quick and unsafe complex operators: they are perfectly safe if the values that may appear in the program are under control. The big difference, however, is that here the default behavior is the unsafe one.

The accuracy of range reduction can be improved by the following pragma [17, Section 5.4]:

```
#pragma import (__use_accurate_range_reduction)
```

The more accurate range reduction is slower and requires more memory (this will be explained in Section 10.2). The ARM mainly focuses on embedded applications such as mobile devices, which are memory-limited.

6.2.5.3 Compiler-specific optimizations

Compilers can have their own directives to provide unsafe optimizations which may be acceptable for most common codes, e.g., assuming that no exceptions or special values occur, that mathematically associative operations can be regarded as associative in floating-point arithmetic, and so on. This is the case of GCC's generic `-ffast-math` option (and other individual options enabled by this one). *Users should use such options with much care.* In particular, using them on a code they have not written themselves is highly discouraged.

6.2.6 Summary: a few horror stories

As pointed out by David Goldberg [214] (in his edited reprint), all these uncertainties make it impossible, in many cases, to figure out the exact semantics of a floating-point C program just by reading its code. As a consequence, portability is limited, and moving a program from one platform to another may involve some rewriting. This also makes the automatic verification of floating-point computations very challenging, as explained by Monniaux [423].

The following section describes a few traps that await the innocent programmer.

6.2.6.1 Printing out a variable changes its value

Even the crudest (and most common) debugging mode of all, printing out data, can be a trap. Consider the program given in Listing 6.4. With GCC

C listing 6.4 Strange behavior caused by spilling data to memory.

```
long double lda = 9007199254740991.0; // 2^53 - 1
double da = 9007199254740991.0;       // ditto
if (lda + 0.5 < 9007199254740992.0)
{
  printf("%.70Lg is strictly less than %.70Lg\n",
         lda + 0.5,
         (long double) 9007199254740992.0);
}
if (da + 0.5 < 9007199254740992.0)
{
  printf("%.70g is strictly less than %.70g\n",
         da + 0.5,
         9007199254740992.0);
}
```

version 7.2.0 on a Linux/x86 platform in 32-bit mode and the default settings, the program will display:

```
9007199254740991.5 is strictly less than 9007199254740992
9007199254740992 is strictly less than 9007199254740992
```

While there is nothing wrong with the first line, the second is a bit more disturbing. For the former we have used the `long double` type, which, on this platform, maps to 80-bit x87 registers. These have enough room to store all the bits of the sum. To no one's surprise, the first test evaluates to `true`. There is also enough room to store all the bits when, to call the `printf()` function, the register holding the sum is spilled out to memory (remember a `long double` is 12 bytes long on this platform). The printed message reflects what happens in the registers.

For the second line, while we are supposed to work with 64 bits, the addition and the test for inequality are also executed in the 80-bit x87 registers. The test evaluates again to `true` since, at register level, we are in the exact same situation. What changes is that, when the sum is spilled to memory, it is rounded to its "actual" 64-bit size. Using the rounding-to-nearest mode and applying the "even rule" to break ties leads us to the 9007199254740992 value, which is eventually printed out. By the way, it has nothing to do (as a suspicious reader might wonder) with the formats used in the `printf()` function. One may be convinced by trying the following:

- on the same platform, add to the command line the flags that require the use of 64-bit SSE registers instead of the 80-bit x87 ones (`-march=pentium4` and `-mfpmath=sse`);
- on the 64-bit corresponding platform, run GCC with the default settings (which are to use the SSE registers and not the x87).

The second line never prints out since the rounding of the sum takes place, this time, in the 64-bit registers *before* the comparison is executed.

6.2.6.2 A possible infinite loop in a sort function

It is difficult to implement a sorting algorithm without the basic hypothesis that if, at some point, two elements have compared as $a < b$, then in the future they will also compare as $b > a$. If this assumption is violated, the programmer of the sorting algorithm cannot be held responsible if the program goes into an infinite loop.

Now let us write a function that compares two `my_type` structures according to their radius.

C listing 6.5 A radius comparison function.

```
int compare_radius (const my_type *a, const my_type *b)
{
  double temp = (a->x*a->x + a->y*a->y) - (b->x*b->x + b->y*b->y);
  if (temp > 0)
    return 1;
  else if (temp < 0)
    return -1;
  else
    return 0;
}
```

We see at least two ways things can go wrong in Listing 6.5.

- If `temp` is computed using an FMA, there are two different ways of computing each side of the subtraction, as we have seen in Section 6.2.3.2.
- Some of the intermediate results in the computation of `temp` may be computed to a wider precision, especially when using an x87 FPU.

In both cases, the net result is that the `compare_radius` function, although written in a symmetrical way, may be compiled into asymmetrical code: it may happen that `compare_radius(a,b)` returns 0 while `compare_radius(b,a)` returns 1, for instance. This is more than enough to break a sorting algorithm.

Getting to the root of such bugs is very difficult for several reasons. First, it will happen extremely rarely. Second, as we have just seen, entering debug mode or adding `printf`s is likely to make the bug vanish. Third, it takes a deep understanding of floating-point issues to catch (and fix) such a bug. We hope that our reader now masters this knowledge.

6.2.7 The CompCert C compiler

The CompCert C compiler, designed by Leroy [383] and, as of 2017, still actively developed, offers a way to avoid some of the troublesome issues related to the use of floating-point arithmetic in C programs. The peculiarity of this compiler is that it comes with a formal semantics of the C language[9] and with a formal proof that, if the source program is free of undefined behaviors, the observable behavior of the compiled program is one of the observable behaviors of the source program. In particular, one does not need to know how the compiler is implemented (e.g., the optimization passes it performs) in order to know how the compiled program will behave.

This *semantics preservation* property would be pointless if the source semantics was so lax that one could not say anything interesting about the behavior of the source program, and thus of the compiled program either. That is why the CompCert C semantics is much more deterministic than the one from the C standard. For instance, Boldo et al. [51] have expressed the behavior of any arithmetic operator on floating-point inputs using the Flocq formalization of the IEEE standard. More generally, the following floating-point properties can be inferred from the semantics and the correctness theorem of the compiler.

- The `float` and `double` types are mapped to the binary32 and binary64 formats, respectively. Extended-precision numbers are not supported: the `long double` type is either unsupported or mapped to binary64, depending on a compiler option.
- Conversions to a floating-point type, either explicit ("type casts") or implicit (at assignment, parameter passing, and function return), always round the floating-point value to the given type, discarding excess precision.
- Reassociation of floating-point operations, or "contraction" of several operations into one (e.g., a multiplication and an addition being contracted into a fused multiply-add) are prohibited. On target platforms that support FMA instructions, CompCert makes them available as compiler built-in functions, so they must be explicitly used by the programmer.
- All intermediate floating-point results in expressions are computed with the precision that corresponds to their static C types, that is, the greater of the precisions of their arguments.

[9]Note, however, that this formalization just comes from an interpretation of the ISO C standard, which is written in plain English. Before version 2.5, CompCert behaved differently from other compilers by not rejecting a lexically invalid program.

- All the computations round to nearest, ties-to-even, except conversions from floating-point numbers to integers, which round toward zero. The CompCert formal semantics make no provisions for programmers to change rounding modes at run-time.
- Floating-point literal constants are also rounded to nearest, ties-to-even.

Note that Section 4.10 presents some of the code sequences generated by CompCert when converting floating-point numbers from and to integers.

6.3 Floating-Point Arithmetic in the C++ Language

6.3.1 Semantics

The semantics of the C++ language is similar to that of the C language with respect to floating-point arithmetic. The parenthesizing order is the same and, as in C, intermediate expressions may use a wider format since as per the C99 standard [273, section 6.3.1.8]:

> The values of the floating operands and the results of floating expressions may be represented in greater precision and range than that required by the type; the types are not changed thereby.

The current C++14 standard explicitly mentions in different occasions (e.g., at 26.3.1 "The floating-point environment") its elder C99 sibling standard as its model for most of the floating-point operations.

6.3.2 Numeric limits

In addition to the macros inherited from C, the C++ standard library provides the template class `std::numeric_limits` in the `<limits>` header file to allow metaprogramming depending on the floating-point capabilities. For a floating-point type `T`, the class `std::numeric_limits<T>` provides the following static members.

1. Format:
   - `int radix`: the radix β, either 2 or 10 usually,
   - `bool is_iec559`: *true* if `T` is an IEEE format and the operations on `T` are compliant.[10]
2. Special values:
   - `bool has_infinity`: *true* if `T` has a representation for $+\infty$,

[10]When `is_iec559` is *true*, the C++ standard mandates that infinities and NaNs are available.

- `bool has_quiet_NaN`: *true* if `T` has a representation for a qNaN,
- `bool has_signaling_NaN`: *true* if `T` has a representation for an sNaN,
- `T infinity()`: representation of $+\infty$,
- `T quiet_NaN()`: representation of a qNaN,
- `T signaling_NaN()`: representation of an sNaN.

3. Range:

- `T min()`: smallest positive normal number,
- `T max()`: largest finite number,
- `T lowest()`: negated `max()`,
- `int min_exponent`: smallest integer k such that β^{k-1} is a normal number, e.g., -125 for binary32,
- `int max_exponent`: largest integer k such that β^{k-1} is a normal number, e.g., 128 for binary32,
- `int min_exponent10`: smallest integer k such that 10^k is a normal number, e.g., -37 for binary32,
- `int max_exponent10`: largest integer k such that 10^k is a normal number, e.g., 38 for binary32.

4. Subnormal numbers:

- `float_denorm_style has_denorm`: `denorm_present`, `denorm_absent`, or `denorm_indeterminate`, depending on whether subnormal numbers are known to be supported or not,
- `bool has_denorm_loss`: *true* if inaccurate subnormal results are detected with an "underflow" exception (Section B.1.4) instead of just an "inexact" exception,
- `T denorm_min()`: smallest positive subnormal number,
- `bool tinyness_before`: *true* if subnormal results are detected before rounding (see Definition 2.1 of Chapter 2).

5. Rounding mode and error:

- `T epsilon()`: the subtraction $1^+ - 1$ with 1^+ the successor of 1 in the format `T`,
- `T round_error()`: biggest relative error for normalized results of the four basic arithmetic operations, with respect to `epsilon()`, hence 0.5 when rounding to nearest,

- `float_round_style round_style`: `round_toward_zero`, `round_to_nearest`, `round_toward_infinity`, `round_toward_neg_infinity`, or `round_indeterminate`, depending on whether the rounding mode is known or not.[11]

6.3.3 Overloaded functions

In C++, functions from the `<cmath>` header have been overloaded to take argument types into account (`float` and `long double`). For instance, while `<math.h>` provides a `float sinf(float)` function for computing sine on `float`, `<cmath>` provides `float sin(float)` in the `std` namespace. Some might think that this refinement boils down to syntactic sugar but it does not. In particular, one should acknowledge that the following piece of code does not have the same semantics in C and in C++ (assuming the `std` namespace is in scope for C++ and `<tgmath.h>` is not included for C):

```
float a = 1.0f;
double b = sin(a);
```

In C, the variable `a` will first be promoted to `double`. The double-precision sine will then be called and the double-precision result will be stored in `b`. In C++, the single-precision sine will be called and its single-precision result will then be promoted to `double` and stored in `b`. Of course, the first approach provides a more accurate result.

The C++14 standard also provides utility functions that replace the C99 macros for classifying or ordering floating-point values:

```
namespace std {
  template <class T> bool signbit(T x);
  template <class T> int fpclassify(T x);
  template <class T> bool isfinite(T x);
  template <class T> bool isinf(T x);
  template <class T> bool isnan(T x);
  template <class T> bool isnormal(T x);

  template <class T> bool isgreater(T x, T y);
  template <class T> bool isgreaterequal(T x, T y);
  template <class T> bool isless(T x, T y);
  template <class T> bool islessequal(T x, T y);
  template <class T> bool islessgreater(T x, T y);
  template <class T> bool isunordered(T x, T y);
}
```

[11] `round_style` is a constant member. Therefore, it may well be set to the default `round_to_nearest` style, even if the architecture allows us to change the rounding direction on the fly.

6.4 FORTRAN Floating Point in a Nutshell

6.4.1 Philosophy

FORTRAN was initially designed as a FORmula TRANslator, and this explains most of its philosophy with respect to floating-point arithmetic. In principle, a FORTRAN floating-point program describes the implementation of a mathematical formula, written by a mathematician, an engineer, or a physicist, and involving real numbers instead of floating-point numbers. This is illustrated by the fact that a floating-point variable is declared with the `REAL` keyword or one of its variants. Compare this with the C `float` keyword which describes a machine implementation of the reals, following the C "close-to-the-metal" philosophy. FORTRAN also draws a clear line between integers and reals, and acknowledges them as fundamentally different mathematical objects.

Therefore, in the compilation of a FORTRAN program, a compiler is free to apply to the source code (formula) any mathematical identity that is valid over the reals, as long as it results in a mathematically equivalent formula. However, it gives little importance to the rounding errors involved in this process. They are probably considered unavoidable, and small anyway.

Here is a biased excerpt of the FORTRAN standard [276] that illustrates this.

> (...) the processor may evaluate any mathematically equivalent expression (...). Two expressions of a numeric type are mathematically equivalent if, for all possible values of their primaries, their mathematical values are equal. However, mathematically equivalent expressions of numeric type may produce different computational results.

Again, integers and reals are distinct objects, as illustrated by the following excerpt:

> Any difference between the values of the expressions `(1./3.)*3.` and `1.` is a computational difference, not a mathematical difference. The difference between the values of the expressions `5/2` and `5./2.` is a mathematical difference, not a computational difference.

Therefore, `(1./3.)*3.` may be quietly replaced by `1.`, but `5/2` and `5./2.` are not interchangeable.

However, the standard acknowledges that a programmer may sometimes want to impose a certain order to the evaluation of a formula. It therefore makes a considerable exception to the above philosophy: when parentheses are given, the compiler should respect them. Indeed, the first sentence of the first excerpt reads in full:

> (...) the processor may evaluate any mathematically equivalent expression, provided that the integrity of parentheses is not violated.

Then, another excerpt elaborates on this:

> In addition to the parentheses required to establish the desired interpretation, parentheses may be included to restrict the alternative forms that may be used by the processor in the actual evaluation of the expression. This is useful for controlling the magnitude and accuracy of intermediate values developed during the evaluation of an expression.

For example, an expression written

$$a/b * c/d$$

may be computed either as

$$(a/b) * (c/d), \tag{6.1}$$

or as

$$(a * c)/(b * d). \tag{6.2}$$

A FORTRAN compiler may choose the parenthesizing it deems the more efficient. These two expressions are mathematically equivalent but do not lead to the same succession of rounding errors, and therefore the results may differ. For instance, if a, b, c, and d are all equal and strictly larger than the square root of the largest representable finite floating-point number Ω, then choosing Equation (6.1) leads to the right result 1, whereas choosing Equation (6.2) leads to the quotient of two infinities, which gives a NaN in an arithmetic compliant with the IEEE 754-2008 standard. During the development of the LHC@Home project [152], such an expression, appearing identically in two points of a program, was compiled differently and—very rarely—gave slightly different results. This led to inconsistencies in the distributed simulation, and was solved by adding explicit parentheses on the offending expression.

Here are some more illustrations of the FORTRAN philosophy, also from the FORTRAN standard [276]. In Tables 6.2 and 6.3, X, Y, Z are of arbitrary numeric types, A, B, C are reals or complex, and I, J are of integer type.

Note how the compiler is free to replace X*Y-X*Z with X*(Y-Z), but not the other way around. This is a strange consequence of Fortran's respecting the parentheses.

The example of the last line of Table 6.2 could be turned into a similar example by replacing the "5.0" by "4.0" and the "0.2" by "0.25." However, it is not possible to design a similar example by replacing the "5.0" by "3.0" because no finite sequence "$0.3333\cdots3$" is exactly equal to $1/3$: FORTRAN only accepts replacing a formula by another formula that is mathematically equivalent.

Expression	*Allowable* alternative
X+Y	Y+X
X*Y	Y*X
-X + Y	Y-X
X+Y+Z	X + (Y + Z)
X-Y+Z	X - (Y - Z)
X*A/Z	X * (A / Z)
X*Y - X*Z	X * (Y - Z)
A/B/C	A / (B * C)
A / 5.0	0.2 * A

Table 6.2: FORTRAN allowable alternatives.

Expression	*Forbidden* alternative	Why
I/2	0.5 * I	integer versus real operation
X*I/J	X * (I / J)	real versus integer division
I/J/A	I / (J * A)	integer versus real division
(X + Y) + Z	X + (Y + Z)	explicit parentheses
(X * Y) - (X * Z)	X * (Y - Z)	explicit parentheses
X * (Y - Z)	X*Y-X*Z	explicit parentheses

Table 6.3: FORTRAN forbidden alternatives.

To summarize, FORTRAN has much more freedom when compiling floating-point expressions than C. As a consequence, the performance of a FORTRAN program is likely to be higher than that of the same program written in C.

6.4.2 IEEE 754 support in FORTRAN

Section 13 of the FORTRAN standard [276], *Intrinsic procedures and modules*, defines a machine model of the real numbers that corresponds to normal floating-point numbers:

The model set for real x is defined by

$$x = \begin{cases} 0 \quad \text{or} \\ s \times b^e \times \sum_{k=1}^{p} f_k \times b^{-k} \end{cases},$$

> where b and p are integers exceeding one; each f_k is a nonnegative integer less than b, with f_1 nonzero (i.e., there are no subnormal numbers); s is $+1$ or -1; and e is an integer that lies between some integer maximum $e_{\max}$ and some integer minimum $e_{\min}$ inclusively. For $x = 0$, its exponent e and digits f_k are defined to be zero. (...) the integer parameters b, p, $e_{\min}$, and $e_{\max}$ determine the set of model floating-point numbers. The parameters of the integer and real models are available for each integer and real type implemented by the processor. The parameters characterize the set of available numbers in the definition of the model. The floating-point manipulation functions (13.5.10) and numeric inquiry functions (13.5.6) provide values of some parameters and other values related to the models.

Numeric inquiry functions are `DIGITS (X)`, `EPSILON (X)`, `HUGE (X)`, `MAXEXPONENT (X)`, `MINEXPONENT (X)`, `PRECISION (X)`, `RADIX (X)`, `RANGE (X)`, `TINY (X)`. Most need no further explanation, but be aware that the significand in the previous model is in $[1/b, 1)$; therefore, $e_{\min}$ and $e_{\max}$ differ from those of the IEEE 754 standard. Some of these functions have strange definitions in 13.5.10: `EPSILON (X)` is defined as *Number that is almost negligible compared to one*, which is not very clear. It becomes clearer in 13.7, *Specifications of the standard intrinsic procedures*, which defines it as b^{1-p} with the notation of the model above. `HUGE (X)` and `TINY (X)` are defined, respectively, as *Largest number of the model* and *Smallest positive number of the model*. For these functions, remember that the model includes neither infinities nor subnormals. Indeed, `TINY (X)` is defined later in 13.7 of the FORTRAN standard as $b^{e_{\min}-1}$.

The following floating-point manipulation functions are available:

- `EXPONENT (X)` Exponent part of a model number;
- `FRACTION (X)` Fractional part of a number;
- `NEAREST (X, S)` Nearest different processor number in the direction given by the sign of S;
- `SPACING (X)` Absolute spacing of model numbers near a given number. This underspecified definition of the ulp is clarified in 3.7 as $b^{\max(e-p, e_{\min}-1)}$;
- `RRSPACING (X)` Reciprocal of the relative spacing of model numbers near a given number;
- `SCALE (X, I)` Multiply a real by its radix to an integer power;
- `SET EXPONENT (X, I)` Set exponent part of a number.

All the previous information is mostly unrelated to the IEEE 754 standard. In addition, the FORTRAN standard dedicates its Section 14 to *Exceptions and IEEE arithmetic*. This section standardizes IEEE 754 support when it is provided, but does not make it mandatory. It provides, in the intrinsic modules `IEEE EXCEPTIONS`, `IEEE ARITHMETIC`, and `IEEE FEATURES`, numerous inquiry functions testing various parts of standard compliance. It also defines read and write access functions to the rounding directions, as well as read and write access functions to the underflow mode (which may be either gradual, i.e., supporting subnormals, or abrupt, i.e., without subnormals). Finally, it defines subroutines for all the functions present in the IEEE 754-1985 standard.

6.5 Java Floating Point in a Nutshell

6.5.1 Philosophy

"Write once, run anywhere (or everywhere)" is the mantra of the Java language evangelists. Reproducibility between platforms is an explicit and essential goal, and this holds for numeric computations as well. In practical terms, this is achieved through *byte-code* compilation (instead of compilation to object code) and interpretation on a *virtual machine* rather than direct execution on the native operating system/hardware combination. In the first versions of the Java platform, this meant poor performance, but techniques such as "Just In Time" or "Ahead Of Time" compilation were later developed to bridge the gap with native execution speed.

The initial language design tried to ensure numerical reproducibility by clearly defining execution semantics, while restricting floating-point capabilities to a subset of formats and operators supported by most processors. Unfortunately, this was not enough to ensure perfect reproducibility, but enough to frustrate performance-aware programmers who could not exploit their possibly available extended precision or FMA hardware [322]. We will see how Java later evolved to try and give to the programmers the choice between reproducibility and performance.

In general terms, Java claims compliance with the IEEE 754-1985 standard, but only implements a subset of it. Let us now look at the details.

6.5.2 Types and classes

Java is an object-oriented language "with a twist": the existence of primitive types and the corresponding wrapper classes. One of the main reasons behind the existence of basic types is a performance concern. Having many small and simple variables incurs a severe access time penalty if created on the heap (as objects are).

As far as floating-point numbers are concerned, there are two basic types:

- `float`: binary32 analogous, `Float` being the wrapper class;
- `double`: binary64 analogous, `Double` being the wrapper class.

As the Java Language Specification puts it, these types are "conceptually associated with the 32-bit single-precision and 64-bit double-precision formats of the IEEE 754 standard."

6.5.2.1 In the virtual machine

Although the virtual machine is supposed to insulate the execution from the peculiarities of the hardware, at some point, floating-point operations have to be performed on the actual processor. Of course, specialized floating-point units could be totally avoided and all instructions could be software-emulated using integer registers, but the speed penalty would be prohibitive.

Looking at the details of the Java Virtual Machine Specification, second edition [390, Section 2.3.2], one may observe that the Java designers have been forced to acknowledge the peculiarity of the x87 hardware, which can be set to round to a 53-bit or 24-bit significand, but always with the 16-bit exponent of the double-extended format. The Java Virtual Machine Specification defines, in addition to the *float* and *double* formats, a *double-extended-exponent* format that exactly corresponds to what is obtained when one sets an x87 FPU to rounding to 53 bits. Unfortunately, from there on, reproducibility vanishes, as the following excerpt illustrates:

> These extended-exponent value sets may, under certain circumstances, be used instead of the standard value sets to represent the values of type float or double.

The first edition of the Java Virtual Machine Specification was much cleaner, with only float and double types, and fewer "may" sentences threatening reproducibility. The only problem was that it could not be implemented efficiently on x86 hardware (or, equivalently, efficient implementations were not strictly compliant with the specification).

Neither the first edition nor the second is fully satisfactory. To be honest, this is not to be blamed on the Java designers, but on the x87 FPU. The fact that it could not round to the standard double-precision format is one of the main flaws of an otherwise brilliant design. It will be interesting to see if Java reverts to the initial specification, now that x86-compatible processors, with SSE2, are turning the x87 page and offering high-performance hardware with straightforward rounding to binary32 and binary64.

6.5.2.2 In the Java language

In Java 2 SDK 1.2, a new keyword (and the corresponding behavior) was introduced to make sure computations were realized as if binary32 and binary64 were actually used all the way (again, at some performance cost). See Program 6.1. Technically, this `strictfp` modifier is translated into a bit associated with each method, down to the virtual machine.

The `strictfp` modifier ensures that all the computations are performed in strict binary32 or binary64 mode, which will almost always imply a performance cost on x86 hardware without SSE2. According to the specification, what non-`strictfp` allows is just an extended exponent range, "at the whim of the implementation" [220].

```
// In this class, all operations in all methods are performed
// in binary32 or binary64 mode.
strictfp class ExampleFpStrictClass {
...
} // End class ExampleFpStrictClass

class NormalClass {
...
  // This particular method performs all its operations in binary32
  // or binary64 mode
  strictfp double aFpStrictMethod(double arg)
  {
    ...
  } // End aFpStrictMethod
} // End class NormalClass
```

Program 6.1: The use of the strictfp keyword.

However, as usual, one may expect compilation flags to relax compliance to the Java specification. Here, the interested reader should look at the documentation not only of the compiler (the `javac` command or a substitute), but also of the runtime environment (the `java` command or a substitute, including just-in-time compilers). There are also Java native compilers such as JET or GCJ, which compile Java directly to machine code (bypassing the virtual machine layer). Some enable extended precision even for the significand.

6.5.3 Infinities, NaNs, and signed zeros

Java has the notion of signed infinities, NaNs, and signed zeros. Infinities and NaNs are defined as constants in their respective wrapper classes (e.g., `java.lang.Double.POSITIVE_INFINITY`, `java.lang.Float.NaN`).

A first pitfall one must be aware of is that `Double` and `double` do not compare the same, as Program 6.2 shows. Here is the output of this program:

```
NaN != NaN
NaN == NaN
aDouble and anotherDouble are different objects.
anotherDouble and aThirdDouble are the same object.
anotherDouble and aThirdDouble have the same value NaN == NaN
```

As one can see, the == operator does not behave the same for basic types and objects.

- For basic types, if the variables hold the same value, the comparison evaluates to `true`, except for the `java.lang.{Float | Double}.NaN` value, in accordance with any version of the IEEE 754 standard.
- For the object types, the == operator evaluates to `true` only if both references point to the same object. To compare the values, one must use the `equals()` method. But as you can see, `NaN` equals `NaN`, which is a bit confusing.

  We are confronted here with a tradeoff between respect for the IEEE 754 standard and the consistency with other Java language elements as maps or associative arrays. If floating-point objects are used as keys, one should be able to retrieve an element whose index is `NaN`.

Also note that in Java, there is only one kind of `NaN`. It is a quiet `NaN` and it has only one possible representation. Therefore, `NaN`s will carry no extra information about why they surfaced during computations and therefore cannot be used for further debugging.

6.5.4 Missing features

The compliance of the Java Virtual Machine [390, Section 2.8] to IEEE 754 remains partial in two main respects:

- It does not support flags or exceptions. The term *exception* must be taken here with the meaning it has in the IEEE 754 standard, not with the one it has in Java (which would more accurately translate into *trap*, in IEEE 754 parlance).
- All operations are performed with rounding to the nearest. As a consequence, some of the algorithms described in this book cannot be implemented. Worse, as this is a virtual machine limitation, there is no possibility of a machine interval arithmetic data type as first class citizen in the Java language. This does not mean that interval arithmetic cannot be done at all in Java; several external packages have been developed for that, but they are much less efficient than operations performed using hardware directed rounding modes. This is all the more surprising as most processors with hardware floating-point support also support directed rounding modes.

```java
import java.io.*;

class ObjectValue {
  public static void main(String args[])
  {
    double adouble = java.lang.Double.NaN;
    double anotherdouble = java.lang.Double.NaN;
    Double aDouble = new Double(java.lang.Double.NaN);
    Double anotherDouble = new Double(java.lang.Double.NaN);
    Double aThirdDouble = anotherDouble;

    if (adouble != anotherdouble){
      System.out.print(adouble);
      System.out.print(" != ");
      System.out.println(anotherdouble);
    }
    if (aDouble.equals(anotherDouble)){
      System.out.print(aDouble.toString());
      System.out.print(" == ");
      System.out.println(anotherDouble.toString());
    }
    if (aDouble != anotherDouble)
      System.out.println("aDouble and anotherDouble are different objects.");
    if (anotherDouble == aThirdDouble)
      System.out.println("anotherDouble and aThirdDouble are the same object.");
    if (anotherDouble.equals(aThirdDouble)){
      System.out.print("anotherDouble and aThirdDouble have the same value ");
      System.out.print(anotherDouble.toString());
      System.out.print(" == ");
      System.out.println(aThirdDouble.toString());
    }
  } // End main
} // End class ObjectValue
```

Program 6.2: Object comparison in Java.

6.5.5 Reproducibility

The `strictfp` keyword enables reproducibility of results computed using basic operations. Expression evaluation is strictly defined and unambiguous, with (among others) left-to-right evaluation, StrictFP compile-time constant evaluation, and widening of the operations to the largest format [220].

However, tightening basic operations is not enough. Until Java 2 SDK 1.3, when a mathematical function of the `java.lang.Math` package was called (sine, exponential, etc.), it was evaluated using the operating system's implementation, and the computed result could change from platform to platform. Java 2 SDK 1.3 was released with the new `java.lang.StrictMath` package. It tried to guarantee the same bit-for-bit result on any platform, again, at the expense of performance. Nevertheless, correct rounding to the last bit was not ensured since, as of Java SE 8 (at time of writing) the `java.lang.StrictMath` package is based on `fdlibm version 5.3` from the `netlib` network library. Eventually, in Java 2 SDK 1.4, the implementation of `java.lang.Math` functions became simple calls to their `java.lang.StrictMath` counterparts.

This enabled numerical consistency on all platforms at last, with two ill effects. Some users observed that the result changed for the same program on the same platform. Users also sometimes experienced a sharp drop in execution speed of their program, and the standard Java platform no longer offers a standard way out for users who need performance over reproducibility. Many tricks (e.g., resorting to JNI for calls to an optimized C library) were tried to gain access again to the speed and precision of the underlying platform.

To summarize this issue, the history of Java shows the difficulty of the "run anywhere with the same results" goal. At the time of writing this book, there are still some inconsistencies. For example, the default choice for elementary function evaluation is reproducibility over performance, while the default choice for expression evaluation (without `strictfp`) is performance over reproducibility.

Things are evolving favorably, however. On the prevalent x86 platform, the gradual phasing out of the historical x87 FPU in favor of the SSE2/AVX units (see Section 3.4) renders `strictfp` increasingly useless. In addition, in the near future, the generalization of correctly rounded elementary functions which are recommended by IEEE 754-2008 (see Section 10.5) could reconcile performance and reproducibility for the elementary functions: the Java virtual machine could again trust the system's optimized mathematical library if it knows that it implements correct rounding. It is unfortunate that the `java.lang.StrictMath` implementation did not make the choice of correct rounding. It remains for Java designers to specify a way to exploit the performance and accuracy advantage of the FMA operator when available [11] without endangering numerical reproducibility.

6.5.6 The BigDecimal package

The Java designers seem to have been concerned by the need for reliable decimal floating-point, most notably for perfectly specified accounting. They provided the necessary support under the form of the `java.math.BigDecimal` package. Although this package predates the IEEE 754-2008 standard and therefore does not exactly match the decimal floating-point specification, it shares many concepts with it.

In particular, the `java.math.MathContext` class encapsulates a notion of precision and a notion of rounding mode. For instance, the preset `MathContext.DECIMAL128` defines a format matching the IEEE 754-2008 decimal128 and the "round to nearest and break ties to even" default rounding mode. Users can define their own kind of MathContext and have, for that purpose, a wide choice of rounding modes.

A MathContext can be used to control how operations are performed, but also to emulate IEEE 754 features otherwise absent from the language, such as the inexact flag. Program 6.3 illustrates this. This program will crash, since we do not catch the exception that would, in IEEE 754 parlance, raise the "inexact status flag," and will print out a message of the following type:

```
 Exception in thread "main" java.lang.ArithmeticException:
Rounding necessary
```

```
import java.math.*;

class  DecimalBig {
  public static void main(String args[])
  {
    // Create a new math context with 7 digits precision, matching
    // that of MathContext.DECIMAL32 but with a different rounding
    // mode.
    MathContext mc = new MathContext(7, RoundingMode.UNNECESSARY);
    BigDecimal a = new BigDecimal(1.0, MathContext.DECIMAL32);
    BigDecimal b = new BigDecimal(3.0, MathContext.DECIMAL32);
    BigDecimal c;

    // Perform the division in the requested MathContext.
    // In this case, if the result is not exact, within the required
    // precision, an exception will be thrown.
    c = a.divide(b, mc);
    // could have been written as
    // c = a.divide(b, 7, RoundingMode.UNNECESSARY)
  } // End main
} // End class DecimalBig
```

Program 6.3: BigDecimal and MathContext.

```
     at java.math.BigDecimal.divide(BigDecimal.java:1346)
     at java.math.BigDecimal.divide(BigDecimal.java:1413)
     at DecimalBig.main(DecimalBig.java:12)
```

This is more awkward and dangerous than the behavior proposed in IEEE 754-2008: in IEEE 754, an inexact computation silently raises a flag and does not interrupt execution. Still, when used with care, this is the closest to floating-point environment control one can find in "out-of-the-box" Java.

A completely different problem is that BigDecimal numbers are objects, not basic types. They incur all the performance overhead associated with objects (in addition to the performance overhead associated with software decimal operations) and require a clumsy object-oriented method syntax instead of the leaner usual infix operators.

6.6 Conclusion

We wish we convinced the reader that, from the floating-point perspective, languages and systems were not "designed equal," and that the designer of numerical programs may save on debugging time by looking carefully at the documentations of both the chosen language and the underlying system.

Obviously, considering the variety of choices made by different systems, there is no perfect solution, in particular because of the performance/reproducibility conflict (where *reproducibility* may be replaced with *portability*, *predictability*, or *numerical consistency*, depending on the programmer's concerns). The perfect solution may be a system which

- is safe by default (favoring portability) so that subtle numerical bugs, such as the infinitely looping sort, are impossible, and
- gives to the programmer the possibility of improving performance when needed, with due disclaimers with respect to numerical predictability.

Even so, the granularity of the programmer's control on this tradeoff is an issue. Compilation flags or operating-system-level behavior control are typically too coarse, while adding pragmas or `strictfp` everywhere in the code may be a lot of work, and may not be possible when external libraries are used.

We have not covered all the existing languages, of course. Some of them are well specified, and some are explicitly underspecified.

Python is a good example. The documentation[12] explicitly warns the user about floating-point arithmetic:

> `numbers.Real (float)`
> These represent machine-level double precision floating-point numbers. You are at the mercy of the underlying machine architecture (and C or Java implementation) for the accepted range and handling of overflow. Python does not support single-precision floating-point numbers; the savings in processor and memory usage that are usually the reason for using these are dwarfed by the overhead of using objects in Python, so there is no reason to complicate the language with two kinds of floating-point numbers.

Note, however, the `numpy` package (part of the SciPy project[13]) provides access to all the IEEE formats and operations for users who need it.

---

[12] For Python3: https://docs.python.org/3/reference/datamodel.html; For Python2: https://docs.python.org/2/reference/datamodel.html.

[13] https://scipy.org/.

So Python, as far as floating-point is concerned, is underspecified, but at least it is clear and honest about it. The same can be said of C#. Its specification has a similarly short paragraph to warn the programmer that the intermediate precision used when evaluating expressions may be larger than the precision of the operand (in this respect, it inherits the philosophy of C).

Let us finally mention JavaScript/ECMAScript (ECMA-262 / ISO/IEC 16262). The language specification[14] is stricter than that of Python, but only mentions binary64:

> The `Number` type has exactly 18437736874454810627 (that is, $2^{64} - 2^{53} + 3$) values, representing the double-precision 64-bit format IEEE 754 values as specified in the IEEE Standard for Binary Floating-Point Arithmetic, except that the 9007199254740990 (that is, $2^{53} - 2$) distinct "Not-a-Number" values of the IEEE Standard are represented in ECMAScript as a single special NaN value. (...) To ECMAScript code, all NaN values are indistinguishable from each other.

Note that JavaScript/ECMAScript does not have integer arithmetic: its integers are embedded in IEEE 754 floating-point numbers. The only difficulty is the integer division, which is commonly implemented as a floating-point division followed by a floor. Developers should be aware that some inputs can yield an incorrect result because of the rounded floating-point division, although in most cases (in particular those encountered in practice), one can prove that the result is correct [377].

All these examples, again, illustrate that all modern languages accurately document their floating-point support... or lack of.

[14]http://www.ecma-international.org/ecma-262/5.1/.

Part III

Implementing Floating-Point Operators

Chapter 7

Algorithms for the Basic Operations

AMONG THE MANY OPERATIONS that the IEEE 754 standards specify (see Chapter 3), we will focus here and in the next two chapters on the five basic arithmetic operations: addition, subtraction, multiplication, division, and square root. We will also study the fused multiply-add (FMA) operator. We review here some of the known properties and algorithms used to implement each of those operators. Chapter 8 and Chapter 9 will detail some examples of actual implementations in, respectively, hardware and software.

Throughout this chapter, the radix β is assumed to be either 2 or 10. Following the IEEE 754-2008 standard [267], we shall further assume that the extremal exponents are related by $e_{\min} = 1 - e_{\max}$. Also, the formats considered here are basic formats only.

7.1 Overview of Basic Operation Implementation

For the five basic operations and the FMA, the IEEE 754-2008 standard requires correct rounding: the result returned must be as if the exact, infinitely precise result was computed, then rounded. The details of the cases that may occur, illustrated in Figure 7.1, are as follows.

- If the result is undefined, a Not a Number (NaN) will be returned.
- Otherwise, let us consider the real number which is the infinitely precise result of the operation.
    - If this real result is exactly representable as a floating-point number, no rounding will be needed. However, there may still be work

J.-M. Muller et al., *Handbook of Floating-Point Arithmetic*,
https://doi.org/10.1007/978-3-319-76526-6_7

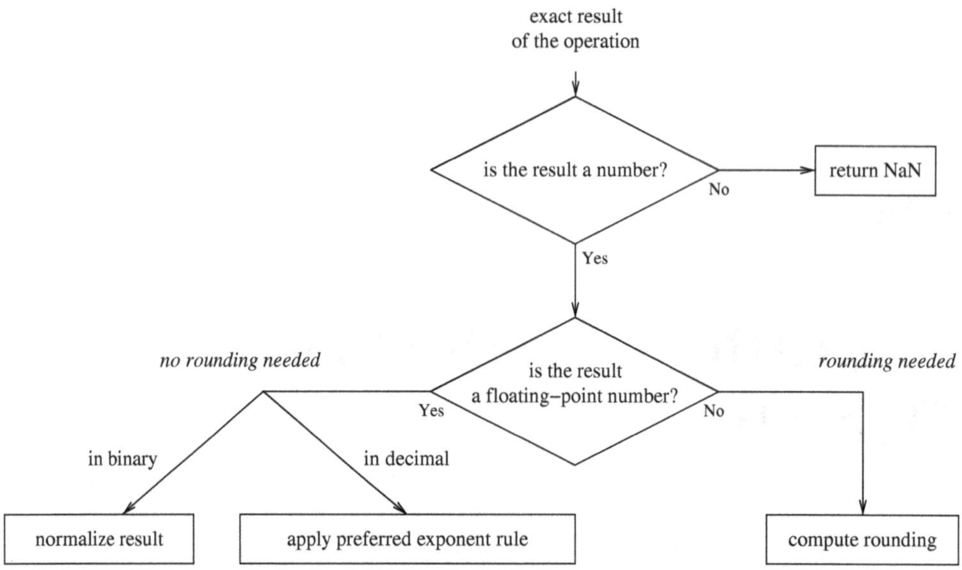

Figure 7.1: Specification of the implementation of a floating-point operation.

to do: a representable result may have several possible representations, and the implementation has to compute which one it returns out of its intermediate representation of the result:

* In binary, there is only one valid representation, which is the one with the smallest possible exponent;
* In decimal, several representations of the result may be valid (they form what is called a *cohort*). The standard precisely defines which member of such a cohort should be returned. For each operation, a *preferred exponent* is defined as a function of the operation's inputs (see Section 3.1.3.3). The implementation has to return the member of the cohort result whose exponent is closest to the preferred exponent;

- If the exact result is not exactly representable as a floating-point number, it has to be rounded to a floating-point number. In decimal, this number may have several possible representations, in which case the returned result is the one with the smallest possible exponent, to get the maximum number of significant digits.

In practice, there are two classes of operations.

- When adding, subtracting, or multiplying two floating-point numbers, or when performing a fused multiply-add (FMA) operation, the infinitely precise result actually always has a finite representation and may be exactly computed by an implementation. Therefore, the previous discussion has a straightforward translation into an architecture or a program.[1]

- When performing a division or computing a square root or an elementary function, the exact result may have an infinite number of digits; consider, for instance, the division $1.0/3.0$. In such cases, other means are used to reduce the rounding problem to a finite computation. For division, one may compute a finite-precision quotient, then the remainder allows one to decide how to round. Similarly, for the square root, one may compute $y \approx \sqrt{x}$, then decide rounding by considering $x - y^2$. Some iterations derived from the Newton-Raphson iteration also allow, when an FMA instruction is available, to directly get a correctly rounded quotient or square root from an accurate enough approximation (see Section 4.7). Correct rounding of the elementary functions will be dealt with in Chapter 10.

Outline of this Chapter

Section 7.2 addresses the general issue of rounding a value (the "compute rounding" box in the bottom right corner of Figure 7.1). The subsequent sections (7.3 to 7.7) specifically address each of the five basic operations as well as the FMA. They focus on homogeneous operators, i.e., operators that have inputs and output in the same format. Indeed, these are the most useful, in particular because language standards rely on them. However, IEEE 754-2008 also states [269, §5.4.1] that implementations shall provide these operations *for operands of all supported arithmetic formats with the same radix as the destination format*. A note then explicits that this can be achieved in software, building upon homogeneous operators. Section 7.8 reviews existing nonhomogeneous operators and some pitfalls when implementing them in software.

7.2 Implementing IEEE 754-2008 Rounding

7.2.1 Rounding a nonzero finite value with unbounded exponent range

Obviously every nonzero finite real number x can be written as

$$x = (-1)^s \cdot m \cdot \beta^e, \tag{7.1}$$

where β is the chosen radix, here 2 or 10, where e is an integer, and where the real m is the (possibly infinitely precise) significand of x, such that $1 \le m < \beta$. We will call the representation (7.1) the *normalized representation* of x (note however that this does *not* mean that x is a normalized floating-point representation in the IEEE 754 sense since x may have no finite radix-β expansion).

[1] In many cases, there are better ways of implementing the operation.

Writing m_i to denote the digit of weight β^{-i} (i.e., the i-th fractional digit) in the radix-β expansion of m, we have

$$m = \sum_{i \geq 0} m_i \beta^{-i} = (m_0.m_1 m_2 \ldots m_{p-1} m_p m_{p+1} \ldots)_\beta,$$

where $m_0 \in \{1, \ldots, \beta - 1\}$ and, for $i \geq 1$, $m_i \in \{0, 1, \ldots, \beta - 1\}$. In addition, an *unbounded exponent range* is assumed from now on, so that we do not have to worry about overflow, underflow, or subnormals (they will be considered in due course).

The result of rounding x to precision p is either the floating-point number

$$x_p = (-1)^s \cdot (m_0.m_1 m_2 \ldots m_{p-1})_\beta \cdot \beta^e,$$

obtained by truncating the significand m after $p - 1$ fractional digits, or

- the floating-point successor of x_p when x is positive,
- the floating-point predecessor of x_p when x is negative.

In other words, writing $\mathrm{Succ}(x)$ for the successor of a floating-point number x, the rounded value of x will always be one of the two following values:

$$(-1)^s \cdot |x_p| \qquad \text{or} \qquad (-1)^s \cdot \mathrm{Succ}(|x_p|),$$

with $|x_p| = (m_0.m_1 m_2 \ldots m_{p-1})_\beta \cdot \beta^e$.

Note that, in practice, rounding essentially reduces to rounding *nonnegative* values because of the following properties (the one for RN being true in particular in the context of IEEE 754-2008: the two considered tie-breaking rules, ties-to-even and ties-to-away, are symmetrical):

$$\mathrm{RN}(-x) = -\mathrm{RN}(x), \qquad \mathrm{RZ}(-x) = -\mathrm{RZ}(x),$$
$$\mathrm{RU}(-x) = -\mathrm{RD}(x). \tag{7.2}$$

7.2.1.1 Computing the successor in a binary interchange format

The reader may check that the binary interchange formats (see [267] and Chapter 3) are built in such a way that *the binary encoding of the successor of a positive floating-point value is the successor of the binary encoding of this value, considered as a binary integer*. This important property (which explains the choice of a biased exponent over two's complement or sign-magnitude) is true for all positive floating-point numbers, including subnormal numbers, from $+0$ to the largest finite number (whose successor is $+\infty$). It also has the consequence that the lexicographic order on the binary representations of positive floating-point numbers matches the order on the numbers themselves.

This provides us with a very simple way of computing $\mathrm{Succ}(|x_p|)$: consider the encoding of $|x_p|$ as an integer, and increment this integer. The possible carry propagation from the significand field to the exponent field will take care of the possible change of exponent.

Example 7.1. *Considering the binary32 format, let* $x_{24} = (2 - 2^{-23}) \cdot 2^{-126}$. *The bit string* $X_{31} \ldots X_0$ *of the 32-bit integer* $X = \sum_{i=0}^{31} X_i 2^i$ *that encodes* x_{24} *is*

$$0 \quad \underbrace{00000001}_{8 \textit{ exponent bits}} \quad \underbrace{11111111111111111111111}_{23 \textit{ fraction bits}}.$$

The successor $\mathrm{Succ}(x_{24})$ *of* x_{24} *is encoded by the 32-bit integer* $X + 1$ *whose bit string is*

$$0 \quad \underbrace{00000010}_{8 \textit{ exponent bits}} \quad \underbrace{00000000000000000000000}_{23 \textit{ fraction bits}}.$$

That new bit string encodes the number $1 \cdot 2^{2-127}$, *which is indeed* $\mathrm{Succ}(x_{24}) = x_{24} + 2^{-23} \cdot 2^{-126} = 2^{-125}$. *Note how the carry in the addition* $X+1$ *has propagated up to the exponent field.*

It is also possible to use floating-point operations instead of integer operations for computing the predecessor and successor of a floating-point number. Algorithms for doing that are described in [533].

7.2.1.2 Choosing between $|x_p|$ and its successor $\mathrm{Succ}(|x_p|)$

As already detailed in Table 2.1 for radix 2, the choice between $|x_p|$ and $\mathrm{Succ}(|x_p|)$ depends on the sign s, the rounding mode, the value of the digit m_p of m (called the *round digit*), and a binary information telling us if there exists at least one nonzero digit among the (possibly infinitely many) remaining digits $m_{p+1}, m_{p+2}, \ldots$.

In radix 2, this information may be defined as the logical OR of all the bits to the right of the round bit, and is therefore named the *sticky bit*. In radix 10, the situation is very similar. One still needs a binary information, which we still call the sticky bit. It is no longer defined as a logical OR, but as follows: its value is 0 if all the digits to the right after m_p are zero, and 1 otherwise.

Let us consider some decimal cases for illustration.

- When rounding x toward zero, the rounded number is always[2] x_p.
- When rounding a positive x toward $+\infty$, the rounded number is $\mathrm{Succ}(x_p)$, except if x was already a representable number, i.e., when both its round digit and sticky bit are equal to zero.
- When rounding to nearest with *roundTiesToEven* a positive decimal number x, if the round digit m_p belongs to $\{0,1,2,3,4\}$, then the rounded number is x_p; if m_p belongs to $\{6,7,8,9\}$, then the rounded number is $\mathrm{Succ}(x_p)$. If m_p is equal to 5, then the sticky bit will decide

[2]In this discussion, we assume that $4.999\ldots 9^{\infty}$ is written 5.0^{∞}; otherwise, this sentence is not true. This remark is academic: a computer will only deal with finite representations, which do not raise this ambiguity.

between $\mathrm{Succ}(x_p)$ (if equal to 1) or a tie (if equal to 0). In case of a tie, the *ties to even* rule considers the last digit (of weight 10^{-p+1}) of the two candidates. The rounded result is the one whose last digit is even. The ties to away rule requires us to choose $\mathrm{Succ}(x_p)$.

Having defined the infinitely accurate normalized representation $x = (-1)^s \cdot m \cdot \beta^e$ of the result allows us to manage flags and exceptional cases as well. However, note first that for some operations, overflow or underflow signaling may be decided by considering the inputs only, before any computation of the results. For example, as we will see later in this chapter, square root can neither underflow nor overflow,[3] and returns NaN if and only if the input is NaN or strictly negative. The possibility of such an *early detection* of exceptional situations will be mentioned when appropriate.

7.2.2 Overflow

As stated in Section 3.1.6, the overflow exception is signaled when the absolute value of the intermediate result is finite and strictly larger than the largest finite number $\Omega = (\beta - \beta^{1-p}) \cdot \beta^{e_{\max}}$. Here, the intermediate result is defined as the infinitely accurate result rounded to precision p with an unbounded exponent range.

For rounding to nearest, this translates to: an overflow is signaled when

$$
\begin{gathered}
(e > e_{\max}) \\
\text{or} \\
\left(e = e_{\max} \quad \text{and} \quad (m_0.m_1m_2\ldots m_{p-1})_\beta = \beta - \beta^{1-p} \quad \text{and} \quad m_p \geq \frac{\beta}{2}\right).
\end{gathered}
$$

Note that in the case $m_p = \frac{\beta}{2}$, when $e = e_{\max}$ and $(m_0.m_1m_2\ldots m_{p-1})_\beta = \beta - \beta^{1-p}$, with *roundTiesToEven* as well as with *roundTiesToAway*, the exact result is rounded to the intermediate result $\beta^{e_{\max}+1}$; therefore, it signals overflow without having to consider the sticky bit.

When rounding a positive number to $+\infty$, an overflow is signaled when

$$
\begin{gathered}
(e > e_{\max}) \\
\text{or} \\
\Big(e = e_{\max} \quad \text{and} \quad (m_0.m_1\ldots m_{p-1})_\beta = \beta - \beta^{1-p} \\
\text{and} \quad (m_p > 0 \quad \text{or} \quad \text{sticky} = 1)\Big).
\end{gathered}
$$

This reminds us that overflow signaling is dependent on the prevailing rounding direction. The other combinations of sign and rounding direction are left as an exercise to the reader.

[3]When the input is $+\infty$, it is the exact result $+\infty$ which must be returned; see [267, §6.1].

7.2.3 Underflow and subnormal results

As stated in Section 3.1.6, the underflow exception is signaled when a nonzero result whose absolute value is strictly less than $\beta^{e_{\min}}$ is computed.[4] This translates to: an underflow is signaled if $e < e_{\min}$, where e is the exponent of the normalized infinitely precise significand.

In such cases, the previous rounding procedure has to be modified as follows: m (the normalized infinitely precise significand) is shifted right by $e_{\min} - e$, and e is set to $e_{\min}$. We thus have rewritten x as

$$x = (-1)^s \cdot m' \cdot \beta^{e_{\min}},$$

with

$$m' = (m'_0.m'_1m'_2 \ldots m'_{p-1}m'_pm'_{p+1}\cdots)_\beta.$$

From this representation, we may define the round digit m'_p, the sticky bit (equal to 1 if there exists a nonzero m'_i for some $i > p$, and 0 otherwise), and the truncated value $|x_p| = (m'_0.m'_1m'_2 \ldots m'_{p-1})_\beta \cdot \beta^{e_{\min}}$ as previously. As the successor function is perfectly defined on the subnormal numbers—and even easy to compute in the binary formats—the rounded value is decided among $(-1)^s \cdot |x_p|$ and $(-1)^s \cdot \mathrm{Succ}(|x_p|)$ in the same way as in the normal case.

One will typically need the implementations to build the *biased exponent* (that is, in binary, what is actually stored in the exponent field), equal to the exponent plus the bias (see Table 3.3). There is one subtlety to be aware of in binary formats: the subnormal numbers have the same exponent as the smallest normal numbers, although their biased exponent is smaller by 1. In general, we may define n_x as the "is normal" bit, which may be computed as the OR of the bits of the exponent field. Its value will be 0 for subnormal numbers and 1 for normal numbers. Then the relation between the value of the exponent e_x and the biased exponent E_x is the following:

$$e_x = E_x - \text{bias} + 1 - n_x. \tag{7.3}$$

This relation will allow us to write exponent-handling expressions that are valid in both the normal and subnormal cases.

In addition, n_x also defines the value of the implicit leading bit: the actual significand of a floating-point number is obtained by prepending n_x to the significand field.

7.2.4 The inexact exception

This exception is signaled when the exact result y is not exactly representable as a floating-point number (i.e., denoting $\circ$ the rounding function, $\circ(y) \neq y$, y

[4]We remind the reader that there remains some ambiguity in the standard, since underflow can be detected before or after rounding. See Sections 2.1.3 and 3.1.6.4 for more on this. Here, we describe underflow detection *before rounding*.

not a NaN). As the difference between $\circ(y)$ and y is condensed in the round digit and the sticky bit, the inexact exception will be signaled unless both the round digit and the sticky bit are equal to 0.

7.2.5 Rounding for actual operations

Actual rounding of the result of an operation involves two additional difficulties.

- Obtaining the intermediate result in normalized form may require some work, all the more as some of the inputs, or the result, may belong to the subnormal range. In addition, decimal inputs may not be normalized (see the definition of cohorts in Section 3.1.1.2).
- For decimal numbers, the result should not always be normalized (see the definition of preferred exponents in Section 3.1.3).

These two problems will be addressed on a per-operation basis.

7.2.5.1 Decimal rounding using the binary encoding

The entire discussion in Section 7.2 assumes that the digits of the infinitely precise significand are available in the radix in which it needs to be rounded. This is not the case for the *binary encoding* of the decimal formats (see Section 3.1.1.2). In this case, one first needs to convert the binary encoding to decimal digits, at least for the digits needed for rounding (the round digit and the digits to its right). Such a conversion is typically done by performing a division by some 10^k (with $k > 0$) with remainder. Cornea et al. [116, 117] have provided several efficient algorithms for this purpose, replacing the division by 10^k with a multiplication by a suitable precomputed approximation to 10^{-k}. They also provide techniques to determine to which precision 10^{-k} should be precomputed.

7.3 Floating-Point Addition and Subtraction

When x or y is nonzero, the addition of $x = (-1)^{s_x} \cdot |x|$ and $y = (-1)^{s_y} \cdot |y|$ is based on the identity

$$x + y = (-1)^{s_x} \cdot \Big(|x| + (-1)^{s_z} \cdot |y|\Big), \qquad s_z = s_x \text{ XOR } s_y \in \{0, 1\}. \tag{7.4}$$

For subtraction, a similar identity obviously holds since $x - y = x + (-y)$. Hence, in what follows we shall consider addition only.

The IEEE 754-2008 specification for $|x| \pm |y|$ is summarized in Tables 7.2 and 7.3. Combined with (7.2) and (7.4), it defines floating-point addition completely provided x or y is nonzero. When both x and y are zero, the standard stipulates to return $+0$ or -0, depending on the operation (addition or subtraction) and the rounding direction attribute, as shown in Table 7.1.

$x \operatorname{op} y$	$\circ(x \operatorname{op} y)$ for $\circ \in \{\text{RN}, \text{RZ}, \text{RU}\}$	$\text{RD}(x \operatorname{op} y)$
$(+0) + (+0)$	$+0$	$+0$
$(+0) + (-0)$	$+0$	-0
$(-0) + (+0)$	$+0$	-0
$(-0) + (-0)$	-0	-0
$(+0) - (+0)$	$+0$	-0
$(+0) - (-0)$	$+0$	$+0$
$(-0) - (+0)$	-0	-0
$(-0) - (-0)$	$+0$	-0

Table 7.1: Specification of addition/subtraction when both x and y are zero. Note that floating-point addition is commutative.

$\lvert x\rvert + \lvert y\rvert$		$\lvert y\rvert$			
		$+0$	(sub)normal	$+\infty$	NaN
$\lvert x\rvert$	$+0$	$+0$	$\lvert y\rvert$	$+\infty$	qNaN
	(sub)normal	$\lvert x\rvert$	$\circ(\lvert x\rvert + \lvert y\rvert)$	$+\infty$	qNaN
	$+\infty$	$+\infty$	$+\infty$	$+\infty$	qNaN
	NaN	qNaN	qNaN	qNaN	qNaN

Table 7.2: Specification of addition for floating-point data of positive sign.

In Tables 7.2 and 7.3, the sum or difference $\circ(|x| \pm |y|)$ of the two positive finite floating-point numbers

$$|x| = m_x \cdot \beta^{e_x} \qquad \text{and} \qquad |y| = m_y \cdot \beta^{e_y}$$

is given by

$$\circ\Big(|x| \pm |y|\Big) = \circ\Big(m_x \cdot \beta^{e_x} \pm m_y \cdot \beta^{e_y}\Big). \tag{7.5}$$

The rest of this section discusses the computation of the right-hand side of the above identity.

Note that for floating-point addition/subtraction, the only possible exceptions are *invalid operation*, *overflow*, *underflow*, and *inexact* (see [187, p. 425]).

In more detail, a sequence of operations that can be used for implementing (7.5) is as follows.

$\|x\| - \|y\|$		$\|y\|$			
		$+0$	(sub)normal	$+\infty$	NaN
$\|x\|$	$+0$	± 0	$-\|y\|$	$-\infty$	qNaN
	(sub)normal	$\|x\|$	$\circ(\|x\| - \|y\|)$	$-\infty$	qNaN
	$+\infty$	$+\infty$	$+\infty$	qNaN	qNaN
	NaN	qNaN	qNaN	qNaN	qNaN

Table 7.3: Specification of subtraction for floating-point data of positive sign. Here ± 0 means $+0$ for "all rounding direction attributes except roundTowardNegative" (i.e., except $\circ = \mathrm{RD}$), and -0 for $\circ = \mathrm{RD}$; see Table 7.1 and [267, §6.3].

- First, the two exponents e_x and e_y are compared, and the inputs x and y are possibly swapped to ensure that $e_x \geq e_y$.

- A second step is to compute $m_y \cdot \beta^{-(e_x - e_y)}$ by shifting m_y right by $e_x - e_y$ digit positions (this step is sometimes called *significand alignment*). The exponent result e_r is tentatively set to e_x.

- The result significand is computed as $m_r = m_x + (-1)^{s_z} \cdot m_y \cdot \beta^{-(e_x - e_y)}$: either an addition or a subtraction is performed, depending on the signs s_x and s_y. Then if m_r is negative, it is negated. This (along with the signs s_x and s_y) determines the sign s_r of the result. At this step, we have an exact sum $(-1)^{s_r} \cdot m_r \cdot \beta^{e_r}$.

- This sum is not necessarily normalized (in the sense of Section 7.2). It may need to be normalized in two cases.

  - There was a carry out in the significand addition ($m_r \geq \beta$). Note that m_r always remains strictly smaller than 2β, so this carry is at most 1. In this case, m_r needs to be divided by β (i.e., shifted right by one digit position), and e_r is incremented, unless e_r was equal to $e_{\max}$, in which case an overflow is signaled as per Section 3.1.6.

  - There was a cancellation in the significand addition ($m_r < 1$). In general, if λ is the number of leading zeros of m_r, then m_r is shifted left by λ digit positions, and e_r is set to $e_r - \lambda$. However, if $e_r - \lambda < e_{\min}$ (the cancellation has brought the intermediate result in the underflow range, see Section 3.1.6), then the exponent is set to $e_{\min}$ and m_r will be shifted left only by $e_r - e_{\min}$.

Note that for decimals, the preferred exponent rule (mentioned in Section 3.1.3) states that *inexact* results must be normalized as just described, but not *exact* results. We will come back to this case.

- Finally, the normalized sum (which again is always finite) is rounded as per Section 7.2.

Let us now examine this algorithm more closely. There are important points.

1. This algorithm never requires more than a p-digit effective addition for the significands. This is easy to see in the case of an addition: the least significant digits of the result are those of m_y, since they are added to zeros. This is also true when y is subtracted, provided the sticky bit computation is modified accordingly.

2. The alignment shift need never be by more than $p + 1$ digits. Indeed, if the exponent difference is larger than $p + 1$, y will only be used for computing the sticky bit, and it doesn't matter that it is not shifted to its proper place.

3. Leading-zero count (LZC) and variable shifting is only needed in case of a cancellation, i.e., when the significands are subtracted and the exponent difference is 0 or 1. But in this case, several things are simpler. The sticky bit is equal to zero and need not be computed. More importantly, the alignment shift is only by 0 or 1 digit.

   In other words, although two possibly large shifts are mentioned in the previous algorithm (one for significand alignment, the other one for normalization in case of a cancellation), they are mutually exclusive. The literature defines these mutually exclusive cases as the *close* case (when the exponents are close) and the *far* case (when their difference is larger than 1).

4. In our algorithm, the normalization step has to be performed before rounding: indeed, rounding requires the knowledge of the position of the round and sticky bits, or, in the terminology of Section 7.2, it requires a *normalized* infinite significand. However, here again the distinction between the close and far cases makes things simpler. In the close case, the sticky bit is zero whatever shift the normalization entails. In the far case, normalization will entail a shift by at most one digit. Classically, the initial sticky bit is therefore computed out of the digits to the right of the $(p + 2)$-nd (directly out of the lower digits of the lesser addend). The $(p + 2)$-nd digit is called the *guard* digit. It will either become the round digit in case of a 1-digit shift, or it will be merged to the previous sticky bit if there was no such shift. The conclusion of this is that the bulk of the sticky bit computation can be performed in parallel with the significand addition.

Let us now detail specific cases of floating-point addition.

7.3.1 Decimal addition

We now come back to the preferred exponent rule (see Section 3.1.3), which states that *exact* results should not be normalized. As the notion of exactness is closely related to that of normalization (a result is exact if it has a normalized representation that fits in p digits), the general way to check exactness is to first normalize X, then apply the previous algorithm.

Exactness of the intermediate result is then determined combinatorially out of the carry-out and sticky bits, and the round and guard digits.

For addition, the preferred exponent is the smallest of the input exponents (in other words, e_y and not e_x). If the result is exact, we therefore need to shift m_r right and reduce e_r to e_y. Therefore, the preferred exponent rule means two large shifts.

In case of a carry out, it may happen that the result is exact, but the result's cohort does not include a member with the preferred exponent. An example is $9.999e0+0.0001e0$ for a $p=4$-digit system. Both input numbers have the same quantum exponent, yet the (exact) value of the result, 10, cannot be represented with the same quantum exponent and must be represented as $1.000e1$.

In practice, the exact case is a common one in decimal applications (think of accounting), and even hardware implementations of decimal floating-point addition distinguish it and try to make this common case fast.

The IBM POWER6 [184] distinguishes the following three cases (from simplest to most complex).

Case 1 Exponents are equal. This is the most common case of accounting: adding amounts of money which have the decimal point at the same place. It is also the simplest case, as no alignment shifting is necessary. Besides, the result is obtained directly with the preferred exponent. It may still require a one-digit normalization shift and one-digit rounding in case of overflow, but again such an overflow is highly rare in accounting applications using decimal64—it would correspond to astronomical amounts of money!

Case 2 Aligning to the operand with the smaller exponent. When the exponent difference is less than or equal to the number of leading zeros in the operand with the larger exponent, this operand can be shifted left to properly align it with the smaller exponent value. Again, after normalization and rounding, the preferred exponent is directly obtained.

Case 3 Shifting both operands. This is the general case that we have considered above.

The interested reader will find in [184] the detail of the operations performed in each case in the POWER6 processor. This leads to a variable number of cycles for decimal addition—9 to 17 for decimal64.

7.3.2 Decimal addition using binary encoding

A complete algorithm for the addition of two decimal floating-point numbers in the binary encoding is presented in [116, 117].

The main issue with the binary encoding is the implementation of the shifts. The number $M_1 \cdot 10^{e_1} + M_2 \cdot 10^{e_2}$, with $e_1 \geq e_2$, is computed as

$$10^{e_2} \cdot \left(M_1 \cdot 10^{e_1 - e_2} + M_2\right).$$

Instead of a shift, the significand addition now requires a multiplication by some 10^k with $0 \leq k \leq p$ (because of remark 2 in the introduction of Section 7.3). There are few such constants, so they may be tabulated, and a multiplier or FMA will then compute the "shifted" values [116, 117]. The full algorithm takes into account the number of decimal digits required to write M_1 and M_2, which is useful to obtain the preferred exponent. This number is computed by table lookup.

For the full algorithm, the interested reader is referred to [117].

7.3.3 Subnormal inputs and outputs in binary addition

Here are the main modifications to the previous addition algorithm to ensure that it handles subnormal numbers properly in the binary case. Let us define, for the inputs x and y, the "is normal" bits n_x and n_y. One may compute n_x (resp. n_y) as the OR of the bits of the exponent field of x (resp. y).

- The implicit leading bit of the significand of x (resp. y) is now set to n_x (resp. n_y).
- If E_x and E_y are the respective biased exponents of the inputs, we now have $e_x = E_x - \text{bias} + 1 - n_x$ and $e_y = E_y - \text{bias} + 1 - n_y$. The exponent difference, used for the alignment shifting, is now computed as $E_x - n_x - E_y + n_y$. Of course, two subnormal inputs are already aligned.
- As already stated in Section 7.2.3, the normalization shift should handle subnormal outputs: the normalization shift distance will be $\min(\lambda, e_x - e_{\min})$ digit positions, where λ is the leading-zero count. The output is subnormal if $\lambda > e_x - e_{\min}$.

7.4 Floating-Point Multiplication

Floating-point multiplication is much simpler than addition. Given $x = (-1)^{s_x} \cdot |x|$ and $y = (-1)^{s_y} \cdot |y|$, the exact product $x \times y$ satisfies

$$x \times y = (-1)^{s_r} \cdot \left(|x| \times |y|\right), \qquad s_r = s_x \text{ XOR } s_y \in \{0, 1\}. \tag{7.6}$$

The IEEE 754-2008 specification for $|x| \times |y|$ is summarized in Table 7.4. Combined with (7.2) and (7.6), it defines floating-point multiplication completely.

$\lvert x\rvert \times \lvert y\rvert$		$\lvert y\rvert$			
		+0	(sub)normal	$+\infty$	NaN
$\lvert x\rvert$	+0	+0	+0	qNaN	qNaN
	(sub)normal	+0	$\circ(\lvert x\rvert \times \lvert y\rvert)$	$+\infty$	qNaN
	$+\infty$	qNaN	$+\infty$	$+\infty$	qNaN
	NaN	qNaN	qNaN	qNaN	qNaN

Table 7.4: Specification of multiplication for floating-point data of positive sign.

In Table 7.4 the product $\circ(|x|\times|y|)$ of the two positive finite floating-point numbers

$$|x| = m_x \cdot \beta^{e_x} \qquad \text{and} \qquad |y| = m_y \cdot \beta^{e_y}$$

is given by

$$\circ(|x| \times |y|) = \circ\big(m_x m_y \cdot \beta^{e_x+e_y}\big). \tag{7.7}$$

The rest of this section discusses the computation of the right-hand side of (7.7).

For floating-point multiplication, the only possible exceptions are *invalid operation, overflow, underflow,* and *inexact* (see [187, p. 438]).

7.4.1 Normal case

Let us first consider the case when both inputs are normal numbers such that $1 \le m_x < \beta$ and $1 \le m_y < \beta$ (this is notably the case for binary normal numbers). It follows that the exact product $m_x m_y$ satisfies $1 \le m_x m_y < \beta^2$. This shows that the significand product has either one or two nonzero digits to the left of the point. Therefore, to obtain the normalized significand required to apply the methods given in Section 7.2, the significand product $m_x m_y$ may need to be shifted right by one position. This is exactly similar to the *far* case of addition, and will be handled similarly, with a guard and a round digit, and a partial sticky bit. Since the product of two p-digit numbers is a $2p$-digit number, this partial sticky computation has to be performed on $p-1$ digits.

The significand multiplication itself is a fixed-point multiplication, and much literature has been dedicated to it; see for instance [187] and the references therein. Hardware implementations, both for binary and decimal, are surveyed in Section 8.2.4, and the issues related to software implementations (for binary) are discussed in Section 9.3.

In binary, the exponent is equal to the biased exponent E_x minus the bias (see Section 3.1.1.1). The exponent computation is therefore $e_x + e_y = E_x - \text{bias} + E_y - \text{bias}$. One may directly compute the biased exponent of the result (before normalization) as $E_x + E_y - \text{bias}$.

7.4.2 Handling subnormal numbers in binary multiplication

We now extend the previous algorithm to accept subnormal inputs and produce subnormal outputs when needed.

Let us define again, for the inputs x and y, the "is normal" bits n_x and n_y. One may compute n_x as the OR of the bits of the exponent field E_x. This bit can be used as the implicit bit to be added to the significand, and also as the bias correction for subnormal numbers since $e_x = E_x - \text{bias} + 1 - n_x$ and $e_y = E_y - \text{bias} + 1 - n_y$.

Let us first note that if both operands are subnormal ($n_x = 0$ and $n_y = 0$), the result will be zero or one of the smallest subnormals, depending on the rounding mode. This case is therefore handled straightforwardly.

Let us now assume that only one of the operands is subnormal. The simplest method is to normalize it first, which will bring us back to the normal case. For this purpose we need to count its leading zeros. Let us call l the number of leading zeros in the significand extended by n_x. We have $l = 0$ for a normal number and $l \geq 1$ for a subnormal number. The subnormal significand is then shifted left by l bit positions, and its exponent becomes $e_{\min} - l$. Obviously, this requires a larger exponent range than what the standard format offers. In practice, the exponent data is only one bit wider.

An alternative to normalizing the subnormal input prior to a normal computation is to normalize the product after the multiplication: indeed, the same multiplication process which computes the product of two p-bit numbers will compute this product exactly if one of the inputs has l leading zeros. The product will then have l or $l+1$ leading zeros, and will need to be normalized by a left shift. However, the advantage of this approach is that counting the subnormal leading zeros can be done in parallel with the multiplication. Therefore, this alternative will be preferred in the hardware implementations presented in the next chapter.

For clarity, we now take the view that both inputs have been normalized with a 1-bit wider exponent range, and focus on producing subnormal outputs. Note that they may occur even for normal inputs. They may be handled by the standard normalization procedure of Section 7.2. It takes an arbitrary shift right of the significand: if $e_x+e_y < e_{\min}$, then shift right by $e_{\min}-(e_x+e_y)$ before rounding.

To summarize, the cost of handling subnormal numbers is: a slightly larger exponent range for internal exponent computations, a leading-zero counting step, a left-shifting step of either the subnormal input or the product, and a right-shifting step before rounding. Section 8.4.4 will show how these additional steps may be scheduled in a hardware implementation in order to minimize their impact on the delay.

7.4.3 Decimal specifics

For multiplication, the preferred exponent rule (see Section 3.1.3) mentions that, for exact results, the preferred quantum exponent is $Q(x) + Q(y)$. We recall the relation $Q(x) = e_x - p + 1$, see Section 3.1.

Again, exactness may be computed by first normalizing the two inputs, then computing and normalizing the product, then observing its round digit and sticky bit.

However, exactness may also be predicted when the sum of the numbers of leading zeros of both multiplicands is larger than p, which will be a very common situation.

To understand why, think again of accounting. Multiplication is used mostly to apply a rate to an account (a tax rate, an interest rate, a currency conversion rate, etc.). After the rate has been applied, the result is rounded to the nearest cent before being further processed. Rounding to the nearest cent can be performed using the *quantize* operation specified by the IEEE 754-2008 standard.

Such rates are 3- to 6-digit numbers (for illustration, the euro official conversion rates with respect to the currencies it replaces were all defined as 6-digit numbers). When using the 16-digit format decimal64, the product of a 6-digit rate with your bank account will be exact, unless your wealth exceeds 10^{10} cents (one hundred million dollars). In the latter case, your bank will be happy to spend a few extra cycles to manage it.

In the common case when the product is exact, the quantum exponent is set to $Q(x) + Q(y)$ (it makes sense to have a zero quantum exponent for the rate, so that the product is directly expressed in cents) and the product needs no normalization. The significand to output is simply the last p digits of the product.

Counting the leading zeros of the inputs is an expensive operation, but it may be performed in parallel to the multiplication.

To summarize, an implementation may compute in parallel the $2p$-digit significand product and the two leading-zero counts of the inputs. If the sum of the counts is larger than p (common case), then the result is exact, and no rounding or shift is required (fast). Otherwise, the result needs to be normalized, and then rounded as per Section 7.2. Note that the result may also be exact in this case, but then the result significand is too large to be representable with the preferred exponent. The standard requires an implementation to return the representable number closest to the result, which is indeed the normalized one.

7.5 Floating-Point Fused Multiply-Add

When computing $\circ(ab + c)$, the product ab is a $2p$-digit number, and needs to be added to the p-digit number c. Sign handling is straightforward: what actually matters is whether the operation will be an effective subtraction or an effective addition.

We base the following analysis on the actual exponent of the input numbers, denoted e_a, e_b, and e_c. Computing $ab+c$ requires first aligning the product ab with the summand c, using the exponent difference

$$d = e_c - (e_a + e_b).$$

In the following figures, the precision used is $p = 5$ digits.

7.5.1 Case analysis for normal inputs

If a and b are normal numbers, one has $|a| = m_a \cdot \beta^{e_a}$ and $|b| = m_b \cdot \beta^{e_b}$, and the product $|ab|$ has at most two digits in front of the point corresponding to $e_a + e_b$:

$|ab| = \beta^{e_a+e_b}.$ 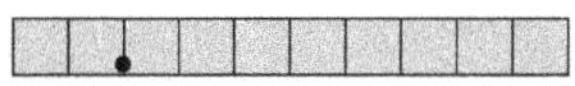 $|c| = \beta^{e_c}.$ 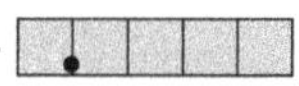

One may think of first performing a 1-digit normalization of this product ab, but this would add to the computation a step which can be avoided, and it only marginally reduces the number of cases to handle. Following most implementations, we therefore choose to base the discussion of the cases on the three input exponents only. We now provide an analysis of the alignment cases that may occur. These cases are mutually exclusive. After any of them, we need to perform a rounding step as per Section 7.2. This step may increment the result exponent again.

7.5.1.1 Product-anchored case

The exponent of the result is almost that of the product, and no cancellation can occur, for $d \leq -2$, as illustrated in Figure 7.2.

In this case, the leading digit may have four positions: the two possible positions of the leading digit of ab, one position to the left in case of effective addition, and one position to the right in case of effective subtraction. This defines the possible positions of the round digit. All the digits lower than the lowest possible position of the round digit may be condensed in a sticky bit. An actual addition is only needed for the digit positions corresponding to the digits of ab (see Figure 7.2); therefore, all the digits shifted out of this range may be condensed into a sticky bit before addition. If $d \leq -2p+1$, all the digits of c will go to the sticky bit. This defines the largest alignment one may need to perform in this case: if $d \leq -2p+1$, a shift distance of $2p-1$ and a sticky computation on all the shifted p digits will provide the required information, namely "Is there a nonzero digit to the right of the round digit?"

To summarize, this case needs to perform a shift of c by at most $2p-1$ with a p-digit sticky computation on the lower digits of the output, a $2p$-digit

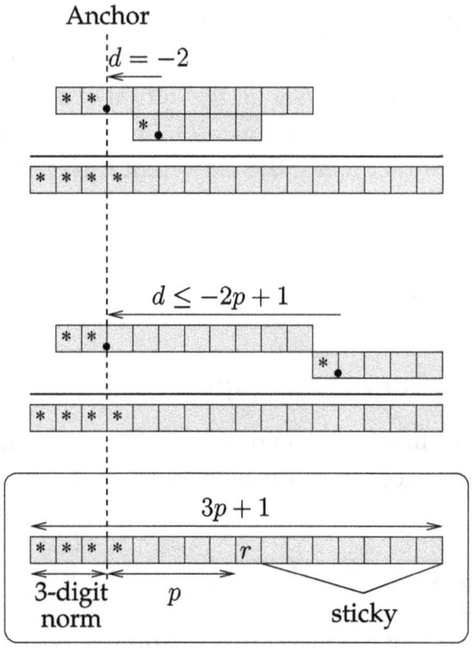

Figure 7.2: Product-anchored FMA computation for normal inputs. The stars show the possible positions of the leading digit.

addition with sticky output of the $p - 3$ lower digits, and a 3-digit normalization (updating the sticky bit). The exponent is set tentatively to $e_a + e_b$, and the 3-digit normalization will add to it a correction in $\{-1, 0, 1, 2\}$.

7.5.1.2 Addend-anchored case

The exponent of the result will be close to that of the addend (and no cancellation can occur) when ($d \geq 3$ or ($d \geq -1$ and EffectiveAdd)), as illustrated in Figure 7.3. In that case, the leading digit may be in five different positions.

The whole of ab may be condensed in a sticky bit as soon as there is a gap of at least two digits between ab and c. One gap digit is needed for the case of an effective subtraction $|c| - |ab|$, when the normalized result exponent may be one less than that of c (for instance, in decimal with $p = 3$, $1.00 - 10^{-100}$ rounded down returns 0.999). The second gap digit is the round digit for rounding to the nearest. A sufficient condition for condensing all of ab in a sticky bit is therefore $d \geq p + 3$ (see Figure 7.3).

To summarize, this case needs to perform a shift of ab by at most $p + 3$ positions, a $(p+2)$-digit addition, a $2p$-digit sticky computation, and a 4-digit normalization (updating the sticky bit). The exponent is set tentatively to e_c, and the 4-digit normalization will add to it a correction in $\{-1, 0, 1, 2, 3\}$.

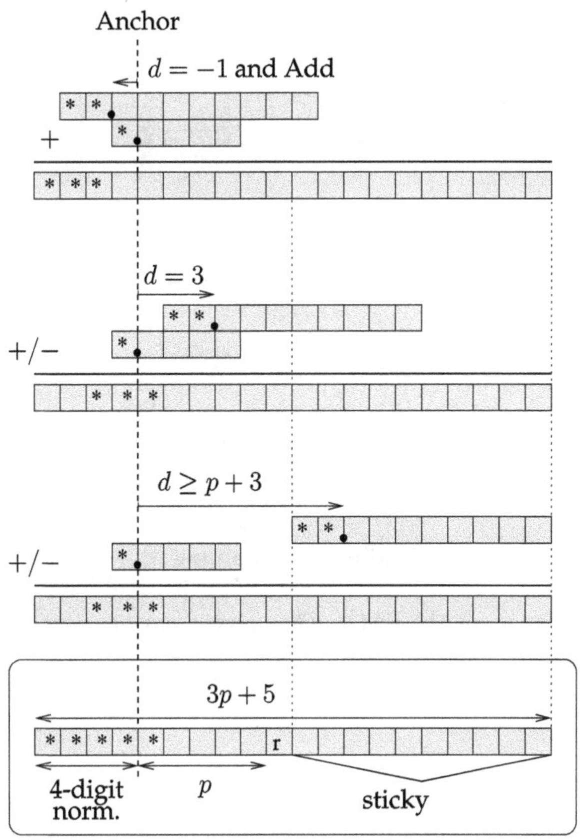

Figure 7.3: Addend-anchored FMA computation for normal inputs. The stars show the possible positions of the leading digit.

7.5.1.3 Cancellation

If $-1 \leq d \leq 2$ and the FMA performs an effective subtraction, a cancellation may occur. This happens for four values of the exponent difference d, versus only three in the case of the floating-point addition, because of the uncertainty on the leading digit of the product. Possible cancellation situations are illustrated in Figure 7.4.

In that case we need a $(2p+1)$-digit addition, and an expensive normalization consisting of leading-zero counting and right shifting, both of size $2p+1$.

Example 7.2. *Consider in radix 2, precision p the inputs $a = b = 1 - 2^{-p}$ and $c = -(1 - 2^{-p+1})$. The FMA should return the exact result $ab + c = 2^{-2p}$. In this example $e_a = e_b = e_c = -1$, $d = 1$, and there is a $(2p-1)$-bit cancellation.*

The result exponent is equal to $e_a + e_b + 3 - \lambda$, where λ is the leading-zero count.

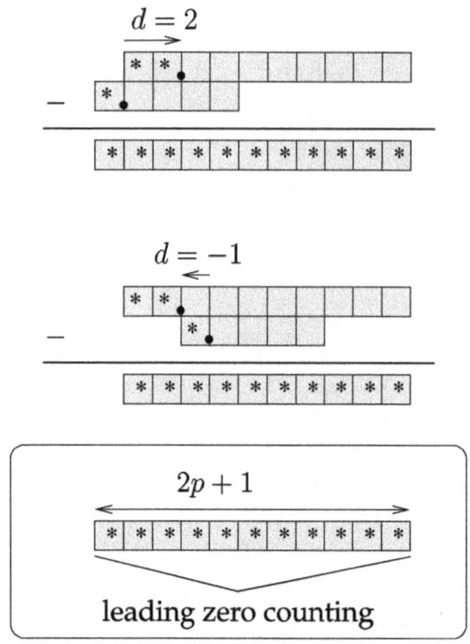

Figure 7.4: Cancellation in the FMA. The stars show the possible positions of the leading digit.

7.5.2 Handling subnormal inputs

To manage subnormal inputs, we define n_a, n_b, and n_c as the "is normal" bits. In binary floating-point, these bits are inserted as leading bits of the significands, and the exponent difference that drives the aligner becomes

$$d = e_c - (e_a + e_b) = E_c - E_a - E_b + \text{bias} - 1 - n_c + n_a + n_b. \tag{7.8}$$

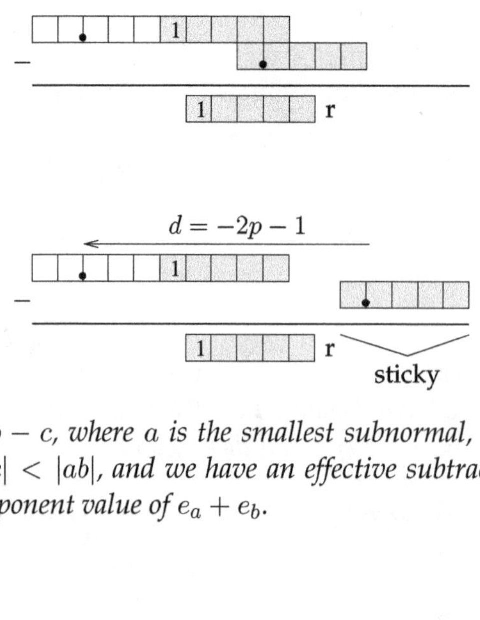

Figure 7.5: FMA $ab - c$, where a is the smallest subnormal, ab is nevertheless in the normal range, $|c| < |ab|$, and we have an effective subtraction. Again, the dot corresponds to an exponent value of $e_a + e_b$.

If both a and b are subnormal, the whole of ab will be condensed in a sticky bit even if c is subnormal. Let us therefore focus on the case when only one of a and b is subnormal. Most of the previous discussion remains valid in this case, with the following changes.

- The significand product may now have up to p leading zeros. Indeed, as the situation where both inputs are subnormals is excluded, the smallest significand product to consider is the smallest subnormal significand, equal to β^{-p+1}, multiplied by the smallest normal significand, equal to 1. This is illustrated in Figure 7.5.
- In the product-anchored case, this requires us to extend the shift by two more digit positions, for the cases illustrated in Figure 7.5. The maximum shift distance is now $-d = 2p + 1$.
- In fact, the product-anchored case is not necessarily product-anchored if one of the multiplicands is subnormal. The leading digit of the result may now come from the addend—be it normal or subnormal. Still, this requires neither a larger shift, nor a larger addition, than that shown in Figure 7.2. However, the partial sticky computation from the lower bits of the product shown on this figure is now irrelevant: one must first determine the leading digit before knowing which digits go to the sticky bit. This requires a leading-zero counting step on p digits. In this respect, the product-anchored case when either a or b is subnormal now closely resembles the cancellation case, although it requires a smaller leading-zero counting (p digits instead of $2p + 1$).
- The addend c may have up to $p - 1$ leading zeros, which is more than what is shown in Figure 7.3, but they need not be counted, as the exponent is stuck to $e_{\min}$ in this case.

7.5.3 Handling decimal cohorts

In decimal, the previous case analysis is valid (we have been careful to always use the word "digit"). The main difference is the handling of cohorts, which basically means that a decimal number may be subnormal for any exponent. As a consequence, there may be more stars in Figures 7.2 to 7.4. In particular, the addend-anchored case may now require up to a $(p+4)$-digit leading-zero count instead of p.

In addition, one must obey the preferred exponent rule: for inexact results, the preferred exponent is the least possible (this corresponds to normalization in the binary case, and the previous analysis applies). For exact results, the preferred quantum exponent is $\min(Q(a)+Q(b), Q(c))$. As for addition and multiplication, this corresponds to avoiding any normalization if possible.

The only available decimal FMA implementation, to our knowledge, is a software one, part of the Intel Decimal Floating-Point Math Library [116, 117]. A hardware decimal FMA architecture is evaluated in Vázquez's Ph.D. dissertation [613].

7.5.4 Overview of a binary FMA implementation

Let us now conclude this section with a complete algorithmic description of an implementation of the binary FMA. Most early hardware FMA implementations chose to manage the three cases evoked above (product-anchored, addend-anchored, and canceling/subnormal) in a single computation path [424, 260]. In the next chapter, more recent, multiple-path implementations [550, 359, 501] will be reviewed.

Here is a summary of the basic single-path implementation handling subnormals. Figure 7.6 represents the data alignment in this case. It is a simplified superimposition of Figures 7.2 to 7.5. The single-path approach is in essence product anchored.

- The "is normal" bits are determined, and added as implicit bits to the three input significands. The exponent difference d is computed as per (7.8).
- The $2p$-digit significand of ab is shifted right by $p+3$ digit positions (this is a constant distance). Appending two zeros to the right, we get a first $(3p+5)$-digit number ab_{shifted}.
- The summand shift distance d' and the tentative exponent e'_r are determined as follows:
  - if $d \leq -2p+1$ (product-anchored case with saturated shift), then $d' = 3p+4$ and $e'_r = e_a + e_b$;
  - if $-2p+1 < d \leq 2$ (product-anchored case or cancellation), then $d' = p+3-d$ and $e'_r = e_a + e_b$;
  - if $2 < d \leq p+2$ (addend-anchored case), then $d' = p+3-d$ and $e'_r = e_c$;
  - if $d \geq p+3$ (addend-anchored case with saturated shift), then $d' = 0$ and $e'_r = e_c$.
- The p-digit significand c is shifted right by d' digit positions. The maximum shift distance is $3p+4$ digits, for instance, 163 bits for binary64.
- The lower $p-1$ digits of the shifted c are compressed into a sticky bit. The leading $3p+5$ digits form c_{shifted}.
- In case of effective subtraction, c_{shifted} is bitwise inverted.

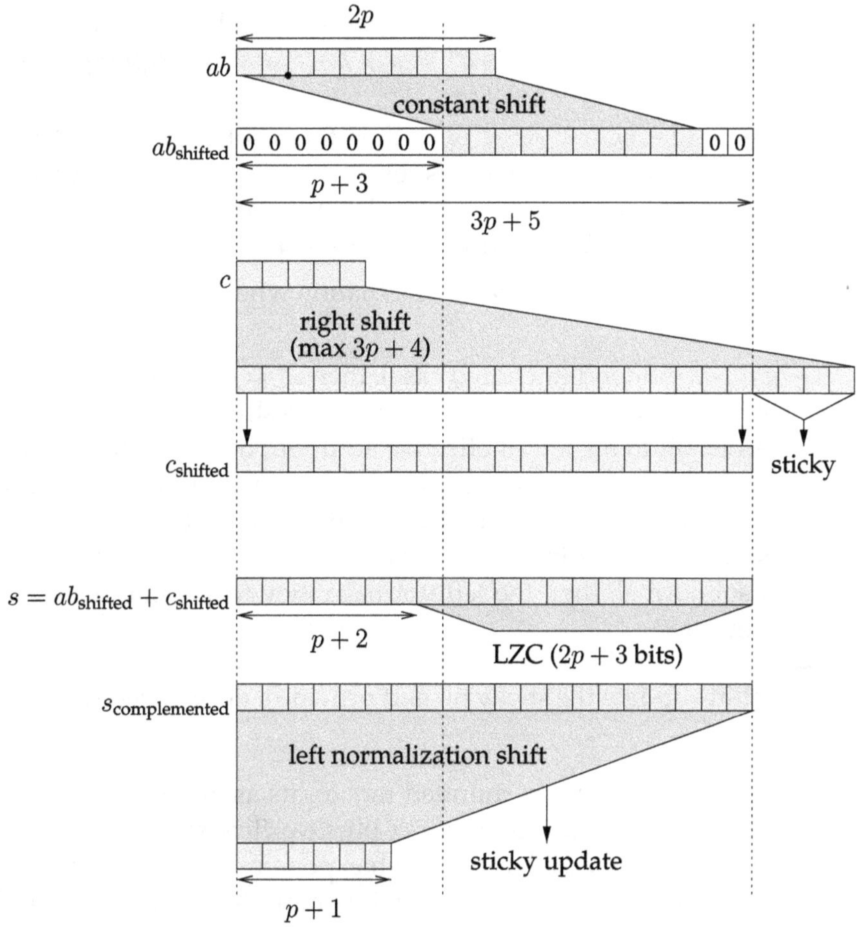

Figure 7.6: Significand alignment for the single-path algorithm.

- The product ab_{shifted} and the possibly inverted addend c_{shifted} are added (with a carry in case of an effective subtraction), leading to a $(3p+5)$-digit number s. This sum may not overflow because of the gap of two zeros in the case $d = p + 3$.

- If s is negative, it is complemented. Note that the sign of the result can be predicted from the signs of the inputs and the exponents, except when the operation is an effective subtraction and $d \in \{0, 1\}$.

- The possibly complemented sum s now needs to be shifted left so that its leading digit is a 1. The value of the left shift distance d'' is determined as follows.

  - if $d \leq 2$ (product-anchored case or cancellation), let l be the number of leading zeros counted in the $2p + 3$ lower digits of s.

  * If $e_a + e_b - l + 2 \geq e_{\min}$, then the result will be normal, the left shift distance is $d'' = p + 2 + l$, and the exponent is set to $e'_r = e_a + e_b - l + 2$.
  * If $e_a + e_b - l + 2 < e_{\min}$, then the result will be a subnormal number, the exponent is set to $e'_r = e_{\min}$, the result significand will have $e_{\min} - (e_a + e_b - l + 2)$ leading zeros, and the shift distance is therefore only $d'' = p + 4 - e_{\min} + e_a + e_b$.

  Note that this case covers the situation when either a or b is a subnormal.

- if $d > 2$ (addend-anchored case), then $d'' = d'$: the left shift undoes the initial right shift. However, after this shift, the leading one may be one bit to the left (if effective addition) or one bit to the right (if effective subtraction), see the middle case of Figure 7.3. The large shift is therefore followed by a 1-bit normalization. The result exponent is accordingly set to one of $e'_r \in \{e_c, e_c - 1, e_c + 1\}$. If c is subnormal, then the left normalization/exponent decrement is prevented.

These shifts update the sticky bit and provide a normalized $(p+1)$-digit significand.

- Finally, this significand is rounded to p digits as per Section 7.2, using the rounding direction and the sticky bit. Overflow may also be handled at this point, provided a large enough exponent range has been used in all the exponent computations (two bits more than the exponent field width are enough).

One nice feature of this single-path algorithm is that subnormal handling comes almost for free.

7.6 Floating-Point Division

7.6.1 Overview and special cases

Given $x = (-1)^{s_x} \cdot |x|$ and $y = (-1)^{s_y} \cdot |y|$, we want to compute:

$$x/y = (-1)^{s_r} \cdot \big(|x|/|y|\big), \qquad s_r = s_x \text{ XOR } s_y \in \{0, 1\}. \tag{7.9}$$

The IEEE 754-2008 specification for $|x|/|y|$ is summarized in Table 7.5 (see [267] and [293]). Combined with (7.2) and (7.9) it specifies floating-point division completely.

$\|x\|/\|y\|$		$\|y\|$			
		+0	(sub)normal	$+\infty$	NaN
$\|x\|$	+0	qNaN	+0	+0	qNaN
	(sub)normal	$+\infty$	$\circ(\|x\|/\|y\|)$	+0	qNaN
	$+\infty$	$+\infty$	$+\infty$	qNaN	qNaN
	NaN	qNaN	qNaN	qNaN	qNaN

Table 7.5: Special values for $|x|/|y|$.

We now address the computation of the quotient of $|x| = m_x \cdot \beta^{e_x}$ and $|y| = m_y \cdot \beta^{e_y}$. We obtain

$$|x|/|y| = m_x/m_y \cdot \beta^{e_x - e_y}. \tag{7.10}$$

In the following, we will assume that m_x and m_y are normal numbers, written $m_x = (m_{x,0}.m_{x,1} \ldots m_{x,p-1})_\beta$ and $m_y = (m_{y,0}.m_{y,1} \ldots m_{y,p-1})_\beta$, with $m_{x,0} \neq 0$ and $m_{y,0} \neq 0$. An implementation may first normalize both inputs (if they are subnormals or decimal numbers with leading zeros) using an extended exponent range. After this normalization, we have $m_x \in [1, \beta)$ and $m_y \in [1, \beta)$, therefore $m_x/m_y \in (\frac{1}{\beta}, \beta)$. This means that a 1-digit normalization may be needed before rounding.

7.6.2 Computing the significand quotient

There are three main families of division algorithms.

- Digit-recurrence algorithms, such as the family of SRT algorithms named after Sweeney, Robertson, and Tocher [511, 602], generalize the paper-and-pencil algorithm learned at school. They produce one digit of the result at each iteration. Each iteration performs three tasks (just like the pencil-and-paper method): determine the next quotient digit, multiply it by the divider, and subtract it from the current partial remainder to obtain a partial remainder for the next iteration.

  In binary, there are only two choices of quotient digits, 0 or 1; therefore, the iteration reduces to one subtraction, one test, and one shift. A binary digit-recurrence algorithm can therefore be implemented on any processor as soon as it is able to perform integer addition.

  Higher radix digit-recurrence algorithms have been designed for hardware implementation, and will be briefly reviewed in Section 8.6. Detailed descriptions of digit-recurrence division theory and implementations can be found in the books by Ercegovac and Lang [186, 187].

  One important thing about digit-recurrence algorithms is that they are exact: starting from fixed-point numbers X and D, they compute at iteration i an i-digit quotient Q_i and a remainder R_i such that the identity

$X = DQ_i + R_i$ holds. For floating-point purposes, this means that all the information needed for rounding the result is held in the pair (R_i, Q_i). In practice, to round to precision p, one needs p iterations to compute Q_p, then possibly a final addition on Q_p depending on a test on R_p.

- Functional iteration algorithms generalize Newton iteration for approximating the function $1/x$. They make sense mostly on processors having a hardware multiplier. The number of iterations is much less than in digit-recurrence algorithms ($\mathcal{O}(\log p)$ versus $\mathcal{O}(p)$), but each iteration involves multiplications and is therefore more expensive.

  Functional iterations are not exact; in particular, they start with an approximation of the inverse, and round their intermediate computations. Obtaining a correctly rounded result therefore requires some care. The last iteration needs to provide at least twice the target precision p, as a consequence of the exclusion lemma, see Lemma 4.14, Chapter 4. In Chapter 4, it has been shown that the FMA provides the required precision. However, AMD processors have used functional iteration algorithms to implement division without an FMA [467]. The iteration is implemented as a hardware algorithm that uses the full $2p$-bit result of the processor's multiplier before rounding. To accommodate double-extended precision ($p = 64$ bits) and cater to the error of the initial approximation and the rounding errors, they use a 76×76-bit multiplier; see [467].

- Polynomial approximation can also be used to evaluate the inverse $1/y$ to the required accuracy [629]. Note that, mathematically speaking, functional iterations evaluate a polynomial in the initial approximation error [118]. Both approaches may be combined; see [492] and [502, §9.5]. It is also possible to directly approximate the ratio x/y by a polynomial, and an example of this approach in a software context will be given in Section 9.4.

Note that each division method has its specific way of obtaining the correctly rounded result.

7.6.3 Managing subnormal numbers

Subnormal inputs are best managed by first normalizing with a wider exponent range.

A subnormal result can be predicted from the exponent difference. As it will have less than p significand digits, it requires less accuracy than a standard computation. In a functional iteration, it suffices to round the high-precision intermediate result to the proper digit position.

In a digit-recurrence implementation, the simplest way to handle rounding to a subnormal number is to stop the iteration after the required number of digits has been produced, and then shift these digits right to their proper place. In this way, the rounding logic is the same as in the normal case.

7.6.4 The inexact exception

In general, the inexact exception is computed as a by-product of correct rounding. Directed rounding modes, as well as round to nearest even in the "ties to even" case, require, respectively, exactness and half-ulp exactness detection. From another point of view, the inexact exception is straightforwardly deduced from the sticky bit. In digit-recurrence algorithms, for instance, exactness is deduced from a null remainder. Methods using polynomial approximations have to compute the remainder to round, and the inexact flag comes at no extra cost.

FMA-based functional iterations are slightly different in that they do not explicitly compute a sticky bit. However, they may be designed in such a way that the final FMA operation raises the inexact flag if and only if the quotient is inexact (see Section 4.7 or [406, page 115]).

7.6.5 Decimal specifics

For division, the preferred exponent rule (see Section 3.1.3) mentions that for exact results the preferred exponent is $Q(x) - Q(y)$.

7.7 Floating-Point Square Root

Although it has similarities with division, square root is somewhat simpler to implement conformally with the IEEE 754 standard. In particular, it is univariate and, as we will recall, it neither underflows nor overflows.

7.7.1 Overview and special cases

If x is a positive (sub)normal floating-point number, then the correctly rounded value $\circ(\sqrt{x})$ must be returned. Otherwise, a special value must be returned conformally with Table 7.6 (see [267] and [292]).

Operand x	$+0$	$+\infty$	-0	less than zero	NaN
Result r	$+0$	$+\infty$	-0	qNaN	qNaN

Table 7.6: Special values for `sqrt`(x).

In the following, we will assume that m_x is a normal number, written $m_x = (m_{x,0}.m_{x,1} \dots m_{x,p-1})_\beta$ with $m_{x,0} \neq 0$. An implementation may first normalize the input (if it is a subnormal or a decimal number having some leading zeros) using an extended exponent range.

After this normalization, we have a number of the form $m_x \cdot \beta^{e_x}$ with $m_x \in [1, \beta)$. Another normalization, by 1 digit, may be needed before taking the square root in order to make the exponent e_x even:

$$m_x \cdot \beta^{e_x} = \begin{cases} m_x \cdot \beta^{e_x} & \text{if } e_x \text{ is even,} \\ (\beta \cdot m_x) \cdot \beta^{e_x - 1} & \text{if } e_x \text{ is odd.} \end{cases}$$

Consequently, for $c \in \{0, 1\}$ depending on the parity of e_x, we have

$$\sqrt{m_x \cdot \beta^{e_x}} = \sqrt{\beta^c \cdot m_x} \cdot \beta^{\frac{e_x - c}{2}},$$

where $(e_x - c)/2 = \lfloor e_x/2 \rfloor$ is an integer. Since $\beta^c \cdot m_x \in [1, \beta^2)$, the significand square root satisfies

$$\sqrt{\beta^c \cdot m_x} \in [1, \beta).$$

7.7.2 Computing the significand square root

The families of algorithms most commonly used are exactly the same as for division, and a survey of these was given by Montuschi and Mezzalama in [425].

- **Digit-recurrence algorithms.** Those techniques are extensively covered in [186] and [187, Chapter 6], with hardware implementations in mind. Here we simply note that the recurrence is typically a little more complicated than for division; see, for example, the software implementation of the restoring method that is described in Section 9.5.3.

- **Functional iteration algorithms.** Again, those methods generalize Newton iteration for approximating the positive real solution $y = \sqrt{x}$ of the equation $y^2 - x = 0$. As for division, such methods are often used when an FMA operator is available. This has been covered in Section 4.8.

- **Evaluation of polynomial approximations.** As for division, these methods consist in evaluating sufficiently accurately a "good enough" polynomial approximation of the function $\sqrt{x}$. Such techniques have been combined with functional iterations in [492] and [502, Section 11.2.3]. It was shown in [291, 292] that, at least in some software implementation contexts, using exclusively polynomials (either univariate or bivariate) can be faster than a combination with a few steps of functional iterations. These approaches are described briefly in Section 9.5.3.

7.7.3 Managing subnormal numbers

As for division, a subnormal input is best managed by first normalizing with a wider exponent range.

Concerning output, the situation is much simpler than for division, since a subnormal result can never be produced. This useful fact is an immediate consequence of the following property.

Property 7.1. *If* $x = m_x \cdot \beta^{e_x}$ *is a positive, finite floating-point number, the real number* $\sqrt{x}$ *satisfies*

$$\sqrt{x} \in \left[\beta^{e_{\min}}, \beta^{\frac{1}{2}e_{\max}}\right).$$

Proof. The floating-point number x is positive, so $\beta^{1-p+e_{\min}} \leq x < \beta^{e_{\max}}$. Since the square root function is monotonically increasing,

$$\beta^{(1-p+e_{\min})/2} \leq \sqrt{x} < \beta^{e_{\max}/2},$$

and the upper bound follows immediately. Using $p \leq 1 - e_{\min}$ (which is a valid assumption for all the formats of [267]), we get further

$$\sqrt{x} \geq \beta^{e_{\min}},$$

which concludes the proof. □

Recalling that for $x = +\infty$ the IEEE 754-2008 standard requires to return $+\infty$ (that is, the exact value of $\sqrt{x}$), we conclude that the floating-point square root neither underflows nor overflows, and that the only exceptions to be considered are *invalid* and *inexact*.

7.7.4 The inexact exception

In digit-recurrence algorithms or polynomial-based approaches, exactness is deduced from a null remainder just as in division—the remainder here is $x - r^2$.

FMA-based functional iterations may be designed in such a way that the final FMA operation raises the inexact flag if and only if the square root is inexact [406, page 115].

7.7.5 Decimal specifics

For square root, the preferred quantum exponent is $\lfloor Q(x)/2 \rfloor$.

7.8 Nonhomogeneous Operators

The typical personal computer claims support of the binary32 and binary64 formats. To fully comply with IEEE 754-2008, however, it must also support all the heterogeneous operations. An example of heterogeneous operation is the addition of a binary32 number a and a binary64 number b, correctly rounded to a binary32 number r.

A naive solution is to first promote a to binary64, then use the homogeneous binary64 addition, then round the binary64 result of this addition to the binary32 format. However, this may lead to a double rounding (see Section 3.2) and is therefore not compliant with IEEE 754-2008.

From a practical point of view, this question may seem rather academic. If, in some language, we have declared `a` and `r` as binary32, and `b` as binary64, then the semantics of `r:=a+b;` will likely be something like:

- cast `a` to binary64 (the larger format),
- then perform the addition in binary64, yielding a binary64 (semantics of +),
- then round the result to a binary32 (semantics of assignment `:=`).

In other words, the semantics of the language will explicitly describe a double rounding—or allow it if it is underspecified; see Chapter 6 for a more detailed discussion of floating-point semantics in programming languages.

This current situation is indeed the result of IEEE 754-1985, which specified only homogeneous operators. The IEEE 754-2008 committee decided that leaving heterogeneous operators unspecified, or potentially doubly rounded, was dangerous.

Note that both hardware binary16 and hardware binary128 are emerging in recent architectures. None of these new formats will be used blindly for all the variables of a program: binary16 because it is not accurate enough, binary128 because it is too expensive. Therefore, these new formats will very likely make mixed-precision programming more frequent. More generally, the idea of selecting the "just right" precision at each point of a program is attracting a lot of attention. This is another motivation for the study of nonhomogeneous operators.

In what follows, we first review in Section 7.8.1 a reasonably efficient purely software approach, proposed by Lutz and Burgess [400], that enables standard-compliant heterogeneous operators, in the case when only the homogeneous operators are available. Then we review in Section 7.8.2 a hardware heterogeneous FMA that has been suggested as a good tradeoff between the accuracy of the larger format, and the performance and resource consumption of the smaller format.

For illustration, we consider in this section the most common case, where we have hardware support for binary32 and binary64, but all the following applies to any pair of IEEE 754-2008 binary formats.

7.8.1 A software algorithm around double rounding

7.8.1.1 Problem statement

A system that claims support for binary32 and binary64 must offer, for each operation, *op*, the following correctly rounded variants:

binary32	*op*	binary32	$\rightarrow$	binary32
binary64	*op*	binary32	$\rightarrow$	binary32
binary32	*op*	binary64	$\rightarrow$	binary32
binary64	*op*	binary64	$\rightarrow$	binary32
binary32	*op*	binary32	$\rightarrow$	binary64
binary64	*op*	binary32	$\rightarrow$	binary64
binary32	*op*	binary64	$\rightarrow$	binary64
binary64	*op*	binary64	$\rightarrow$	binary64

For the 3-input FMA, there is a similar table with twice as many lines.

In this table, we assume that the hardware provides only the first and last variants (that is, the two homogeneous operators), and we now discuss the implementation of all the other lines.

7.8.1.2 The easy cases

A first key observation is that conversion from a smaller IEEE format to a wider IEEE format is always exact. Indeed, the set of representable numbers in the wider format is always a superset of the set of representable numbers in the smaller one. Besides, this conversion is easy to implement in hardware, and an instruction to this effect will likely exist in most instruction sets.

Therefore, the operations in the second half of the table (returning the larger format and possibly inputting an operand in the smaller format) are easily implemented by first converting all the input operands to the larger format, then using the homogeneous operation in the larger format.

Similarly, all the variants in the first half of the table can be implemented if we know how to implement the single format-reducing operation:

binary64 *op* binary64 $\rightarrow$ binary32

For the FMA and its variants, all we need is an FMA that inputs three binary64 and returns a correctly rounded binary32.

One option would be to provide these instructions in hardware. The following shows how to emulate them in software (albeit low-level software, since it requires bit manipulations and access to the inexact flag).

7.8.1.3 Implementing the format-reducing operation

A second key observation (detailed by case analysis in [400]) is that we have a double-rounding problem only when rounding to nearest, and only when the intermediate result (before the second rounding) is exactly between two numbers in the smaller format. Indeed, this case (and only this case) triggers the tie-breaking rule, which takes an arbitrary rounding decision that may not reflect the exact result of the operation.

The solution proposed in Algorithm 7.1 is therefore to detect this case and take specific measures when it happens:

Algorithm 7.1 Implementing a format-reducing operation with correct rounding to nearest

```
y ← homogeneous operation in the larger format
if y is a tie-breaking case for the smaller format then
  clear inexact flag
  y ← homogeneous operation in the larger format, but in RZ mode
  if inexact flag is set then
    set the least-significant bit of y to 1
  else
    y is exact, there will be no double-rounding problem
  end if
end if
round y to the smaller format in round to nearest mode
```

Here are a few explanations on this algorithm.

- The test for the tie-breaking situation is best performed by bit manipulations. For instance, for the binary64/binary32 case, it consists in testing if the lower 29 bits of the significand are exactly 10000000000000000000000000000.
- If the operation in RZ mode is exact (in which case the initial RN operation was exact, too), there is no double-rounding problem since an exact operation does not involve a rounding.
- Otherwise, since we were in the RN tie-breaking case, the significand in the larger format may only end with
  - 0111...111, in which case setting the least significant bit does not change it, and the final rounding will round down;
  - 1000...000, in which case setting the least significant bit converts this significand into 1000...001: this takes it out of the tie-breaking case, and it will round up.

A final remark on this algorithm is that the most expensive part (if we are in the tie-breaking case) is taken rarely.

7.8.2 The mixed-precision fused multiply-and-add

Implementations of all the heterogeneous variants of the FMA are mandated by the standard, and can be achieved by the previous algorithm. Among these, one is particularly useful: It computes $r = \circ(a \cdot b + c)$ where the addend c and the result r are of a larger format, for instance binary64 (double precision), while the multiplier inputs a and b are of a smaller format, for instance binary32 (single precision). Such a mixed-precision FMA will be denoted MPFMA in the following.

7.8.3 Motivation

For sum-of-product applications, the MPFMA provides the accumulation accuracy of the larger format at a cost that has been shown in [83] to be only one third more than that of a classical FMA in the smaller format. With minor modifications, it is also able to perform the standard FMA in the smaller format, and the standard addition in the larger format.

Besides, the MPFMA operator is perfectly compliant to the standards governing the languages C and Fortran. For instance, consider the following C code, archetypal of many computing kernels (including matrix operations, finite impulse response (FIR) filters, fast Fourier transforms (FFT), etc.).

C listing 7.1 Sum of products using a mixed-precision FMA

```
float A[], B[];    /* binary32 */
double C;          /* binary64 */
C=0;
for(i=0; i< N; i++)
   C = C + A[i]*B[i];
```

The reader is invited to check that, according to Section 6.2, using an MPFMA for the line `C = C + A[i]*B[i]` is compliant with the C standard.

Here are three use cases for such mixed-precision FMAs.

- For embedded applications, a mixed binary32/binary64 FMA will enable binary64 computing where it is most needed, at a small cost overhead with respect to current binary32 FMAs, and with fewer data transfers, hence lower power than a pure binary64 approach. This application context motivated the first implementation of an MPFMA [83, 84].
- In high-end processors, very large scale computing applications push towards binary128. There, a hardware FMA that accumulates products of binary64 into a binary128 could provide an adequate solution, at a much lower cost than full hardware binary128 support.
- To double their peak Teraflop/s performance on deep learning applications, the latest generations of GPUs provide full hardware support for the binary16 format. However, precision is very quickly lost in binary16 accumulations. Here again, accumulating in a binary32 will provide a cost-effective solution to scale binary16 computations beyond their precision limit.

7.8.4 Implementation issues

Let us write p for the precision of the multiplier operands (smaller format), and q the precision of the addend and result (larger format).

When one draws the alignment cases [83], the intermediate computation width (which was $3p + 5$ in Figure 7.3) becomes $q + 2p + 5$ bits. This is larger than $3p + 5$ bits (for an FMA in the smaller format), but smaller than $3q + 5$ (for an FMA in the larger format).

For all the pairs of formats mentioned above, we have $q \geq 2p + 2$ (see Table 3.1). In other words, the significand product of $a \times b$ is smaller than the significand of c. Therefore, another point of view is that this intermediate computation width is, roughly, the one of the floating-point addition in the larger format. The interested reader will find more details in [83].

Subnormal handling has some specificities.

- Contrary to the homogeneous case, if either a, or b, or even both are subnormals, the product ab nevertheless belongs in the normal range of the result format. This is easily checked using the values of the smallest subnormals given in Table 3.5. Managing these cases therefore resumes to normalizing this product, i.e., bringing its leading one in the leftmost position. This corresponds to a shift of up to $2p$ bits. The shift distance is the sum of the leading-zero counts (LZC) on the significands of a and b. These LZCs can be performed in parallel with the multiplication.
- Managing subnormal values of c has no overhead at all. If c is subnormal, then either $ab = 0$ and the result is c, or $ab \neq 0$ and then it is very far from the subnormal range of the (larger) result format (again see Table 3.5). In this last case the whole of c will be accounted for as a sticky bit.

This chapter was devoted to the algorithms for the basic operations. Now, we are going to focus on the implementation of these algorithms, in hardware (Chapter 8) and in software (Chapter 9).

Chapter 8

Hardware Implementation of Floating-Point Arithmetic

Chapter 7 has shown that operations on floating-point numbers are naturally expressed in terms of integer or fixed-point operations on the significand and the exponent. For instance, to obtain the product of two floating-point numbers, one basically multiplies the significands and adds the exponents. However, obtaining the correct rounding of the result may require considerable design effort and the use of nonarithmetic primitives such as leading-zero counters and shifters. This chapter details the implementation of these algorithms in hardware, using digital logic.

Describing in full detail all the possible hardware implementations of the needed integer arithmetic primitives is much beyond the scope of this book. The interested reader will find this information in the textbooks on the subject [345, 483, 187]. After an introduction to the context of hardware floating-point implementation in Section 8.1, we just review these primitives in Section 8.2, discuss their cost in terms of area and delay, and then focus on wiring them together in the rest of the chapter.

8.1 Introduction and Context

We assume in this chapter that inputs and outputs are encoded according to the IEEE 754-2008 Standard for Floating-Point Arithmetic.

8.1.1 Processor internal formats

Some systems, although compatible with the standard from a user point of view, may choose to use a different data format internally to improve perfor-

J.-M. Muller et al., *Handbook of Floating-Point Arithmetic*,
https://doi.org/10.1007/978-3-319-76526-6_8

mance. These choices are related to processor design issues that are out of the scope of this book. Here are a few examples.

- Many processors add *tag* bits to floating-point numbers. For instance, a bit telling if a number is subnormal saves having to detect it by checking that all the bits of the exponent field are zero. This bit is set when an operand is loaded from memory, or by the arithmetic operator if the number is the result of a previous computation in the floating-point unit: each operator has to determine if its result is subnormal anyway, to round it properly. Other tags may indicate other special values such as zero, infinities, and NaNs. Such tags are stored in the register file of the processor along with the floating-point data, which may accordingly not be fully compliant with the standard. For instance, if there is a tag for zero, there is no need to set the data to the full string of zeros in this case.

- The fused multiply-add (FMA) of the IBM POWER6 has a short-circuit feedback path which sends results back to the input. On this path, the results are not fully normalized, which reduces the latency on dependent operations from 7 cycles to 6. They can be normalized as part of the first stage of the FMA.

- An internal data format using a redundant representation of the significands has been suggested in [198].

- Some AMD processors have a separate "denormalization unit" that formats subnormal results. This unit receives data in a nonstandard format from the other arithmetic units, which alone do not handle subnormals properly.

- ARM implements a custom binary16 (half precision) format which trades encoding for special values (NaNs and infinites) to extend the exponent range. ARM architecture also implements an IEEE compliant binary16 format.

8.1.2 Hardware handling of subnormal numbers

In early processors, it was common to trap to software for the handling of subnormals. The cost could be several hundreds of cycles, which sometimes made the performance collapse each time subnormal numbers would appear in a computation. Conversely, most recent processors have fixed-latency operators that handle subnormals entirely in hardware. This improvement is partly due to very large-scale integration (VLSI): the overhead of managing subnormal numbers is becoming negligible with respect to the total area of a processor. In addition, several architectural improvements have also made the delay overhead acceptable.

An intermediate situation was to have the floating-point unit (FPU) take more cycles to process subnormal numbers than the standard case. The solution, already mentioned above, used in some AMD processors is a "denormalizing" unit that takes care of situations when the output is a subnormal number. The adder and multiplier produce a normalized result with a larger exponent. If this exponent is in the normal range, it is simply truncated to the standard exponent. Otherwise, that result is sent to the denormalizing unit which, in a few cycles, will shift the significand to produce a subnormal number. This can be viewed as a kind of "trap," but one that is managed in hardware. An alternative approach, used in some IBM processors, saves the denormalizing unit by sending the number to be denormalized back to the shifter of the adder. The problem is then to manage conflicts with other operations that might be using this shifter.

The 2005 state of the art concerning subnormal handling in hardware was reviewed by Schwarz, Schmookler, and Trong [549]. They showed that subnormal numbers can be managed with relatively little overhead, which explains why most recent FPUs in processors handle subnormal numbers in hardware. This is now even the case in graphics processing units (GPUs), the latest of which provide binary64 standard-compatible hardware. We will present, along with each operator, some of the techniques used. The interested reader is referred to [549] and references therein for more details and alternatives.

8.1.3 Full-custom VLSI versus reconfigurable circuits (FPGAs)

Most floating-point architectures are implemented as full-custom VLSI in processors or GPUs. There has also been a lot of interest in the last decade in floating-point acceleration using reconfigurable hardware, with field-programmable gate arrays (FPGAs) replacing or complementing processors. The feasibility of floating-point arithmetic on FPGA was studied long before it became a practical possibility [556, 386, 389]. At the beginning of the century, several libraries of floating-point operators were published almost simultaneously (see [464, 374, 388, 512] among others). The increase of capacity of FPGAs soon meant that they could provide more floating-point computing power than a processor in binary32 [464, 388, 512], then in binary64 [642, 179, 159, 392, 251, 30]. FPGAs also revived interest in hardware architectures for the elementary functions [178, 172, 171, 174, 154, 141, 144, 141, 364, 143, 601, 366] and other coarser or more exotic operators [643, 69, 155, 227]. This will be reviewed in Section 8.8.

Current Intel FPGAs include embedded floating-point hardware [365], and high-level FPGA design tools support floating-point computing [559].

We will survey floating-point implementations for both full-custom VLSI and FPGA. The performance metrics of these targets may be quite different (they will be reviewed in due course), and so will be the best implementation of a given operation.

By definition, floating-point implementation on an FPGA is application-specific. The FPGA is programmed as an "accelerator" for a given problem, and the arithmetic operators will be designed to match the requirements of the problem but no more. For instance, most FPGA implementations are parameterized by exponent and significand sizes, not being limited to those specified by the IEEE 754 standard. In addition, FPGA floating-point operators are designed with optional subnormal support, or no support at all. If tiny values appear often enough in an application to justify subnormal handling, the application can often be fixed at a much lower hardware cost by adding one bit to the exponent field. This issue is still controversial, and subnormal handling is still needed for some applications, including those which require bit-exact results with respect to a reference software.

8.1.4 Hardware decimal arithmetic

Most of the research so far has focused on *binary* floating-point arithmetic. It is still an open question whether it is worth implementing decimal arithmetic in hardware [122, 193, 121, 613, 628], or if a software approach [116, 117], possibly with some minor hardware assistance, is more economical. This chapter covers both hardware binary and hardware decimal, but the space dedicated to hardware decimal reflects the current predominance of binary implementations.

As exposed in detail in Section 3.1.1.2, the IEEE 754-2008 standard specifies two encodings for the decimal numbers, corresponding to the two main competing implementations of decimal arithmetic at the time IEEE 754-2008 was designed. The *binary* encoding allows for efficient software operations, using the native binary integer operations of a processor. It is probable that processor instruction sets will be enriched to offer hardware assistance to this software approach. The *decimal* encoding, also known as *densely packed decimal* or *DPD*, was designed to make a hardware implementation of decimal floating-point arithmetic as efficient as possible.

In the DPD format, the significand is encoded as a vector of radix-1000 digits, each encoded in 10 bits (declets). This encoding is summarized in Tables 3.9, 3.10, and 3.11. It was designed to facilitate the conversions: all these tables have a straightforward hardware implementation, and can be implemented in three gate levels [184]. Although decimal numbers are stored in memory in the DPD format, hardware decimal arithmetic operators internally use the simpler binary coded decimal (BCD) representation, where each decimal digit is straightforwardly encoded as a 4-bit number.

Considering that the operations themselves use the BCD encoding, should the internal registers of a processor use DPD or BCD? On one hand, having the registers in BCD format saves the time and power of converting the input operands of an operation to BCD, then converting the result back to DPD—to be converted again to BCD in a subsequent operation. On the other hand, as pointed out by Eisen et al. [184], it is unlikely that an application will intensively use binary floating-point and decimal floating-point at the same time; therefore, it makes sense to have a single register file. The latter will contain 64-bit or 128-bit registers, which is a strong case for accepting DPD numbers as inputs to a decimal FPU.

8.1.5 Pipelining

Most floating-point operator designs are pipelined. There are three characteristics of a pipelined design:

- its frequency, which is the inverse of the cycle time;
- its depth or latency, which is the number of cycles it takes to obtain the result after the inputs have been presented to the operator;
- its silicon area.

IEEE 754-compliant operations are combinatorial (memory-less) functions of their inputs. Therefore, one may in principle design a combinational (unpipelined) operator of critical path delay T, then insert n register levels to convert it into a pipelined one of latency n. If the logical depth between each register is well balanced along the critical path, the cycle time will be close to T/n—it will always be larger due to the delay incurred by the additional registers. These additional registers also add to the area. This defines a technology-dependent tradeoff between a deeply pipelined design and a shallower one [564]. For instance, studies suggested an optimal delay of 6 to 8 fanout-of-4 (FO4) inverter delays per stage if only performance is considered [262], but almost double if power is taken into consideration [649].

Besides, some computations will not benefit from deeper pipelines. An addition using the result of another addition must wait for this result. This is called a *data dependency*. The deeper the pipeline, the more cycles during which the processor must wait. If there are other operations that can be launched during these cycles, the pipeline will be fully utilized, but if the only possible next operation is the dependent operation, the processor will be idle. In this case, the pipeline is not used to its full capacity, and this inefficiency is worse for a deeper pipeline. Very frequently, due to data dependencies, a deeper pipeline will be less efficient, in terms of operations per cycle, than a shallower one. This was illustrated by the transition from the P6 microarchitecture (used among others in the Pentium III) to the NetBurst microarchitecture (used among others in the Pentium 4). The latter had roughly twice as deep a pipeline as the former, and its frequency could be almost twice as high, but it was not uncommon that a given piece of floating-point code would need almost twice as many cycles to execute [148], negating the performance advantage of the deeper pipeline. Considering this, recent Intel microprocessors (most notably the Core microarchitecture) have stepped back to shallower pipelines, which have additional advantages in terms of power consumption and design complexity. Relatedly, an ARM paper [399] suggests that it is worth splitting a complex FMA architecture into two shallower pipelines (one for the multiplication, one for addition). This speeds up dependent computations by lowering the latency of elementary operations.

In the design of a processor, the target frequency is decided early, which imposes a limit on the logic depth delay in each pipeline level. One then tries to keep the number of cycles as close as possible to the theoretical optimal (obtained by dividing the critical path delay by the cycle time). In general-purpose processors, the area of the arithmetic operators is not as much of a problem. Thanks to Moore's law, transistors are cheap, and the floating-point units account for a few percent of the area of a processor anyway. It is not uncommon to replicate computations if this decreases the critical path. In the following pages, we will see many examples.

In GPUs or reconfigurable circuits, the context is slightly different. The applications being targeted to GPUs and FPGAs will expose enough parallelism so that the issue of operation latency will not be significant: due to very cheap context switches, the cycles between dependent operations will be filled with other computations. On the other hand, one wants as many operators as possible on a chip to maximally exploit this parallelism. Therefore, floating-point operators on GPUs and FPGAs tend to favor small area over small latency.

8.2 The Primitives and Their Cost

Let us first review the primitives which will be required for building most operations. These include integer arithmetic operations (addition/subtraction, multiplication, integer division with remainder) but also primitives used for significand alignment and normalization (shifters and leading-zero counters). Small tables of precomputed values may also be used to accelerate some computations.

These primitives may be implemented in many different ways, usually exposing a range of tradeoffs between speed, area, and power consumption. The purpose of this section is to expose these tradeoffs.

8.2.1 Integer adders

Integer addition deserves special treatment because it appears in virtually all floating-point operators, though with very different requirements. For instance, the significand addition in a floating-point adder needs to be as fast as possible, while operating on numbers of fairly large width. In contrast, exponent addition in a floating-point multiplier operates on smaller integers, and does not even need to be fast, since it can be performed in parallel with the significand product. As a third example, integer multiplication can be

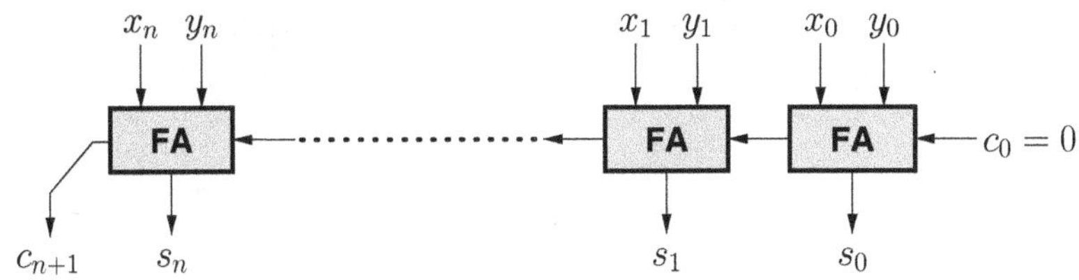

Figure 8.1: Carry-ripple adder.

expressed in terms of iterated additions, and adders may be designed specifically for this context.

Fortunately, a wide range of adder architectures has been proposed and allows us to address each of these needs. We present each adder family succinctly. The interested reader will find details in the bibliographic references (e.g. [187, 341, 446, 503, 582, 645, 554, 85, 610, 486]).

8.2.1.1 Carry-ripple adders

The simplest adder is the *carry-ripple adder*, represented in Figure 8.1 for binary inputs and in Figure 8.2 for decimal BCD inputs. Carry-ripple addition is simply the paper-and-pencil algorithm learned at school. A carry-ripple adder has $\mathcal{O}(n)$ area and $\mathcal{O}(n)$ delay, where n is the size in digits of the numbers to be added.

The building block of the binary carry-ripple adder is the *full adder* (FA), which outputs the sum of three input bits x_i, y_i, and z_i (this sum is between 0 and 3) as a 2-bit binary number $c_i s_i$. Formally, it implements the equation $2c_i + s_i = x_i + y_i + z_i$. The full adder can be implemented in many ways with a two-gate delay [187]. Typically, one wants to minimize the delay on the carry propagation path (horizontal in Figure 8.1). Much research has been dedicated to implementing full adders in transistors; see, for instance, Zimmermann [645] for a review. A clean CMOS implementation requires 28 transistors, but many designs have been suggested with as few as 10 transistors (see [641, 1, 554, 85] among others). These transistor counts are given for illustration only: smaller designs have limitations, for instance they cannot be used for building carry-ripple adders of arbitrary sizes. The best choice of a full-adder implementation depends much on the context in which it is used.

Carry-ripple adders can be built in any radix β (take $\beta = 10$ for illustration). The basic block DA (for digit addition) now computes the sum of an input carry (0 or 1) and two radix-β digits (between 0 and $\beta - 1$). This sum is between 0 and $2\beta - 1$ and can therefore be written in radix β as $c_i s_i$, where c_i is an output carry (0 or 1) and s_i is a radix-β digit.

Useful radices for building hardware floating-point operators are 2, small powers of 2 (in which case the DA block is simply a binary adder as

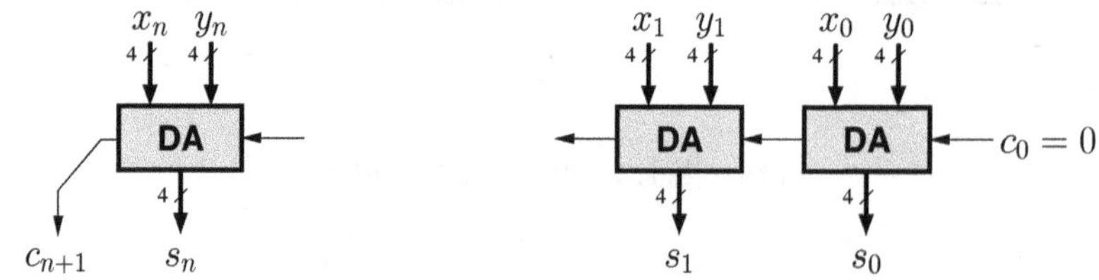

Figure 8.2: Decimal addition. Each decimal digit is coded in BCD by 4 bits.

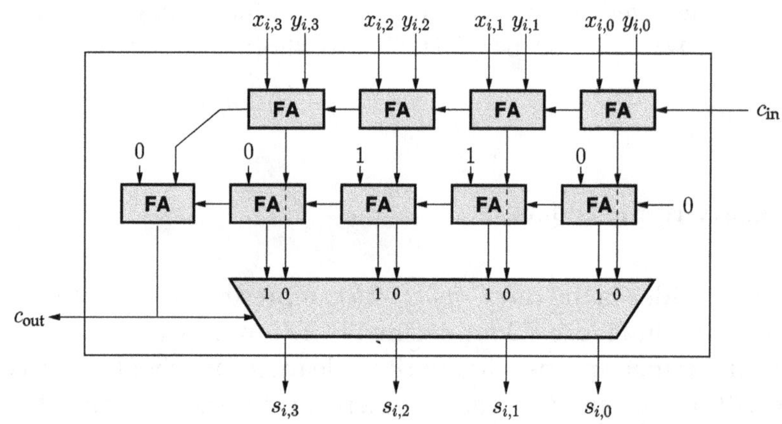

Figure 8.3: An implementation of the decimal DA box.

shown in Figure 8.4), 10, and small powers of 10. Figure 8.2 gives the example of a radix-10 ripple-carry adder. The implementation of a typical decimal DA block is depicted by Figure 8.3. It first computes $x_i + y_i + z_i$ using a binary 4-bit adder. To detect if the sum is larger than 10, a second 5-bit adder adds 6 to this sum. The carry out of this adder is always 0 (the sum is at most $9 + 9 + 1 + 6 = 25 < 32$) and can be ignored. The bit of weight 2^4 is the decimal carry: it is equal to 1 iff $x_i + y_i + z_i + 6 \geq 16$, i.e., $x_i + y_i + z_i \geq 10$. The sum digit is selected according to this carry. Many low-level optimizations can be applied to this figure, particularly in the constant addition. However, comparing Figures 8.4 and 8.3 illustrates the intrinsic hardware overhead of decimal arithmetic over binary arithmetic.

Many optimizations can also be applied to a radix-2^p adder, such as the one depicted in Figure 8.4, particularly to speed up the carry-propagation path from c_{in} to c_{out}. In a carry-skip adder [187], each radix-2^p DA box first computes, from its x_i and y_i digit inputs (independently of its c_i input, therefore in parallel for all the digits), two signals g_i (for generate) and p_i (for propagate), such that $c_{i+1} = g_i$ OR (p_i AND c_i). Carry propagation then takes only 2 gate delays per radix-2^p digit instead of $2p$ gate delays when the DA box is implemented as per Figure 8.4. The drawback is that the DA box is

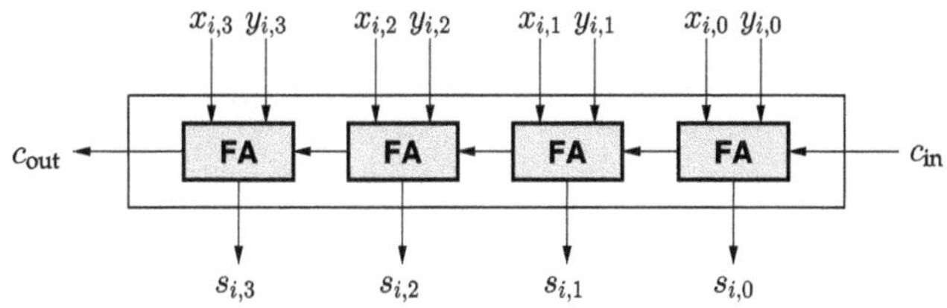

Figure 8.4: An implementation of the radix-16 DA box.

now larger: this family of adders exposes a tradeoff between area and adder delay.

8.2.1.2 Parallel adders

The hardware of a carry-ripple adder can be used to perform *carry-save* addition in $\mathcal{O}(n)$ area and $\mathcal{O}(1)$ time, as shown in Figure 8.5. The catch is that the result is obtained in a nonstandard redundant number format: Figure 8.5 shows a carry-save adder that adds three binary numbers and returns the sum as two binary numbers: $R = \sum_{i=0}^{n+1}(c_i + s_i)2^i$.

The cost of converting this result to standard representation is a carry propagation, so this is only useful in situations when many numbers have to be added together. This is the case, e.g., for multiplication.

As shown in Figure 8.6 for radix 2^{32}, this idea works for any radix. A radix-β carry-save number is a vector of pairs (c_i, s_i), where c_i is a bit and s_i a radix-β digit, representing the value $\sum_{i=0}^{n}(c_i + s_i)\beta^i$. A radix-$\beta$ carry-save adder adds one radix-β carry-save number and one radix-β number and returns the sum as a radix-β carry-save number.

Radix-2^p carry-save representation is also called *partial carry save*. It allows for the implementation of very large adders [640] or accumulators [354, 155] working at high frequency in a pipelined way. It has also been proposed that its use as the internal format for the significands inside a processor floating-point unit would reduce the latency of most operations [198]. Conversions between the IEEE 754-2008 interchange formats presented in Section 3.1.1 and this internal format would be performed only when a number is read from or written to memory.

8.2.1.3 Fast adders

It is usual to call *fast adders* adders that take their inputs and return their results in standard binary representation, and compute addition in $\mathcal{O}(\log n)$ time instead of $\mathcal{O}(n)$ for carry-ripple addition. They require more area ($\mathcal{O}(n \log n)$ instead of $\mathcal{O}(n)$). In this case, the constants hidden behind the $\mathcal{O}$ notation are important.

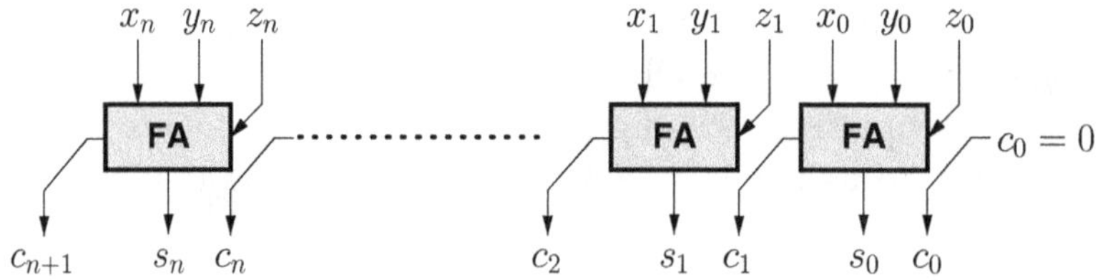

Figure 8.5: Binary carry-save addition.

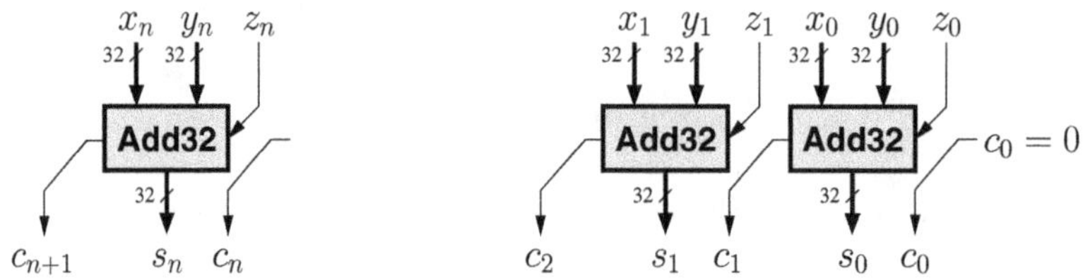

Figure 8.6: Partial carry-save addition.

Here is a primer on *prefix tree adders* [187], a very flexible family of fast adders. For any block $i : j$ of consecutive bits of the addition with $i \geq j$, one may define a *propagate* signal $P_{i:j}$ and a *generate* signal $G_{i:j}$. These signals are defined from the input bits of the range $i : j$ only, and have the following meanings: $P_{i:j} = 1$ iff the block will propagate its input carry, whatever its value; $G_{i:j} = 1$ iff the block will generate an output carry, regardless of its input carry. For a 1-bit block we have $G_{i:i} = a_i$ AND b_i and $P_{i:i} = a_i$ XOR b_i. The key observation is that generate and propagate signals can be built in an associative way: if $i \geq j \geq k$, we have

$$(G_{i:k}, P_{i:k}) = (G_{i:j} \text{ OR } (P_{i:j} \text{ AND } G_{j:k}),\ P_{i:j} \text{ AND } P_{j:k}),$$

which is usually noted $(G_{i:k}, P_{i:k}) = (G_{i:j}, P_{i:j}) \bullet (G_{j:k}, P_{j:k})$. The operator $\bullet$ is associative: for any value of the bits $a_1, b_1, a_2, b_2, a_3, b_3$, we have

$$((a_1, b_1) \bullet (a_2, b_2)) \bullet (a_3, b_3) = (a_1, b_1) \bullet ((a_2, b_2) \bullet (a_3, b_3)).$$

Because of this associativity, it is possible to compute in parallel in logarithmic time all the $(G_{i:0}, P_{i:0})$, using a parallel-prefix tree. The sum bits are then computed in time $\mathcal{O}(1)$ as $s_i = a_i$ XOR b_i XOR $G_{i:0}$. This recurrence was first studied by Kogge and Stone in [344] with a recurrence generalizing the adder problem.

The family of prefix tree adders has the advantage of exposing a wide range of tradeoffs, with delay between $\mathcal{O}(\log n)$ and $\mathcal{O}(n)$ and area between $\mathcal{O}(n \log n)$ and $\mathcal{O}(n)$.

In addition, the tradeoff can be expressed in other terms such as fan-out (the number of inputs connected to the output of a gate) or in terms of wire length. With submicrometer VLSI technologies, the delay and power consumption are increasingly related to wires. Some of the fastest theoretical adder variants of parallel prefix (often named Kogge-Stone) cannot be implemented in practice due to the overwhelming fan-out and wire routing constraints. See [486] for a survey on prefix adder tradeoff.

Another advantage of prefix adders is that computing $A + B + 1$ on top of the computation of $A + B$ comes at the price of $\mathcal{O}(n)$ additional hardware and $\mathcal{O}(1)$ additional delay. Once the $(G_{i:0}, P_{i:0})$ are available, the bits of $A + B + 1$ are defined by setting the input carry to 1 as follows: $s'_i = a_i \text{ XOR } b_i \text{ XOR } (G_{i:0} \text{ OR } P_{i:0})$ [610]. A fast adder designed to compute both $A+B$ and $A+B+1$ is called a *compound adder* [552]. It is useful for floating-point operators, typically because the rounded significand is either a sum or the successor of this sum.

A variation of the compound adder computes $|A - B|$ when A and B are two positive numbers (e.g., significands). In that case, using two's complement, one needs to compute either $A - B = A + \overline{B} + 1$ or $B - A = \overline{A + \overline{B}}$, depending on the most significant bit s_{n-1} of $A - B$ which is set if $A - B$ is negative (see Figure 8.12). This approach is sometimes referred to as an *end-around carry adder*. As this variation can be used to compute the absolute value of a subtraction, it is also very useful when implementing a floating-point operation: as floating-point formats use a sign-magnitude encoding, operations often require computing the absolute value of the result significand.

To summarize, the design of an adder for a given context typically uses a mixture of all the techniques presented above. See [640] for an example.

8.2.1.4 Fast addition in FPGAs

In reconfigurable circuits, the area is measured in terms of the elementary operation, which is typically an arbitrary function of 4 to 6 inputs implemented as a look-up table (LUT). As the routing is programmable, it actually accounts for most of the delay: in a circuit, you just have a wire, but in an FPGA you have a succession of wire segments and programmable switches [138]. However, all current FPGAs also provide *fast-carry* circuitry. This is simply a direct connection of each LUT to one of its neighbors which allows carry propagation to skip the slow generic routing.

The net consequence is that fast (logarithmic) adders are irrelevant to FPGAs up to very large sizes [638, 462]. For illustration, the peak practical frequency of a high-end FPGA is roughly defined by the delay of traversing one register, one LUTs, and a few programmable wires. In this cycle time, one may typically also perform a 32-bit carry-propagation addition using the fast-carry logic. A simple carry select adder (Figure 8.7) runs at the peak practical frequency for additions up to 64 bits. The fast-carry logic can also be abused to implement short-latency adders of larger sizes [462].

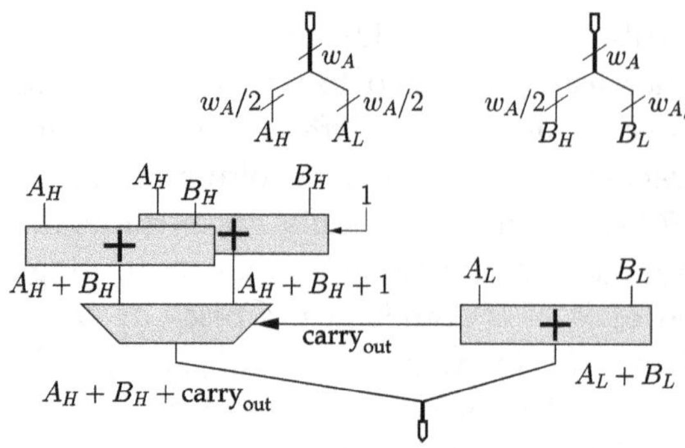

Figure 8.7: Carry-select adder.

8.2.2 Digit-by-integer multiplication in hardware

The techniques learned at school for paper-and-pencil multiplication and division require the computation of the product of a digit by a number. This operation also occurs in hardware versions of these algorithms.

Multiplying an n-bit number by a bit simply takes n AND gates operating in parallel in $\mathcal{O}(1)$ time. However, as soon as the radix β is larger than two, the product of two digits in $\{0, \ldots, \beta - 1\}$ does not fit on a single digit, but on two. Computing the product of a digit by a number then takes a carry propagation, and time is now $\mathcal{O}(n)$, still for $\mathcal{O}(n)$ area with the simplest operators (the previously seen methods for fast addition can readily be adapted). Similar to addition, if it is acceptable to obtain the product in nonstandard redundant format, this operation can be performed in $\mathcal{O}(1)$ time.

For small radices, digit-by-integer multiplication in binary resumes to a few additions and constant shifts. For instance, if the radix is a power of 2, $2X = X \ll 1$ (X shifted left by one bit), $3X = X + (X \ll 1)$, etc. All the multiplications up to $10X$ can be implemented by a single addition/subtraction and constant shifts. These additions require a carry propagation. We now see a general method, *recoding*, that can avoid these carry propagations.

8.2.3 Using nonstandard representations of numbers

Note that the radix here is not necessarily 2 nor 10. For instance, considering a binary number as a radix-4 number simply means considering its digits two by two. An operation expressed in radix 4 will have fewer digit operations than the same operation expressed in radix 2. However, each digit operation will be more complex. The choice of a larger radix may allow one to reduce the number of cycles for an operation while maximizing the amount of computation done within one cycle. For instance, Intel reduced the latency of division in their x86 processors by replacing the radix-4 division algorithm with a radix-16 algorithm [431].

The digit set to be used in radix β is not necessarily $\{0, \ldots, \beta - 1\}$. Any set of at least β consecutive digits including 0 allows for representation of any number. Furthermore, it is common to use a *redundant* digit set, i.e., a digit set with more than β digits. The value of a string of p digits $(d_{p-1}d_{p-2}\cdots d_0)$ in radix β is still

$$N = \sum_{i=0}^{p-1} d_i \beta^i \quad .$$

In such a system, some numbers will have more than one representation.

A digit set including *negative* digits makes it possible to represent negative numbers (these redundant number systems with negative digits are sometimes called *signed-digit* number systems).

Example 8.1 (Radix 10 with digit set {-5,-4,-3,-2,-1,0,1,2,3,4,5}). *Let us note the negative digits with an overline, e.g.,* $\overline{5} = -5$. *Here are some examples of number representations in this system:*

$$\begin{aligned} 0 &= 0 \\ 9 &= 1\overline{1} \quad (= 10 - 1) \\ -8 &= \overline{1}2 \quad (= -10 + 2) \\ 17 &= 2\overline{3} \\ 5 &= 05 = 1\overline{5} \quad (= -10 + 2) \quad . \end{aligned}$$

The main point of redundant digit sets is that they provide more algorithmic freedom. For instance, carry propagation can be prevented by choosing, among two possible representations of the sum, the one that will be able to consume an incoming carry. This is the trick behind Avizienis' signed-digit addition algorithm [21]. Another example where this freedom of choice is useful is SRT division, introduced in Section 8.6.

Nonstandard digit sets also have other advantages. A trick commonly used to speed up binary multiplication is *modified Booth recoding* [57]. It consists in rewriting one of the operands of the multiplication as a radix-4 number using the digit set $\{\overline{2}, \overline{1}, 0, 1, 2\}$. The advantage is that multiplication of the other operand by any of these recoded digits still consists of a row of AND as in binary, and a possible shift. As the digits are in radix 4, they are twice as few. Compared to using standard radix 4 with digits in $\{0, 1, 2, 3\}$, we no longer have the carry propagation that was needed to compute $3X = X + 2X$.

Booth recoding can be done in $\mathcal{O}(1)$ time using $\mathcal{O}(n)$ area. In a first step, the initial radix-4 digits in $\{0, 1, 2, 3\}$ are rewritten in parallel as follows: 0 and 1 are untouched, 2 is rewritten $4 + \overline{2}$, 3 is rewritten $4 + \overline{1}$. Here the 4 corresponds to a carry in radix 4, sent one digit position to the left. In a second step, these carries are added, again in parallel, to the digits from the previous step, which belonged to $\{\overline{2}, \overline{1}, 0, 1\}$. The resulting digits indeed belong to $\{\overline{2}, \overline{1}, 0, 1, 2\}$. Again, thanks to redundancy, there was no carry propagation.

Recoding can also be used for decimal multiplication and will be surveyed in Section 8.2.5.

8.2.4 Binary integer multiplication

There are many ways of implementing integer multiplication with a range of area-time tradeoffs [187]. We focus here on high-performance implementations. They are typically performed in three steps, illustrated in Figure 8.8.

- First, one of the operands is Booth recoded, and each of its digits is multiplied by the other operand. This results in a *partial product array* of $(n+1) \times n/2$ weighted bits. The result of the multiplication will be the sum of these weighted bits. This step can be performed in constant time.

- A second step reduces this array to only two lines using several carry-save adders, or more generally, *compressors*. For instance, the 3:2 compressor is the carry-save adder of Figure 8.5, a 4:2 compressor takes 4 binary numbers and writes their sum as two binary numbers (i.e., as a carry-save number). A 4:2 compressor can be implemented as two carry-save adders, but it may also be implemented more efficiently (using fewer transistors). It has been argued that a good 4:2 compressor implementation may perform the same function as Booth recoding in less time using less resources [618].

  This step can be performed in $\mathcal{O}(\log n)$ time, using a tree of compressors. Compression tree schemes have been extensively studied, for example by Wallace [624], Dadda [126, 127] and others [88, 605].

- Finally, the carry-save result of the previous step is summed using a fast adder in $\mathcal{O}(\log n)$ time.

This multiplier scheme is quite flexible. It easily accommodates signed integers at no extra cost. More important for floating-point, rounding can be performed at almost no cost by adding a few bits in the partial product array, as detailed in Section 8.4. Finally, computing a multiply-and-add $a \times b + c$ adds only one more line of bits (corresponding to c) to the initial partial product array of $a \times b$ depicted in Figure 8.8. In practice, the requirement of correct rounding of the FMA makes the overall data path much more complex; see Section 8.5.

8.2.5 Decimal integer multiplication

The basic scheme presented above can be used for decimal multiplication. The design of a carry-save decimal adder is straightforward. The partial product array may be built by computing all the digit-by-digit products in parallel, as illustrated by Figure 8.9. Notice that this figure is different from the one obtained using the paper-and-pencil algorithm, which needs a carry propagation to compute the product of one number by one digit.

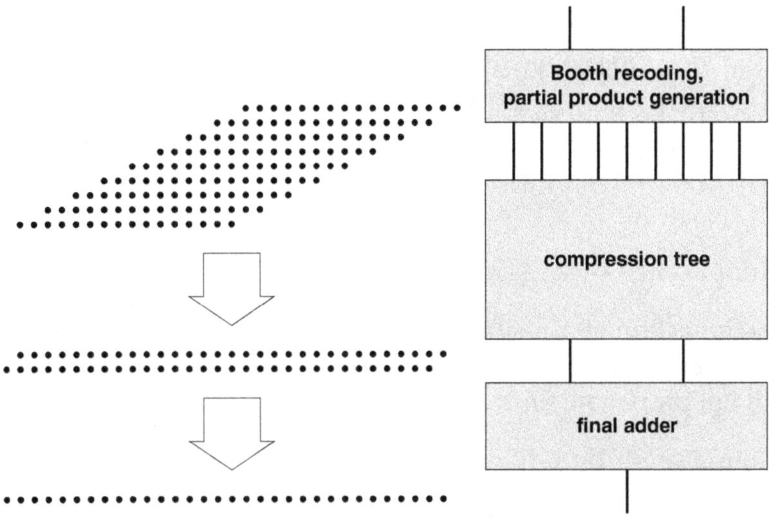

Figure 8.8: Binary integer multiplication.

```
        3 8 2
   ×    9 7 6
   ----------
        8 8 2
      1 4 1
      1 6 4
    2 5 1
    7 2 8
  2 7 1
  -----------
  3 7 2 8 3 2
```

Figure 8.9: Partial product array for decimal multiplication. One digit product, $9 \times 2 = 18$, is highlighted.

Here also, recoding can help in several ways. Recoding the decimal digits to $d_l \in \{-5, \ldots, 5\}$, for instance, and handling signs separately, reduces the number of digit-by-digit products to compute from 100 to 36, with a corresponding reduction in the complexity of the digit-by-digit multiplier [194]. Interestingly enough, Cauchy already had that idea (of course, in a different context!) in 1840 [91]. In [360], each decimal digit is recoded as the sum of two digits, $d_h + d_\ell$, with $d_h \in \{0, 5, 10\}$ and $d_\ell \in \{-2, -1, 0, 1, 2\}$. Multiplying a decimal input by such a recoded digit still leads to two lines in the partial product array, as in Figure 8.9, but the computation of these two lines requires very little hardware.

- Multiplication by 0, ± 1, and ± 10 is almost "for free."
- Multiplication by 2 resumes to a shift and a 4-bit addition. A carry may be produced to the next digit position, but, added to an even digit, it will not propagate further.
- Multiplication by 5 is performed as multiplication by 10 (digit shift) then division by 2. The latter is performed digit-wise. For odd digits, there is a fractional part equal to 0.5. It is sent as a $+5$ to the digit position to the right. Similar to the previous case, this "anti-carry" may not propagate any further.

In [614], the standard BCD code is named BCD-8421 (indicating the weights of the 4 bits of a digit), and different codings of decimal digits are suggested: BCD-4221 or BCD-5211. The advantage of these codes is that they are slightly redundant, which in practice enables faster decimal carry-save addition. More alternatives can be found in Vásquez' PhD dissertation [613].

Other variations on decimal multiplication may be found in [191, 628]. Earlier decimal architectures were more sequential, more or less like the paper-and-pencil algorithm [87, 190]. In this case, a first step may be to compute multiples of the multiplicand and store them in registers.

In [453], Neto and Véstias implement decimal multiplication in FPGAs using a radix-1000 binary format. This is motivated by the availability of embedded multipliers able to perform the product of two 10-bit numbers and return the exact product. They also propose addition-based radix conversion algorithms that exploit the fast addition fabric of FPGAs.

8.2.6 Shifters

Shifters are used for two purposes in a floating-point operation: aligning operands (e.g., in an addition or FMA) and normalization. Normalizing a binary floating-point result means bringing its leading "1" to the first position of the significand (i.e., the most significant bit). If this "1" was preceded by a string of zeros, an idea is to count these zeros first, and then feed this count to a shifter. Note that a simple way of handling subnormal inputs is to normalize them to an internal format with a few extra exponent bits. This concerns practically all operations that need to manage subnormal inputs.

A hardware shifter (commonly called a *barrel shifter*) takes one input x of size n (the number to be shifted) and one input d of size $\lceil \log_2 n \rceil$ (the shift distance). It consists of $\lceil \log_2 n \rceil$ stages. The i-th stage considers the i-th bit of d, say d_i, shifts by 2^i if $d_i = 1$ and does nothing otherwise, using a 2:1 multiplexer. Ignoring fan-in and fan-out issues, such a multiplexer for a data of n bits has area $\mathcal{O}(n)$ and delay $\mathcal{O}(1)$. Therefore, the area of a complete barrel shifter is $\mathcal{O}(n \log n)$, and its delay is $\mathcal{O}(\log n)$.

Shifting to a constant distance is cheaper, as it reduces to wires and possibly a multiplexer, in $\mathcal{O}(n)$ area and $\mathcal{O}(1)$ time.

8.2.7 Leading-zero counters

If an intermediate result may produce an arbitrary number of leading zeros (a typical case is a cancellating subtraction), it is necessary to count these leading zeros. A leading-zero counter (LZC) provides the shift value that will allow one to normalize the result.

Of course, one may derive from an LZC an architecture that counts leading ones. A leading-zero-or-one counter (LZOC) may also be useful: if an operation may return a negative number and the latter is available in two's complement, it will have to be complemented and normalized. In this case, one may perform the leading-zero counting in parallel with the complementing: what needs to be counted is the numbers of bits identical to the most significant bit (MSB) of the two's complement value.

8.2.7.1 Tree-based leading-zero counter

In general, counting the leading zeros in a number of size n can be performed in a binary tree manner in $\mathcal{O}(\log n)$ time and $\mathcal{O}(n)$ area. The basic algorithm is the following [473]. The first level considers pairs of adjacent bits and associates to each pair a 2-bit number that counts the leading zeros in this pair. The second level considers groups of 4 consecutive bits and builds 3-bit numbers that count the leading zeros in each group, and so on. Each computing node at level i considers groups of 2^i consecutive bits and outputs the leading-zero count for such a group as an $(i+1)$-bit number. A node is able to build this count using simple multiplexers out of the counts of the previous level, because of the following observation: the MSB of such a count is 1 if and only if the group only consists of zeros. Algorithm 8.1 makes explicit the operation of a node.

Algorithm 8.1 One node of the i-th level of an LZC tree.

Inputs: two i-bit numbers L and R
$l \leftarrow$ most significant bit of L
$r \leftarrow$ most significant bit of R
if $l = 0$ (there is a 1 on the left group) **then**
  $D \leftarrow 0L$ (ignore the right group)
else if $l = 1$ and $d = 1$ (both groups are all zeros) **then**
  $D \leftarrow 2^{i+1}$ (written $10\ldots0$)
else if $l = 1$ and $d = 0$ (all zeros on the left) **then**
  $D \leftarrow 1R$ (add 2^i to the count of the right group)
end if
Returns: the $(i+1)$-bit number D

Many variations of this algorithm, e.g., using coarser levels, can be designed to match the constraints or requirements of a given operator.

8.2.7.2 Leading-zero counting by monotonic string conversion

A completely different approach to leading-zero counting consists in first converting the input into a monotonic string, i.e., a bit string of the same size, which has the same leading zeros as the input, but only ones after the leading one. This is a very simple prefix-OR computation that may be performed efficiently in hardware in several ways [543]. By ANDing this string with itself negated and shifted by one bit position, one gets a string $S = s_{n-1}\ldots s_1s_0$ consisting of only zeros, except at the position of the leading one. From this last string, each bit of the count R can be computed as an $n/2$-wide OR. This approach is also well suited to *leading-zero anticipation* techniques, which will be presented in Section 8.3.3.

8.2.7.3 Combined leading-zero counting and shifting for FPGAs

Algorithm 8.2 combines leading-zero counting and shifting, which is what is needed for floating-point normalization.

Algorithm 8.2 Combined leading-zero counting and shifting.

Inputs: an n-bit number x
$k \leftarrow \lceil \log_2 n \rceil$
$x_k \leftarrow x$
for $i = k - 1$ downto 0 **do**
  if there are 2^i leading zeros in x_{i+1} **then**
    $d_i \leftarrow 1$
    $x_i \leftarrow x_{i+1}$, shifted left by 2^i
  else
    $d_i \leftarrow 0$
    $x_i \leftarrow x_{i+1}$
  end if
end for
Returns: (d, x_0)

The theoretical delay of this algorithm is worse than that of an LZC followed by a shifter (both in $\mathcal{O}(\log n)$). Indeed, the test

if there are 2^i leading zeros in x_{i+1}

is itself an expensive operation: implemented as a binary tree, its delay is in $\mathcal{O}(i)$. The total delay would therefore be $\mathcal{O}((\log n)^2)$.

However, in recent FPGAs, the dedicated hardware accelerating carry-propagation addition can also be used to compute the required wide OR operation. Thus, the practical delay of leading-zero counting in Algorithm 8.2 is completely hidden in that of the shifter for input sizes up to roughly 30 bits. If the input is larger, pipeline levels may have to be inserted to reach the peak FPGA frequency, but this only concerns the first few stages. As this algorithm also leads to a compact implementation, it will be preferred over an LZC followed by a shifter.

8.2.8 Tables and table-based methods for fixed-point function approximation

There are several situations where an approximation to a function must be retrieved from a table. In particular, division and square root by functional iteration require an initial approximation to the reciprocal, square root, or reciprocal square root of a significand [482, 130, 131, 132, 574, 350]. Many other applications exist, especially in the field of reconfigurable computing. For instance, table-based architectures for floating-point elementary functions are presented in [172] (exponential and logarithm), [171] (sine and cosine), and [181] (power function). Multiplication by a constant may also resort to precomputed tables, as will be detailed in Section 8.8.2. All these tables make sense in an architecture targeting FPGAs, because these circuits include a huge number of memory elements (from the LUT to larger configurable RAMs).

8.2.8.1 Plain tables

A table with an n-bit address may hold 2^n values. If each of these values is stored as m bits, the total number of bits stored in the table is $m \cdot 2^n$. Such a table is typically implemented in VLSI technology as a ROM (read-only memory) of area $\mathcal{O}(m \cdot 2^n)$. The delay is that of the address decoding, in $\mathcal{O}(n)$. Some tables, for instance with very small n or very redundant content, may be best implemented directly as Boolean combinatorial circuits.

In an FPGA, a LUT-based table will have a LUT cost of $\mathcal{O}(m \cdot 2^{n-k})$ (where k is the number of inputs to the elementary LUT, currently from 4 to 6). However, a table may also be implemented using one or several configurable RAM blocks—for more details on the RAM blocks available in a given FPGA circuit, refer to vendor documentation.

8.2.8.2 Table-based methods

If the function to be tabulated is continuous and derivable with well-bounded derivatives on the interval of interest (this is the case of the reciprocal on $[1, 2)$, for instance), it is possible to derive an architecture that may replace a table of the function, at a much lower hardware cost. Such an architecture is based on a piecewise linear approximation to the function, where both the constant term and the product are tabulated in two smaller tables (typically each with $2n/3$ address bit). The resulting *bipartite* architecture consists of an addition that sums the output of the two tables.

To our knowledge, this idea was first published for the sine function [581], then rediscovered independently for the reciprocal function [131]. It was later generalized to arbitrary functions and improved gradually to more than two tables [574, 438, 156, 263]. The generalized multipartite method in [156] builds an architecture with several tables and several additions which guarantees faithful rounding of the result while optimizing a given cost function. For instance, consider a 16-bit approximation to the reciprocal on $[1, 2)$, with value in $(\frac{1}{2}, 1]$. As there is one constant input bit and one constant output bit, the function to tabulate is actually $f(x) = \frac{2}{x+1} - 1$. A 15 bits in, 15 bits out faithful multipartite architecture for such a function is presented in Figure 8.10. It consists of three tables of 18.2^9, 9.2^9, and 6.2^8 bits respectively, and two additions. This represents 30 times less storage than using a plain table with 15-bit addresses. Moreover, as the three tables are accessed in parallel, the delay of a multipartite architecture is very small—in an FPGA, it is smaller than the delay of the plain table, because the smaller tables are also accessed faster, which more than compensates the delay due to the adders.

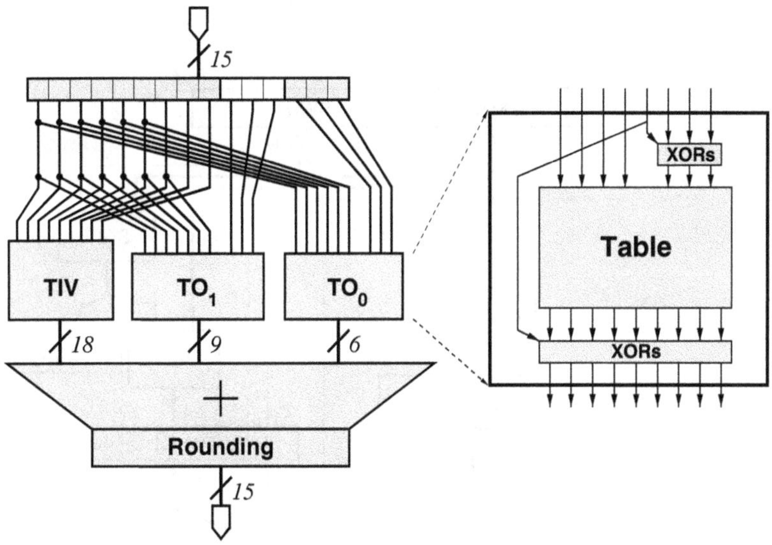

Figure 8.10: A multipartite table architecture for the initial approximation of $1/x$. The XORs implement changes of sign to exploit the symmetry of a linear approximation segment with respect to its center [574].

In this architecture, the largest table could even be decomposed further as the sum of two smaller ones [263]. The multipartite method has been generalized to higher-degree approximations, with architectures involving small multipliers [170] which further reduce the area at the expense of the delay.

8.2.8.3 Conclusion

The previous sections introduced the basic blocks useful to build floating-point operators. The next sections will describe how to assemble these blocks into the main floating-point operations: addition, multiplication, FMA, and division.

8.3 Binary Floating-Point Addition

A binary floating-point adder basically follows the generic algorithm given in Section 7.3. We focus here on the implementation issues.

8.3.1 Overview

Let us call x and y the floating-point inputs. First, the exponent difference $e_x - e_y$ is computed, and the inputs are possibly swapped to ensure that $e_y \leq e_x$. Then comes a possibly large shift to align the significands. In case of effective subtraction, one of the significands is possibly complemented using

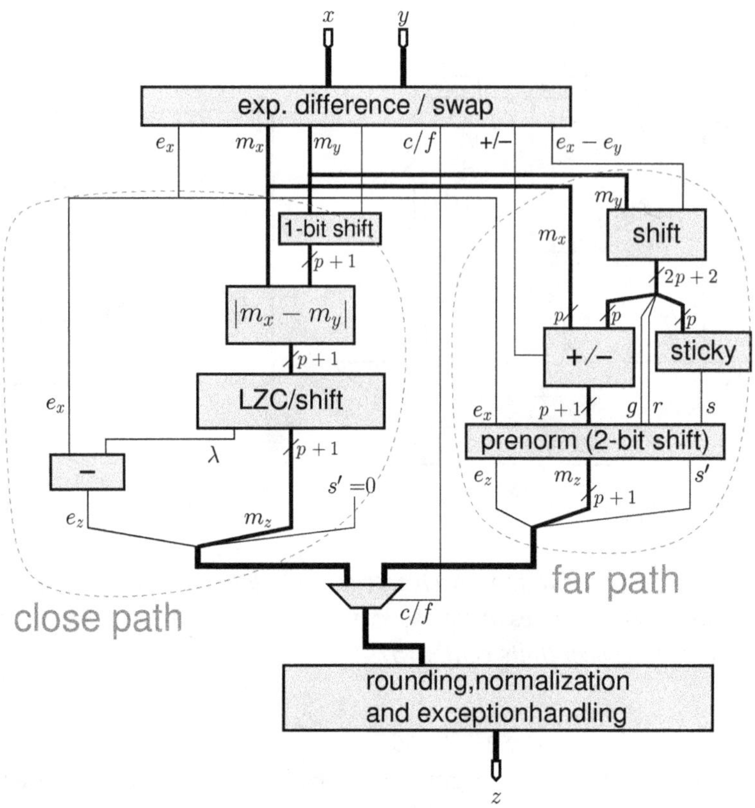

Figure 8.11: A dual-path floating-point adder.

1's complement (the increment required for complete negation is injected as addition carry). The addition itself is then performed. Finally, the result may be normalized. In the case of a cancellation, this normalization may require an LZC and a shifter.

8.3.2 A first dual-path architecture

As remarked already in the previous chapter, the two large shifts occur in situations that are exclusive: a cancellation may occur only if the input exponents differ by at most one, but in this case there is no need for a large alignment shift. Besides, the decision of which path to choose depends only on the input exponents and can be made early, typically in the same time as the initial exponent comparison. Therefore, virtually all the recent floating-point adders are variations of the *dual-path* architecture presented in Figure 8.11. The *close* path is for situations where a massive cancellation (more than 1 bit) may occur: effective subtractions of inputs with exponents that differ by at most 1. The *far* path is for distant exponents (their difference is at least 2).

Here are a few remarks in Figure 8.11.

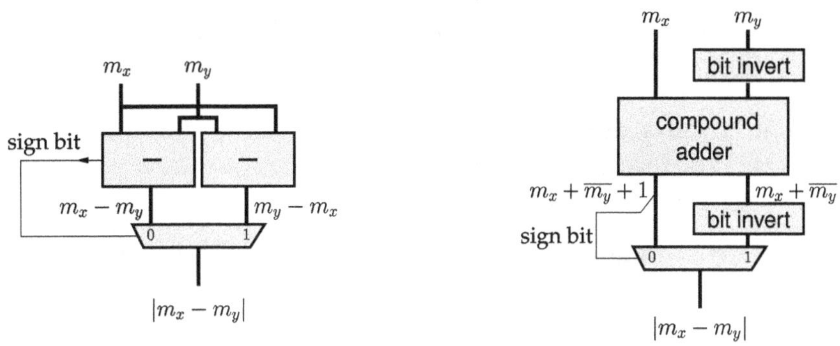

Figure 8.12: Possible implementations of significand subtraction in the close path.

- Both paths are relatively balanced in terms of critical path delay; however, the close path has slightly more work to do.
- The far path may perform either an addition or a subtraction. The sign of the result is known in advance; it will be the sign of the number whose exponent is larger.
- The close path always performs a significand subtraction; however, the sign of the result is unknown. If it is negative, it needs to be negated to obtain a positive significand (changing the sign bit of the result). This change of sign is represented after the subtraction in Figure 8.11. A better latency is obtained by computing in parallel $m_x - m_y$ and $m_y - m_x$ and selecting the one that is positive, as shown in Figure 8.12.
- The final normalization unit inputs an exponent e_z, a $(p+1)$-bit significand m_z whose MSB is 1 and whose least-significant bit (LSB) is a round bit, and a sticky bit s' (the sticky bit coming from the close path is always 0). This information suffices to perform rounding according to the rules defined in Chapter 7. The integer increment trick described in Section 7.2.1.1 may be used: a $(p + w_E)$-bit integer is built by removing the leading 1 of the significand and concatenating the exponent to its left, then the rounding rules simply decide (from the round and sticky bits) if this integer should be incremented or not. This provides a normalized representation of the result even when the increment entails an overflow.
- The LZC/shift box of the close path, and the prenorm box of the far path, prepare the information for the final normalization unit.
- The prenorm box in the far path inputs the $(p+1)$-bit significand sum (including a possible carry out), and three more bits from the shifted significand, classically called g for *guard bit*, r for *round bit*, and s a sticky bit computed from the p least significand bits of the shifted sum.

The prenorm box is essentially a multiplexer controlled by the two leading bits of significand sum (including the carry out):

- if 1x (carry out), this carry will be the leading 1 of the p-bit significand of the result. In this case the exponent is $e_z = e_x + 1$ and the sticky out is $s' = g$ OR r OR s ;
- if 01, the significand sum is still aligned with m_x. The $(p+1)$-bit significand m_z is obtained by removing the leading zero of the sum and concatenating g as its LSB. The exponent is $e_z = e_x$ and the sticky out is $s' = r$ OR s;
- if 00 (an effective subtraction leads to a one-bit cancellation), since the shift was at least by two bit positions, it is easy to prove that the third bit is a 1. In this case the exponent is $e_x - 1$ and the sticky out is $s' = s$.

Compared to a single-path architecture, the area overhead of the dual-path architecture is limited. The only duplicated hardware are the significand additions, and the far path prenormalization, which may be considered as a 2-bit combined LZC/shifter. Besides there is also the cost of the multiplexer selecting the results from the close or far path, and in a pipelined implementation the cost of registers synchronizing both paths.

Still, the dual-path design makes sense even when area is a concern, as in FPGA implementations [374, 173]. The design of Figure 8.11 is indeed suitable for an FPGA implementation since the delay of leading-zero counting can be hidden in the shift delay.

Now in a standard full-custom VLSI technology, implementing Figure 8.11 will lead to a larger latency in the close path than in the far path. The far path still involves only two logarithmic steps in sequence (a shifter and an adder), whereas the close path has three: the subtracter (if implemented as shown in Figure 8.12), an LZC, and a shifter. The latter cannot be overlapped: the LZC provides the most significant bits of the shift in its last levels, whereas the shifter needs them in its first levels.

The solution is *leading-zero anticipation*, a technique that allows us to compute an approximate of the leading-zero count in parallel with the significand subtraction.

8.3.3 Leading-zero anticipation

A leading-zero anticipator (LZA) is a block that computes or estimates the number of leading zeros from the inputs of the significand adder, instead of having to wait for its output. Figure 8.13 describes an adder architecture using an LZA. Note that some authors call the LZA a leading-one predictor or LOP.

An LZA works in two stages.

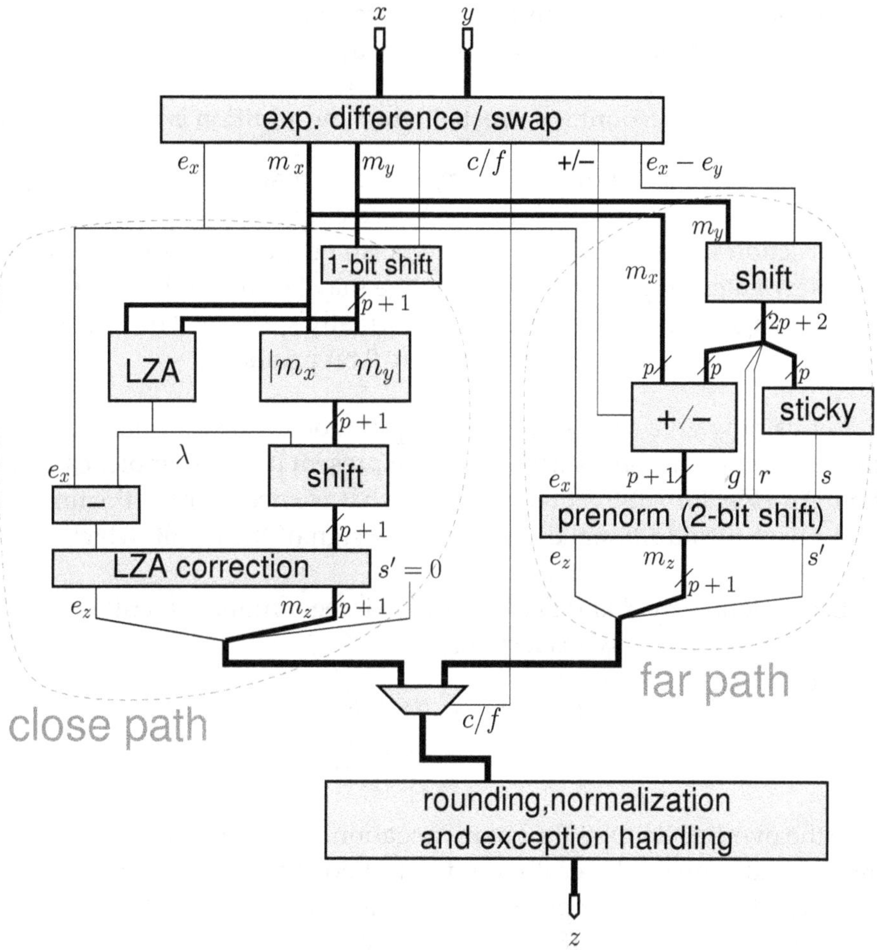

Figure 8.13: A dual-path floating-point adder with LZA.

- The first stage recodes in parallel the two p-bit inputs m_x and m_y to be subtracted into a single p-bit *indicator string* f. This string has the following property: if its leading one is in position i, then the leading one of the sum is at position i or $i+1$. This stage has a constant delay of a few gates.

- A standard LZC on the string f then provides an estimation of the leading-zero count of the result of the subtraction, which may be off by at most 1.

The reason why the result may be off by 1 is that the LZA computes the count from the MSB to the LSB, therefore it ignores a possible carry propagation coming in the other direction. Some LZA designs ([460, 270]) implement a built-in correction but according to [543] this induces extra complexity and delays making them less interesting than aftermath correction. Such a correction may be implemented in two possible ways:

- either the LZA count is corrected when the addition is finished;
- or the shifter is followed by a multiplexer that implements the needed possible 1-bit shift. This is the LZA correction box of Figure 8.13. Its construction is very similar to the prenorm box, and it also has a delay of a few gates. In practice, this last stage may be efficiently fused with the last stage of the shifter: this stage normally shifts by 0 or 1 bit position, and may be modified to shift also by 2 bit positions.

Let us now give an example of computation of the indicator string (see [543] for a comprehensive survey). We first present it in the simpler case of addition. An LZA for subtraction, which is what we need for the floating-point adder, raises the additional difficulty of the sign of the result, which we will address later.

Let us denote by A and B two positive integer numbers written on n bits $a_{n-1} \dots a_0$ and $b_{n-1} \dots b_0$, respectively.

Let

$$\begin{cases} p_i &= a_i \text{ XOR } b_i \\ g_i &= a_i \text{ AND } b_i \\ k_i &= \overline{a_i} \text{ AND } \overline{b_i} \end{cases},$$

where the overline denotes the binary negation. These are the propagate, generate, and kill signals classically used in fast adders (see Section 8.2.1.3). If p_i is set, the i-th bit position propagates the incoming carry, whatever it is, to the $(i+1)$-th bit position. If g_i is set, a carry is generated at the i-th bit position regardless of the incoming carry. If k_i is set, no carry will exit the i-th bit position, regardless of the incoming carry.

One may then check that a leading-zero string on the sum corresponds:

- either to a leftmost sequence of consecutive bit positions such that k_i is set (this is often noted k^*, using a notation inspired by regular expressions);
- or to a sequence (from left to right) of zero or more consecutive bit positions with p_i set, followed by one position with g_i set, followed by zero or more consecutive positions with k_i set (in regular expression notation, a sequence p^*gk^*). Note that this case corresponds to an addition of A and B with overflow (carry out). Here we ignore the carry out in the leading-zero count: this will be justified as we use such an addition for two's complement subtraction.

The trick is that these two sequences can be encoded as the indicator string defined by

$$f_i = p_i \text{ XOR } \overline{k_{i-1}}$$

(the proof is by enumeration).

The important thing here is that the computation of f_i is a Boolean function of only two bits of A and two bits of B: it may be computed with a few gates and in a very small delay.

Let us now consider an LZA for the close path of a floating-point adder. As the addition is actually a subtraction, A will be $2^p \cdot m_x$ with $n = p+1$, and B will be the binary complement of either $2^p \cdot m_y$ or $2^{p-1} \cdot m_y$. If the result is positive, we have our estimation of the leading-zero count. However, if the result turns out to be negative, we should have counted the leading ones instead.

A conceptually simple solution is to duplicate the LZA hardware, just as the addition itself is duplicated in Figure 8.12. This approach means duplicating the LZC as well. Another solution is to derive an indicator string that works for leading zeros as well as for leading ones. This approach does not need to duplicate the LZC, but the computation of the indicator string is now more complex, as it needs to examine three bits of each input [79]. A discussion of this tradeoff is given in [543].

As a conclusion, the best implementation of leading-zero anticipation is very technology dependent. It also depends on the pipeline organization and on the implementation of the adder itself. For a comprehensive review of these issues, see the survey by Schmookler and Nowka [543]. Many patents are still filed each year on this subject (see for example [460, 270]).

8.3.3.1 Subnormal handling in addition

Binary floating-point addition has a nice property (Theorem 4.2): any addition that returns a subnormal number is exact. A subnormal either results from the addition or subtraction of two subnormals, or from a cancellation (in the close path of Figure 8.13). In the first case, no shift is needed. In the second case, the shift amount has to be limited to the one that brings the exponent to $e_{\min}$: the remaining leading zeros will be left in the subnormal significand. In both cases, we do not have to worry about rounding the result, since it is exact.

The only issue is to ensure that the detection of subnormal inputs on one side and the limitation of the shift amount in the close path on the other side do not degrade the critical path of the overall operator.

The shift at the beginning of the addition far path is now driven by the exponent difference $E_x - n_x + E_y - n_y$, where n_x and n_y are the "is normal" bits. To hide the delay of the computation of these bits (n_x is computed as the OR of the bits of E_x), one may compute in parallel n_x, n_y, and the three alternative differences $e_x - e_y$, $e_x - e_y - 1$, and $e_x - e_y + 1$, and select the correct one depending on n_x and n_y. This adds just one level of multiplexers to the critical path. Besides, n_x and n_y are the implicit bits of the significand, and are available before the significand shift, so there is no increase of the critical path delay here.

It is more difficult to limit to $e_x - e_{\min}$ the normalizing shift in the close path without increasing the close path critical path. The shift distance is now $\min(\lambda, e_x - e_{\min})$, where λ is now the tentative number of leading zeros output by the LZA. The value $e_x - e_{\min}$ may be computed early, but the delay of the comparison with λ, from the LZA output, to decide shifting is added to the critical path. A solution is to perform the two shifts (by λ bits and by $e_x - e_{\min}$ bits) tentatively while comparing these two numbers and to select the valid one when the comparison is available. This adds a lot to the area.

Another solution would be to inject a dummy leading 1 in the proper position ($e_x - e_{\min}$) before leading-zero counting. Again, $e_x - e_{\min}$ may be computed in parallel with exponent difference, but bringing a 1 in this position is a shift. It is difficult to obtain this value before the leading-zero count.

To summarize, we have to add at least the delay of a few multiplexers to both the far path and the close path, and this may cost much area. Or, a more area-efficient solution may be preferred, and it will add to the critical path the delay of an exponent addition, which remains small. The best solution in a given adder will depend on whether the critical path delay is in the close path or in the far path.

8.3.4 Probing further on floating-point adders

We have not discussed the technology-dependent decomposition of the adder in pipeline stages. The pipeline depth is typically 2 [552], 3 [468], or 4 [448] cycles for binary64 operands. In [468], the latency is variable from 1 to 3 cycles, depending on the operands.

Even and Seidel suggest a nonstandard two-path architecture in [552], and compare it to implementations in Sun and AMD processors. Their motivation is the minimization of the overall logic depth of floating-point addition [551].

Pillai et al. present a three-path approach for lower power consumption in [491].

Fahmy, Liddicoat, and Flynn suggest using an internal redundant format for floating-point numbers within processors [198]. This enables significand addition with $\mathcal{O}(1)$ delay. Numbers have to be converted back to the standard format only when writing back to memory.

As floating-point adders are complex designs, there have been several attempts to prove their correctness using formal methods [536, 37]. A formal approach to the design is also advocated by Even and Paul [195].

8.4 Binary Floating-Point Multiplication

8.4.1 Basic architecture

The architecture depicted by Figure 8.14 directly follows the algorithm given in Section 7.4. It does not illustrate sign or overflow handling, which is straightforward. Underflow handling is discussed in Section 8.4.4. Here are a few comments on this architecture.

- The computation in parallel of $e_x + e_y - b$ and $e_x + e_y - b + 1$ consumes little area, as it may use small and slow carry-propagate adders: the exponents are not needed before the end of the significand product, which takes much more time.
- The sticky bit computation may in principle be performed as a by-product of the multiplication, in such a way that the sticky bit is available at the same time as the result of the significand multiplexer.
- The increment in the final normalization unit may change the exponent a second time. One idea could be to use a single adder using a carry propagation from the significand to the exponent, as explained in Section 7.2.1.1. However, note that even rounded up, the product of two significands in $[1, 2)$ never reaches 4. Indeed, the largest possible significand is $2-2^{-p+1}$, whose square is $4-2^{-p+2}+2^{-2p+2}$. This is the largest possible significand product, and it is rounded up to $4-2^{-p+2}+2^{-p+1} < 4$. As a consequence, the exponent is never incremented twice. In other words, there will only be a carry out from the normalization incrementer if the exponent was $e_x + e_y - b$, in which case it is changed to $e_x + e_y - b + 1$.
- Not shown in Figure 8.14 are sign and exception handling, which are straightforward, except for subnormal numbers, which are managed using methods discussed in Section 8.4.4.

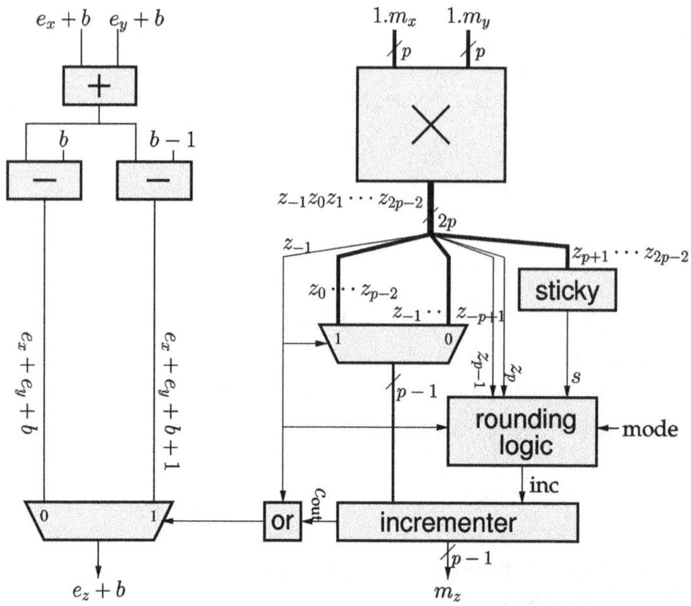

Figure 8.14: Basic architecture of a floating-point multiplier without subnormal handling.

8.4.2 FPGA implementation

The architecture of Figure 8.14 is well suited for FPGA implementations.

- Subnormal handling is not a strong requirement for applications using FPGA floating-point accelerators. The floating-point format used in these accelerators can be nonstandard, and in particular can have an ad hoc exponent range.

- Significand multiplication can be performed efficiently using the small integer multipliers embedded in the FPGA fabric of high-performance FPGAs. These multipliers are typically able to perform 18×18-bit products, and recent FPGAs have increased this size to 25×18-bit to facilitate the implementation of binary32 arithmetic. For larger significand sizes, several of these multipliers have to be grouped together; for instance, a 36×36-bit product can be implemented using four 18×18-bit multipliers and a few adders. In recent FPGAs, the embedded multipliers are tightly coupled to specific adders. The main purpose of these blocks is efficient multiply-and-accumulate operations for digital signal processing (DSP), but they also allow for building larger multipliers [153].

- Embedded multipliers are not able to compute the sticky bit as a by-product. However, a wide OR can be computed using the fast-carry circuitry. As soon as more than one embedded multiplier is needed, the higher part of the result comes from an addition, and the sticky computation can be overlapped with this addition.

- The increment in the final rounding unit can be performed by an area-efficient carry-propagate adder through the fast-carry circuitry.

8.4.3 VLSI implementation optimized for delay

Let us analyze the delay of the previous design when implemented in full-custom VLSI. The critical path is in the significand computation.

- One of the operands is Booth recoded (see Section 8.2.4), and the partial product array is generated. This step takes a small constant delay.
- A compression tree in $\log p$ time compresses the partial product array into a carry-save representation of the product.
- A fast adder, also in $\log p$ time, converts this carry-save product to standard representation.
- The normalization unit needs to wait for the MSB of the result of this addition to decide where to increment. This increment takes another fast adder, again with a $\log p$ delay.

It is possible to reduce the two adder latencies to only one (and a few more gate delays). Let us start from the output of the compression tree, which is a carry-save number, i.e., a pair of vectors $s_{-1}s_0s_1\cdots s_{2p-2}$ and $c_{-1}c_0c_1\cdots c_{2p-2}$.

An easy case is round to zero, which comes down to a truncation. In this mode, all one needs is a $2p$-bit fast adder inputting $s_{-1}s_0s_1\cdots s_{2p-2}$ and $c_{-1}c_0c_1\cdots c_{2p-2}$ and outputting $y_{-1}y_0y_1\cdots y_{2p-2}$. Then, a multiplexer selects $y_{-1}y_0\cdots y_{p-2}$ if $y_{-1}=1$, and $y_0y\cdots y_{p-1}$ if $y_{-1}=0$. It also selects the exponent among e_a+e_b and e_a+e_b+1.

The reason why we need a second large adder in the previous architecture is the possible increment of the result, which never happens in round-to-zero mode.

The trick will therefore be to reduce the other rounding modes to the round-to-zero mode. For this purpose, we add to the partial product array a few bits which depend only on the rounding mode. These *rounding injection* bits [196] may, in principle, add at most one gate delay to the delay of the reduction tree. In practice, however, the tree is designed to fit in one or two pipeline cycles, and the injection bits do not increase its latency.

The injection bits are defined as follows:

$$\mathrm{inj} = \begin{cases} 0 & \text{if round to zero} \\ 2^{-p} & \text{if round to nearest} \\ 2^{-p+1} - 2^{-2p+2} & \text{if round to infinity.} \end{cases}$$

If the significand product is in $[1,2)$, we indeed have $\circ(z) = RZ(z + \text{inj})$ except in one case: the round to nearest thus implemented is *round to nearest up*, not round to nearest even. In case of a tie, it always returns the larger number. This will be easy to fix by simply considering the LSB z_{p-1} of the rounded result. If $z_{p-1} = 0$, the result is even and there is nothing to do. If $z_{p-1} = 1$, the nearest even number is obtained by simply pulling z_{p-1} to 0, which takes no more than one gate delay.

To summarize so far, the injection bits allow us to obtain the correctly rounded result in any rounding mode using only one $2p$-bit adder and a few extra gates, in the case when the significand product is in $[1,2)$. The correctly rounded result is obtained by truncation as the bits $z_0 \cdots z_{p-1}$, with the bit z_{p-1} pulled down in case of a tie for RN mode. Detecting a tie requires a sticky bit computation as usual, which can be performed in parallel to the sum of the lower bits.

Let us now consider the case when the result will be in $[2,4)$. In this case, the injection bits have been added one bit to the right of their proper position: the injection should have been

$$\text{inj}_{[2,4)} = \begin{cases} 0 & \text{if round to zero} \\ 2^{-p+1} & \text{if round to nearest} \\ 2^{-p+2} - 2^{-2p+2} & \text{if round to infinity.} \end{cases}$$

To correct this injection, if the result turns out to be in $[2,4)$, we have to add a correction defined by

$$\text{corr} = \text{inj}_{[2,4)} - \text{inj} = \begin{cases} 0 & \text{if round to zero} \\ 2^{-p} & \text{if round to nearest} \\ 2^{-p+1} & \text{if round to infinity.} \end{cases}$$

The problem is that we do not know if this correction should be applied before the end of the significand adder. Of course, we don't want this final correction to involve another carry propagation delay over the full significand. The solution is to compute in parallel the two versions of the higher bits, one with the correction, and one without.

The sum of the lower bits of the carry-save product, $s_p \cdots s_{2p-2} + c_p \cdots c_{2p-2}$, is condensed into a sticky bit, a sum bit s'_p, and a carry-out bit c'_{p-1} (see Figure 8.15). This carry out will be the carry in to the sum of the higher part of the carry-save product. As the correction may add one bit at position $p-1$, a row of half-adders is used to reduce $s_{-1} \cdots s_{p-1} + c_1 \cdots c_{p-1}$ to $s'_{-1} \cdots s'_{p-1} + c'_{-1} \cdots c'_{p-2}$. Thus, if the correction is 2^{-p+1} (round to infinity), it will be added to at most two bits at position $p-1$ (s'_{p-1} and c'_{p-1}) and generate a carry at position $p-2$. If the correction is 2^{-p}, it will be added to one bit of the same weight (s'_p), and again a carry may propagate to the higher bits.

Thus, the two versions of the higher sum to compute in parallel are $z_{-1} \cdots z_{p-2} = s'_{-1} \cdots s'_{p-2} + c'_{-1} \cdots c'_{p-2}$, and the same with a carry in in case

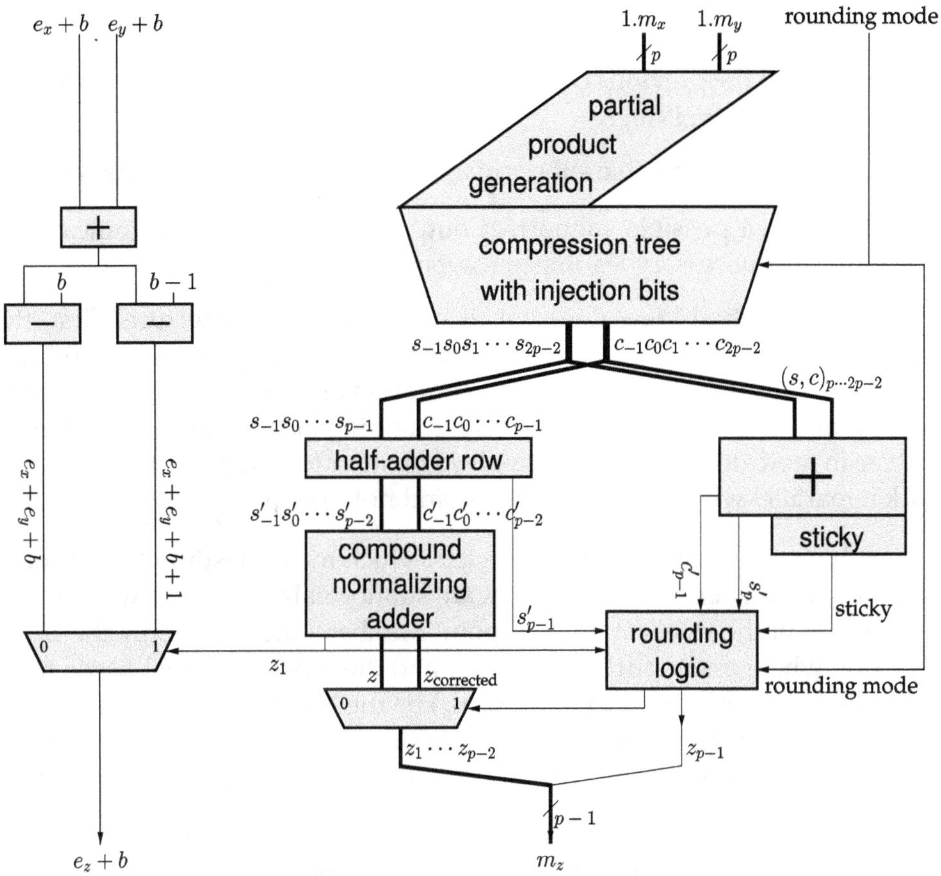

Figure 8.15: A floating-point multiplier using rounding by injection, without subnormal handling.

of correction. These two sums can be computed by a compound adder. The carry in—which will decide which of these results is valid—is a combinatorial function of s'_{p-1}, s'_p, c'_p, the rounding mode which defines the correction to apply, and of course the value of z_{-1} which decides normalization, and hence whether a correction should be applied. This is the z_{-1} of the sum, not of the incremented sum: there will be a carry in to the compound adder only if the sum is in $[2, 4)$, in which case we know that the incremented sum is also in $[2, 4)$.

The lower bit of the normalized result is another combinatorial function from the same information, plus the sticky bit.

The resulting architecture is shown in Figure 8.15. The compound adder on this figure outputs the normalized versions of z and $z_{\text{corrected}}$, i.e., $z_0 \cdots z_{p-3}$ if $z_{-1} = 1$ and $z_1 \cdots z_{p-2}$ if $z_{-1} = 0$.

The full details of the Boolean equations involved can be found in [196], which also compares this rounding algorithm to two other approaches.

8.4.4 Managing subnormals

As explained in the previous chapter, handling subnormals conceptually consists in two additional steps:

- normalizing a possible subnormal input using a wider exponent range;
- detecting a possible subnormal output and shifting its significand to insert the necessary leading zeros, before rounding.

This conceptual view does not fit our previous architecture. First, the two steps added are long-latency ones, and they are sequential. Second, we perform rounding by injection in the compression tree, and if the output is subnormal, we will have performed the rounding at the wrong place. Let us analyze in more detail the issues involved in order to schedule this additional work in parallel with the existing steps, and not in sequence.

- Subnormal inputs must be detected, which means testing the exponent field for all zeros. If both inputs are subnormals, the result will be zero or one of the smallest floating-point numbers; therefore, only the situation where one input is subnormal and the other is normal needs to be managed with care. Let n_x (resp n_y) be the "is normal" bit, a flag bit of value 1 if x (resp. y) is normal, and 0 if it is subnormal. This bit can be used as the implicit bit to be added to the significand, and also the bias correction for subnormals.
- The implicit bit affects one row and one column of partial products. The simplest solution here is to delay the compression of these partial products long enough so that it begins when the implicit bit has been determined. This does not add to the critical path.
- The leading zeros of the subnormal input need to be counted. Define λ as the number of leading zeros of the significand with n_x appended as the leading bit.
- The subnormal input need not be normalized before multiplication. Indeed, the existing multiplier which computes the product of two p-bit numbers will compute this product as well if one of the inputs has leading zeros. The product will then have leading zeros, too, so it needs to be normalized, but this gives us the time to compute λ in parallel with the multiplication tree. This latency reduction comes at a hardware cost: we now have to shift two $2p$-bit strings (c and s) instead of one p-bit string.
- After this shift we have, in carry-save form (c, s), a significand in $[1, 4)$ whose exponent is $e_x + e_y - \lambda$. The result will be subnormal if $e_x + e_y + z_1 - \lambda < e_{\min}$, where z_1 is the leading bit of the (yet to be computed) sum of c and s. In this case, the result will have to be shifted right by $e_{\min} - (e_x + e_y + z_1 - \lambda)$ bit positions before rounding.

- We now need to preprocess the injected rounding bits to anticipate this shift. Ignoring z_{-1}, which will be handled by the same correction as previously, the injection has to be shifted left by $\max\{e_{\min} - (e_x + e_y - \lambda), 0\}$ bit positions. If we are able to compute this shift value and perform the shift in a delay shorter than that of the compression tree, the shifted injection will be ready to be added in a late stage of the compression tree, and this will not add to the delay.

- Finally, the two previous shifts may be combined in a single one, but this requires changing the injection again. Finally, the injection becomes:

$$\text{inj} = \begin{cases} 0 & \text{if round to zero} \\ 2^{-p-l+\max(e_{\min}-(e_x+e_y-l),0)} & \text{if round to nearest} \\ 2^{-p+1-l+\max(e_{\min}-(e_x+e_y-l),0)} - 2^{-2p+2} & \text{if round to infinity.} \end{cases}$$

- To summarize, the delay overhead related to subnormal handling is the delay of one large significand shift. All the other computations can be hidden by the delay of the compression tree.

- Note that no correction will occur in the case of a subnormal output as z_{-1} will be 0, which means that there is nothing to change in the correction logic.

8.5 Binary Fused Multiply-Add

Let us now discuss the hardware implementation of a fused multiply-add (FMA) binary operator. Most of the algorithm is explained in Section 7.5.4. We first present the most classical architectures, then discuss several propositions of alternative architectures.

8.5.1 Classic architecture

The first widely available FMA was that of the IBM RS/6000 [424, 260]. Its architecture closely follows the algorithm sketched in Section 7.5.4 and is depicted in Figure 8.16. The data width and alignments are identical. The main differences one may observe on this figure with respect to the algorithm are the following.

- As in the floating-point multiplier, the intermediate significand product is produced in carry-save representation. The "is normal" bits are injected in late stages of the compression tree.

- The shift of the addend is performed in parallel with the product compression tree.

- A carry-save adder inputs the carry-save product and the $2p$ lower bits of the shifted summand, and produces their sum in carry-save representation.
- This sum is completed with the higher bits of the shifted summand and a carry in that completes the bit inversion in case of effective subtraction.
- A fast adder transforms this carry-save sum in standard representation, and complements it if negative. This step may use an end-around carry adder, or two adders in parallel, with the proper result being selected depending on the sign bit.
- In parallel, an LZA determines λ, the number of leading zeros needed in case of cancellation or subnormal input. In some implementations, in cases when the result will be subnormal, a 1 is injected before the LZA to limit the count to the proper value [549].
- The normalization box is mostly a large shifter, but it also performs the case analysis described in Section 7.5.4 and determines the exponent before rounding.
- The rounding box performs the rounding and the possible subsequent 1-bit normalization, and handles overflow and sign.

This architecture is typically pipelined in 3 to 7 cycles. In the POWER6 FMA implementation, the latency is reduced from 7 to 6 cycles for dependent operations, as the final rounding is performed at the beginning of the next operation [607].

8.5.2 To probe further

An overview of the issues related to subnormal handling is given in [549].

Several authors [550, 501] have investigated multiple-path FMA implementations.

Lang and Bruguera described an FMA architecture that anticipates the normalization step before the addition [359]. This enables a shorter-latency rounding-by-injection approach. They also suggested an FMA architecture that reduces the latency in case of an addition by bypassing the multiplier stages [80].

Quinnell et al. [501] propose a bridge FMA architecture that adds FMA functionality to an existing floating-point unit by reusing parts of the adder and of the multiplier.

An example of a recent deep submicrometer, high-frequency implementation is the POWER6 FMA described in [607, 640]. To minimize wiring, the pipeline of this FMA is laid out as a U with the floating-point registers on the top.

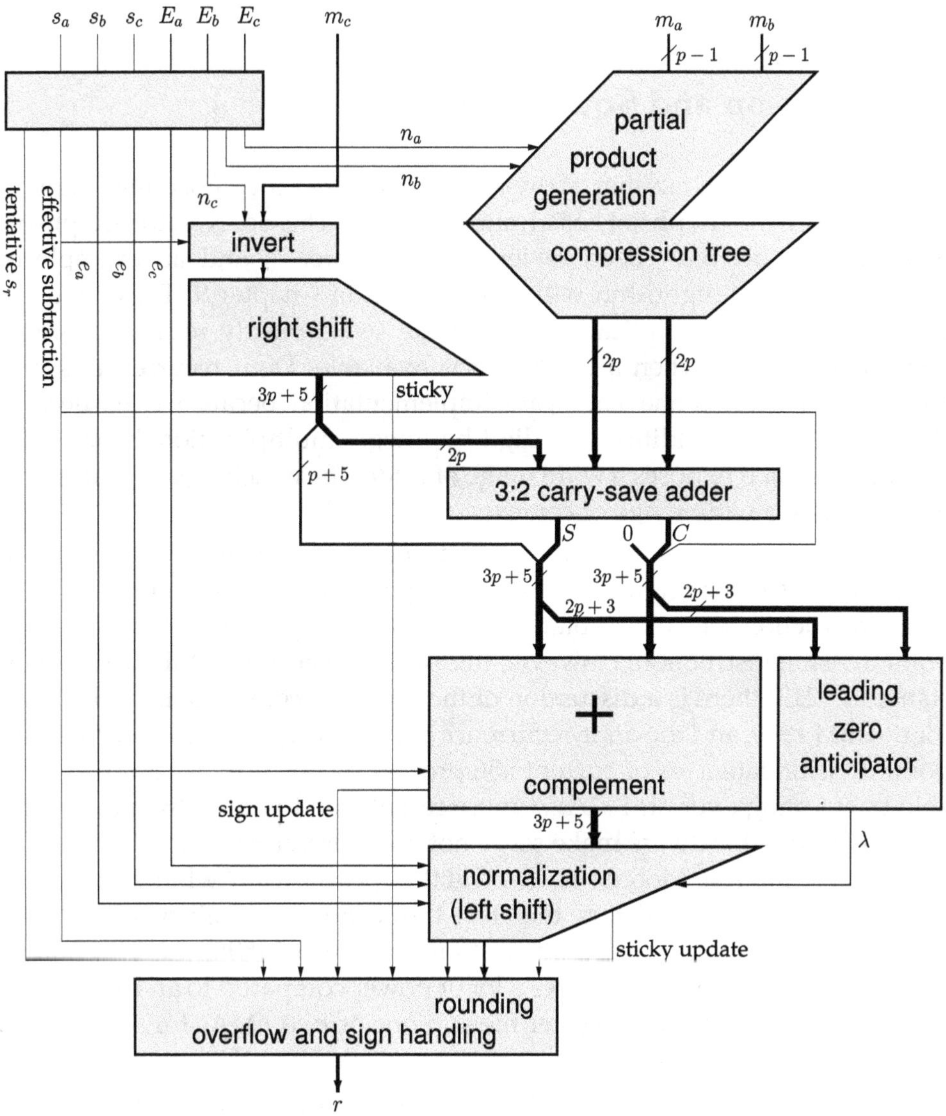

Figure 8.16: The classic single-path FMA architecture.

Finally, Lutz suggested [399] to decompose the FMA into a 4-cycle multiplier followed by a 4-cycle adder, which can be used independently or fused into the FMA. He showed that this arrangement performs consistently better on application code than a monolithic 7-cycle FMA, thanks to the reduced latency in case of dependent FMA operations (for instance a sum of products).

8.6 Division and Square Root

There are three main families of division and square root algorithms: digit recurrence, Newton–Raphson based, and polynomial based. Newton–Raphson-based algorithms have been reviewed in Chapter 4, and an example of polynomial based algorithm will be presented in Chapter 9. These two approaches rely on multiplication, and make sense mostly in a software or microcode context, when a multiplier is available. Digit recurrence is preferred for a stand-alone hardware implementation because it reduces to simpler primitives: addition and digit-by-integer multiplication. In addition, we will see that it exposes a wide range of tradeoffs, which means that it may be adapted to a wide range of contexts.

In microprocessors, there has been a trend not to include a division operator, and instead compute divisions using the FMA as shown in Chapter 4. Some instruction sets, most notably IA-64, were designed without a floating-point division instruction. However, this design decision is still open. For instance, in [213] there is a discussion of the pros and cons of both approaches for the IBM z990, and the digit-recurrence approach is preferred. The floating-point division latencies of current x86 processors [431] do not seem achievable by iterating over an FMA. A more recent trend is that discrete divider and square root hardware make sense not only for performance, but also for power consumption. Moore's law has hit the "power wall" while still providing more and more transistors, therefore the economics of silicon on a chip is moving from "saving transistors" to "saving power" [596]. In this context, dedicated divider hardware saves a lot of power compared to an FMA-based division that exercises the register file and one or two FMAs for several tens of cycles.

In FPGA-accelerated applications, a stand-alone hardware divider often makes sense. Earlier implementations employed digit recurrences, but this choice has been reassessed considering the availability of embedded multipliers. Multiplication-based dividers and square roots for FPGAs have been published without [492, 629] and with [146, 485] correct rounding. They are similar to the approaches surveyed in Section 4.7, with the difference that the iterations are unrolled for a pipelined design, and each multiplication can be computed by an ad-hoc multiplier tailored to the precision required at each step.

Let us now focus on the mainstream technique used to implement dedicated hardware floating-point dividers: digit-recurrence algorithms. More details on digit-recurrence division theory and implementations can be found in textbooks by Ercegovac and Lang [186, 187].

8.6.1 Digit-recurrence division

Figure 8.17 describes the architecture of a digit-recurrence divider. The sign and special cases may been handled as per Section 7.6. This include the normalization of subnormal inputs, using a larger exponent range (hence the intermediate exponents extended by one bit in Figure 8.17). We refer to Section 7.6 for more details of sign and exponent handling.

It will be interesting to use a large radix for the intermediate computation, as this will lead to fewer iterations. To this purpose, a p-bit normalized significand can be considered as a radix-2^k number simply by grouping its bits k by k. For instance, hexadecimal is a radix-16 representation of binary numbers. The SRT4 algorithm depicted in Figure 8.17 uses radix 4. In all generality, we may therefore consider our significands as n-digit numbers in radix-β.

We now focus on dividing the significand

$$X = x_0.x_1 \cdots x_{n-1}$$

by the significand:

$$D = d_0.d_1 \cdots d_{n-1},$$

with $x_0 \neq 0$ and $d_0 \neq 0$. We have $X \in [1, \beta)$ and $D \in [1, \beta)$; therefore, $X/D \in (1/\beta, \beta)$. For clarity, in the rest of this section, upper-case letters denote multi-digit numbers, and lower-case letters denote digits.

In digit-recurrence approaches, the division is expressed using the Euclidean equation

$$X \quad = \quad QD + R \quad \text{with} \quad 0 \leq R < D\,\text{ulp}(Q). \tag{8.1}$$

This equation corresponds to rounding the quotient down. For the other rounding directions, a final correction will be needed.

Algorithm 8.3 is a generic radix-β digit recurrence that computes two numbers $Q = q_0.q_1 \cdots q_{n-1}$ and R satisfying

$$X = QD + R.$$

Note that this algorithm depicts the paper-and-pencil division. In this algorithm, β may indeed be 10, or 2, or small powers of 2. We will come back to the choice of number representation.

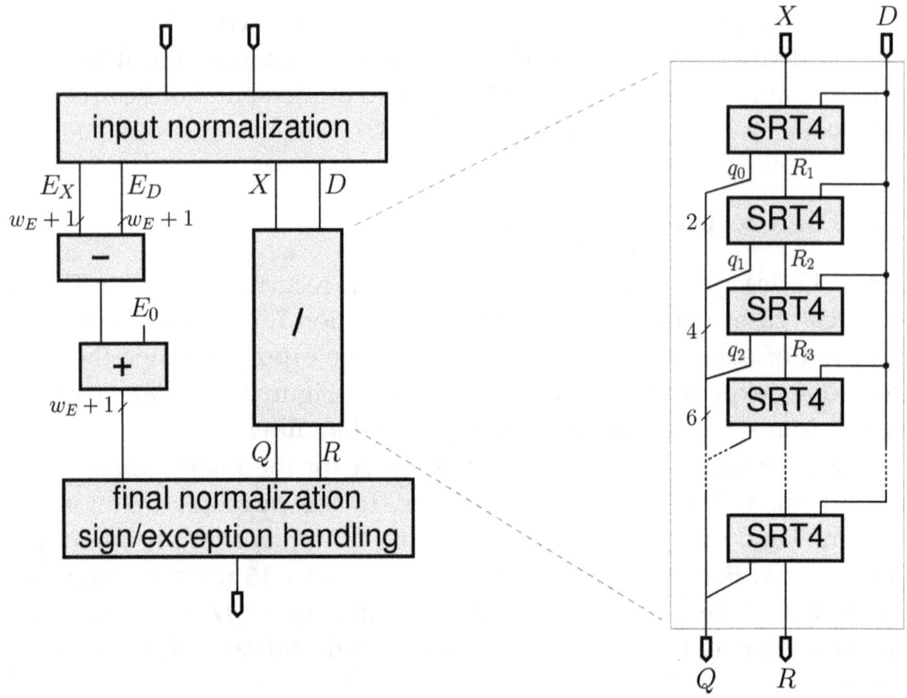

Figure 8.17: An unrolled floating-point divider using a digit recurrence, here SRT4. Note that an alternative, often used in processors, is to iterate on a single SRT stage. This means smaller area, but divisions can no longer be pipelined.

Algorithm 8.3 Generic digit-recurrence algorithm.

Inputs: a n-digit significand X, a n-digit divider D (both in radix β)
$R^{(0)} \leftarrow X$
$Q^{(0)} \leftarrow 0$
for $j = 0$ to $n-1$ **do**
  $q_j \leftarrow \text{Sel}(R^{(j)}, D)$
  $R^{(j+1)} \leftarrow \beta R^{(j)} - q_j D$
end for
$R \leftarrow R^{(n)}$
$Q \leftarrow q_0.q_1 \cdots q_{n-1}$
Returns: (Q, R);

The reader may check that, if we call $Q^{(j)} = q_0.q_1 \cdots q_j$ the quotient computed at iteration j, this algorithm maintains the invariant

$$X = Q^{(j)} D + R^{(j+1)}.$$

Inside the loop, the second line is composed of a subtraction and a digit-by-significand multiplication. Both operations are much simpler than significand-by-significand multiplication, having an $\mathcal{O}(n)$ digit-level operation cost, and possibilities of $\mathcal{O}(1)$ computation time when using redundant number systems. Also, note that it is possible to precompute, before starting the iteration, all the q_jD for all the possible digit values. They may even be computed in parallel. Then, the iteration no longer involves any product.

The first line of the loop is the selection of the next quotient digit. The choice has to be made in order to ensure the convergence of the algorithm towards a final residual such that $0 \le R < D\,\mathrm{ulp}(Q)$. This is obtained by ensuring this condition at each iteration; thus, the invariant of the algorithm becomes

$$\begin{cases} X = Q^{(j)}D + R^{(j+1)}, \\ 0 \le R^{(j+1)} < D\,\mathrm{ulp}(Q^{(j)}). \end{cases}$$

In the paper-and-pencil method, the Sel function is "implemented" by intuition. The human operator performs an approximation of the computation of the second line, using leading digits only. It fails very rarely, when βR^j is very close to a nontrivial multiple of D such as $7D$. In this case, the human computes the next residual $R^{(j+1)}$, observes that it is too large or negative, and backtracks with a neighboring value of q_j.

In a hardware implementation, a form of backtracking can be used in the binary case ($\beta = 2$) as follows. The subtraction is performed as if $q_j = 1$:

$$R_1^{(j+1)} = 2R^{(j)} - D.$$

Then, if $R_1^{(j+1)} < 0$, it means that the correct choice was $q_j = 0$; therefore, the next partial remainder is set to the current one, shifted: $R^{(j+1)} = 2R^{(j)}$. Otherwise $R^{(j+1)} = R_1^{(j+1)}$. Here, backtracking means simply restoring the previous partial remainder and does not involve a new computation. This algorithm is called the *restoring division algorithm*. A classical variant, called the nonrestoring algorithm, uses one register less [186].

In higher radices, backtracking is not a desirable option as it would need a recomputation of $R^{(j+1)}$ and thus lead to a variable-latency operator. A better alternative is the use of a redundant digit set for the quotient Q. Redundancy will give some freedom in the choice of q_j. Thus, in the difficult cases (again, when βR^j is close to a nontrivial multiple of D), there will be two valid choices of q_j. For both choices, the iteration will converge. This in turn means that the selection function $\mathrm{Sel}(R^{(j)}, D)$ need not consider the full $2p$-digit information $(R^{(j)}, D)$ to make its choice. As the human operator in the paper-and-pencil approach, it may consider the leading digits of $R^{(j)}$ and D only, but it will do so in a way that never requires backtracking.

The family of division algorithms obtained this way is called *SRT division*, after the initials of the names of Sweeney, Robertson, and Tocher, who invented it independently [511, 602] . Its theory and a comprehensive survey of implementations are available in the book by Ercegovac and Lang [187]. We now briefly overview performance and cost issues.

The choice of the radix β has an obvious impact on performance. In binary, we have to choose $\beta = 2^k$. Each iteration then produces k bits of the quotient, so the total number of iterations is roughly p/k. Notice, thus, that here β is not necessarily the radix of the floating-point system: it is a power of that radix. Intel recently reduced the latency of division in their x86 processors by replacing a radix-4 division algorithm with a radix-16 algorithm [431]. However, each iteration is also more complex for larger k, in particular for the product $q_j D$. In a processor, the choice of radix is limited by the target cycle time.

A second factor that has an impact on the cost is the choice of the digit set used for the quotient. Most implementations use a symmetrical digit set:

$$q_j \in \{-\alpha, \dots, \alpha\} \quad \text{with} \quad \lceil \beta/2 \rceil \leq \alpha < \beta.$$

The choice of α is again a tradeoff. A larger α brings in more redundancy and therefore eases the implementation of the selection function. However it also means that more digit-significand products must be computed.

With a quotient using a redundant digit set, the selection function may be implemented as a table indexed by the leading digits of R^j and D. Other choices are possible; see [187] for a review.

Another variation on digit-recurrence algorithms is *prescaling*. The idea is that if the divider D is closer to 1, the selection function is easier to implement. To understand it, think again of the paper-and-pencil algorithm: when dividing for instance by 1.0017, one gets a very quick intuition of the next quotient digit by simply looking at the first digit of the partial remainder. In SRT algorithms, it is possible to formally express the benefit of prescaling. In practice, prescaling consists in multiplying both X and D by an approximation to the inverse of D. This multiplication must remain cheap, typically equivalent to a few additions. Prescaling has also been used to implement complex division [188].

8.6.2 Decimal division

An implementation of a decimal Newton–Raphson iteration was proposed by Wang and Schulte [627].

The SRT scheme is well suited to decimal implementations. The implementation of a digit recurrence decimal divider in the POWER6 processor is described by Eisen et al. in [184]. They use a radix-10 implementation with digit set $\{-5, \dots, 5\}$ and prescaling to get the scaled divisor in $[1, 1.11)$. This is probably the only decimal division architecture actually in operation at the time of writing this book.

A complete architecture for decimal SRT division with the same digit set, but with D recoded in the BCD-5421 code (see Section 8.2.5), is also presented in Vásquez's PhD dissertation [613].

8.7 Beyond the Classical Floating-Point Unit

8.7.1 More fused operators

With the fused multiply-add (FMA) now standard, it is natural to look beyond and see if other operators can be fused. As for the FMA, there are two motivations: accuracy and performance. Besides, there is a new architectural opportunity: in a processor designed for an FMA, the register file is able to feed three inputs to an operator. This is a motivation to study other three-input operators.

Among these, the most useful seems to be the fused sum of three inputs. It has been extensively studied by Tenca [597], then Tao et al [595], who also showed that supporting fused 3-input addition only adds one third to the area of an FMA [592].

The decision to fuse additions in software is not without consequences in terms of accuracy: see the discussion of Section 6.1.1. However, the availability of hardware 3-input addition with one single rounding would considerably speed up triple-word arithmetic (see Section 14.1.2). It would also offer a very simple way of expressing the sum of two floating-point numbers as a double-word: Algorithms 2Sum and Fast2Sum, presented in Section 4.3, would become useless.

Another fused operator of interest is the combined sum and difference, which computes at the same time $a + b$ and $a - b$. It would in particular be very useful in the computation of fast Fourier transforms (FFT). It is smaller than the juxtaposition of two standard adders, since only one close path is needed [593]. However, its integration in a processor datapath may be more problematic, as this operation produces two results.

Section 8.7.3 will review another class of useful fused hardware operators: those which return both the rounded result of an operation, and the rounding error.

Finally, the fused sum of two products [538, 594] and sum of squares [421] have also been studied. They improve complex arithmetic in both performance and accuracy. The floating-point unit of Kalray processors supports the fused sum of two products in binary32.

Designing architectures for reconfigurable hardware (FPGAs) opens a new opportunity: the fusion of operators for a specific context. It will be studied further in Section 8.8. When floating point is implemented in software, fused operators are also very relevant: this will be studied in Section 9.6.

8.7.2 Exact accumulation and dot product

As already explained in Chapter 5, summing many independent terms is a very common operation. Scalar product, matrix-vector, and matrix-matrix products are defined as sums of products. Summations also appear in digital filters, integration, Monte Carlo simulations, etc.

The main issue with large summations, as already exposed in Chapter 5, is that of *accuracy*. To address it, Kulisch [353, 354, 355] advocated building into a floating-point unit a very large register (or accumulator) that would allow sums of products to be performed exactly, without any rounding:

- Instead of rounding the result of a multiplication to p bits, all the $2p$ bits of the exact significand product are kept. This is what an FMA also does.

- Instead of adding the $2p$-bit product to a p-bit significand (as in an FMA), it is added to the large accumulator.

This is an old idea: many earlier (fixed-point) desk calculators offered a wider-than-displayed accumulator. Wider-than-product accumulator are also commonly used in fixed-point DSP processors. In floating point, if the accumulator is large enough to cover all the possible significand products, the addition will entail no rounding either. The minimum size of such an accumulator is deduced from the format parameters, and given in Table 8.1.

format	$e_{\min}$	$e_{\max}$	p	min. accumulator width ($2e_{\max} - 2e_{\min} + 2p$)
binary16	-14	15	11	80 bits
binary32	-126	127	24	554 bits
binary64	-1022	1023	53	4,196 bits
binary128	-16382	16383	113	65,756 bits

Table 8.1: Minimum size of an exact accumulator for the main IEEE formats

The value held in the large accumulator cannot, in general, be represented as a precision-p floating-point number. One accumulation or sum of product therefore entails one rounding when converting this value back to standard floating point. This conversion can also overflow or underflow, since the exponent range of the accumulator is twice as large as the exponent range of a floating-point number.

Kulisch also suggested that at least 60 bits should be added to the large accumulator. This would avoid intermediate spurious overflow of sums of up to 2^{60} products (well beyond what can be computed in several years at current processor speeds). This protection, of course, is only useful if the result itself is representable, thanks to later cancellations.

There are several difficulties in implementing the exact accumulator concept, especially if one aims at performance levels comparable to standard floating-point units (one accumulation per cycle at GHz frequency). Firstly, there is the intrinsic area of the large accumulator. Secondly, shifting the product to the proper place is very expensive. Finally, potential carry propagations along thousands of bits must be managed at high speed.

Several academic implementations of the exact accumulator have been demonstrated [340, 445, 598, 343]. So far, however, processor manufacturers always considered this idea too costly to implement. This is changing with the binary16 format: at this precision, an exact accumulator is barely larger than an FMA [82] while being vastly more accurate.

8.7.3 Hardware-accelerated compensated algorithms

The basic blocks of compensated algorithms are the 2Sum and 2Mult operations described in Sections 4.3 and 4.4 and illustrated in Figure 8.18.

Although their software implementation is expensive, they are relatively straightforward to implement in hardware: the lower part of the result is ac-

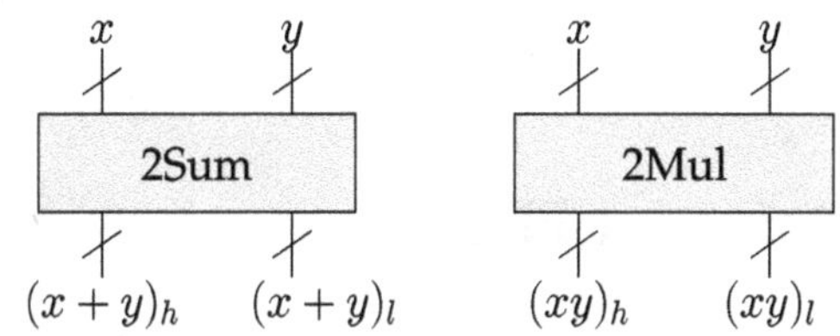

Figure 8.18: The 2Sum and 2Mult operators.

tually computed by the classical operator, but used only to determine correct rounding.

Dieter et al. [175] suggested the following modification to floating-point units. First, an additional p-bit register receives the residual bits that, in classical implementations of addition and multiplication, are used only to compute the sticky bits. Second, an additional instruction converts this residual register into a normal floating-point register. Filling the residual register involves no additional work with respect to a classical floating-point operator. Specifically, a floating-point multiplier has to compute the full $2p$-bit significand product to determine rounding. A floating-point adder does not perform the full addition, but the lower bits are untouched by the addition, so no additional computation is needed to recover them.

In this proposal, the residual register holds an unnormalized significand. It is normalized, only when needed, by the (specific) copy instruction, which uses the LZC and shifter available in the adder. Actually, the residual register also has to store a bit of information saying whether the rounded significand was the truncated significand or its successor, as this information propagates to the residual and may change its sign. Consider for example (in decimal for clarity), the product of the 4-digit numbers 6.789 and 5.678. The product $3.8547942 \cdot 10^1$ may be represented by the unevaluated sum $3.854 \cdot 10^1 + 7.942 \cdot 10^{-3}$, but 2Sum and 2Mul return in their higher part the correct rounding of the exact result. Therefore, the product returned should be $3.855 \cdot 10^1 - 2.058 \cdot 10^{-3}$ (the reader can check that the sum is the same). The details may be found in [175].

With this approach, 2Sum and 2Mul now only cost two instructions each, for an area overhead of less than 10%. The authors of [175] were concerned with the binary32 units of GPUs, but the same approach could be applied to general purpose processors.

Manoukian and Constantinides [404] and then Kadric, Gurniak, and DeHon [312] implement the 2Sum operator in Figure 8.18 and observe (on FPGA) that its delay is identical to that of a standard adder while the size is only 50% larger. The extra area corresponds to the LZC and shifter needed to normalize the error term.

Nathan et al. [450] suggest adding to the instruction set one new 3-operand instruction that computes the rounding error from adding the first two operands and adds this rounding error to the third operand. No implementation details are given.

8.8 Floating-Point for FPGAs

Processors need "one size fits all" operators. A good illustration is the attempt to have a single operator, the FMA, replacing the adder, the multiplier, and even the divider and square root.

The situation is exactly opposite in the domain of FPGA-based floating-point computing. There, the circuit of an operator can be optimized differently in different contexts. For instance, when programming an FPGA, you don't have to make a dramatic choice between a large and fast divider or a small and slow one. One will suit one application, and the other will suit another application. You could even use both for different tasks at the same time in one application. Similarly, you don't have to make a dramatic choice between small but inaccurate binary32 or slow but accurate binary64. You may use a home-made "binary42" format if it represents the soft spot in performance/accuracy for your application. Most floating-point libraries targeted for FPGAs are parameterized in precision.

This section is a survey of the architectural opportunities offered by the "one size need not fit all" feature of the FPGA target with respect to floating point. The FPGA is the near-term target of the operators prospected here, and indeed some of these operators are so context-specific that they make no sense at all in a processor. However, they may also be used in application-specific circuits.

8.8.1 Optimization in context of standard operators

Optimization in context can be more radical than just a change of precision. Consider a common example, the computation of the Euclidean norm of a three-dimensional floating-point vector, $\sqrt{x^2 + y^2 + z^2}$.

- The multipliers can be optimized in this context. On one side, computing the significand square m_x^2 requires in principle half the work of an arbitrary significand multiplication $m_x m_y$, due to symmetries in the partial product array. On the other side, the exponent (before normalization) of x^2 is $2e_x$, which saves one addition.

- The adders can also be optimized in this context. The squares are positive, so these additions are effective additions, which means that the close path of Figure 8.11 can be removed altogether [173].

These optimizations preserve a bit-exact result when compared to unoptimized operators in sequence. If one lifts this requirement, one may go further.

- Spurious overflows, i.e., overflows occurring in the intermediate computation of $x^2 + y^2 + z^2$ although the result $\sqrt{x^2 + y^2 + z^2}$ is in the floating-point range, may be prevented by using two more exponent bits in the intermediate results, or by dividing in such cases the three inputs by a power of two, the result being then multiplied by the same value. When implemented in hardware, both solutions have a very small area and latency overhead. The scaling idea can be used in software, but its relative overhead (several tests and additional operations) is much higher.

- The alignment of the three significand squares can be computed in parallel before adding them, reducing the critical path.

- Then, a 3-input significand adder is more efficient than two 2-input ones in sequence.

- Intermediate normalizations and roundings can be saved or relaxed. For instance, it is possible to build a Euclidean norm operator [153] that returns a faithful result (for which the absolute error is thus strictly smaller than one ulp of the exact result). This is not correct rounding, but is nevertheless more accurate than the combination of correctly rounded operators, at a much lower hardware cost than this combination. Such a fused operator may have other advantages, for instance, symmetry in its three inputs.

Going even further, Takagi and Kuwahara [587] have described an optimized architecture for the Euclidean norm that fuses the computation of the square root with that of the sum of squares. This approach could even be extended to the inverse square root [586] and other useful algebraic combinations such as

$$\frac{x}{\sqrt{x^2 + y^2}}.$$

The Euclidean norm is but one example of a coarser operator that is of general use and in the context of which the basic operators can be optimized. Let us first focus on an important case of operator optimization in context: the case of a constant operand.

8.8.2 Operations with a constant operand

A typical floating-point program involves many constants. Addition and subtraction with a constant operand do not allow for much optimization of the operator, but multiplication and division by a constant do.

By definition, a constant has a fixed exponent, therefore the floating point is of little significance here: all the research that has been done on integer constant multiplication [94, 375, 150, 636, 230, 176] can be used straightforwardly. Let C be an integer constant and X an input integer. Chapman's approach [94, 636] is FPGA specific: it exploits the structure of FPGAs, based on k-input LUTs, by writing the input X in radix 2^k: $X = \sum_{i=0}^{n}(2^k)^i x_i$. Then

$$CX = \sum_{i=0}^{n}(2^k)^i C x_i,$$

where the Cx_i may be efficiently tabulated in k-input LUTs. Most other recent approaches consider the binary expression of the constant. Let X be a p-bit integer. The product is written $CX = \sum_{i=0}^{k} 2^i c_i X$, and by only considering the nonzero c_i, it is expressed as a sum of $2^i X$; for instance, $17X = X + 2^4 X$. In the following, we will note this using the shift operator $\ll$, which has higher priority than $+$ and $-$; for instance, $17X = X + X{\ll}4$. As in standard multipliers, recoding the constant using signed bits allows us to reduce the number of nonzero bits, replacing some of the additions with subtractions; for instance, $15X = X \ll 4 - X$. Finally, one may reuse sub-constants that have already been computed; for instance, $4369X = 17X + (17X){\ll}8$. One needs to resort to heuristics to find the best (or a nearly best) decomposition of a large constant into shifts and additions [375, 61, 230, 621, 357, 356]. When applied to hardware, the cost function of a decomposition must take into account not only the number of additions, but also the sizes of these additions [168, 305, 7, 69]. Other approaches are possible; for instance, the complexity of multiplication by a constant was shown to be sublinear in the size of the constant [176], using an algorithm that is not based on the binary decomposition of the input, but is inefficient in practice.

To summarize, it is possible to derive an architecture for a multiplier by an integer constant that is smaller than that of a standard multiplier. How much smaller depends on the constant. A full performance comparison with operators using embedded multipliers or Digital Signal Processing (DSP) blocks remains to be done, but when these DSP blocks are a scarce resource, the multiplications to be implemented in logic should be the constant ones.

The architecture of a multiplier by a floating-point constant of arbitrary size is a straightforward specialization of the usual floating-point multiplier [69]. There are extreme cases, such as the multiplication by the constant 2.0, which reduces to one addition on the exponent.

We may also define constant multipliers that are much more accurate than what can be obtained with a standard floating-point multiplier. For instance, consider the irrational constant π. It cannot be stored on a finite number of bits, but it is nevertheless possible to design an operator that provably always returns the correctly rounded result of the (infinitely accurate) product πx [69], using techniques similar to those presented in Section 4.11. The

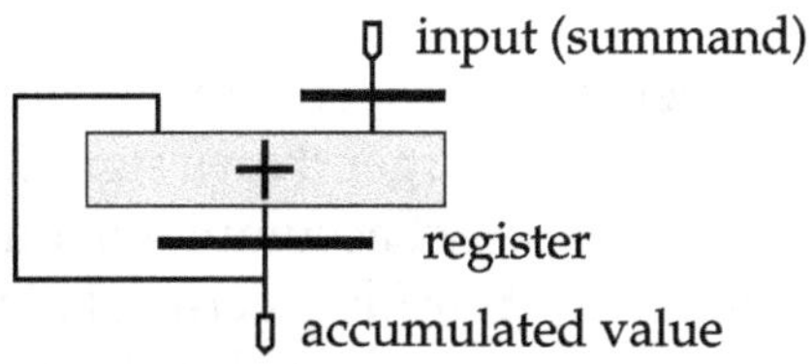

Figure 8.19: Iterative accumulator.

number of bits of the constant that is needed depends on the constant, and may be computed using continued fraction arguments [74]. Although one needs to use typically $2p$ bits of the constant to ensure correct rounding, the resulting constant multiplier may still be smaller than a generic one.

A practical application is division by a floating-point constant. It may be implemented as a multiplication by the inverse, but to obtain a bit-exact result (the same result as using a divider), one must compute the inverse to a sufficiently high accuracy. In software, one may check that the working precision is enough to ensure this accuracy [75]. On FPGAs, one may tailor the working precision to this purpose [180, 139].

In the case of the division by a small integer constant, a specialization of the Euclidean division algorithm may be used [140, 611]. This technique produces a quotient and a small remainder, which can be used to decide correct rounding [140].

8.8.3 Computing large floating-point sums

The issue of accumulating many floating-point terms in an FPGA has received a lot of attention. For a few simple terms, one may build trees of adders. However, when one has to add an arbitrary number of terms, one needs the iterative accumulator depicted by Figure 8.19.

One problem is that the floating-point adder is pipelined: Its latency l is typically 3 to 5 cycles in a processor, but 6 to 12 cycles for FPGA high-frequency implementations. Either the pipeline will actually work one cycle out of l, or one needs to compute l independent sub-sums and add them together at the end. Many architectural techniques have been proposed to design single-cycle accumulators [398, 480].

Another problem is accuracy. A first idea, to accumulate more accurately, is to use a standard floating-point adder with a larger significand. However, this leads to several inefficiencies. In particular, this large significand will have to be shifted, sometimes twice (first to align both operands and then to normalize the result). These shifts are in the critical path loop of the sum (see Figure 8.11).

8.8.3.1 Application-specific accurate accumulator

A better solution may therefore be to perform the accumulation in a large fixed-point register, typically much larger than a significand (see Figure 8.21). This is a variation on the Kulisch exact accumulator, but tailored to the problem at hand. It removes all the shifts from the critical path of the loop, as illustrated by Figure 8.20. The loop is now a fixed-point accumulation for which current FPGAs are highly efficient. Fast-carry logic enables high frequencies for medium-sized accumulators, and larger ones may use partial carry save (see Figure 8.6).

The shifters now only concern the summand (see Figure 8.20) and can be pipelined as deeply as required by the target frequency.

The normalization of the result may be performed at each cycle, also in a pipelined manner. However, most applications will not need all the intermediate sums: they will output the contents of the fixed-point accumulator (or only some of its MSBs), and the final normalization may be performed offline in software, once the sum is complete, or in a single normalizer shared by several accumulators (case of matrix operations). Therefore, it makes sense to provide this final normalizer as a separate component, as shown by Figure 8.20.

For clarity, implementation details are missing from these figures—the interested reader will find them in [155]. For example, the accumulator stores a two's complement number, so the shifted summand has to be sign-extended. The normalization unit also has to convert back from two's complement to sign-magnitude, and perform a carry propagation in case the accumulator holds a partial carry-save value. None of this is on the critical path of the loop.

In many applications, the proposed accumulator may be designed to be much more accurate than the floating-point adder it replaces. Indeed, it will even be exact (entailing no roundoff error whatsoever) if the accumulator size is large enough so that its LSB is smaller than that of all the inputs, and its MSB is large enough to ensure that no overflow may occur. Figure 8.21 illustrates this idea, showing the significands of the summands, and the accumulator itself.

When accelerating a specific application using an FPGA, instead of a huge generic accumulator, one may choose the size that matches the requirements of the application. There are five parameters in Figures 8.20 and 8.21: w_E and p are the exponent and significand size of the summands; MSB_A and LSB_A are the respective weights of the most and least significant bits of the accumulator (the size in bits of the accumulator is $w_A = \text{MSB}_A - \text{LSB}_A + 1$), and MaxMSB_X is the maximum expected weight of the MSB of a summand. By default MaxMSB_X will be equal to MSB_A, but sometimes the designer is able to tell that each summand is much smaller in magnitude than the final sum. For example, when integrating a function that is known to be positive,

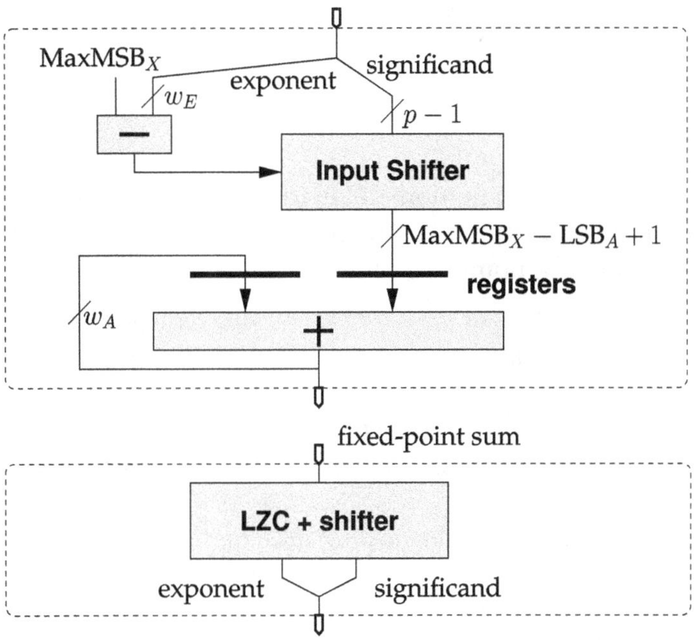

Figure 8.20: The proposed accumulator (top) and post-normalization unit (bottom). Only the registers on the accumulator itself are shown. The rest of the design is combinatorial and can be pipelined arbitrarily.

the size of a summand could be bounded by the product of the integration step and the max of the function. In this case, providing $\text{MaxMSB}_X < \text{MSB}_A$ will save hardware in the input shifter.

Defining these parameters requires some trial-and-error, or (better) some error analysis which will involve one more hidden parameter, the number N of summands to be accumulated. In most cases, the application dictates an *a priori* bound on either MaxMSB_X or MSB_A. LSB_A may be viewed as controlling the absolute accuracy, which still depends on N. The error analysis can be loose. For instance, adding to the maximum expected value of the result a margin of three orders of magnitude means adding only 10 bits to the accumulator.

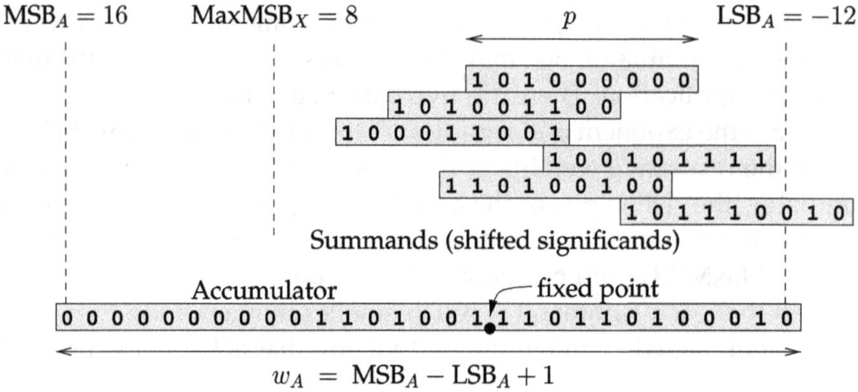

Figure 8.21: Accumulation of floating-point numbers into a large fixed-point accumulator.

This issue is surveyed in more detail in [155], where it is also shown that such an accumulator is much better in terms of area and latency than one using standard floating-point operators.

Example 8.2. *In [123], FPGAs are used to accelerate the computation of the inductance of a set of coils. This inductance is computed by the integration of inductances from elementary wire segments. Physical expertise tells us that the sum will be less than* 10^5 *(using arbitrary units), while profiling of a software version of the computation showed that the absolute value of an elementary inductance was always between* 10^{-2} *and* 2.

Converting to bit positions and adding two orders of magnitude (or 7 bits) for safety in all directions, this defines $MSB_A = \lceil \log_2(10^2 \times 10^5) \rceil = 24$, $MaxMSB_X = 8$ *and* $LSB_A = -p - 14$ *where* p *is the significand width of the summands. For* $p = 24$ *(binary32), we conclude that an accumulator stretching from* $LSB_A = -24 - 14 = -38$ *(least significant bit) to* $MSB_A = 24$ *(most significant bit) will be able to absorb all the additions without any rounding error: according to our profiling, no summand will ever add bits lower than* 2^{-38}, *and the accumulator is large enough to ensure it never overflows. The accumulator size should therefore be* $w_A = 24 + 38 + 1 = 63$ *bits.*

Of course, this is an optimistic view: profiling does not guarantee that no summand will ever be smaller than 10^{-4}, *so in practice this accumulator will not be exact. However, profiling does show that such very small summands are extremely rare, and the accumulation of errors due to them is very likely to have no measurable effect.*

Remark that in this application, only LSB_A *depends on* p, *since the other parameters (*MSB_A *and* $MaxMSB_X$*) are related to physical quantities, regardless of the precision used to simulate them. This illustrates that* LSB_A *is the parameter that allows one to manage the accuracy/area tradeoff for an accumulator.*

8.8.3.2 Parallel summations

DeHon and Kapre [157] have designed a technique to evaluate a large sum in parallel and obtain the sequential sum. If determinism is important, it is probably better to aim at faithful or correct rounding with respect to the exact mathematical sum [162].

Even on FPGAs, the compensated algorithms reviewed in Section 5.3.2 have several advantages with respect to application-specific accumulators. They require less error analysis, they scale better to problem sizes unknown *a priori*, and they are compatible with software.

With the 2Mul and 2Sum operators reviewed in Section 8.7.3, it is also possible to implement a parallel version of the compensated sum algorithms reviewed in Section 5.3.2. For instance, Kadric, Gurniak, and DeHon iterate over a tree of hardware 2Sum operators to compute the correctly rounded sum [312]. This expensive solution has the advantage of being generic.

8.8.4 Block floating point

When an input floating-point vector is to be multiplied by another constant one (as it happens in filters, Fourier transforms, etc.), one may use block floating point, a technique first used in the 1950s, when floating-point arithmetic was implemented in software, and more recently applied to FPGAs [10]. The technique consists in an initial alignment of all the input significands to the largest one, which brings them all to the same exponent (hence the phrase "block floating point"). After this alignment, all the computations (multiplications by constants and accumulation) can be performed in fixed point, with a single normalization at the end. Compared with the same computation using standard operators, this approach saves the renormalization shifts of the intermediate results. The argument is that the information lost in the initial shifts would have been lost in later shifts anyway. As seen in Section 5.3, this argument may be disputable in some cases. In practice, however, a typical block floating-point implementation will accumulate the dot product in a fixed-point register slightly larger than the input significands, thus ensuring a better accuracy than that achieved using standard operators.

8.8.5 Algebraic operators

We have already mentioned Euclidean norm [587], inverse square root [586], dot products [395], and sums of squares [153]. Other useful classes of algebraic operators are operations on complex numbers, and polynomial or rational evaluators [145].

8.8.6 Elementary and compound functions

Interest in hardware implementations of elementary functions was revived by its relevance to reconfigurable computing [174]. Floating-point elementary functions for FPGAs were proposed as soon as the FPGAs were large enough to accommodate them: exp and log in binary32 [178, 172] then up to binary64 [174, 366], trigonometric functions in binary32 [476, 171, 364], power function in binary32 [181] then up to binary64 [141]. Thomas designed a generator for arbitrary functions based on piecewise polynomial approximation [601]. Most of the recent implementations are parameterized in precision and exponent range.

The goal of such works is to design specific, combinatorial architectures based on fixed-point computations for such functions. These architectures can then be pipelined arbitrarily to run at the typical frequency of the target FPGA, with latencies that can be quite long (up to several tens of cycles for binary64), but with a throughput of one elementary function per cycle. The area of the state-of-the-art architecture is comparable to that of a few multipliers. As these pipelined architectures compute one result per cycle, the performance here is one order of magnitude better than that of a processor.

8.9 Probing Further

The reader wishing to examine actual designs will find several open source implementations of hardware FPUs in the OpenCores project.[1] The VFLOAT project from Northeastern University[2] is an open-source library of parameterized operators designed for FPGAs. The FloPoCo project[3] was started at ENS Lyon and is now supported by a wider community. It also provides the standard operators, but its main goal is to explore nonstandard ones such as those listed in Section 8.8.

[1]http://www.opencores.org/.

[2]http://www.coe.neu.edu/Research/rcl/projects/floatingpoint/index.html.

[3]http://flopoco.gforge.inria.fr/.

Chapter 9

Software Implementation of Floating-Point Arithmetic

THE PREVIOUS CHAPTER has presented the basic paradigms used for implementing floating-point arithmetic in hardware. However, some processors may not have such dedicated hardware, mainly for cost reasons. When it is necessary to handle floating-point numbers on such processors, one solution is to implement floating-point arithmetic in software. The goal of this chapter is to describe a set of techniques for writing an efficient implementation of the five floating-point operators $+, -, \times, \div$, and $\sqrt{\cdot}$ by working exclusively on integer numbers.

We will focus on algorithms for radix-2 floating-point arithmetic and provide some C codes for the binary32 format (see Chapter 3 for details), for which the usual floating-point parameters β (radix), k (width), p (precision), and $e_{\max}$ (maximum exponent), are

$$\beta = 2, \qquad k = 32, \qquad p = 24, \qquad e_{\max} = 127.$$

Details will be given on how to implement some key specifications of the IEEE 754 standard for each arithmetic operator. How should we implement special values? How should we implement correct rounding? For simplicity, we restrict ourselves here to "rounding to nearest, ties to even" (called roundTiesToEven in IEEE 754-2008 [267, §4.3.1]), but other rounding direction attributes can be implemented using very similar techniques. Note also that our codes will not handle exceptions.

Except for Listings 9.24 and 9.29, the codes we give in this chapter are taken from the FLIP software library [300]. In the last section, we also illustrate the benefits of going beyond the five basic operations with the development of so-called custom operators.

J.-M. Muller et al., *Handbook of Floating-Point Arithmetic*,
https://doi.org/10.1007/978-3-319-76526-6_9

For other software implementations for radix 2, we refer especially to Hauser's SoftFloat library [248].[1] Some software for radix-10 floating-point arithmetic has also been released. It will not be covered here, but the interested reader may refer to [116, 117].

9.1 Implementation Context

9.1.1 Standard encoding of binary floating-point data

Assume that the width of the floating-point format is k, and that the binary floating-point numbers are represented according to one of the basic formats specified by the IEEE 754 standard. As explained in Chapter 3, the encoding chosen for the standard allows one to compare floating-point numbers as if they were integers. We will make use of that property, and frequently manipulate the floating-point representations as if they were representations of integers. This is the "standard encoding" into k-bit unsigned integers [267, §3.4].

For a nonzero finite binary floating-point number

$$x = (-1)^{s_x} \cdot m_x \cdot 2^{e_x},$$

the standard encoding is the integer X represented by the bit string $X_{k-1} \dots X_0$ such that

$$X = \sum_{i=0}^{k-1} X_i 2^i. \tag{9.1}$$

Here the X_i are defined from the following splitting of the bit string into three fields:

- **Sign bit:** $X_{k-1} = s_x$;
- **Biased exponent:**

$$\begin{aligned} \sum_{i=0}^{k-p-1} X_{i+p-1} 2^i &= E_x \\ &= e_x - e_{\min} + n_x, \end{aligned} \tag{9.2}$$

  where

$$e_{\min} = 1 - e_{\max}$$

  and where n_x is the "is normal" bit[2] of x:

$$n_x = \begin{cases} 1 & \text{if } x \text{ is normal,} \\ 0 & \text{if } x \text{ is subnormal;} \end{cases} \tag{9.3}$$

[1]SoftFloat is available at http://www.jhauser.us/arithmetic/SoftFloat.html.

[2]That bit is not actually stored in the floating-point representation. It is easily obtained from the exponent field, and we will use it many times.

- **Fraction bits:** $X_i = m_{x,p-i-1}$ for $i = 0, \ldots, p-2$.

Notice that E_x is indeed nonnegative, for $e_{\min} \le e_x \le e_{\max}$ and $n_x \ge 0$. Also, this encoding allows us to represent zeros, infinities, and Not a Number data (NaNs). For the binary32 format, this is made clear in Table 9.1. The table displays the integer value (or range) of the encoding as well as its bit string. In this chapter, both will be used heavily, leading to fast code sequences. For similar implementation tricks in a different context (using SSE2 instructions on the Intel IA-32 architecture) see [14].

Of course, the same encoding will be used for the other operand y (if any) and for the result r of an operation involving x and/or y. Consequently, implementing, say, a multiplication operator for $\circ$ = RN and the binary32 format means writing a C function of the form

```
uint32_t RN_binary32_mul(uint32_t X, uint32_t Y) { ... }
```

which returns an unsigned 32-bit integer R encoding the multiplication result r. The goal of Section 9.3 will be precisely to show how the core `{ ... }` of this function can be implemented in C using exclusively 32-bit integer arithmetic.

9.1.2 Available integer operators

Concerning the operations on input or intermediate variables, we assume available the basic C arithmetic and logical operators

```
+,  -,  <<,  >>,  &,  |,  ^,
```

etc. We assume further that we have a fast way of computing the following functions:

- **Maximum or minimum of two unsigned integers:** $\max(A, B)$ and $\min(A, B)$ for $A, B \in \{0, \ldots, 2^k - 1\}$. These functions will be written

  ```
  maxu,  minu
  ```

  in all our C codes.

- **Maximum or minimum of two signed integers:** $\max(A, B)$ and $\min(A, B)$ for $A, B \in \{-2^{k-1}, \ldots, 2^{k-1} - 1\}$. These functions will be written

  ```
  max,  min
  ```

  in all our C codes.

- **Number of leading zeros of an unsigned integer:** This function counts the number of leftmost zeros of the bit string of $A \in \{0, \ldots, 2^k - 1\}$. (Note the similarity with the LZC (leading-zero count) function used in Chapters 7 and 8.)

Value or range of integer X	Floating-point datum x	Bit string $X_{31} \ldots X_0$
0	$+0$	$00000000000000000000000000000000$
$(0, 2^{23})$	positive subnormal number	$000000000X_{22} \ldots X_0$ with some $X_i = 1$
$[2^{23}, 2^{31} - 2^{23})$	positive normal number	$0\,\underbrace{X_{30}X_{29}X_{28}X_{27}X_{26}X_{25}X_{24}X_{23}}_{\text{not all ones, not all zeros}}\,X_{22} \ldots X_0$
$2^{31} - 2^{23}$	$+\infty$	$01111111100000000000000000000000$
$(2^{31} - 2^{23}, 2^{31} - 2^{22})$	sNaN	$0111111110X_{21} \ldots X_0$ with some $X_i = 1$
$[2^{31} - 2^{22}, 2^{31})$	qNaN	$0111111111X_{21} \ldots X_0$
2^{31}	-0	$10000000000000000000000000000000$
$(2^{31}, 2^{31} + 2^{23})$	negative subnormal number	$100000000X_{22} \ldots X_0$ with some $X_i = 1$
$[2^{31} + 2^{23}, 2^{32} - 2^{23})$	negative normal number	$1\,\underbrace{X_{30}X_{29}X_{28}X_{27}X_{26}X_{25}X_{24}X_{23}}_{\text{not all ones, not all zeros}}\,X_{22} \ldots X_0$
$2^{32} - 2^{23}$	$-\infty$	$11111111100000000000000000000000$
$(2^{32} - 2^{23}, 2^{32} - 2^{22})$	sNaN	$1111111110X_{21} \ldots X_0$ with some $X_i = 1$
$[2^{32} - 2^{22}, 2^{32})$	qNaN	$1111111111X_{21} \ldots X_0$

Table 9.1: Binary32 datum x and its encoding into integer $X = \sum_{i=0}^{31} X_i 2^i$ (see [292]).

- **Lower half of the product of two unsigned integers:** This function computes $AB \bmod 2^k$, for $A, B \in \{0, \ldots, 2^k - 1\}$.

- **Upper half of the product of two unsigned integers:** This function computes $\lfloor AB/2^k \rfloor$, where $\lfloor \cdot \rfloor$ denotes the usual floor function.

The last three functions will be written, respectively,

`nlz`, `*`, `mul`

in all our C codes.

Some typical C99 implementations of the `maxu`, `minu`, `max`, `min`, `mul`, and `nlz` operators are given in Listings 9.1 through 9.4 for $k = 32$.

C listing 9.1 Implementation of the `maxu` and `minu` operators for $k = 32$.

```
uint32_t maxu(uint32_t A, uint32_t B)
{ return A > B ? A : B; }

uint32_t minu(uint32_t A, uint32_t B)
{ return A < B ? A : B; }
```

C listing 9.2 Implementation of the `max` and `min` operators for $k = 32$.

```
int32_t max(int32_t A, int32_t B)
{ return A > B ? A : B; }

int32_t min(int32_t A, int32_t B)
{ return A < B ? A : B; }
```

C listing 9.3 Implementation of the `mul` operator for $k = 32$.

```
uint32_t mul(uint32_t A, uint32_t B)
{
  uint64_t t0, t1, t2;

  t0 = A;
  t1 = B;
  t2 = (t0 * t1) >> 32;
  return t2;
}
```

C listing 9.4 Implementation of the `nlz` operator for $k = 32$.

```
uint32_t nlz(uint32_t X)
{
  uint32_t Z = 0;

  if (X == 0) return 32;
  if (X <= 0x0000FFFF) {Z = Z +16; X = X <<16;}
  if (X <= 0x00FFFFFF) {Z = Z + 8; X = X << 8;}
  if (X <= 0x0FFFFFFF) {Z = Z + 4; X = X << 4;}
  if (X <= 0x3FFFFFFF) {Z = Z + 2; X = X << 2;}
  if (X <= 0x7FFFFFFF) {Z = Z + 1;}
  return Z;
}
```

Therefore, in principle all that is needed to run our implementation examples is a C99 compiler that supports 32-bit integer arithmetic. However, optimizations (to achieve low latency) depend on the features of the targeted architectures and compilers. Some of these features will be described in Section 9.1.4.

Before that, we shall give in the next section a few examples that show how one can implement some basic building blocks (needed in several places later in this chapter) by means of some of the integer operators we have given above.

9.1.3 First examples

For $k = 32$, given an integer X as in (9.1), we consider here three sub-tasks. How does one deduce E_x as in (9.2)? How shall we deduce n_x as in (9.3)? Assuming that x is (sub)normal and defining λ_x as the number of leading zeros of the binary expansion of the significand m_x, that is,

$$m_x = [\underbrace{0.0\ldots0}_{\lambda_x \text{ zeros}}1m_{x,\lambda_x+1}\cdots m_{x,23}], \tag{9.4}$$

how does one compute that number λ_x?

9.1.3.1 Extracting the exponent field

Since by (9.2) the bit string of E_x consists of the bits $X_{30}, \ldots, X_{23}$, a left shift by one position (to remove the sign bit X_{31}) followed by a right shift by 24 positions (to remove in particular the fraction bits $X_{22}, \ldots, X_0$) will give us what we want. This can be implemented as in Listing 9.5.

C listing 9.5 Computation of the biased exponent E_x of a binary32 datum x.

```
uint32_t Ex;

Ex = (X << 1) >> 24;
```

Of course, an alternative to the left shift would be to mask the sign bit by taking the bitwise AND of X and

$$2^{31} - 1 = (0\underbrace{1111111111111111111111111111111}_{31 \text{ ones}})_2 = (7FFFFFFF)_{16}.$$

This alternative is implemented in Listing 9.6. In what follows we will often interpret the fact of masking the sign bit as taking the "absolute value" of X (even if X encodes a NaN), and we will write $|X|$ for the corresponding integer (which is in fact simply $X \bmod 2^{31}$) and absX for the corresponding variable. This variable absX will turn out to be extremely useful for implementing addition, multiplication, and division (see Sections 9.2, 9.3, and 9.4).

C listing 9.6 Another way of computing the biased exponent E_x of a binary32 datum x.

```
uint32_t absX, Ex;

absX = X & 0x7FFFFFFF;
Ex = absX >> 23;
```

9.1.3.2 Computing the "is normal" bit

Assume that X encodes a (sub)normal binary floating-point number x and recall that the "is normal" bit of x is defined by (9.3). Once absX has been computed (see Listing 9.6), we deduce from Table 9.1 that x is normal if and only if absX is at least $2^{23} = (800000)_{16}$. Hence the code in Listing 9.7.

C listing 9.7 For a (sub)normal binary32 number x, computation of its "is normal" bit n_x.

```
uint32_t absX, nx;

absX = X & 0x7FFFFFFF;
nx = absX >= 0x800000;
```

9.1.3.3 Computing the number of leading zeros of a significand

Assume again that X encodes a (sub)normal binary floating-point number $x = (-1)^{s_x} \cdot m_x \cdot 2^{e_x}$ and recall that the number of leading zeros of m_x is $\lambda_x \geq 0$ as in (9.4).

Let us define L_X as the number of leading zeros of $|X|$. Clearly, $L_X \geq 1$ since the leading bit of $|X|$ is zero. Besides, x is nonzero, therefore $|X|$ is nonzero as well, and $L_X \leq 31$. More interestingly, one can show (see [292]) that λ_x is related to L_X as follows:

$$\lambda_x = M_X - w, \qquad M_X = \max(L_X, w),$$

where $w = k - p$ is the bit width of the biased exponent field. For the binary32 format, it follows from $k = 32$ and $p = 24$ that $w = 8$.

Consequently, one can implement the computation of λ_x by using the integer operators of Section 9.1.2. This is shown in Listing 9.8.

C listing 9.8 For a (sub)normal binary32 number x, computation of the number λ_x of leading zeros of the significand of x.

```
uint32_t absX, MX, lambdax;

absX = X & 0x7FFFFFFF;
MX = maxu(nlz(absX),8);
lambdax = MX - 8;
```

The number λ_x appears in some formulas used for computing the exponent of a product or a quotient in Sections 9.3 and 9.4. However, we will see there that these formulas also involve a constant term which can be updated in order to carry the -8 of identity $\lambda_x = M_X - 8$. Consequently, in such cases, λ_x itself is not needed and computing M_X will be enough.

9.1.4 Design choices and optimizations

We have already seen with the examples of Listings 9.5 and 9.6 that even a simple expression can be implemented in a variety of ways. Which one is best depends on the goal to achieve (low latency or high throughput, for example), as well as on some features of the targeted architecture and compiler.

The C codes given in the following sections have been written so as to expose some instruction-level parallelism (ILP). In fact, we will see that the algorithms used for implementing addition, multiplication, division, and square root very often lead in a fairly natural way to some code with relatively high ILP. For example, it is clear that multiplying two floating-point numbers can be done by, roughly, multiplying the significands and *simultaneously* adding the exponents. However, we shall show further that some ILP can already be exposed for performing the exponent addition itself.

Our codes have also been written with some particular architectural and compiler features in mind. These features are those of the ST231 VLIW[3] processor and compiler (see [509, §1.1] and [291, §II]). Concerning the processor, these features include the following.

[3]VLIW is an acronym for *very long instruction word*.

- Four instructions can be launched simultaneously at every cycle, with some restrictions (for example, at most two of them can be $*$ or `mul`).
- The latency of 32-bit integer arithmetic and logic is 1, except for $*$ and `mul`, whose latency is 3.
- Registers consist in 64 general-purpose registers and 8 condition registers.
- It is possible to encode immediate operands up to 32 bits.
- An efficient branch architecture is available, with multiple condition registers.
- Execution is predicated through `select` instructions.

The compiler features include:

- if-conversion optimization [78]: it generates mostly straight-line code by emitting efficient sequences of `select` instructions instead of costly control flow;
- the linear assembly optimizer (LAO) [137]: it generates a schedule for the instructions that is often very close to the optimal.

In practice, the codes we propose in the following sections are well suited to these features in the sense that when compiling for the ST231 architecture, the obtained latencies are indeed low (from 20 to 30 cycles, depending on the operator). For more performance results in this context, we refer to [291, 293, 292, 509] and to the FLIP software library. [4]

9.2 Binary Floating-Point Addition

An implementation of the method of Section 7.3 for computing $\circ(x + y)$ will be described here for $\circ = \mathrm{RN}$ and the binary32 format:

$$\beta = 2, \qquad k = 32, \qquad p = 24, \qquad e_{\max} = 127.$$

The case where either x or y is a special datum (like ± 0, $\pm\infty$, or NaN) is discussed in Section 9.2.1, while the case where both x and y are (sub)normal numbers is described in Sections 9.2.2 through 9.2.4.

[4]See http://flip.gforge.inria.fr/.

Value or range of integer X	Floating-point datum x
0	$+0$
$(0, 2^{k-1} - 2^{p-1})$	positive (sub)normal number
$2^{k-1} - 2^{p-1}$	$+\infty$
$(2^{k-1} - 2^{p-1}, 2^{k-1} - 2^{p-2})$	sNaN
$[2^{k-1} - 2^{p-2}, 2^{k-1})$	qNaN

Table 9.2: Some floating-point data encoded by X.

9.2.1 Handling special values

In the case of addition, the input (x, y) is considered a *special input* when x or y is ± 0, $\pm\infty$, or NaN. For each possible case the IEEE 754-2008 standard requires that a special value be returned. When both x and y are zero, these special values are given by Table 7.1; when x or y is nonzero, they follow from Tables 7.2 and 7.3 by adjoining the correct sign, using

$$x + y = (-1)^{s_x} \cdot \Big(|x| + (-1)^{s_z} \cdot |y|\Big), \qquad s_z = s_x \text{ XOR } s_y. \tag{9.5}$$

(Notice that the standard does not specify the sign of a NaN result; see [267, §6.3].)

As said above, considering $|x|$ means setting the sign bit of X to zero, that is, considering $X \bmod 2^{k-1}$ instead of X. Recall that we use the notation

$$|X| := X \bmod 2^{k-1}. \tag{9.6}$$

9.2.1.1 Detecting that a special value must be returned

Special inputs can be filtered out very easily using Table 9.2, which is a direct consequence of the standard binary encoding [267, §3.4]. Indeed, we deduce from this table that (x, y) is a special input if and only if

$$|X| \text{ or } |Y| \text{ is in } \{0\} \cup [2^{k-1} - 2^{p-1}, 2^{k-1}). \tag{9.7}$$

An equivalent formulation of (9.7) that allows us to use the `max` operator is

$$\max\Big(\big(|X| - 1\big) \bmod 2^k, \big(|Y| - 1\big) \bmod 2^k\Big) \geq 2^{k-1} - 2^{p-1} - 1. \tag{9.8}$$

For example, when $k = 32$ and $p = 24$, an implementation of (9.8) is given by lines 3, 4, 5 of Listing 9.9.

C listing 9.9 Special value handling in a binary32 addition operator.

```
 1  uint32_t absX, absY, absXm1, absYm1, Max, Sx, Sy;
 2
 3  absX = X & 0x7FFFFFFF; absY = Y & 0x7FFFFFFF;
 4  absXm1 = absX - 1;     absYm1 = absY - 1;
 5  if (maxu(absXm1, absYm1) >= 0x7F7FFFFF)
 6  {
 7     Max = maxu(absX, absY); Sx = X & 0x80000000; Sy = Y & 0x80000000;
 8     if (Max > 0x7F800000 ||
 9         (Sx != Sy && absX == 0x7F800000 && absY == 0x7F800000))
10       return 0x7FC00000 | Max;          // qNaN with payload encoded in
11                                         // the last 22 bits of X or Y
12     if (absX > absY) return X;
13     else if (absX < absY) return Y;
14     else return X & Y;
15  }
```

9.2.1.2 Returning special values as recommended or required by IEEE 754-2008

Once our input (x, y) is known to be special, one must return the corresponding result as specified in Tables 7.2 and 7.3. First, a quiet NaN (qNaN) must be returned as soon as one of the following two situations occurs:

- If $|X|$ or $|Y|$ encodes a NaN, that is, according to Table 9.2, if

$$\max(|X|, |Y|) > 2^{k-1} - 2^{p-1}; \tag{9.9}$$

- If $s_z = 1$ and if both $|X|$ and $|Y|$ encode $+\infty$, that is, if

$$X_{k-1} \text{ XOR } Y_{k-1} = 1 \quad \text{and} \quad |X| = |Y| = 2^{k-1} - 2^{p-1}. \tag{9.10}$$

When $k = 32$ and $p = 24$, we have $2^{k-1} - 2^{p-1} = 2^{31} - 2^{23} = (7F800000)_{16}$. Therefore, the conditions in (9.9) and (9.10) can be implemented as in lines 7, 8, 9 of Listing 9.9.

Let us raise here a few remarks about the qNaN we return at line 10 of Listing 9.9. Since $(7FC00000)_{16} = 2^{31} - 2^{22}$, the bit string of this qNaN has the form

$$0\underbrace{111111111}_{9\text{ ones}}\underbrace{Z_{21}\dots Z_0}_{\text{payload}},$$

where the string $Z_{21}\dots Z_0$ is either $X_{21}\dots X_0$ or $Y_{21}\dots Y_0$ (we feel free to set the sign bit to zero because, as stated above, the standard does not specify the sign of a NaN result). That particular qNaN carries the last 22 bits of one of the operand encodings. So, in a sense, this follows the IEEE 754-2008 recommendation that, in order "to facilitate propagation of diagnostic information contained in NaNs, as much of that information as possible should be preserved in NaN results of operations" (see [267, §6.2]). More precisely, if only one of the inputs is NaN, then its payload is propagated (by means of the bitwise OR involving the variable `Max` at line 10), as recommended in [267, §6.2.3]); if both inputs x and y are NaN, then the payload of one of them is propagated, again as recommended in [267, §6.2.3]).

Assume now that the case of a qNaN result has been handled, and let us focus on the remaining special values by inspecting three cases:

- If $|X| > |Y|$ then, using Tables 7.2 and 7.3, together with (9.5), we must return $(-1)^{s_x} \cdot |x| = x$.
- If $|X| < |Y|$ then we return $(-1)^{s_x} \cdot |y|$ if $s_z = 0$, and $(-1)^{s_x} \cdot (-|y|)$ if $s_z = 1$, that is, y in both cases.
- If $|X| = |Y|$ then both x and y are either zero or infinity. However, since at this stage qNaN results have already been handled, the only possible inputs to addition are $(\pm 0, \pm 0)$, $(+\infty, +\infty)$, and $(-\infty, -\infty)$. Using Table 7.1 for $\mathrm{RN}(x + y)$ shows that for all these inputs the result is given by the bitwise AND of X and Y.

An example of implementation of these three cases is given for the binary32 format at lines 12, 13, 14 of Listing 9.9.

9.2.2 Computing the sign of the result

We assume from now on that the input (x, y) is not *special; that is, both x and y are finite nonzero (sub)normal numbers.*

Recalling that $s_z = s_x$ XOR s_y, we have

$$x + y = (-1)^{s_x} \cdot \Big(|x| + (-1)^{s_z} \cdot |y|\Big) = (-1)^{s_y} \cdot \Big((-1)^{s_z} \cdot |x| + |y|\Big). \qquad (9.11)$$

Hence, the sign of the exact result $x + y$ is s_x if $|x| > |y|$, and s_y if $|x| < |y|$. What about the particular case where $|x| = |y|$ and $s_x \neq s_y$?[5] In that case $x + y = \pm 0$, and the standard prescribes that $\circ(x + y) = +0$ "in all rounding-direction attributes except roundTowardNegative; under that attribute, the sign of an exact zero sum (or difference) shall be -0." (See [267, §6.3].)

Recalling that $\mathrm{RN}(x) \geq 0$ if and only if $x \geq 0$, we conclude that the sign s_r to be returned is given by

$$s_r = \begin{cases} s_x & \text{if } |x| > |y|, \\ s_y & \text{if } |x| < |y|, \\ s_x = s_y & \text{if } |x| = |y| \text{ and } s_z = 0, \\ 0 & \text{if } |x| = |y| \text{ and } s_z = 1. \end{cases}$$

[5] We remind the reader that we have assumed at the beginning of this section that x and y are nonzero.

Now, because of the standard encoding of binary floating-point data, the condition $|x| > |y|$ is equivalent to $|X| > |Y|$. Consequently, the computation of s_r for the binary32 format can be implemented as shown in Listing 9.10, assuming s_x, s_y, $|X|$, and $|Y|$ are available (all of them have been computed, e.g., in Listing 9.9 when handling special values due to special inputs).

C listing 9.10 Sign computation in a binary32 addition operator, assuming rounding to nearest ($\circ = \mathrm{RN}$) and that s_x, s_y, $|X|$, and $|Y|$ are available.

```
uint32_t Sr;

if (absX > absY)
    Sr = Sx;
else if (absX < absY)
    Sr = Sy;
else
    Sr = minu(Sx,Sy);
```

In fact, Listing 9.10 can be used even if the input (x, y) is special. Hence, if special values are handled *after* the computation of s_r, one can reuse the variables `Sr` and `Max`, and replace lines 12, 13, 14 of Listing 9.9 with

```
return Sr | Max;
```

9.2.3 Swapping the operands and computing the alignment shift

As explained in Section 7.3, one may classically compare the exponents e_x and e_y and then, if necessary, swap x and y in order to ensure $e_x \geq e_y$. It turns out that in radix 2 comparing $|x|$ and $|y|$ is enough (and is cheaper in our particular context of a software implementation), as we will see now.

Recalling that $\mathrm{RN}(-x) = -\mathrm{RN}(x)$ (symmetry of rounding to nearest) and that $s_z = s_x$ XOR s_y, we deduce from the identities in (9.11) that the correctly rounded sum satisfies

$$\begin{aligned}\mathrm{RN}(x+y) &= (-1)^{s_x} \cdot \mathrm{RN}\Big(|x| + (-1)^{s_z} \cdot |y|\Big) \\ &= (-1)^{s_y} \cdot \mathrm{RN}\Big((-1)^{s_z} \cdot |x| + |y|\Big).\end{aligned}$$

Thus, it is natural to first ensure that

$$|x| \geq |y|,$$

and then compute only one of the two correctly rounded values above, namely,

$$\mathrm{RN}\Big(|x| + (-1)^{s_z} \cdot |y|\Big).$$

Since $|x| = m_x \cdot 2^{e_x}$ and $|y| = m_y \cdot 2^{e_y}$, this value is in fact given by

$$r := \mathrm{RN}\Big(|x| + (-1)^{s_z} \cdot |y|\Big) = \mathrm{RN}(m_r \cdot 2^{e_x}),$$

where

$$m_r = m_x + (-1)^{s_z} \cdot m_y \cdot 2^{-\delta} \tag{9.12}$$

and

$$\delta = e_x - e_y. \tag{9.13}$$

Interestingly enough, the condition $|x| \geq |y|$ implies in particular that $e_x \geq e_y$. More precisely, we have the following property.

Property 9.1. *If* $|x| \geq |y|$ *then* $m_r \geq 0$ *and* $\delta \geq 0$.

Proof. Let us show first that $e_x < e_y$ implies $|x| < |y|$. If $e_x < e_y$ then $e_y \geq e_x + 1 > e_{\min}$. Thus, y must be a normal number, which implies that its significand satisfies $m_y \geq 1$. Since $m_x < 2$, one obtains $m_y \cdot 2^{e_y} \geq 2^{e_x+1} > m_x \cdot 2^{e_x}$.

Let $\delta = e_x - e_y$. We can show that $m_x - m_y \cdot 2^{-\delta}$ is nonnegative by considering two cases:

- If $e_x = e_y$ then $\delta = 0$, and $|x| \geq |y|$ is equivalent to $m_x \geq m_y$.
- If $e_x > e_y$ then $\delta \geq 1$ and, reasoning as before, x must be normal. On the one hand, $\delta \geq 1$ and $m_y \in (0, 2)$ give $m_y \cdot 2^{-\delta} \in (0, 1)$. On the other hand, x being normal, we have $m_x \geq 1$. Therefore, $m_x - m_y \cdot 2^{-\delta} \geq 0$, which concludes the proof.

□

9.2.3.1 Operand swap

To ensure that $|x| \geq |y|$, it suffices to replace the pair $(|x|, |y|)$ by the pair $(\max(|x|, |y|), \min(|x|, |y|))$. Using the `maxu` and `minu` operators and assuming that $|X|$ and $|Y|$ are available, an implementation for the binary32 format is straightforward, as shown at line 3 of Listing 9.11.

9.2.3.2 Alignment shift

For $|x| \geq |y|$, let us now compute the nonnegative integer δ in (9.13) that is needed for shifting the significand m_y right by δ positions (in order to align it with the significand m_x as in (9.12)). Recall that n_x and n_y are the "is normal" bits of x and y (so that $n_x = 1$ if x is normal, and 0 if x is subnormal). Recall also that the biased exponents E_x and E_y of x and y satisfy

$$E_x = e_x - e_{\min} + n_x \qquad \text{and} \qquad E_y = e_y - e_{\min} + n_y.$$

Therefore, the shift δ in (9.13) is given by

$$\delta = (E_x - n_x) - (E_y - n_y).$$

For the binary32 format, this expression for δ can be implemented as shown at lines 4, 5, 6 of Listing 9.11. Note that E_x, n_x, E_y, and n_y can be computed in parallel, and that the differences $E_x - n_x$ and $E_y - n_y$ can then be computed in parallel too.

C listing 9.11 Operand swap and alignment shift computation in a binary32 addition operator, assuming $|X|$ and $|Y|$ are available.

```
1 uint32_t Max, Min, Ex, Ey, nx, ny, delta;
2
3 Max = maxu(absX, absY);     Min = minu(absX, absY);
4 Ex = Max >> 23;             Ey = Min >> 23;
5 nx = Max >= 0x800000 ;      ny = Min >= 0x800000;
6 delta = (Ex - nx) - (Ey - ny);
```

9.2.4 Getting the correctly rounded result

It remains to compute the correctly rounded value $r = \mathrm{RN}(m_r \cdot 2^{e_x})$, where m_r is defined by (9.12). Recalling that n_x and n_y are "is normal" bits, the binary expansions of m_x and $m_y \cdot 2^{-\delta}$ are, respectively,

$$\begin{aligned} m_x &= (n_x.m_{x,1} \ldots\ldots\ldots m_{x,p-1} \quad 0 \ldots\ldots\ldots 0)_2 \\ m_y \cdot 2^{-\delta} &= (\underbrace{0.0\ldots0}_{\delta \text{ zeros}} n_y m_{y,1} \ldots m_{y,p-1-\delta} m_{y,p-\delta} \ldots m_{y,p-1})_2. \end{aligned}$$

Therefore, the binary expansion of m_r must have the form

$$m_r = (c s_0.s_1 \ldots s_{p-1} s_p \ldots s_{p+\delta-1})_2, \quad c \in \{0, 1\}. \tag{9.14}$$

We see that m_r is defined by at most $p + \delta + 1$ bits. Note that $c = 1$ is due to a possible carry propagation when adding $m_y \cdot 2^{-\delta}$ to m_x. If no carry propagates during that addition, or in the case of subtraction, then $c = 0$.

9.2.4.1 A first easy case: $x = -y \neq 0$

This case, for which it suffices to return $+0$ when $\circ = \mathrm{RN}$, will not be covered by the general implementation described later (the exponent of the result would not always be set to zero). Yet, it can be handled straightforwardly once $|X|$, $|Y|$, and s_z are available.

C listing 9.12 Addition for the binary32 format in the case where $x = -y \neq 0$ and assuming rounding to nearest ($\circ = \mathrm{RN}$).

```
if (absX == absY && Sz == 1) return 0;
```

9.2.4.2 A second easy case: both x and y are subnormal numbers

When $|x| \geq |y|$, this case occurs when $n_x = 0$. Indeed, if $n_x = 0$ then x is subnormal and therefore $|x| < 2^{e_{\min}}$. Now, the assumption $|x| \geq |y|$ clearly implies $|y| < 2^{e_{\min}}$ as well, which means that y is subnormal too.

In this case, x and y are such that

$$|x| = (0.m_{x,1} \dots m_{x,p-1})_2 \cdot 2^{e_{\min}} \qquad \text{and} \qquad |y| = (0.m_{y,1} \dots m_{y,p-1})_2 \cdot 2^{e_{\min}},$$

so that

$$\delta = 0$$

and

$$m_r = (0.m_{x,1} \dots m_{x,p-1})_2 \pm (0.m_{y,1} \dots m_{y,p-1})_2.$$

This is a fixed-point addition/subtraction and m_r can thus be computed exactly. Furthermore, we have $m_r \in [0, 2)$ and $e_x = e_{\min}$, so that $r = \mathrm{RN}(m_r \cdot 2^{e_x})$ is in fact given by

$$r = m_r \cdot 2^{e_{\min}},$$

with

$$m_r = (m_{r,0}.m_{r,1} \dots m_{r,p-1})_2.$$

Note that the result r can be either normal ($m_{r,0} = 1$) or subnormal ($m_{r,0} = 0$).

The code in Listing 9.13 shows how this can be implemented for the binary32 format. Here we assume that $|x| \geq |y|$ and that the variables `Max` and `Min` encode, respectively, the integers $|X|$ and $|Y|$. We also assume that n_x (the "is normal" bit of x) and the operand and result signs s_x, s_y, s_r are available. (These quantities have been computed in Listings 9.9, 9.10, and 9.11.)

C listing 9.13 Addition for the binary32 format in the case where both operands are *subnormal* numbers.

```
uint32_t Sz, Mx, My, compMy, Mr;
// Assume x and y are non-special and such that |x| >= |y|.
if (nx == 0)
{
   Sz = (Sx != Sy);
   compMy = (My ^ (0 - Sz)) + Sz;
   Mr = Mx + compMy;
   return Sr | Mr;
}
```

Since both x and y are subnormal, the bit strings of `Mx` and `My` in Listing 9.13 are

$$[0\,\underbrace{00000000}_{8 \text{ zeros}}\,m_{x,1} \dots m_{x,23}] \qquad \text{and} \qquad [0\,\underbrace{00000000}_{8 \text{ zeros}}\,m_{y,1} \dots m_{y,23}].$$

Recalling that $\delta = 0$, we see that it suffices to encode $(-1)^{s_z} \cdot m_y$. This is achieved by computing `compMy`, using two's complement in the case of effective subtraction: `compMy` is equal to `My` if $s_z = 0$, and to `~My + 1` if $s_z = 1$, where "~" refers to the Boolean bitwise NOT function.

Notice also that the bit string of the returned integer is

$$[s_r \underbrace{0000000}_{\text{7 zeros}} m_{r,0} m_{r,1} \dots m_{r,23}].$$

Therefore, the correct biased exponent has been obtained automatically as a side effect of concatenating the result sign with the result significand. Indeed,

- if $m_{r,0} = 1$, then r is normal and has exponent $e_{\min}$, whose biased value is $e_{\min} + e_{\max} = 1$;

- if $m_{r,0} = 0$, then r is subnormal, and its biased exponent will be zero.

9.2.4.3 Implementation of the general case

Now that we have seen two preliminary easy cases, we can describe how to implement, for the binary32 format, the general algorithm recalled in Section 7.3 (note that our general implementation presented here will in fact cover the previous case where both x and y are subnormal numbers).

In the general case, because of significand alignment, the exact sum m_r in (9.14) has $p + \delta - 1$ fraction bits (which can be over 200 for the binary32 format). However, it is not necessary to compute m_r exactly in order to round correctly, and using $p - 1$ fraction bits together with three additional bits (called guard bits) is known to be enough (see, for example, [187, §8.4.3] as well as [502, §5]).

Thus, for the binary32 format, we can store m_x and m_y into the following bit strings of length 32 (where the last three bits will be used to update the guard bits):

$$[0\,0\,0\,0\,0\,n_x m_{x,1} \dots m_{x,23}\,0\,0\,0]$$

and

$$[0\,0\,0\,0\,0\,n_y m_{y,1} \dots m_{y,23}\,0\,0\,0].$$

This corresponds to the computation of integers `mx` and `my` at lines 3 and 4 of Listing 9.14.

C listing 9.14 Addition/subtraction of (aligned) significands in a binary32 addition operator.

```
1  uint32_t mx, my, highY, lowY, highR;
2
3  mx = (nx << 26) | ((Max << 9) >> 6);
4  my = (ny << 26) | ((Min << 9) >> 6);
5
6  highY = my >> minu(27, delta);
7
8  if (delta <= 3)
9     lowY = 0;
10 else if (delta >= 27)
11    lowY = 1;
12 else
13    lowY = (my << (32 - delta)) != 0;
14
15 if (Sz == 0)
16    highR = mx + (highY | lowY);
17 else
18    highR = mx - (highY | lowY);
```

The leading bits of $m_y \cdot 2^{-\delta}$ are then stored into the 32-bit integer `highY` obtained by shifting `my` right by δ positions:

$$\texttt{highY} = [\underbrace{000\ldots000}_{5+\delta \text{ zeros}} n_y m_{y,1} \ldots m_{y,26-\delta}].$$

We see that if $\delta \geq 27$, then all the bits of `highY` are zero: hence, the use of the `minu` instruction at line 6 of Listing 9.14. This prevents us from shifting by a value greater than 31, for which the behavior of the bitwise shift operators >> and << is undefined.

The bits $m_{y,27-\delta}, \ldots, m_{y,23}$ of m_y that have been discarded by shifting are used to compute the sticky bit T. In fact, all we need is to know whether all of them are zero or not. This information about the lower part of $m_y \cdot 2^{-\delta}$ can thus be collected into an integer `lowY` as follows.

- If $\delta \leq 3$ then, because of the last three zeros of the bit string of `my`, none of the bits of m_y has been discarded and thus `lowY` is equal to 0.
- If $\delta \geq 27$ then all the bits of m_y have been discarded and, since at least one of them is nonzero because $m_y \neq 0$, we must have `lowY` equal to 1.
- If $\delta \in \{4, \ldots, 26\}$ then the last $\delta - 3$ bits of m_y can be extracted from the integer `my` by shifting it left by $32 - \delta$ positions.

A possible implementation of `lowY` is thus as in lines 8–13 of Listing 9.14.

Once the value 0 or 1 of `lowY` has been obtained, it is used to update the last bit (sticky bit position) of `highY`. Only then can the effective operation (addition if $s_z = 0$, subtraction if $s_z = 1$) be performed. Since the bit string of `lowY` consists of 31 zeros followed by either a zero or a one, a possible implementation is as in lines 15–18 of Listing 9.14.

The bit string of the result `highR` is

$$\texttt{highR} = [0000\, c\, r_0 r_1 \ldots r_{23} r_{24} r_{25} r_{26}].$$

As stated in Section 7.3, normalization may now be necessary, either because of *carry propagation* ($c = 1$) or because of *cancellation* ($c = 0$ and $r_0 = \cdots = r_i = 0$ for some $i \geq 0$). Such situations, which further require that the tentative exponent e_x be adjusted, can be detected easily using the `nlz` instruction to count the number n of leading zeros of the bit string of `highR` shown above:

- If $n = 4$ then $c = 1$. In this case, the guard bit and the sticky bit are, respectively,

  $$G = r_{23} \qquad \text{and} \qquad T = \text{OR}(r_{24}, r_{25}, r_{26}).$$

  Consequently, $r = \text{RN}(m_r \cdot 2^{e_x})$ is obtained as

  $$r = \Big((1.r_0 \ldots r_{22})_2 + B \cdot 2^{-23} \Big) \cdot 2^{e_x + 1}, \tag{9.15}$$

  where, for rounding to nearest, the bit B is defined as

  $$B = G \text{ AND } \big(r_{22} \text{ OR } T \big),$$

  (see for example [187, page 425]). An implementation of the computation of B from the integer `highR` is detailed in Listing 9.15. Notice that G and T can be computed in parallel (line 7). Note also that since the `&` operator is a *bitwise* operator and since G is either 0 or 1, the bit r_{22} need not be extracted explicitly. Instead, the whole significand

  $$M = (1.r_0 \ldots r_{22})_2 \cdot 2^{23} \tag{9.16}$$

  can be used to produce $B \in \{0, 1\}$ (line 8). The value of the integer M can also be computed simultaneously with the values of G and T.

C listing 9.15 Computation of the rounding bit in a binary32 addition operator, in the case of carry propagation ($c = 1$) and rounding to nearest ($\circ = \text{RN}$).

```
1  uint32_t n, M, G, T, B;
2
3  n = nlz(highR);
4
5  if (n == 4)
6  {
7     M = highR >> 4;     G = (highR >> 3) & 1;     T = (highR << 29) != 0;
8     B = G & (M | T);
9  }
```

- If $n = 5$ then $c = 0$ and $r_0 = 1$. In this case, the guard bit and the sticky bit are, respectively,

$$G = r_{24} \qquad \text{and} \qquad T = \text{OR}(r_{25}, r_{26}).$$

Consequently, $r = \text{RN}(m_r \cdot 2^{e_x})$ is obtained as

$$r = \left((1.r_1 \dots r_{23})_2 + B \cdot 2^{-23}\right) \cdot 2^{e_x}, \tag{9.17}$$

where, for rounding to nearest, the bit B is now defined as

$$B = G \text{ AND } \left(r_{23} \text{ OR } T\right).$$

An implementation can be obtained in the same way as the one of Listing 9.15. Note here that, unlike the previous case ($n = 4$), no normalization is necessary. Indeed, the tentative exponent e_x had to be adjusted to $e_x + 1$ in (9.15), while it is kept unchanged in (9.17) since neither carry propagation nor cancellation has occurred.

- If $n \geq 6$ then $c = r_0 = \cdots = r_i = 0$ for some $i \geq 0$. Hence, normalization is required, by shifting left the bits of `highR` until either the leading 1 reaches the position of r_0, or the exponent obtained by decrementing e_x reaches $e_{\min}$. Again, this can be implemented in the same fashion as in Listing 9.15. However, when $n \geq 7$, a simplification occurs since, as stated in Section 7.3, δ must be either 0 or 1 in this case, implying $G = T = 0$.

At this stage, the truncated normalized significand M and the rounding bit B have been computed, and only the result exponent is missing. It turns out that it can be deduced easily from e_x for each of the preceding cases: $n = 4$, $n = 5$, and $n \geq 6$.

For example, an implementation for the case $n = 4$ is given at line 5 of Listing 9.16. When $n = 4$, we can see in (9.15) that the result exponent before rounding is

$$d = e_x + 1.$$

Note that $d > e_{\min}$, so that the result r is either infinity or a normal number, for which the bias is $e_{\max}$. Since the integer M in (9.16) already contains the implicit 1, we will in fact not compute $D = d + e_{\max}$ but rather

$$D - 1 = e_x + e_{\max},$$

and then add M to $(D - 1) \cdot 2^{23}$. Recalling that $E_x = e_x - e_{\min} + n_x$ and that $e_{\min} + e_{\max} = 1$, the value $D - 1$ can be obtained as shown at line 5 and stored in the unsigned integer `Dm1`. Notice the parenthesizing, which allows us to reuse the value of $E_x - n_x$, already needed when computing the shift δ in Listing 9.11. Then two situations may occur.

- If Dm1 is at least $2e_{\max} = 254 = (FE)_{16}$, then $e_x + 1 > e_{\max}$ because $e_{\min} = 1 - e_{\max}$. Consequently, $\pm\infty$ must be returned in this case (line 8).
- Otherwise, $e_x + 1$ is at most $e_{\max}$ and one can round and pack the result. The line 10 follows from (9.15), (9.16), and the standard encoding of binary32 data, and the considerations about rounding/computing a successor in Section 7.2. More precisely, if the addition of B to M propagates a carry up to the exponent field, then, necessarily, $B = 1$ and $M = 2^{24} - 1$. In this case, $(D-1) \cdot 2^{23} + M + B$ equals $(D+1) \cdot 2^{23}$ and the result exponent is not $e_x + 1$ but $e_x + 2$. (In particular, overflow will occur if $e_x = 126$.) Else, $M + B$ fits in 24 bits and then $(D-1) \cdot 2^{23} + M + B$ encodes the floating-point number whose normalized representation is given in (9.15). One may check that in both cases the bit string of $(D-1) \cdot 2^{23} + M + B$ has the form $[0 * \cdots *]$ and thus has no overlap with the bit string $[* 0 \cdots 0]$ of Sr.

C listing 9.16 Computation of the correctly rounded result in a binary32 addition operator, in the case of carry propagation ($c = 1$) and rounding to nearest ($\circ = \text{RN}$).

```
1  uint32_t Dm1;
2
3  if (n == 4)
4  {
5     Dm1 = (Ex - nx) + 1;
6
7     if ( Dm1 >= 0xFE ) // overflow
8        return Sr | 0x7F800000;
9     else
10       return ((Sr | (Dm1 << 23)) + M) + B;
11 }
```

9.3 Binary Floating-Point Multiplication

We now turn to the implementation of the method of Section 7.4, for computing $\circ(x \times y)$, again for $\circ = \text{RN}$ and the binary32 format:

$$\beta = 2, \qquad k = 32, \qquad p = 24, \qquad e_{\max} = 127.$$

The case where either x or y is a special datum (like ± 0, $\pm\infty$, or NaN) is described in Section 9.3.1, while the case where both x and y are (sub)normal numbers is discussed in Sections 9.3.2 through 9.3.4.

9.3.1 Handling special values

In the case of multiplication, the input (x, y) is considered a *special input* when x or y is ± 0, $\pm\infty$, or NaN. For each possible case the IEEE 754-2008 standard requires that a special value be returned. These special values follow from those given in Table 7.4 by adjoining the correct sign, using

$$x \times y = (-1)^{s_r} \cdot \Big(|x| \times |y| \Big), \qquad s_r = s_x \text{ XOR } s_y. \tag{9.18}$$

We remind the reader that the standard does not specify the sign of a NaN result (see [267, §6.3]).

9.3.1.1 Detecting that a special value must be returned

Special inputs can be filtered out exactly in the same way as for addition (see Section 9.2.1.1); hence, when $k = 32$ and $p = 24$, lines 3, 4, 5 of Listing 9.17. Here and hereafter $|X|$ will have the same meaning as in (9.6).

C listing 9.17 Special value handling in a binary32 multiplication operator.

```
1  uint32_t absX, absY, Sr, absXm1, absYm1, Min, Max, Inf;
2
3  absX = X & 0x7FFFFFFF; absY = Y & 0x7FFFFFFF; Sr = (X ^ Y) & 0x80000000;
4  absXm1 = absX - 1;      absYm1 = absY - 1;
5  if (maxu(absXm1, absYm1) >= 0x7F7FFFFF)
6  {
7     Min = minu(absX, absY); Max = maxu(absX, absY); Inf = Sr | 0x7F800000;
8     if (Max > 0x7F800000 || (Min == 0 && Max == 0x7F800000))
9        return Inf | 0x00400000 | Max;      // qNaN with payload encoded in
10                                           // the last 22 bits of X or Y
11    if (Max != 0x7F800000) return Sr;
12    return Inf;
13 }
```

9.3.1.2 Returning special values as recommended or required by IEEE 754-2008

Once the input (x, y) is known to be special, one must return the corresponding result as specified in Table 7.4. First, that table shows that a qNaN must be returned as soon as one of the following two situations occurs:

- if $|X|$ or $|Y|$ encodes a NaN, that is, according to Table 9.2, if

$$\max\left(|X|, |Y|\right) > 2^{k-1} - 2^{p-1}; \tag{9.19}$$

- if $(|X|, |Y|)$ encodes either $(+0, +\infty)$ or $(+\infty, +0)$, that is, if

$$\min(|X|, |Y|) = 0 \quad \text{and} \quad \max(|X|, |Y|) = 2^{k-1} - 2^{p-1}. \tag{9.20}$$

When $k = 32$ and $p = 24$, the conditions in (9.19) and (9.20) can be implemented as in lines 7 and 8 of Listing 9.17. Here, we must raise two remarks.

- The condition in (9.19) is the same as the one used in (9.9) for handling NaN results in binary floating-point addition. However, the condition in (9.20) is specific to multiplication.

- The qNaN returned at line 9 of Listing 9.17 enjoys the same nice properties as for binary floating-point addition. Roughly speaking, just as for addition, our code for multiplication returns a qNaN that keeps as much information on the input as possible, as recommended by IEEE 754-2008 (see [267, §6.2]).

Once the case of a qNaN output has been handled, it follows from Table 7.4 that a special output must be ± 0 if and only if neither x nor y is $\pm\infty$, that is, according to Table 9.2, if and only if

$$\max(|X|, |Y|) \neq 2^{k-1} - 2^{p-1}.$$

For $k = 32$ and $p = 24$, the latter condition is implemented at line 11 of Listing 9.17. Finally, the remaining case, for which one must return $(-1)^{s_r}\infty$, is handled by line 12.

9.3.2 Sign and exponent computation

We assume from now on that the input (x, y) is not *special; that is, both x and y are finite nonzero (sub)normal numbers.*

The sign s_r of the result is straightforwardly obtained by taking the XOR of the sign bits of X and Y. It has already been used in the previous section for handling special values (for an example, see variable `Sr` at line 3 of Listing 9.17).

Concerning the exponent of the result, let us first recall that using (9.18) together with the symmetry of rounding to nearest gives

$$\mathrm{RN}(x \times y) = (-1)^{s_r} \cdot \mathrm{RN}\Big(|x| \times |y|\Big).$$

Second, defining λ_x and λ_y as the numbers of leading zeros of the significands m_x and m_y, and defining further

$$m'_x = m_x \cdot 2^{\lambda_x} \qquad \text{and} \qquad m'_y = m_y \cdot 2^{\lambda_y}, \tag{9.21}$$

and

$$e'_x = e_x - \lambda_x \qquad \text{and} \qquad e'_y = e_y - \lambda_y, \tag{9.22}$$

we obtain a product expressed in terms of *normalized* significands:

$$\mathrm{RN}\Big(|x| \times |y|\Big) = \mathrm{RN}\Big(m'_x m'_y \cdot 2^{e'_x + e'_y}\Big), \qquad m'_x,\, m'_y \in [1, 2).$$

Third, taking

$$c = \begin{cases} 0 & \text{if } m'_x m'_y \in [1, 2), \\ 1 & \text{if } m'_x m'_y \in [2, 4), \end{cases} \tag{9.23}$$

we have

$$\mathrm{RN}\Big(|x| \times |y|\Big) = \mathrm{RN}(\ell \cdot 2^d),$$

with

$$\ell = m'_x m'_y \cdot 2^{-c} \qquad \text{and} \qquad d = e'_x + e'_y + c. \tag{9.24}$$

Here c allows one to ensure that ℓ lies in the range $[1, 2)$. Thus, computing the exponent should, in principle, mean computing the value of d as above. There are in fact two possible situations:

- if $d \geq e_{\min}$ then the real number $\ell \cdot 2^d$ lies in the normal range or in the overflow range, and therefore $\mathrm{RN}(\ell \cdot 2^d) = \mathrm{RN}(\ell) \cdot 2^d$;
- if $d < e_{\min}$, which may happen since both e'_x and e'_y can be as low as $e_{\min} - p + 1$, the real $\ell \cdot 2^d$ falls in the subnormal range. As explained in Section 7.4.2, in this case d should be increased up to $e_{\min}$ by shifting ℓ right by $e_{\min} - d$ positions.

For simplicity, here we will detail an implementation of the first case only:

$$d \geq e_{\min}. \tag{9.25}$$

The reader may refer to the FLIP software library for a complete implementation that handles both cases.

To compute the exponent d of the result, we shall as usual manipulate its biased value, which is the integer D such that

$$D = d + e_{\max}. \tag{9.26}$$

And yet, similar to the implementation of binary floating-point addition (see for example Listing 9.16), we will in fact compute $D - 1$ and then let this tentative (biased) exponent value be adjusted automatically when rounding and packing. Recalling that $e_{\min} = 1 - e_{\max}$ and using (9.25), we have

$$D - 1 = d - e_{\min} \geq 0.$$

9.3.2.1 Computing the nonnegative integer $D - 1$

Using (9.22) and (9.24) together with the fact that

$$E_x = e_x - e_{\min} + n_x,$$

one may check that

$$D - 1 = (E_x - n_x) + (E_y - n_y) - (\lambda_x + \lambda_y - e_{\min}) + c.$$

The identity

$$\lambda_x + \lambda_y - e_{\min} = M_X + M_Y - (2w + e_{\min}),$$

where $2w + e_{\min}$ is a constant defined by the floating-point format, shows further that computing λ_x and λ_y is in fact not needed.

In practice, among all the values involved in the preceding expression of $D-1$, the value of condition $c = [m'_x m'_y \geq 2]$ may be the most expensive to determine. On the other hand, n_x, n_y, E_x, and E_y require 2 instructions, while M_X and M_Y require 3 instructions. Hence, recalling that $2w + e_{\min}$ is known *a priori*, a scheduling for the computation of $D-1$ that exposes some ILP is given by the following parenthesizing:

$$D-1 = \Big(\big[(E_x - n_x) + (E_y - n_y)\big] - \big[(M_X + M_Y) - (2w + e_{\min})\big]\Big) + c. \quad (9.27)$$

For example, $w = 8$ and $e_{\min} = -126$ for the binary32 format. Then $2w + e_{\min} = -110$, and an implementation of the computation of $D-1$ that uses (9.27) is thus as shown in Listing 9.18.

C listing 9.18 Computing $D-1$ in a binary32 multiplication operator, assuming $d \geq e_{\min}$ and that $|X|$, $|Y|$, and c are available.

```
uint32_t Ex, Ey, nx, ny, MX, MY, Dm1;

Ex = absX >> 23;                Ey = absY >> 23;
nx = absX >= 0x800000;          ny = absY >= 0x800000;
MX = maxu(nlz(absX), 8);        MY = maxu(nlz(absY), 8);

Dm1 = (((Ex - nx) + (Ey - ny)) - ((MX + MY) + 110)) + c;
```

Notice that in Listing 9.18 n_x, n_y, E_x, E_y, M_X, and M_Y are independent of each other. Consequently, with unbounded parallelism and a latency of 1 for all the instructions involved in the code, the value of $D-1-c$ can be deduced from X and Y in 6 cycles. Then, once c has been computed, it can be added to that value in 1 cycle.

9.3.3 Overflow detection

As stated in Chapter 7, overflow can occur for multiplication only when $d \geq e_{\min}$. In this case we have $|x| \times |y| = \ell \cdot 2^d$, with $\ell \in [1, 2)$ given by (9.24).

9.3.3.1 Overflow before rounding

A first case of overflow is when $d \geq e_{\max} + 1$, since then the exact product $|x| \times |y|$ is larger than the largest finite number $\Omega = (2 - 2^{1-p}) \cdot 2^{e_{\max}}$, and so will be its rounded value. This first case can be detected easily from the value of $D-1$, applying (9.26):

$$d \geq e_{\max} + 1 \quad \text{if and only if} \quad D - 1 \geq 2e_{\max}.$$

For the binary32 format, $2e_{\max} = 254 = (FE)_{16}$ and this first case of overflow can be implemented as shown in Listing 9.19.

C listing 9.19 Returning $\pm\infty$ when $d \geq e_{\max} + 1$ in the binary32 format.

```
if (Dm1 >= 0xFE) return Inf;                    // Inf = Sr | 0x7F800000;
```

Notice that `Inf` has already been used for handling special values (see line 7 of Listing 9.17). Notice also the similarity with binary floating-point addition (see lines 7 and 8 of Listing 9.16).

9.3.3.2 Overflow after rounding

A second case of overflow is when $d = e_{\max}$ and $\mathrm{RN}(\ell) = 2$. This case is handled automatically when rounding the significand and packing the result viá the integer addition

$$(D-1)\cdot 2^{p-1} + \mathrm{RN}(\ell)\cdot 2^{p-1}$$

(see the end of Section 9.3.4). For example, for the binary32 format, when $D - 1 = 2e_{\max} - 1 = 253 = (11111101)_2$ and $\mathrm{RN}(\ell) = 2$, the bit string of $(D-1)\cdot 2^{p-1}$ is

$$[0\,11111101\,\underbrace{00000000000000000000000}_{p-1=23\text{ bits}}],$$

and the bit string of $\mathrm{RN}(\ell)\cdot 2^{p-1} = 2\cdot 2^{p-1}$ is

$$[0\,00000010\,\underbrace{00000000000000000000000}_{p-1=23\text{ bits}}].$$

Summing them gives the integer that encodes $+\infty$, and it remains to concatenate its bit string with the correct sign. The result will then be a correctly signed infinity, as required.

9.3.4 Getting the correctly rounded result

It remains to compute the correctly rounded value $r = \mathrm{RN}(\ell \cdot 2^d)$. As in the previous section, assume for simplicity that $d \geq e_{\min}$. Then

$$r = \mathrm{RN}(\ell)\cdot 2^d$$

and, since a biased value $D - 1$ of d has already been computed (see Listing 9.18), we are left with the computation of $\mathrm{RN}(\ell)$, with

$$\ell = m'_x m'_y \cdot 2^{-c}$$

as in (9.24). The following paragraphs explain how to implement this for the binary32 format.

9.3.4.1 Computing the normalized significands m'_x and m'_y

Recall that the integer λ_x is defined as the number of leading zeros of the binary expansion of the significand m_x:

$$m_x = [\underbrace{0.0\ldots0}_{\lambda_x \text{ zeros}} 1 m_{x,\lambda_x+1} \ldots m_{x,23}].$$

Consequently, we start by storing the bits of m'_x into the following string of length 32:

$$[1\, m_{x,\lambda_x+1} \cdots m_{x,23}\, \underbrace{000\ldots000}_{\lambda_x+8 \text{ zeros}}]. \tag{9.28}$$

If $\lambda_x = 0$ then this bit string is deduced from the bit string of $|X|$ by shifting it left by 8 positions and simultaneously introducing the implicit 1. If $\lambda_x > 0$ then it suffices to shift $|X|$ left by $8 + \lambda_x$ positions. In both cases, the amount of the shift is

$$M_X = \max(\texttt{nlz}(|X|), 8).$$

An implementation can be found at line 3 of Listing 9.20, where the bit string of the variable `mpX` is as in (9.28). The same is done for storing m'_y by means of variable `mpY`.

9.3.4.2 Computing the product $m'_x m'_y$ exactly

Since m'_x and m'_y can each be represented using at most 24 bits, their exact product can be represented using at most 48 bits. Those bits fit into two 32-bit integers, which are called `highS` and `lowS`, and are defined as follows:

$$m'_x m'_y = (\underbrace{c s_0.s_1 \ldots s_{30}}_{=:\, \texttt{highS}}\, \underbrace{s_{31} \ldots s_{46}\, \overbrace{000\ldots000}^{16 \text{ zeros}}}_{=:\, \texttt{lowS}})_2. \tag{9.29}$$

Here $c = 0$ if $m'_x m'_y < 2$, and $c = 1$ if $m'_x m'_y \geq 2$. In the latter case, a carry has been propagated; thus, normalization will be necessary to obtain ℓ in $[1, 2)$.

The computation of the integers `highS`, `lowS`, and c is done at lines 5 and 6 of Listing 9.20, using our basic multiply instructions `mul` and $*$ as well as the fact that $c = 1$ if and only if the integer `highS` is at least $2^{31} = (80000000)_{16}$.

9.3.4.3 Computing the guard and sticky bits needed for rounding correctly

Once the product $m'_x m'_y$ (and thus ℓ as well) is known exactly, one can deduce $\mathrm{RN}(\ell)$ in the classical way, by means of a guard bit G and a sticky bit T. We now discuss how to compute those bits efficiently. We start by considering the cases $c = 0$ and $c = 1$ separately, and then propose a single code for handling both of them.

- If $c = 0$ then, since $m'_x m'_y \geq 1$, the bit s_0 in (9.29) must be equal to 1. Thus, combining (9.24) and (9.29) gives

$$\ell = (1.s_1 \dots s_{46})_2,$$

from which it follows that the guard bit and the sticky bit are, respectively,

$$G = s_{24} \qquad \text{and} \qquad T = \mathrm{OR}(s_{25}, \dots, s_{46}).$$

Finally, the correctly rounded value $\mathrm{RN}(\ell)$ is obtained as

$$\mathrm{RN}(\ell) = (1.s_1 \dots s_{23})_2 + B \cdot 2^{-23},$$

where, for rounding to nearest, the bit B is defined as

$$B = G \text{ AND } \big(s_{23} \text{ OR } T\big)$$

(see, for example, [187, page 425]). Using (9.29) we see that G can be obtained by shifting `highS` $= [01s_1 \dots s_{24}s_{25} \dots s_{30}]$ right by 6 positions and then masking with 1:

```
G = (highS >> 6) & 1;
```

On the other hand, $T = \mathrm{OR}(s_{25}, \dots, s_{30}, \texttt{lowS})$ by definition of `lowS`. Therefore, the value of T can be obtained as follows:

```
T = ((highS << 26) != 0) | (lowS != 0);
```

Clearly, the two comparisons to zero that appear in the computation of T can be done in parallel. Besides, given `highS` and `lowS`, the computation of T itself can be performed in parallel with that of G.

- If $c = 1$ then (9.24) and (9.29) give

$$\ell = (1.s_0 s_1 \dots s_{46})_2.$$

It follows that $\mathrm{RN}(\ell) = (1.s_0 \dots s_{22})_2 + B \cdot 2^{-23}$, where, for rounding to nearest, the bits B, G, and T are now given by

$$B = G \text{ AND } \big(s_{22} \text{ OR } T\big), \qquad G = s_{23}, \qquad T = \mathrm{OR}(s_{24}, \dots, s_{46}).$$

The computation of G and T can be implemented as before, the only difference being the values by which we shift:

```
G = (highS >> 7) & 1;        T = ((highS << 25) != 0) | (lowS != 0);
```

It is now clear that the two cases $c = 0$ and $c = 1$ that we have just analyzed can be handled by a single code fragment, which corresponds to lines 8, 10, 12 of Listing 9.20. As in Listing 9.15 for binary floating-point addition, computing B does not require us to extract the last bit (s_{23} if $c = 0$, s_{22} if $c = 1$) of the significand (stored in the variable `M`). Instead, we work directly on M.

Also, since `highT` will generally be more expensive to obtain than `lowT` and `M`, one can expose more ILP by ORing `lowT` and `M` during the computation of `highT`. This is done at line 10 of Listing 9.20, and is just one example of the many ways of optimizing this kind of code. The FLIP software library implements some other tricks that allow one to expose even more ILP and thus, in some contexts, to save a few cycles.

9.3.4.4 Rounding the significand and packing the result

Once the sign, the (biased) exponent, the significand, and the rounding bit have been computed (and are available in the integers `Sr`, `Dm1`, `M`, and `B`, respectively), one can round the significand and pack the result. This is done in the same way as in Listing 9.16 for binary floating-point addition; see line 14 of Listing 9.20.

C listing 9.20 Computation of the rounding bit and of the correctly rounded result in a binary32 multiplication operator, in the case of rounding to nearest ($\circ = \mathrm{RN}$) and assuming $|X|$, $|Y|$, M_X, and M_Y are available.

```
1   uint32_t mpX, mpY, highS, lowS, c, G, M, highT, lowT, M_OR_lowT, B;
2
3   mpX = (X << MX) | 0x80000000;          mpY = (Y << MY) | 0x80000000;
4
5   highS = mul(mpX, mpY);                 lowS = mpX * mpY;
6   c = highS >= 0x80000000;               lowT = (lowS != 0);
7
8   G = (highS >> (6 + c)) & 1;            M = highS >> (7 + c);
9
10  highT = (highS << (26 - c)) != 0;      M_OR_lowT = M | lowT;
11
12  B = G & (M_OR_lowT | highT);
13
14  return ((Sr | (Dm1 << 23)) + M) + B;
```

9.4 Binary Floating-Point Division

This section describes how to implement the floating-point division of binary32 data. The main steps of the algorithm have been recalled in Section 7.6 and we refer to [509, §4] for a complete description of a possible implementation.

The case where either x or y is a special datum (like ± 0, $\pm\infty$, or NaN) is described in Section 9.4.1, while the case where both x and y are (sub)normal numbers is discussed in Sections 9.4.2 through 9.4.4.

9.4.1 Handling special values

Similar to addition and multiplication, the input (x, y) to division is considered a *special input* when x or y is ± 0, $\pm\infty$, or NaN. For each possible case the IEEE 754-2008 standard requires that a special value be returned by the division operator. These special values are obtained from those in Table 7.5 by adjoining the correct sign, using

$$x/y = (-1)^{s_r} \cdot \left(|x|/|y|\right), \qquad s_r = s_x \text{ XOR } s_y. \tag{9.30}$$

We remind the reader that the standard does not specify the sign of a NaN result; see [267, §6.3].

9.4.1.1 Detecting that a special value must be returned

For division, a special input (x, y) can be filtered out in the same way as for addition and multiplication. When $k = 32$ and $p = 24$, such a filter can be implemented as shown at lines 3, 4, 5 of Listing 9.21.

C listing 9.21 Special value handling in a binary32 division operator.

```
1  uint32_t absX, absY, Sr, absXm1, absYm1, Max, Inf;
2
3  absX = X & 0x7FFFFFFF; absY = Y & 0x7FFFFFFF; Sr = (X ^ Y) & 0x80000000;
4  absXm1 = absX - 1;      absYm1 = absY - 1;
5  if (maxu(absXm1, absYm1) >= 0x7F7FFFFF)
6  {
7     Max = maxu(absX, absY); Inf = Sr | 0x7F800000;
8     if (Max > 0x7F800000 || absX == absY)
9        return Inf | 0x00400000 | Max;      // qNaN with payload encoded in
10                                            // the last 22 bits of X or Y
11    if (absX < absY) return Sr;
12    return Inf;
13 }
```

9.4.1.2 Returning special values as recommended or required by IEEE 754-2008

Here and hereafter $|X|$ will have the same meaning as in (9.6). Once our input (x, y) is known to be special, one must return the corresponding result as specified in Table 7.5. From that table we see that a qNaN must be returned as soon as one of the following two situations occurs:

- if $|X|$ or $|Y|$ encodes a NaN, that is, according to Table 9.2, if

$$\max\left(|X|, |Y|\right) > 2^{k-1} - 2^{p-1}; \tag{9.31}$$

- if $(|X|, |Y|)$ encodes either $(+0, +0)$ or $(+\infty, +\infty)$, that is, since (x, y) is known to be special and assuming the case where x or y is NaN is already handled by (9.31), if

$$|X| = |Y|. \tag{9.32}$$

When $k = 32$ and $p = 24$, the conditions in (9.31) and (9.32) can be implemented as in lines 7 and 8 of Listing 9.21. One can raise the following remarks:

- The condition in (9.31) is the same as the one used for addition and multiplication (see (9.9) and (9.19)). On the contrary, the condition in (9.32) is specific to division.
- The qNaN returned at line 9 of Listing 9.21 enjoys the same nice properties as for binary floating-point addition and binary floating-point multiplication: in a sense, it keeps as much information on the input as possible, as recommended by IEEE 754-2008 (see [267, §6.2]).

Once the case of a qNaN output has been handled, it follows from Table 7.5 that a special output must be ± 0 if and only if $|x| < |y|$, that is, according to Table 9.2, if and only if

$$|X| < |Y|.$$

For $k = 32$ and $p = 24$, the latter condition is implemented at line 11 of Listing 9.21. Finally, the remaining case, for which one must return $(-1)^{s_r}\infty$, is handled by line 12.

9.4.2 Sign and exponent computation

We assume from now on that the input (x, y) is not *special; that is, that both x and y are finite nonzero (sub)normal numbers.*

Exactly as for multiplication, the sign s_r of the result is straightforwardly obtained by taking the XOR of the sign bits of X and Y. It has already been used in the previous section for handling special values (for an example, see variable `Sr` at line 3 of Listing 9.21).

Concerning the exponent of the result, the approach is also very similar to what we have done for multiplication in Section 9.3.2. First, using (9.30) together with the symmetry of rounding to nearest, we obtain

$$\mathrm{RN}(x/y) = (-1)^{s_r} \cdot \mathrm{RN}\Big(|x|/|y|\Big).$$

Then, using the same notation as in Section 9.3.2, we can express the fraction in terms of *normalized* significands:

$$\mathrm{RN}\Big(|x|/|y|\Big) = \mathrm{RN}\Big(m'_x/m'_y \cdot 2^{e'_x - e'_y}\Big), \qquad m'_x, m'_y \in [1, 2).$$

Finally, as in multiplication, we introduce a parameter c that will be used for normalizing the fraction m'_x/m'_y. Indeed, $1 \le m'_x, m'_y < 2$ implies that $m'_x/m'_y \in (1/2, 2)$. Hence, $2m'_x/m'_y \in (1, 4) \subset [1, 4)$ and so $2m'_x/m'_y \cdot 2^{-c} \in [1, 2)$, provided c satisfies

$$c = \begin{cases} 0 & \text{if } m'_x < m'_y, \\ 1 & \text{if } m'_x \ge m'_y. \end{cases} \tag{9.33}$$

It follows that the binary floating-point number $\mathrm{RN}(|x|/|y|)$ is given by

$$\mathrm{RN}\Big(|x|/|y|\Big) = \mathrm{RN}(\ell \cdot 2^d),$$

with

$$\ell = 2m'_x/m'_y \cdot 2^{-c} \qquad \text{and} \qquad d = e'_x - e'_y - 1 + c. \tag{9.34}$$

Since by definition $\ell \in [1, 2)$, the result exponent will thus be set to d.

As in multiplication, the following two situations may occur:

- if $d \ge e_{\min}$ then the real number $\ell \cdot 2^d$ lies in the normal range or the overflow range, and therefore

$$\mathrm{RN}(\ell \cdot 2^d) = \mathrm{RN}(\ell) \cdot 2^d; \tag{9.35}$$

- if $d < e_{\min}$, which may happen since e'_x can be as low as $e_{\min} - p + 1$ and e'_y can be as large as $e_{\max}$, the real number $\ell \cdot 2^d$ falls in the subnormal range. As explained in Section 7.6.3, several strategies are possible in this case, depending on the method chosen for the approximate computation of the rational number ℓ.

For simplicity, here we detail an implementation of the first case only:

$$d \ge e_{\min}. \tag{9.36}$$

The reader may refer to the FLIP software library for a complete implementation that handles both cases by means of polynomial approximation.

Note, however, that unlike multiplication the binary floating-point number $\mathrm{RN}(\ell)$ is always strictly less than 2. This fact is a direct consequence of the following property, shown in [293]:

Property 9.2. *If* $m'_x \ge m'_y$ *then* $\ell \in [1, 2 - 2^{1-p}]$, *else* $\ell \in (1, 2 - 2^{1-p})$.

When $d \geq e_{\min}$ it follows from this property that $\mathrm{RN}(\ell) \cdot 2^d$ is the normalized representation of the correctly rounded result sought.[6] Consequently, d is not simply a tentative exponent that would require updating after rounding, but is already the exponent of the result. We will come back to this point in Section 9.4.3 when considering the overflow exception.

Let us now compute d using (9.34). As usual, what we actually compute is the integer $D-1$ such that $D = d + e_{\max}$ (biased value of d). Since $e_{\min} = 1 - e_{\max}$, the assumption (9.36) gives, as in binary floating-point multiplication,

$$D - 1 = d - e_{\min} \geq 0. \tag{9.37}$$

9.4.2.1 Computing the nonnegative integer $D - 1$

The approach is essentially the same as for multiplication, with d now defined by (9.34) instead of (9.24). Recalling that $E_x = e_x - e_{\min} + n_x$ and that $\lambda_x = M_X - w$, one obtains

$$D - 1 = (E_x - n_x) - (E_y - n_y) - (M_X - M_Y) + c - e_{\min} - 1. \tag{9.38}$$

Before parenthesizing this expression for $D - 1$, let us consider the computation of c. Interestingly enough, the value of c is in fact easier to obtain for division than for multiplication. Indeed, unlike (9.23), its definition in (9.33) does not involve the product $m'_x m'_y$, and it suffices to compute the normalized significands m'_x and m'_y, and to compare them to deduce the value of c. For the binary32 format, a possible implementation is as shown in Listing 9.22.

C listing 9.22 Computation of the normalized significands m'_x and m'_y and of the shift value c in a binary32 division operator.

```
uint32_t absX, absY, MX, MY, mpX, mpY, c;

absX = X & 0x7FFFFFFF;                absY = Y & 0x7FFFFFFF;
MX = maxu(nlz(absX), 8);              MY = maxu(nlz(absY), 8);
mpX = (X << MX) | 0x80000000;         mpY = (Y << MY) | 0x80000000;

c = mpX >= mpY;
```

With unbounded parallelism and a latency of 1 for each of the instructions in Listing 9.22, we see that c can be obtained in 6 cycles. In comparison, the shift value c for binary floating-point multiplication (obtained at line 6 of Listing 9.20) costs 6 cycles plus the latency of `mul`. For example, on the STMicroelectronics ST200 processor, which has four issues and where `mul` has a

[6]Since ℓ is upper bounded by the largest binary floating-point number strictly less than 2, this is *not* specific to rounding to nearest ties to even.

latency of 3 cycles, computing c costs 9 cycles in the case of binary floating-point multiplication, but only 6 cycles in the case of binary floating-point division.

Computing c as fast as possible is important, for c is used in the definition of the fraction ℓ in (9.34), whose approximate computation usually lies on the critical path. Therefore, every cycle saved for the computation of c should, in principle, allow one to reduce the latency of the whole division operator.

Let us now turn to the computation of $D - 1$ as given by (9.38). Still assuming unbounded parallelism and with a latency of 6 cycles for c, one can keep the parenthesizing proposed at line 7 of Listing 9.18 since there the expression to be added to c can also be computed in 6 cycles. The only modification consists in replacing the constant $-(2w + e_{\min})$ with $-e_{\min} - 1$. For the binary32 format, $-e_{\min} - 1 = 125$, and the corresponding code is given by Listing 9.23.

C listing 9.23 Computing $D - 1$ in a binary32 division operator, assuming $d \geq e_{\min}$ and that $|X|$, $|Y|$, and c are available.

```
uint32_t Ex, Ey, nx, ny, MX, MY, Dm1;

Ex = absX >> 23;                Ey = absY >> 23;
nx = absX >= 0x800000;          ny = absY >= 0x800000;
MX = maxu(nlz(absX), 8);        MY = maxu(nlz(absY), 8);

Dm1 = (((Ex - nx) + (Ey - ny)) - ((MX + MY) - 125)) + c;
```

With unbounded parallelism, the code given above thus produces $D - 1$ in 7 cycles. Of course, this is only one of the many possible ways of parenthesizing the expression of $D-1$. Other schemes may be better suited, depending on the algorithm chosen for approximating ℓ and on the actual degree of parallelism of the target processor. Examples of other ways of computing $D - 1$ have been implemented in FLIP, with the STMicroelectronics ST200 processor as the main target.

9.4.3 Overflow detection

Because of Property 9.2, binary floating-point division will never overflow because of rounding, but only when

$$d \geq e_{\max} + 1.$$

This is different from the case of binary floating-point multiplication (see Section 9.3.3).

The situation where $d \geq e_{\max} + 1$ clearly requires that our assumption $d \geq e_{\min}$ be true and thus, using (9.37), it is characterized by the condition

$$D - 1 \geq 2e_{\max}.$$

This characterization has already been used for multiplication in Section 9.3.3 and one can reuse the implementation provided by Listing 9.19.

9.4.4 Getting the correctly rounded result

As in the previous sections, assume for simplicity that $d \geq e_{\min}$ (the general case is handled in the FLIP library). Recall from (9.35) that in this case the correctly rounded result to be returned is $\mathrm{RN}(\ell) \cdot 2^d$. Since the values of c and $D-1 = d - e_{\min}$ have already been computed in Listing 9.22 and Listing 9.23, we are left with the computation of

$$\mathrm{RN}(\ell) = (1.r_1 \dots r_{p-1})_2,$$

where ℓ is defined by (9.34) and thus *a priori* has an infinite binary expansion of the form

$$\ell = (1.\ell_1 \dots \ell_{p-1}\ell_p \dots)_2. \tag{9.39}$$

As explained in Section 7.6, many algorithms exist for such a task, and they belong to one of the following three families: digit-recurrence algorithms (SRT, etc.), functional iteration algorithms (Newton-Raphson iteration, Goldschmidt iteration, etc.), and polynomial approximation-based algorithms.

We will now show how to implement in software two examples of such algorithms. The first one is the *restoring division* algorithm, which is the simplest digit-recurrence algorithm (see, for example, the Sections 1.6.1 and 8.6.2 of [187] for a detailed presentation of this approach). The second one consists in computing approximate values of a function underlying ℓ by means of polynomial evaluation [293, 509]. Note that this second approach has already been used for implementing division (and square root) in software in a different context, namely when a floating-point FMA instruction is available; see, for example, [3].

As we will see, our first example leads to a highly sequential algorithm, while our second example allows us to expose much instruction-level parallelism (ILP). Hence those examples can be seen as two extremes of a wide range of possible implementations.

9.4.4.1 First example: restoring division

Restoring division is an iterative process, where iteration i produces the value of the bit ℓ_i in (9.39).

In order to see when to stop the iterations, recall first that since we have assumed that $d \geq e_{\min}$, the rational number ℓ lies in the normal range or the overflow range. Then in this case it is known that ℓ cannot lie exactly halfway between two consecutive normal binary floating-point numbers (see for example [187, page 452] or [118, page 229]).[7] Consequently, the binary

[7]This is not the case when $d < e_{\min}$ since in that case ℓ lies in the subnormal range; see the FLIP library on how to handle such cases.

floating-point number $\mathrm{RN}(\ell)$ can be obtained with only one guard bit (the sticky bit is not needed, in contrast to addition and multiplication):

$$\mathrm{RN}(\ell) = (1.\ell_1 \ldots \ell_{p-1})_2 + \ell_p \cdot 2^{1-p}. \tag{9.40}$$

We conclude from this formula that p iterations suffice to obtain $\mathrm{RN}(\ell)$.

Let us now examine the details of one iteration and how to implement it in software using integer arithmetic. To do so, let us first write the binary expansions of the numerator and denominator of $\ell = 2m'_x \cdot 2^{-c}/m'_y$ as follows:

$$2m'_x \cdot 2^{-c} = (N_{-1}N_0.N_1 \ldots N_{p-1})_2$$

and

$$m'_y = (01.M_1 \ldots M_{p-1})_2.$$

Notice that $N_{-1} = 1 - c$, where c is given by (9.33). It follows from the two identities above that the positive integers

$$N = \sum_{i=0}^{p} N_{p-1-i}2^i$$

and

$$M = 2^{p-1} + \sum_{i=0}^{p-2} M_{p-1-i}2^i$$

satisfy

$$\ell = N/M.$$

Next, define for $i \geq 0$ the pair (q_i, r_i) by $q_i = (1.\ell_1 \ldots \ell_i)_2$ and $N = q_i \cdot M + r_i$. Clearly, $r_i \geq 0$ for all $i \geq 0$, and (q_i, r_i) tends to $(\ell, 0)$ when i tends to infinity. Defining further $Q_i = q_i \cdot 2^i$ and $R_i = r_i \cdot 2^i$, we arrive at

$$2^i N = Q_i \cdot M + R_i, \qquad i \geq 0. \tag{9.41}$$

Since $q_i > 0$ has at most i fractional bits, Q_i is indeed a positive integer. Besides, since the identity $R_i = 2^i N - Q_i \cdot M$ involves only integers, R_i is an integer as well. Note also that since $\ell - q_i$ is, on the one hand, equal to $2^{-i}R_i/M$ and, on the other hand, equal to $(0.0 \ldots 0\ell_{i+1}\ell_{i+2} \ldots)_2 \in [0, 2^{-i})$, we have

$$0 \leq R_i < M. \tag{9.42}$$

This tells us that Q_i and R_i are, respectively, the quotient and the remainder in the Euclidean division of integer $2^i N$ by integer M. Note the scaling of N by 2^i.

The preceding analysis shows that, given

$$Q_0 = 1 \qquad \text{and} \qquad R_0 = N - M,$$

computing the first p fractional bits of ℓ iteratively simply reduces to deducing (Q_i, R_i) from (Q_{i-1}, R_{i-1}), for $1 \le i \le p$. Using $Q_i = q_i \cdot 2^i$ and $q_i = q_{i-1} + \ell_i \cdot 2^{-i}$, it is not hard to verify that

$$Q_i = 2Q_{i-1} + \ell_i. \tag{9.43}$$

Then, the identity in (9.41) gives

$$\begin{aligned} 2^i N &= (2Q_{i-1} + \ell_i) \cdot M + R_i \quad \text{using (9.43)} \\ &= 2Q_{i-1}M + \ell_i \cdot M + R_i \\ &= 2(2^{i-1}N - R_{i-1}) + \ell_i \cdot M + R_i \quad \text{using (9.41) with } i-1. \end{aligned}$$

Hence, after simplification,

$$R_i = 2R_{i-1} - \ell_i \cdot M. \tag{9.44}$$

Equations (9.43) and (9.44) lead to Algorithm 9.1. For $i \ge 1$, one computes a tentative remainder $T = 2R_{i-1} - M$; that is, we do as if $\ell_i = 1$. Then two situations may occur:

- if T is negative, then ℓ_i is in fact equal to zero. In this case $Q_i = 2Q_{i-1}$ and we restore the correct remainder $R_i = 2R_{i-1}$ by adding M to T;
- if $T \ge 0$, then $R_i = T$ satisfies (9.42), so that $\ell_i = 1$ and $Q_i = 2Q_{i-1} + 1$.

Algorithm 9.1 The first p iterations of the restoring method for binary floating-point division.

```
(Q_0, R_0) ← (1, N − M)
for i ∈ {1, ..., p} do
  T ← 2R_{i−1} − M
  if T < 0 then
    (Q_i, R_i) ← (2Q_{i−1}, T + M)
  else
    (Q_i, R_i) ← (2Q_{i−1} + 1, T)
  end if
end for
```

An implementation of Algorithm 9.1 for the binary32 format, for which $p = 24$, is detailed in Listing 9.24.

Recalling that the bit strings of m'_x and m'_y are stored in the variables `mpX` and `mpY` (see (9.28) as well as line 5 of Listing 9.22), one can deduce the integers N and M simply by shifting to the right. The initial values $Q_0 = 1$ and $R_0 = N - M \in [0, \dots, 2^{p+1})$ can then be stored exactly into the 32-bit unsigned integers `Q` and `R`. These integers will be updated during the algorithm so as to contain, respectively, Q_i and R_i at the end of the i-th iteration. Obviously, the multiplications by two are implemented with shifts by one position to the left.

C listing 9.24 First 24 iterations of the restoring method, rounding and packing for a binary32 division operator. Here we assume that $d \geq e_{\min}$ and that `Sr`, `Dm1`, `mpX`, `mpY`, and `c` are available.

```
1  uint32_t N, M, Q, R, i;
2  int32_t T;
3
4  N = mpX >> (7 + c);        M = mpY >> 8;
5  Q = 1;                     R = N - M;
6
7  for (i = 1; i < 25; i++)
8  {
9    T = (R << 1) - M;
10
11   if (T < 0)
12   {
13     Q = Q << 1;            R = T + M;
14   }
15   else
16   {
17     Q = (Q << 1) + 1; R = T;
18   }
19 }
20
21 return (Sr | (Dm1 << 23)) + ((Q >> 1) + (Q & 1));
```

At the end of iteration 24, the bit string of `Q` contains the first 24 fraction bits of ℓ:

$$\texttt{Q} = [\underbrace{0000000}_{7 \text{ zeros}} 1\ell_1 \ldots \ell_{23}\ell_{24}].$$

Applying (9.40) with $p = 24$, we have $\mathrm{RN}(\ell) = (1.\ell_1 \ldots \ell_{23})_2 + \ell_{24} \cdot 2^{-23}$. Consequently, the 32-bit unsigned integer

$$S_r \cdot 2^{31} + (D-1) \cdot 2^{23} + \mathrm{RN}(\ell) \cdot 2^{23}$$

that encodes the result is thus obtained as in line 21 of Listing 9.24.

From this code it is clear that the restoring method is sequential in nature. Although at iteration i both variables `Q` and `R` can be updated independently from each other, 24 iterations are required. A rough (but reasonable) count of at least 3 cycles per iteration thus yields a total latency of more than 70 cycles, and this only for the computation of $\ell_1, \ldots, \ell_{24}$. Adding to this the cost of computing c, handling special values, computing the sign and exponent, rounding and packing, will result in an even higher latency.

To reduce latency, one may use higher radix digit recurrence algorithms. For example, some implementations of radix-4 and radix-512 SRT algorithms have been studied in [502, §9.4], which indeed are faster than the restoring method on some processors of the ST200 family. In the next paragraph, we present a third approach that allows one to obtain further speed-ups by expressing even more ILP. In the ST200 processor family, the latency is then typically reduced to less than 30 cycles for the *complete* binary floating-point division operator (see [293] and the division code in FLIP).

9.4.4.2 Second example: division by polynomial evaluation

This example summarizes an approach first introduced in [292] for square root, and then adapted to division in [293]. For ℓ as in (9.34), this approach uses two main steps to produce $\mathrm{RN}(\ell)$. First, for each input (x, y) we compute a value v which is representable with at most k bits and which approximates ℓ from above as follows:

$$-2^{-p} < \ell - v \leq 0. \tag{9.45}$$

Then, we can deduce from v the correctly rounded value $\mathrm{RN}(\ell)$ by implementing a method essentially similar to the method outlined in [187, page 460].

Computing the one-sided approximation v

Although a functional iteration approach (the Newton-Raphson iteration or one of its variants; see Section 4.7 and Section 7.6.2) could be used to compute v, more ILP can be exposed by evaluating a suitable polynomial that approximates the exact quotient ℓ. Instead of (9.45) we shall in fact ensure the slightly stronger condition

$$\left|(\ell + 2^{-p-1}) - v\right| < 2^{-p-1}. \tag{9.46}$$

This form is more symmetric and more natural to attain than (9.45). First, the rational number $\ell + 2^{-p-1}$ can be considered as the exact value at some particular point (σ, τ) of the function

$$F(s,t) = 2^{-p-1} + \frac{s}{1+t}.$$

Recalling that $\ell = 2m'_x/m'_y \cdot 2^{-c}$, a typical choice is

$$(\sigma, \tau) = (2m'_x \cdot 2^{-c}, m'_y - 1).$$

Then F is approximated over a suitable domain $\mathcal{S} \times \mathcal{T}$ (that contains all the possible values for (σ, τ) when x and y vary) by a polynomial P of the form

$$P(s,t) = 2^{-p-1} + s \cdot a(t), \quad \text{with} \quad a(t) = \sum_{i=0}^{\delta} a_i t^i. \tag{9.47}$$

Finally, the polynomial P is evaluated at (σ, τ) and the obtained value is assigned to v.

This process introduces two kinds of errors: an *approximation error* which quantifies the fact that $P \approx F$ over $\mathcal{S} \times \mathcal{T}$ and, since the evaluation of P at (σ, τ) is done by a finite-precision program, an *evaluation error* which quantifies the fact that $v \approx P(\sigma, \tau)$.

In [293] some sufficient conditions on these two errors have been given in order to ensure that (9.46) holds. Then, a suitable polynomial $a(t)$ has been obtained under such conditions. By "suitable" we mean a polynomial of smallest degree δ and for which each coefficient absolute value $|a_i|$ fits in a 32-bit unsigned integer. Such a polynomial was obtained as a "truncated minimax" polynomial approximation,[8] using the software environment *Sollya*.[9] More precisely, the polynomial $a(t)$ has the form

$$a(t) = \sum_{i=0}^{10} (-1)^i \cdot A_i \cdot 2^{-32} \cdot t^i, \tag{9.48}$$

where $A_i \in \{0, \ldots, 2^{32} - 1\}$ for $0 \leq i \leq 10$. *Sollya* has further been used to guarantee, using interval arithmetic, that the approximation error induced by this choice of polynomial is small enough for our purposes.

A minimal degree δ has been sought for obvious speed reasons (intuitively, the smaller the polynomial a is, the faster our division code will be). However, once δ has been fixed, there are still many ways of evaluating the arithmetic expression $P(s,t)$ for P as in (9.47) and a as in (9.48). A classical way is Horner's rule:

$$P(s,t) = 2^{-p-1} + s \cdot \Big(a_0 + t \cdot \big(\cdots + t \cdot (a_9 + a_{10} \cdot t) \ldots \big)\Big).$$

This parenthesizing of $P(s,t)$ involves 11 additions and 11 multiplications. Assuming a latency of 1 for an addition and 3 for multiplication (of type `mul`; see Section 9.1.2), one can, in principle, deduce from s and t an approximate value for $P(s,t)$ in 44 cycles. This already seems faster than the rough estimate of 70 cycles used by the 24 iterations in the restoring method (see Listing 9.24).

And yet, Horner's rule is fully sequential (just like the restoring method), and other parenthesizings of $P(s,t)$ exist that make it possible to expose much more ILP. In such cases, provided the degree of parallelism is high enough, much lower latencies can be expected. An example of such a parenthesizing, generated automatically, has been given in [293]. (For more about such generation of polynomial evaluation schemes, we refer to [433].) It reduces the 44-cycle latency of Horner's rule shown above to only 14 cycles,

[8]We have used a minimax-like approximation and not, for example, a simpler Taylor-like polynomial $1 - t + t^2 - t^3 + \cdots$ because, as detailed in [293], the domain $\mathcal{T}$ on which $a(t)$ should approximate $1/(1+t)$ is $[0, 1 - 2^{1-p}]$ and thus contains values that are close to 1.

[9]See Section 10.3.4 of this book, as well as http://sollya.gforge.inria.fr/ and Lauter's Ph.D. dissertation [371].

provided one can launch at least 3 operations + or `mul` simultaneously, at most 2 of them being `mul`. Note that this latency of 14 cycles comes from viewing $P(s,t)$ as a bivariate polynomial, and could not be obtained by first evaluating the univariate polynomial $a(t)$ as fast as possible, and then applying the final Horner step $2^{-p-1} + s \cdot a(t)$.

Once a fast/highly parallel evaluation scheme has been found, it remains to implement it using 32-bit integer arithmetic and to check the numerical accuracy of the resulting program (evaluation error). An example of a C program that implements the 14-cycle evaluation scheme just mentioned is given in Listing 9.25. The variables S, T, V used in Listing 9.25 are integer

C listing 9.25 Polynomial evaluation program used in a binary32 division operator (Listing 1 in [293]).

```c
uint32_t S, T, V,
         r0, r1, r2, r3, r4, r5, r6, r7, r8, r9, r10, r11, r12, r13,
         r14, r15, r16, r17, r18, r19, r20, r21, r22, r23, r24, r25;

r0  = mul( T , 0xffffe7d7 );
r1  = 0xffffffe8 - r0;
r2  = mul( S , r1 );
r3  = 0x00000020 + r2;
r4  = mul( T , T );
r5  = mul( S , r4 );
r6  = mul( T , 0xffbad86f );
r7  = 0xfffbece7 - r6;
r8  = mul( r5 , r7 );
r9  = r3 + r8;
r10 = mul( r4 , r5 );
r11 = mul( T , 0xf3672b51 );
r12 = 0xfd9d3a3e - r11;
r13 = mul( T , 0x9a3c4390 );
r14 = 0xd4d2ce9b - r13;
r15 = mul( r4 , r14 );
r16 = r12 + r15;
r17 = mul( r10 , r16 );
r18 = r9 + r17;
r19 = mul( r4 , r4 );
r20 = mul( T , 0x1bba92b3 );
r21 = 0x525a1a8b - r20;
r22 = mul( r4 , 0x0452b1bf );
r23 = r21 + r22;
r24 = mul( r19 , r23 );
r25 = mul( r10 , r24 );
V   = r18 + r25;
```

representations of σ, τ, v, respectively:

$$S = \sigma \cdot 2^{30}, \qquad T = \tau \cdot 2^{32}, \qquad V = v \cdot 2^{30}.$$

Also, each hexadecimal constant corresponds to a particular A_i as in (9.48). It turns out that the variables `r1, ..., r25, V` are indeed in the range $\{0,\ldots,2^{32}-1\}$ of 32-bit unsigned integers, and that $v = V \cdot 2^{-30}$ satisfies (9.45), as desired. Checking such properties by paper-and-pencil calculations, if possible, would surely be long and error-prone. In [293] they have thus been verified mechanically using the Gappa software.[10]

Rounding to nearest ties to even

For some details on how to effectively implement the move from v to $\mathrm{RN}(\ell)$, we refer to the second example of a square root operator in Section 9.5.3.[11] A complete implementation of division, including subnormal numbers as well as all the IEEE 754-2008 rounding direction attributes, can be found in the FLIP library and is described in [509, §4].

9.5 Binary Floating-Point Square Root

Now, let us deal with the software implementation, for the binary32 format and $\circ = \mathrm{RN}$, of a square root operator. Our exposition follows [291, 292].

The case where x is a special datum (like ± 0, $\pm\infty$, or NaN) is described in Section 9.5.1, while the case where x is a (sub)normal is discussed in Sections 9.5.2 and 9.5.3.

As we will see, our analysis and code have many similarities with those for division in the previous section. However, things are simpler for at least two obvious reasons: first, $\sqrt{x}$ is univariate while x/y was bivariate; and second, $\sqrt{x} \geq 0$ over the reals, so that the sign is known in advance.[12] A third reason is that binary floating-point square roots never underflow or overflow, and an efficient implementation should take advantage of this.

9.5.1 Handling special values

For the square root, the input x is considered a *special input* when it is either ± 0, $\pm\infty$, NaN, or a binary floating-point number that is finite, nonzero, and negative. The special output values that the IEEE 754-2008 standard mandates in such situations have been listed in Table 7.6.

9.5.1.1 Detecting that a special value must be returned

Using Table 9.2 we see that x is a special input if and only if $X \notin (0, 2^{k-1} - 2^{p-1})$; that is, if and only if

$$(X-1) \bmod 2^k \geq 2^{k-1} - 2^{p-1} - 1. \tag{9.49}$$

[10]http://gappa.gforge.inria.fr/ and Section 13.3.

[11]For the square root, the situation will be a little simpler since the output cannot be a subnormal number.

[12]Note however that the IEEE 754 standards require that $\sqrt{-0} = -0$, a special case which is covered in Section 9.5.1.

For example, when $k = 32$ and $p = 24$, an implementation of (9.49) is given by lines 3 and 4 of Listing 9.26.

C listing 9.26 Special value handling in a binary32 square root operator.

```
1  uint32_t Xm1;
2
3  Xm1 = X - 1;
4  if (Xm1 >= 0x7F7FFFFF)
5  {
6     if (X <= 0x7F800000 || X == 0x80000000)
7        return X;
8     else
9        return 0x7FC00000 | X;        // qNaN with payload encoded in
10                                     // the last 22 bits of X
11 }
```

9.5.1.2 Returning special values as recommended or required by IEEE 754-2008

Once x is known to be special, we deduce from Table 7.6 that there are only two cases to be considered: the result must be either x itself, or a qNaN. Since here x is known to be *not* a positive (sub)normal number, this first case, which is $x \in \{+0, +\infty, -0\}$, is characterized by

$$X \leq 2^{k-1} - 2^{p-1} \qquad \text{or} \qquad X = 2^{k-1}.$$

When $k = 32$ and $p = 24$, an implementation of this condition is given at line 6 of Listing 9.26, using

$$2^{31} - 2^{23} = (7F800000)_{16} \qquad \text{and} \qquad 2^{31} = (80000000)_{16}.$$

Returning a qNaN is done in the same way as for addition, multiplication, or division, by masking with the constant $(7FC00000)_{16}$, whose bit string is

$$[0\,\underbrace{111111111}_{9\text{ ones}}\,\underbrace{0000000000000000000000}_{22\text{ zeros}}].$$

Again, this kind of mask keeps as much of the information of the input as possible, and, in particular, the payload of x, as recommended by the IEEE 754-2008 standard (see [267, §6.2]).

9.5.2 Exponent computation

We assume from now on that the input x is not *special; that is, x is a positive finite nonzero (sub)normal number.*

9.5.2.1 Formula for the exponent of the result

In order to see how to compute the exponent of our correctly rounded square root, let us first rewrite $x = m_x \cdot 2^{e_x}$ as

$$x = m'_x \cdot 2^{e'_x},$$

with m'_x as in (9.21) and e'_x as in (9.22). Let us also introduce the shift value c defined as

$$c = \begin{cases} 0 & \text{if } e'_x \text{ is even,} \\ 1 & \text{if } e'_x \text{ is odd.} \end{cases} \tag{9.50}$$

Rewriting x as $x = (2^c \cdot m'_x) \cdot 2^{e'_x - c}$, we obtain $\sqrt{x} = \ell \cdot 2^d$, so that, after rounding,

$$\mathrm{RN}(\sqrt{x}) = \mathrm{RN}(\ell \cdot 2^d),$$

where

$$\ell = \sqrt{2^c \cdot m'_x} \qquad \text{and} \qquad d = \frac{e'_x - c}{2}. \tag{9.51}$$

Notice that, by definition of c, the real ℓ lies in $[1, 2)$. Moreover, as recalled in Section 7.7, floating-point square roots never underflow or overflow, and thus

$$e_{\min} \le d \le e_{\max}. \tag{9.52}$$

Hence, the correctly rounded value of the square root of x is given by the following product:

$$\mathrm{RN}(\sqrt{x}) = \mathrm{RN}(\ell) \cdot 2^d. \tag{9.53}$$

In general, applying a rounding operator $\circ(\cdot)$ to a real number $\ell \in [1, 2)$ yields the weaker enclosure $\circ(\ell) \in [1, 2]$. However, in the case of square root with rounding to nearest, we have the following nice property (see [292] for a proof).

Property 9.3. $\mathrm{RN}(\ell) \in [1, 2)$.

From Property 9.3 and the inequalities in (9.52) we conclude that

$$\mathrm{RN}(\ell) \cdot 2^d$$

in (9.53) is already the *normalized representation* of $\mathrm{RN}(\sqrt{x})$. In particular, neither overflow detection nor exponent update is needed.[13]

In conclusion, the result exponent is indeed d as in (9.51). Since c is either zero or one, and since $e'_x = e_x - \lambda_x$, we arrive at the following formula:

$$d = \left\lfloor \frac{e_x - \lambda_x}{2} \right\rfloor, \tag{9.54}$$

where $\lfloor \cdot \rfloor$ denotes the usual floor function. Note that the shift value c in (9.50) does *not* appear explicitly in the above formula for d. In practice, this means that c and d can be computed independently of each other.

[13] The latter is of course still true for $\circ \in \{\mathrm{RZ}, \mathrm{RD}\}$, but not for $\circ = \mathrm{RU}$, as shown in [292].

9.5.2.2 Implementation for the binary32 format

Let us now implement (9.54). As for other basic operations, we shall in fact compute the biased value

$$D - 1 = d + e_{\max} - 1.$$

Combining (9.54) with $E_x = e_x - e_{\min} + n_x$ and $\lambda_x = M_X - w$ and $e_{\min} = 1 - e_{\max}$, one can deduce that

$$D - 1 = \left\lfloor \frac{E_x - n_x - M_X + w - e_{\min}}{2} \right\rfloor.$$

For the binary32 format, $w - e_{\min} = 8 + 126 = 134$, and thus $D - 1$ can be implemented as shown in Listing 9.27.

C listing 9.27 Computing $D - 1$ in a binary32 square root operator.

```c
uint32_t Ex, nx, MX, Dm1;

Ex = X >> 23;      nx = X >= 0x800000;     MX = max(nlz(X), 8);

Dm1 = ((Ex - nx) + (134 - MX)) >> 1;
```

Notice that in the above code x is assumed to be nonspecial, which in the case of square root implies $X = |X|$. Hence, computing $|X|$ is not needed, and the biased exponent value E_x can be deduced simply by shifting X.

9.5.3 Getting the correctly rounded result

In (9.53), once d has been obtained through the computation of $D - 1$, it remains to compute the binary floating-point number

$$\mathrm{RN}(\ell) = (1.r_1 \ldots r_{p-1})_2,$$

where

$$\ell = \sqrt{2^c \cdot m'_x} = (1.\ell_1 \ldots \ell_{p-1}\ell_p \ldots)_2. \tag{9.55}$$

Exactly as for division in Section 9.4.4,

- in general, the binary expansion of ℓ is infinite;
- we will give two examples of how to implement the computation of $\mathrm{RN}(\ell)$: the first uses the classical *restoring* method for square rooting (see for example [187, §6.1] for a description of the basic recurrence). The second one, which comes from [292], is based on polynomial approximation and evaluation.

Before that, we shall see how to implement the computation of the shift value c defined in (9.50). Since the value of c was not explicitly needed for obtaining the square root exponent d, it has not been computed yet. This is in contrast to binary floating-point multiplication and division, where the corresponding c value was already used for deducing $D-1$ (see line 7 in Listing 9.18 and line 7 in Listing 9.23).

9.5.3.1 Computation of the shift value c

From (9.50) we see that c is 1 if and only if e'_x is odd. Again, combining the facts that $E_x = e_x - e_{\min} + n_x$ and $\lambda_x = M_X - w$ with the definition $e'_x = e_x - \lambda_x$, we have for the binary32 format (where $e_{\min} = -126$ and $w = 8$)

$$e'_x = E_x - n_x - M_X - 118.$$

Therefore, e'_x is odd if and only if $E_x - n_x - M_X$ is odd.

For the binary32 format, we have seen in Listing 9.27 how to deduce E_x, n_x, and M_X from X. Hence, we use the code in Listing 9.28 for deducing the value of c. Since M_X is typically more expensive to obtain than E_x and n_x, one can compute M_X in parallel with $E_x - n_x$. Of course, other implementations are possible (for example, using logic exclusively), but this one reuses the value $E_x - n_x$ that appears in Listing 9.27.

C listing 9.28 Computing the shift value c in a binary32 square root operator. Here we assume that Ex, nx, and MX are available.

```
uint32_t c;

c = ((Ex - nx) - MX) & 1;
```

9.5.3.2 First example: restoring square root

Similar to restoring division (which we have recalled in detail in Section 9.4.4), restoring square root is an iterative process whose iteration number $i \geq 1$ produces the bit ℓ_i of the binary expansion of ℓ in (9.55).

Exactly as for division, it is known (see, for example, [118, page 242] and [187, page 463]) that the square root of a binary floating-point number cannot be the exact middle of two consecutive (normal) binary floating-point numbers. Therefore, the correctly rounded result $\mathrm{RN}(\ell)$ can be obtained using the same formula as the one used for division in (9.40), and p iterations are again enough. Compared to division, the only specificity of the restoring square root method is the definition of the iteration itself, which we will recall now.

Let $q_0 = 1$. The goal of iteration $1 \leq i \leq p$ is to produce ℓ_i, or, equivalently, to deduce $q_i = (1.\ell_1 \ldots \ell_i)_2$ from $q_{i-1} = (1.\ell_1 \ldots \ell_{i-1})_2$. For $0 \leq i \leq p$, one way of encoding the rational number q_i into a k-bit unsigned integer ($k > p$) is through the integer

$$Q_i = q_i \cdot 2^p, \tag{9.56}$$

whose bit string is

$$[\underbrace{000\ldots000}_{k-p-1 \text{ zeros}}\, 1\, q_1 \ldots q_i\, \underbrace{000\ldots000}_{p-i \text{ zeros}}].$$

We have $0 \leq \ell - q_i < 2^{-i}$, and thus $q_i \geq 0$ approximates ℓ from below. Hence, q_i^2 approximates ℓ^2 from below as well, and one can define the remainder $r_i \geq 0$ by

$$\ell^2 = q_i^2 + r_i. \tag{9.57}$$

Now, $\ell^2 = 2^c \cdot m_x'$ has at most $p-1$ fraction bits, and

$$q_i^2 = \left((1.q_1 \ldots q_i)_2\right)^2$$

has at most $2i$ fraction bits. Since $0 \leq i \leq p$, both ℓ^2 and q_i^2 have at most $p+i$ fraction bits and we conclude that the following scaled remainder is a *nonnegative integer*:

$$R_i = r_i \cdot 2^{p+i}. \tag{9.58}$$

The following property shows further that the integer R_i can be encoded using at most $p+2$ bits. For example, for the binary32 format, $p+2 = 26 \leq 32$, and so R_i will fit into a 32-bit unsigned integer.

Property 9.4. *For $0 \leq i \leq p$, the integer R_i satisfies $0 \leq R_i < 2^{p+2}$.*

Proof. We have already seen that R_i is an integer such that $0 \leq R_i$. Let us now show the upper bound. From $0 \leq \ell - q_i < 2^{-i}$ it follows that

$$\ell^2 - q_i^2 = (\ell - q_i)(2q_i + (\ell - q_i)) \leq 2^{-i}(2q_i + 2^{-i}).$$

Now, $q_i = (1.\ell_1 \ldots \ell_i)_2 \leq 2 - 2^{-i}$, so that $2q_i + 2^{-i} \leq 4 - 2^{-i} < 4$. From this we conclude that

$$R_i = (\ell^2 - q_i^2) \cdot 2^{p+i} < 4 \cdot 2^{-i} \cdot 2^{p+i} = 2^{p+2}.$$

□

Combining the definition of Q_i in (9.56) with the facts that $q_0 = 1$ and, for $1 \leq i \leq p$, that $q_i = q_{i-1} + \ell_i \cdot 2^{-i}$, we get

$$Q_0 = 2^p \tag{9.59}$$

and, for $1 \leq i \leq p$,

$$Q_i = Q_{i-1} + \ell_i \cdot 2^{p-i}. \tag{9.60}$$

Then, using (9.59) and (9.60) together with the invariant in (9.57) and the definition of R_i in (9.58), one may check that

$$R_0 = m'_x \cdot 2^{p+c} - 2^p \tag{9.61}$$

and, for $1 \le i \le p$,

$$R_i = 2R_{i-1} - \ell_i \cdot (2Q_{i-1} + \ell_i \cdot 2^{p-i}). \tag{9.62}$$

Note that the recurrence relation (9.62) is, up to a factor of 2^p, analogous to Equation (6.12) in [187, page 333].

As for division, we deduce from the above recurrence relations for Q_i and R_i a *restoring* algorithm (see Algorithm 9.2), which works as follows. A tentative remainder T is computed at each iteration, assuming $\ell_i = 1$ in the recurrence (9.62) that gives R_i. Then two situations may occur:

- if $T < 0$ then the true value of ℓ_i was in fact 0. Hence, by (9.60) and (9.62), $Q_i = Q_{i-1}$ while R_i is equal to $2R_{i-1}$;
- if $T \ge 0$ then $\ell_i = 1$ was the right choice. In this case, (9.60) and (9.62) lead to $Q_i = Q_{i-1} + 2^{p-i}$ and $R_i = T$.

Algorithm 9.2 The first p iterations of the restoring method for binary floating-point square root.

$(Q_0, R_0) \leftarrow (2^p, m'_x \cdot 2^{p+c} - 2^p)$
for $i \in \{1, \ldots, p\}$ **do**
  $T \leftarrow 2R_{i-1} - (2Q_i + 2^{p-i})$
  if $T < 0$ **then**
    $(Q_i, R_i) \leftarrow (Q_{i-1}, 2R_{i-1})$
  else
    $(Q_i, R_i) \leftarrow (Q_{i-1} + 2^{p-i}, T)$
  end if
end for

An implementation of Algorithm 9.2 for the binary32 format, for which $p = 24$, is detailed in Listing 9.29. Here are a few remarks about lines 4 to 20:

- The variable `Q` is initialized with

$$Q_0 = 2^{24} = (1000000)_{16}$$

and, at the end of iteration $1 \le i \le 24$, contains the integer Q_i.

- Similarly, the variable `R` is initialized with R_0 and then contains the successive values of R_i. Note that the initialization of `R` requires the knowledge of both m'_x and c. These are assumed available in variables `mpX` and c, respectively. Concerning mpX, we may compute it exactly as for division (see for example Listing 9.22). Concerning c, we have seen how to compute it in Listing 9.28. The right shift of $7 - c$ positions at line 4 of Listing 9.29 comes from the fact that the bit string of mpX in Listing 9.22 has the form

$$[1\underbrace{* \cdots *}_{23 \text{ bits}}\underbrace{00000000}_{8 \text{ bits}}],$$

and from the fact that the bit string of $m'_x \cdot 2^{p+c}$ has the form

$$[\underbrace{0000000}_{7 \text{ bits}}1\underbrace{* \cdots *}_{23 \text{ bits}}0] \qquad \text{if } c = 0,$$

and

$$[\underbrace{000000}_{6 \text{ bits}}1\underbrace{* \cdots *}_{23 \text{ bits}}00] \qquad \text{if } c = 1.$$

- The tentative remainder T is stored in variable T, while S is initialized with 2^{24} and contains the integer 2^{24-i} at the end of the i-th iteration.

C listing 9.29 First 24 iterations of the restoring method, rounding and packing for a binary32 square root operator. Here we assume that Dm1, mpX, and c are available.

```
1   uint32_t Q, R, S, i;
2   int32_t T;
3
4   Q = 0x1000000;                  R = (mpX >> (7 - c)) - Q;
5   S = 0x1000000;
6
7   for (i = 1; i < 25; i++)
8   {
9      S = S >> 1;
10     T = (R << 1) - ((Q << 1) + S);
11
12     if (T < 0)
13     {
14        R = R << 1;
15     }
16     else
17     {
18        Q = Q + S; R = T;
19     }
20  }
21
22  return (Dm1 << 23) + ((Q >> 1) + (Q & 1));
```

Final rounding of the significand and packing with the result exponent is done at line 22 of Listing 9.29. The variable Dm1 which encodes the (biased) result exponent has been computed in Listing 9.23.

The only difference with division (see line 21 of Listing 9.24) is that the sign bit of $\mathrm{RN}(\sqrt{x})$ is known to be zero when x is not special.

9.5.3.3 Second example: square root by polynomial evaluation

For $\ell = \sqrt{2^c \cdot m'_x}$, a way of computing $\mathrm{RN}(\ell)$ that is much faster on some parallel architectures was proposed in [291, 292]. Its generalization to division has already been outlined in the second example of Section 9.4.4, following [293], and extensions to other algebraic functions have been investigated in [509, §3].

Computing the one-sided approximation v

As for division, $\mathrm{RN}(\ell)$ is obtained by first computing a one-sided approximation v to ℓ that satisfies (9.46) and is representable using at most k bits. The main differences concern the choice of function F to be approximated, the domain $\mathcal{S} \times \mathcal{T}$ on which it is approximated,[14] the fact that the evaluation point (σ, τ) is not exactly representable in finite precision, and the sufficient conditions on the approximation and evaluation errors. In particular, for the binary32 format those conditions are simpler and, above all, much easier to verify, than for division (see [293] and [292] for the details).

To summarize, this first step is aimed at producing a code with high ILP, in the same spirit as Listing 9.25 and whose output is a 32-bit unsigned integer V such that $v = V \cdot 2^{-30}$ satisfies

$$\left|(\ell + 2^{-25}) - v\right| < 2^{-25}. \tag{9.63}$$

Rounding to nearest ties to even

Let us now see how to deduce $\mathrm{RN}(\ell)$ from v. The binary expansion of v has the form

$$v = (1.v_1 \ldots v_{30})_2.$$

Therefore, truncating v after 24 fraction bits and using (9.63) gives a value

$$w = (1.v_1 \ldots v_{23}v_{24})_2$$

such that

$$0 \le v - w < 2^{-24}. \tag{9.64}$$

By combining (9.63) and (9.64), we conclude that

$$|\ell - w| < 2^{-24}. \tag{9.65}$$

[14]Although the square root is a univariate function, the real $\ell = \sqrt{2^c \cdot m'_x}$ depends on c and m'_x. Hence, $\ell + 2^{-25}$ may be considered as the exact value of a suitable function $F : \mathcal{S} \times \mathcal{T} \to \mathbb{R}$.

Given this "truncated approximation" w, one can now deduce the correctly rounded value $\mathrm{RN}(\ell)$ fairly easily by means of Algorithm 9.3.

Algorithm 9.3 Deducing $\mathrm{RN}(\ell)$ from the truncated approximation w.

if $w \geq \ell$ **then**
  $\mathrm{RN}(\ell) \leftarrow$ truncate w after 23 fraction bits
else
  $\mathrm{RN}(\ell) \leftarrow$ truncate $w + 2^{-24}$ after 23 fraction bits
end if

It was shown in [292] that Algorithm 9.3 indeed returns $\mathrm{RN}(\ell)$. In order to implement it, all we need is a way of evaluating the condition $w \geq \ell$. This can be done because of the following property [292, Property 4.1].

Property 9.5. *Let W, P, and Q be the 32-bit unsigned integers such that*

$$w = W \cdot 2^{-30},$$

$$P = \mathtt{mul}(W, W),$$

and

$$\ell^2 = Q \cdot 2^{-28}.$$

Then $w \geq \ell$ if and only if $P \geq Q$.

As an immediate consequence of this property, we obtain the code in Listing 9.30 that implements the computation of the correctly rounded significand $\mathrm{RN}(\ell)$ and its packing with the biased exponent $D - 1$.

C listing 9.30 Truncating a one-sided approximation, rounding, and packing for a binary32 square root operator. Here we assume that `V`, `mpX`, `c`, and `Dm1` are available.

```
uint32_t W, P, Q;

W = V & 0xFFFFFFC0;

P = mul(W, W);          Q = mpX >> (3 - c);

if (P >= Q)
   return (Dm1 << 23) + (W >> 7);
else
   return (Dm1 << 23) + ((W + 0x40) >> 7);
```

9.6 Custom Operators

The previous sections in this chapter detailed how to implement the five basic operations in software, using only integer arithmetic. Once these generic operators have been optimized for a given environment, achieving further performance may be possible by considering a set of *non-generic* or *custom* operators as well. These are similar to those studied in hardware in Sections 8.7 and 8.8.

Indeed, many benchmarks and application codes involve operations which are either specializations or generalizations of the five basic ones. For example, evaluating the Euclidean norm of an n-dimensional vector needs not a general multiplier as the one presented in Section 9.3, but only a squarer; on the other hand, radix-2 FFT computations and complex arithmetic typically involve two-dimensional dot-products (DP2), for which we may desire an implementation that is more efficient than the naive one (via two multiplications and an addition, or one multiplication and an FMA) and at least as accurate.

Of course, to benefit from such custom operators, work is required not only at the library level (to implement the operators themselves) but also at the compiler level (so that it becomes able to select each of those operators). In the context of the FLIP library and the ST231 processor, this approach has been studied in [301, 509, 290, 289, 288, 311], leading to significant performance improvements in practice. Specifically, three groups of operators have been considered:

1. *Specialized operators*, which will be substituted to generic operators when some properties about their floating-point arguments can be detected at compile time. Those include
   - mul2 (multiplication by two): $2x$;
   - div2 (multiplication by one half): $\frac{1}{2}x$;
   - scaleB (multiplication by an integer power of two): $x \cdot 2^n$ with n a 32-bit signed integer;
   - square (squaring): x^2;
   - addnn (addition of two nonnegative numbers): $x + y$ with $x \geq 0$ and $y \geq 0$;
   - inv (inversion): $1/x$.
2. *Fused operators*, which replace several basic operations by a single one (thus with a single rounding). For example,
   - FMA (fused multiply-add): $xy + z$;
   - FSA (fused square-add): $x^2 + y$ with $y \geq 0$;

- DP2 (two-dimensional dot product): $xy + zt$;
- SOS (sum of two squares): $x^2 + y^2$.

3. *Paired operators*, for performing simultaneously two operations on the same input. A typical case, which appears in FFT computations, is

   - addsub (simultaneous addition and subtraction): $(x + y, x - y)$.

This list could of course be extended depending on the application needs, for example with more multipliers by some specific constants or with fused operations like $x^{-1/2}$ (reciprocal square root) and $\sqrt{x^2 + y^2}$ (hypotenuse), whose correctly rounded evaluation is already recommended by the IEEE 754-2008 standard. (For various roots and their reciprocals, this is detailed in [509, §3].)

Remark also that while the FMA cannot really be considered as a custom operator (since it is now fully specified by this standard), it does go beyond the five basic operations studied in the previous sections and so we have included it here.

Finally, note that another simultaneous operation has been studied in the context of the FLIP library, namely, the evaluation of sine and cosine of a given floating-point argument in the reduced range $[-\pi/4, \pi/4]$. We do not mention it here, since it pertains more to so-called elementary functions (which are the subject of the next chapter), but interesting gains have been obtained in this case as well; see for example [288].

Table 9.3 indicates the speedups obtained on the ST231 processor with the custom operators listed above, for the binary32 format with rounding to nearest and full support of subnormal numbers. Here,

$$\text{speedup} = \frac{\text{latency of the original code}}{\text{latency of the customized code}},$$

and by "original code" we mean the straightforward implementation of the given operation using only the optimized generic operations of FLIP.

We see in Table 9.3 that some operators can be up to 4x faster than their generic counterpart. In all cases, this has required a complete re-design of the generic algorithms seen in the previous sections. For example, for mul2, a speedup of only 1.17 would be obtained by a naive specialization, that is, by simply replacing y with the floating-point number 2 everywhere in the existing code for xy, and then recompiling this code.

Finally, combining such custom operators can also bring interesting improvements for various numerical kernels and applications. For example, for the UTDSP Benchmark Suite (which aims to evaluate the efficiency of C com-

Custom operator	Speedup
mul2	4.2
div2	3.5
scalb	1.4
square	1.75
addnn	1.73
inv	1.36
FSA	2.14
FMA	1.02
SOS	2.62
DP2	1.24
addsub	1.86

Table 9.3: Speedups provided by various custom operators (compared with their original implementation using generic operators) for binary32 arithmetic with rounding to nearest and full subnormal support.

pilers on typical DSP codes), it was shown in [289] that FFTs in dimensions 256 and 1024 can be made 1.59x faster when selecting the two custom operators DP2 and addsub.

Chapter 10

Evaluating Floating-Point Elementary Functions

THE ELEMENTARY FUNCTIONS (the right term is "elementary transcendental functions") are the most common mathematical functions: sine, cosine, tangent, and their inverses, exponentials and logarithms of radices e, 2, or 10, etc. They appear everywhere in scientific computing. Therefore, being able to evaluate them quickly and accurately is important for many applications. Many very different methods have been used for evaluating them: polynomial or rational approximations, shift-and-add algorithms, table-based methods, etc. The choice of the method greatly depends on whether the function will be implemented in hardware or in software, on the target precision, and on the required performance (in terms of speed, accuracy, memory consumption, size of code, etc.). Some methods are better for average performance, some are better for worst-case performance.

The goal of this chapter is to present a short survey of the main methods and to introduce the basic techniques. More details can be found in books dedicated to elementary function algorithms, such as [118, 406, 442].

10.1 Introduction

10.1.1 Which accuracy?

As in almost all domains of numerical computing, judicious tradeoffs must be found between accuracy and speed. When designing special-purpose software or hardware, the targeted applications may have latency, throughput and/or accuracy constraints that somehow impose a family of solutions. For general-purpose hardware or software, however, the choices are not so clear. Assume we wish to implement function f (which is a function over the reals).

J.-M. Muller et al., *Handbook of Floating-Point Arithmetic*,
https://doi.org/10.1007/978-3-319-76526-6_10

Denote by F the function (over the floating-point numbers) being actually computed. We wish F to be as close as possible to f. Ideally, we would like to provide correct rounding, i.e., if $\circ$ is the chosen rounding function, to have $F(x) = \circ(f(x))$ for all finite floating-point numbers x.

However, some functions are specified to be much less accurate, to enable higher-performance implementations. For instance, the OpenCL SPIR-V environment specification defines several profiles, some of which allow an error bound up to $8192\,\mathrm{ulp}$ in binary32 [336, Section 8.5]. Some profiles are defined more tightly, with an error bound of 2 to $4\,\mathrm{ulp}$. This error bound is related to the bound on $\left|\frac{F(x)-f(x)}{f(x)}\right|$, and correct rounding would entail a bound of $0.5\,\mathrm{ulp}$.

Even accurate elementary function libraries often report an error bound very slightly above $0.5\,\mathrm{ulp}$. Indeed, the goal of correctly rounded elementary functions is sometimes difficult to reach at a reasonable cost, because of a difficulty known as the *Table maker's dilemma* (TMD), a term coined by Kahan. Let us quote him [321]:

> *Why can't Y^W be rounded within half an ulp like SQRT? Because nobody knows how much computation it would cost to resolve what I long ago christened "The Table-Maker's Dilemma" $(\cdots)$. No general way exists to predict how many extra digits will have to be carried to compute a transcendental expression and round it correctly to some preassigned number of digits. Even the fact (if true) that a finite number of extra digits will ultimately suffice may be a deep theorem.*

We will present the Table maker's dilemma in Section 10.5, and show that in some cases, there is a solution to this problem.

10.1.2 The various steps of function evaluation

After a preliminary filtering of the "special cases" (input values that are infinities or NaNs, or that lie outside the domain of the function), evaluating an elementary function is usually done in several steps: range reduction, evaluation of an approximation valid in a small domain, and possibly reconstruction of the final result. Let us quickly describe the main steps.

10.1.2.1 Range reduction

The usual approximation algorithms (e.g., polynomial approximation—see below) allow one to approximate a function f in a given bounded domain D, which is generally rather small. Hence, the first step when evaluating f at a given floating-point input argument x consists in finding an intermediate argument $y \in D$, such that $f(x)$ can be deduced from $f(y)$ (or, sometimes, from $g(y)$, where g is a related function). A typical example is the function

$\sin(x)$: assume we have algorithms that evaluate $\sin(y)$ and $\cos(y)$ for $y \in [0, \pi/2]$. The computation of $\sin(x)$ can be decomposed in three steps:

- compute y and k such that $y \in [0, \pi/2]$ and $y = x - k\pi/2$;
- compute
$$g(y,k) = \begin{cases} \sin(y) & \text{if} \quad k \bmod 4 = 0 \\ \cos(y) & \text{if} \quad k \bmod 4 = 1 \\ -\sin(y) & \text{if} \quad k \bmod 4 = 2 \\ -\cos(y) & \text{if} \quad k \bmod 4 = 3; \end{cases} \tag{10.1}$$
- obtain $\sin(x) = g(y,k)$.

When high accuracy is at stake, y may be represented by two (or more) floating-point numbers, in a way similar to what is described in Section 14.1.

The first step (computation of y and k from x) is called *range reduction* (or, sometimes, *argument reduction*). In this example, the range reduction is *additive*: y is equal to x plus or minus a multiple of some constant (here, $\pi/2$). Range reduction can also be *multiplicative*, for instance in

$$\cos(x) = 2\cos^2\left(\frac{x}{2}\right) - 1$$

or

$$\ln(x \cdot 2^k) = \ln(x) + k \cdot \ln(2).$$

Let us consider the previous example of additive range reduction. When x is large or very close to an integer multiple of $\pi/2$, computing y by the naive method, i.e., actually subtracting from x the result of the floating-point operation $k \times \mathrm{RN}(\pi/2)$, leads to a *catastrophic cancellation*: the result obtained may be very poor. Consider the following example in the binary64/double precision format of the IEEE 754-2008 standard, assuming round to nearest even. We wish to evaluate the sine of $x = 5419351$. The value of k is 3450066. The binary64 number that is nearest to $\pi/2$ is

$$\begin{aligned} P &= \frac{884279719003555}{2^{49}} \\ &= 1.5707963267948965579989817342720925807952880859375. \end{aligned}$$

First, assume that a fused multiply-add (FMA) instruction is not used. The value of kP computed in the binary64 format will be $5818983827636183/2^{30}$. This will lead to a computed value of $x - kP$ (namely, $\mathrm{RN}(x - \mathrm{RN}(kP))$) equal to

$$\frac{41}{2^{30}} = 0.000000038184225559234619140625,$$

whereas the exact value of $x - k\pi/2$ is

$$0.000000038200475070896611209301164704279487763080326 14\ldots.$$

This means that if we use that naive range reduction for evaluating $\sin(x)$, our final result will only have two significant decimal digits. If an FMA instruction is available, the computed reduced argument (namely, $\mathrm{RN}(x - kP)$) will be

$$\frac{10811941}{2^{48}} = 0.0000000384117306850839668186381459236145019531 25,$$

which, surprisingly enough,[1] is even worse. And yet, x is not a huge number: one can construct worse cases.

This shows that range reduction is a critical step, which should not be overlooked if one wishes to build good function software.

We present some algorithms for accurate range reduction in Section 10.2.

10.1.2.2 Function approximation

Once range reduction has been performed, we need to evaluate some function at the reduced argument. For this purpose, we need to use an approximation to the function. The first algorithms used for approximating elementary functions were based on truncated power series expansions. Another family of *shift-and-add* algorithms was designed for early calculators. They work in fixed-point (indeed, one benefit of range reduction is that the reduced argument can be represented in fixed point), and only use additions, multiplications by a power of the radix—that is, digit shifts—and small tables. A shift-and-add algorithm of considerable historical importance is CORDIC (COrdinate Rotation DIgital Computer), introduced in 1959 by Volder [619, 620], and generalized in 1971 by Walther [625, 626]. This algorithm allowed one to evaluate square roots, sines, cosines, arctangents, exponentials, and logarithms. Its versatility made it attractive for hardware implementations. CORDIC (or variants of CORDIC) has been implemented in the Hewlett-Packard 9100 desktop calculator, built in 1968, in pocket calculators (for instance, Hewlett-Packard's HP 35 [103]) and in arithmetic co-processors such as the Intel 8087 [451] and some of its successors. And still, a handful of papers presenting new variants of this algorithm are published every year. The *Journal of VLSI Signal Processing* devoted a special issue to CORDIC (June 2000), and Meher et al. gave a survey in 2009 for the 50th birthday of CORDIC [414]. Although the original CORDIC algorithm was designed for radix-2 arithmetic, there exist radix-10 variants [542, 539].

Power series are still of interest for multiple-precision evaluation of functions, and shift-and-add methods may still be attractive for hardware-

[1] In general, the result of the naive reduction will be *slightly* better with an FMA.

oriented implementations.[2] And yet, mainstream current implementations of floating-point elementary functions are written in software, and use polynomial approximations.

For this reason, we focus, in Section 10.3, on polynomial approximations to functions. Let us now give more details on the various steps of elementary function evaluation.

10.2 Range Reduction

10.2.1 Basic range reduction algorithms

In the following, we deal with *additive* range reduction. Let us denote by x the initial floating-point argument, and by C the constant of the range reduction (typically, for trigonometric functions, C is π divided by some power of the radix of the floating-point format). We are looking for an integer k and a real number $y \in [0, C]$ or $[-C/2, +C/2]$ such that

$$y = x - kC.$$

10.2.1.1 Cody and Waite's reduction algorithm

Cody and Waite [107, 105] tried to improve the accuracy of naive range reduction by representing C more accurately, as the sum of two floating-point numbers. More precisely,

$$C = C_1 + C_2,$$

where the significand of C_1 has sufficiently many zeros on its right part, so that when multiplying C_1 by an integer k whose absolute value is not too large (say, less than k_{max}, where $k_{max}C$ is the order of magnitude of the largest input values for which we want the reduction to be accurate), the result is exactly representable as a floating-point number.[3] The reduction operation consists in computing

$$y = \text{RN}(\text{RN}(x - \text{RN}(kC_1)) - \text{RN}(kC_2)), \tag{10.2}$$

which corresponds, in the C programming language, to the line:

```
y = (x - k*C1) - k*C2
```

[2]As we write this book, for general-purpose systems, software implementations that use polynomial approximations are used on all platforms of commercial significance [406, 118]. This does not mean that CORDIC-like algorithms are not interesting for special-purpose circuits. The choice also depends on the function and target technology. For instance, on FPGAs, recent studies indicate that CORDIC is better than polynomial-based techniques for the arctangent [143], but worse for sine/cosine [144].

[3]This algorithm was designed before the time of the FMA, hence the need for $\text{RN}(kC_1)$ to be computed exactly. Note also that contracting pairs of multiplication and subtraction into FMA does not invalidate the algorithm.

In (10.2), the product kC_1 is exact (provided that $|k| \leq k_{max}$). The subtraction $x - kC_1$ is also exact, by Sterbenz's lemma (Lemma 4.1 in Chapter 4). Hence, the only sources of error are the computation of the product kC_2 and the subtraction of this product from $x - kC_1$: we have somehow simulated a larger precision. Let us continue our previous example (sine of $x = 5419351$ in binary64 arithmetic, with $C = \pi/2$, which gives $k = 3450066$), with the two binary64 floating-point numbers:

$$\begin{aligned} C_1 &= \frac{1686629713}{2^{30}} \\ &= 1.1001001000011111101101010100001 \quad \text{(binary)} \\ &= 1.570796326734125614166259765625 \quad \text{(decimal)} \end{aligned}$$

and

$$\begin{aligned} C_2 &= \frac{4701928774853425}{2^{86}} \\ &= 1.0000101101000110000100011010011000\,1 \\ &\quad 0011000110011000 1 \times 2^{-34} \quad \text{(binary)} \\ &= 6.077100506506192249319157691538431 13 \\ &\quad 331570080504206998739391565322875976 5625 \times 10^{-11} \quad \text{(decimal)}. \end{aligned}$$

C_1 is exactly representable with only 31 bits in the significand. The product of C_1 with any integer less than or equal to 5340353 is exactly representable (hence, exactly computed) in binary64 arithmetic. In the case considered, the computed value of the expression kC_1 will be exactly equal to kC_1; that is,

$$\begin{aligned} &\frac{2909491913705529}{2^{29}} \\ &= 10100101011000101010110.11111111111100100100000111001 \quad \text{(binary)} \\ &= 5419350.99979029782116413116455078125 \quad \text{(decimal)}, \end{aligned}$$

and the computed value of the expression $(x - kC_1)$ will be exactly equal to $x - kC_1$,

$$\begin{aligned} &\frac{112583}{2^{29}} \\ &= 1.1011011111000111 \times 2^{-13} \quad \text{(binary)} \\ &= 0.000209702178835868835449 21875 \quad \text{(decimal)}. \end{aligned}$$

Therefore, the computed value y of the expression $(x - kC_1) - kC_2$ will be

$$\begin{aligned}
&\frac{704674387127}{2^{64}} \\
&= 1.0100100000100011101110111010100101101111 \times 2^{-25} \quad \text{(binary)} \\
&= 3.8200475070899238966282140950 17 \cdots \times 10^{-8} \quad \text{(decimal)},
\end{aligned}$$

which is much more accurate than what we got using the naive method (recall that the exact result is $3.820047507089661120930116 47 \cdots \times 10^{-8}$).

Cody and Waite's idea of splitting the constant C into two floating-point numbers can be generalized. For instance, in the library CRlibm[4] [129], this idea is used with a constant $C = \pi/256$ for small arguments of trigonometric functions ($|x| < 6433$, this threshold value being obtained by error analysis of the algorithm). However, the reduced argument must be known with an accuracy better than 2^{-53}. To this purpose, the final subtraction is performed using the Fast2Sum procedure (introduced in Section 4.3.1). For larger arguments ($6433 \leq |x| < 13176794$), the constant is split into three floating-point numbers, again with a Fast2Sum at the end.

Note that when an FMA instruction is available, one can use a very accurate generalization of Cody and Waite's algorithm, due to Boldo, Daumas, and Li [47, 442].

10.2.1.2 Payne and Hanek's algorithm

Payne and Hanek [487] designed a range reduction algorithm for the trigonometric functions that becomes very interesting when the input arguments are large. Assume we wish to compute

$$y = x - kC, \tag{10.3}$$

where $C = 2^{-t}\pi$ is the constant of the range reduction. Typical values of C that appear in current programs are between $\pi/256$ and $\pi/4$. Assume x is a binary floating-point number of precision p, and define e_x as its exponent and X as its integral significand, so that $x = X \cdot 2^{e_x - p + 1}$. Equation (10.3) can be rewritten as

$$y = 2^{-t}\pi \left(\frac{2^{t+e_x-p+1}}{\pi} X - k \right). \tag{10.4}$$

The division by π can be implemented as a multiplication by $1/\pi$. Let us therefore denote by

$$0.v_1 v_2 v_3 v_4 \cdots$$

[4]CRlibm was developed by the Arénaire/AriC team of CNRS, INRIA, and ENS Lyon, France. It is available at https://gforge.inria.fr/scm/browser.php?group_id=5929&extra=crlibm.

the infinite binary expansion of $1/\pi$. The core idea of Payne and Hanek's algorithm is that the implementation of Equation (10.4) only needs a small subset of these bits in practice. The most significant bits will not be needed due to the periodicity of the trigonometric functions, and the least significant bits will not be needed due to the absolute error bound given in the input. The conditions are detailed as follows:

- The bit v_i is of weight 2^{-i}. The product $2^{t+e_x-p+1}X \cdot v_i 2^{-i}$ is therefore an integral multiple of $2^{t+e_x-p+1-i}$. When multiplied, later on, by $2^{-t}\pi$, this value will become an integral multiple of $2^{e_x-p+1-i}\pi$. Therefore, as soon as $i \leq e_x - p$, the contribution of bit v_i in Equation (10.4) will result in an integral multiple of 2π: it will have no influence on the trigonometric functions. So, in the product
$$\frac{2^{t+e_x-p+1}}{\pi}X,$$
we do not need to use the bits of $1/\pi$ of rank i less than or equal to $e_x - p$;
- Since $|X| \leq 2^p - 1$, the contribution of bits $v_i, v_{i+1}, v_{i+2}, v_{i+3}, \ldots$ in the reduced argument is less than
$$2^{-t}\pi \times 2^{t+e_x-p+1} \times 2^{-i+1} \times 2^p < 2^{e_x-i+4};$$
therefore, if we want the reduced argument with an absolute error less than $2^{-\ell}$, we can ignore the bits of $1/\pi$ of rank i larger than or equal to $4 + \ell + e_x$.

Altogether, to obtain y using (10.4) with an absolute error less than $2^{-\ell}$, it is enough to compute the product $\frac{2^{t+e_x-p+1}}{\pi}X$ using a window of $3 + p + \ell$ bits of $1/\pi$: these bits range from $4+\ell+e_x+1$ to $e_x - p - 1$. The position of this window is defined by e_x. Contrary to Cody and Waite, this range reduction can be accurate on the whole floating-point range: the bits of $1/\pi$ that need to be stored are defined by the extremal values that e_x can take.

10.2.2 Bounding the relative error of range reduction

Payne and Hanek's algorithm (as well as Cody and Waite's algorithm or similar methods) allows one to easily bound the absolute error in the reduced argument. In some situations, we are more interested in bounding *relative errors* (or errors in ulps). To do this, we need to know the smallest possible value of the reduced argument. That is, we need to know what is the smallest possible value of
$$x - k \cdot C,$$
where x is a floating-point number of absolute value larger than C.

If X is the integral significand of x and e_x is its exponent, the problem becomes that of finding very good rational approximations to C, of the form

$$\frac{X \cdot 2^{e_x - p + 1}}{k},$$

which is a typical *continued fraction* problem (see Section A.1). This problem is solved using a continued fraction method due to Kahan [316] (a very similar method is also presented by Smith in [560]). See [442, 459] for details.

Program 10.1, written in Maple, computes the smallest possible value of the reduced argument and the input value for which it is attained. For instance, by entering

```
worstcaseRR(2,53,0,1023,Pi/2,400);
```

we get

```
The worst case occurs
 for x = 6381956970095103 * 2 ^ 797,
The corresponding reduced argument is:
 .46871659242546276111225828019638843989495... e-18
whose base 2 logarithm is -60.88791794
```

which means that, for binary64 (double-precision) inputs and $C = \pi/2$, a reduced argument will never be of absolute value less than $2^{-60.89}$. This has interesting consequences:

- to make sure that the relative error on the reduced argument will be less than η, it suffices to make sure that the absolute error is less than $\eta \times 2^{-60.89}$;
- since the sine and tangent of $2^{-60.89}$ are much larger than $2^{e_{\min}} = 2^{-1022}$, the sine, cosine, and tangent of a double-precision/binary64 floating-point number larger than $2^{e_{\min}}$ cannot underflow;
- since the tangent of $\pi/2 - 2^{-60.89}$ is much less than $\Omega \approx 1.798 \times 10^{308}$, the tangent of a binary64 floating-point number cannot overflow.

Table 10.1, extracted from [442], gives the input values for which the smallest reduced argument is reached, for various values of C and various formats. These values were obtained using Program 10.1.

10.2.3 More sophisticated range reduction algorithms

The most recent elementary function libraries sometimes use techniques that are more sophisticated than the Cody and Waite or the Payne and Hanek methods presented above. First, several reduction steps may be used. Many current implementations derive from P. Tang's *table-based methods* [588, 589,

```
worstcaseRR := proc(Beta,p,efirst,efinal,C,ndigits)
  local epsilonmin,powerofBoverC,e,a,Plast,r,Qlast,
        Q,P,NewQ,NewP,epsilon,
      numbermin,expmin,ell;
      epsilonmin := 12345.0 ;
      Digits := ndigits;
      powerofBoverC := Beta^(efirst-p)/C;
      for e from efirst-p+1 to efinal-p+1 do
          powerofBoverC := Beta*powerofBoverC;
          a := floor(powerofBoverC);
          Plast := a;
          r := 1/(powerofBoverC-a);
          a := floor(r);
          Qlast := 1;
          Q := a;
          P := Plast*a+1;
          while Q < Beta^p-1 do
              r := 1/(r-a);
              a := floor(r);
              NewQ := Q*a+Qlast;
              NewP := P*a+Plast;
              Qlast := Q;
              Plast := P;
              Q := NewQ;
              P := NewP
          od;
          epsilon :=
              evalf(C*abs(Plast-Qlast*powerofBoverC));
          if epsilon < epsilonmin then
              epsilonmin := epsilon; numbermin := Qlast;
              expmin := e
          fi
      od;
      printf("The worst case occurs\n for x = %a * %a ^ %a,\n",
         numbermin,Beta,expmin);
      printf("The corresponding reduced argument is:\n %a\n",
         epsilonmin);
      ell := evalf(log(epsilonmin)/log(Beta),10);
      printf("whose base %a logarithm is %a",Beta,ell)
  end:
```

Program 10.1: This Maple program, extracted from [442], gives the worst cases for range reduction, for constant C, assuming the floating-point systems has radix `Beta`*, precision* `p`*, and that we consider input arguments of exponents between* `efirst` *and* `efinal`*. It is based on Kahan's algorithm [316]. Variable* `ndigits` *indicates the radix-10 precision with which the Maple calculations must be carried out. A good rule of thumb is to choose a value slightly larger than* $(\texttt{efinal} + 1 + 2\texttt{p}) \log(\texttt{Beta})/\log(10) + \log(\texttt{efinal} - \texttt{efirst} + 1)/\log(10)$.

β	p	C	efinal	Worst case	$-\log_r(\epsilon)$
2	24	$\pi/2$	127	$16367173 \times 2^{+72}$	29.2
2	24	$\pi/4$	127	$16367173 \times 2^{+71}$	30.2
2	24	$\ln(2)$	127	8885060×2^{-11}	31.6
2	24	$\ln(10)$	127	9054133×2^{-18}	28.4
10	10	$\pi/2$	99	$8248251512 \times 10^{-6}$	11.7
10	10	$\pi/4$	99	$4124125756 \times 10^{-6}$	11.9
10	10	$\ln(10)$	99	$7908257897 \times 10^{+30}$	11.7
2	53	$\pi/2$	1023	$6381956970095103 \times 2^{+797}$	60.9
2	53	$\pi/4$	1023	$6381956970095103 \times 2^{+796}$	61.9
2	53	$\ln(2)$	1023	$5261692873635770 \times 2^{+499}$	66.8

Table 10.1: Worst cases for range reduction for various floating-point systems and reduction constants C [442]. The corresponding reduced argument is ϵ.

590, 591]. Second, the range reduction and polynomial evaluation steps are not fully separated, but somewhat interleaved. Also, very frequently, the reduced argument y is not represented by just one floating-point number: it is generally represented by two or three floating-point values, whose sum approximates y (this is called a double-word or triple-word number, see Chapter 14). Another important point is the following: although in the first implementations, the reduced arguments would lie in fairly large intervals (typically, $[-\pi/4, +\pi/4]$ for trigonometric functions, $[0, 1]$ or $[0, \ln(2)]$ for the exponential function), some current implementations use reduced arguments that lie in much smaller intervals. This is due to the fact that the degree of the smallest degree polynomial that approximates a function f in an interval I within some given error decreases with the width of I (the speed at which it decreases depends on f, but it is very significant). This is illustrated by Table 10.2, extracted from [442].

a	arctan	exp	$\ln(1+x)$
10	19	16	15
1	6	5	5
0.1	3	2	3
0.01	1	1	1

Table 10.2: Degrees of the minimax polynomial approximations that are required to approximate certain functions with error less than 10^{-5} on the interval $[0, a]$. When a becomes small, a very low degree suffices [442].

Other range reduction methods are presented in [442]. We now give two examples of reduction methods used in the CRlibm library of correctly rounded elementary functions [129, 371]. We assume binary64 (double-precision) input values.

10.2.4 Examples

10.2.4.1 An example of range reduction for the exponential function

The natural way to deal with the exponential function is to perform an additive range reduction, and later on to perform a multiplicative "reconstruction." First, from the input argument x, we find z, $|z| < \ln(2)$, so that

$$e^x = e^{E \cdot \ln(2) + z} = \left(e^{\ln(2)}\right)^E \cdot e^z = 2^E \cdot e^z, \tag{10.5}$$

where E is a signed integer.

The use of such an argument reduction implies a multiplication by the transcendental constant $\ln(2)$. This means that the reduced argument will not be exact, and the reduction error has to be taken into account when computing a bound on the overall error of the implemented function.

A reduced argument obtained by the reduction shown above is generally still too large to be suitable for a polynomial approximation of reasonable degree. A second reduction step is therefore necessary. By the use of table look-ups, the following method can be employed to implement this second step. It yields smaller reduced arguments. Let ℓ be a small positive integer (a typical value is around 12). Let w_1 and w_2 be positive integers such that $w_1 + w_2 = \ell$. Let

$$k = \left\lfloor z \cdot \frac{2^\ell}{\ln(2)} \right\rceil,$$

where $\lfloor u \rceil$ is the integer closest to u. We compute the final reduced argument y, and integers M, $i_1 < 2^{w_1}$, and $i_2 < 2^{w_2}$ such that

$$\begin{aligned} y &= z - k \cdot \frac{\ln(2)}{2^\ell}, \\ k &= 2^\ell \cdot M + 2^{w_1} \cdot i_2 + i_1, \end{aligned} \tag{10.6}$$

which gives

$$\begin{aligned} e^z &= e^y \cdot 2^{k/2^\ell} \\ &= e^y \cdot 2^{M + i_2/2^{w_2} + i_1/2^\ell}. \end{aligned} \tag{10.7}$$

The two integers w_1 and w_2 are the widths of the indices to two tables T_1 and T_2 that store

$$t_1 = 2^{i_2/2^{w_2}}$$

and

$$t_2 = 2^{i_1/2^\ell}.$$

A polynomial approximation will be used for evaluating e^y. From (10.5) and (10.7), we deduce the following "reconstruction" step:

$$e^x = 2^E \cdot e^z = 2^{M+E} \cdot t_1 \cdot t_2 \cdot e^y.$$

This argument reduction ensures that

$$|y| \leq \ln(2)/2^\ell < 2^{-\ell}.$$

This magnitude is small enough to allow for efficient polynomial approximation.

Typical values currently used are $\ell = 12$ and $w_1 = w_2 = 6$.

The subtraction in $y = z - k \cdot \mathrm{RN}\left(\ln(2)/2^\ell\right)$ can be implemented exactly, but it leads to a catastrophic cancellation that amplifies the absolute error of the potentially exact multiplication of k by the approximated $\ln(2)/2^\ell$. A careful implementation must take this into account.

10.2.4.2 An example of range reduction for the logarithm

The range reduction algorithm shown below is derived from one due to Wong and Goto [637] and discussed further in [442]. The input argument x is initially reduced in a straightforward way, using integer arithmetic, in order to get an integer E' and a binary64 floating-point number m so that

$$x = 2^{E'} \cdot m$$

with $1 \leq m < 2$.

When x is a normal number, E' is the exponent of x and m is its significand. When x is subnormal, the exponent can be adjusted so that $1 \leq m < 2$ even in this case.

This first argument reduction corresponds to the equation

$$\ln(x) = E' \cdot \ln(2) + \ln(m). \tag{10.8}$$

Using (10.8) directly would lead to a catastrophic cancellation in the case $E' = -1$ and $m \approx 2$. To overcome this problem, an adjustment is necessary. An integer E and a binary64 floating-point number y are defined as follows:

$$E = \begin{cases} E' & \text{if} \quad m \leq \sqrt{2} \\ E' + 1 & \text{if} \quad m > \sqrt{2} \end{cases}, \tag{10.9}$$

and

$$y = \begin{cases} m & \text{if} \quad m \leq \sqrt{2} \\ m/2 & \text{if} \quad m > \sqrt{2} \end{cases}. \tag{10.10}$$

The test $m \leq \sqrt{2}$ need not be performed exactly—of course, the very same test must be performed for (10.9) and (10.10). Here, the bound $\sqrt{2}$ is somewhat arbitrary. Indeed, software implementations perform this test using integer arithmetic on the high order bits of m, whereas hardware implementations have used 1.5 instead of $\sqrt{2}$ to reduce this comparison to the examination of two bits [174].

The previous reduction step can be implemented exactly, as it mainly consists in extracting the exponent and significand fields of a floating-point number, and multiplying by powers of 2. The evaluation of the logarithm becomes:

$$\ln(x) = E \cdot \ln(2) + \ln(y), \tag{10.11}$$

where y is in some reduced domain. If (10.9) and (10.10) could use the exact value $\sqrt{2}$, the reduced domain would be $[-\frac{\ln(2)}{2}, +\frac{\ln(2)}{2}]$. In practice it will be slightly larger.

This reduced domain is still large. It will not be possible to approximate $\ln(y)$ by a polynomial of reasonably small degree with very good accuracy. A second range reduction step is therefore performed, using a table with 2^ℓ entries as follows:

- the high order ℓ bits of y are used as an index i into a table that produces a very good approximation r_i of $\frac{1}{y}$;
- a new reduced argument z is defined as

$$z = y \cdot r_i - 1; \tag{10.12}$$

- the logarithm is evaluated as

$$\ln(y) = \ln(1 + z) - \ln(r_i) \tag{10.13}$$

  where the value $\ln(r_i)$ is also read from a table indexed by i.

Since y and $1/r_i$ are very close to each other, the magnitude of the final reduced argument z is now small enough (roughly speaking, $|z| < 2^{-\ell}$) to allow good approximation of $\ln(1+z)$ by a polynomial $p(z)$ of reasonably small degree. Current implementations use values of ℓ between 6 and 8, leading to tables of 64 to 256 entries.

It is important to note that (10.12) can be implemented exactly. However, this requires the reduced argument z to be represented either in a format wider than the input format (a few more bits are enough), or as an unevaluated sum of two floating-point numbers.

From Equations (10.11) to (10.13), we easily deduce the reconstruction step:

$$\ln(x) \approx E \cdot \ln(2) + p(z) - \ln(r_i).$$

This range reduction process can be repeated one more time [284, 372], or more: it can even be iterated until the argument z is small enough that the approximation $\ln(z) \approx z$ is accurate enough [185, 174].

10.3 Polynomial Approximations

After the range reduction step, our problem is reduced to the problem of evaluating a polynomial approximation to function f in a rather small interval $[a, b]$. Let us first see how such an approximation is found. The theory of the polynomial approximation to functions was developed mainly in the nineteenth and the early twentieth centuries. It makes it possible to build polynomials that better approximate a given function, in a given interval, than truncated power series.

Figure 10.1 (resp. 10.2) shows the difference between the natural logarithm ln function in $[1, 2]$ and its degree-5 Taylor at point 1.5 (resp. minimax) approximation. One immediately sees that the minimax approximation is much better than the truncated Taylor series.

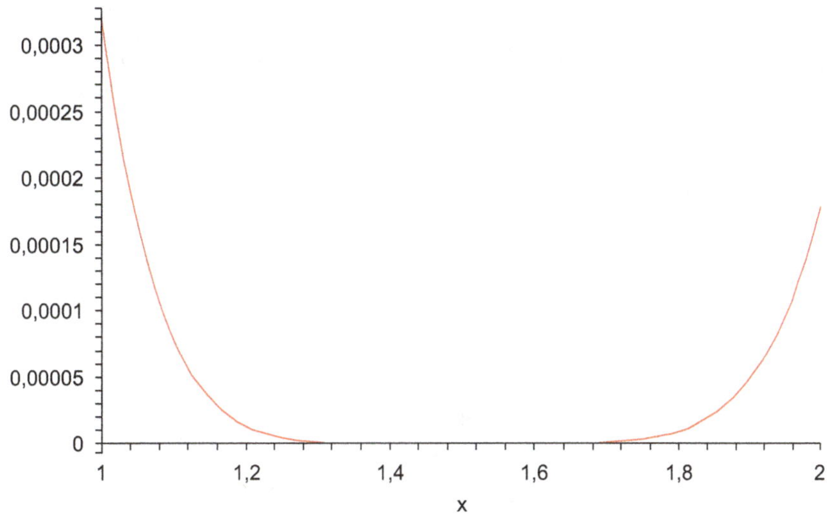

Figure 10.1: The difference between the logarithm function and its degree-5 Taylor approximation (at $x = 1.5$) in the interval $[1, 2]$.

When trying to build a good polynomial approximation p to function f, we will consider two cases, depending on whether we wish to minimize

$$||f - p||_{L^2,[a,b],w} = \int_a^b w(x)\,(f(x) - p(x))^2\,dx,$$

(this is the L^2 *case*), where w is a positive *weight* function, or

$$||f - p||_{\infty,[a,b]} = \sup_{x \in [a,b]} |f(x) - p(x)|$$

(this is the L^∞, *or minimax, case*).

We will first recall the classical methods for getting polynomial approximations. Then, we will deal with the much more challenging problem of

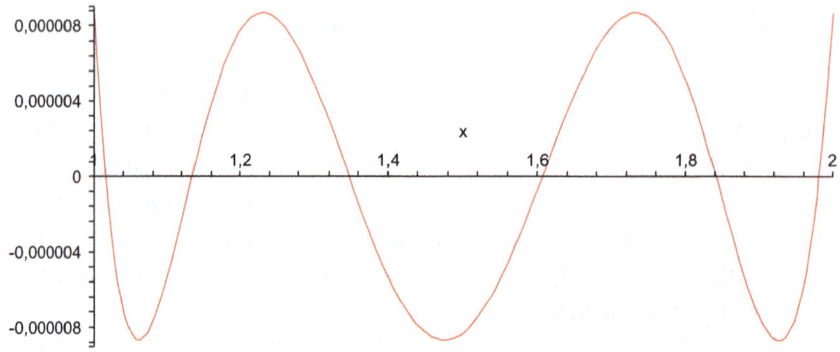

Figure 10.2: The difference between the logarithm function and its degree-5 minimax approximation in the interval $[1, 2]$. *By comparing with Figure 10.1, we see that the minimax approximation is much better than the truncated Taylor series.*

getting approximations with constraints. Examples of constraints that we wish to use are requiring all coefficients to be exactly representable in the floating-point system, or requiring some of them to be zero.

10.3.1 L^2 case

Finding an L^2 approximation is done by means of a *projection*. More precisely,

$$\langle f, g \rangle = \int_a^b w(x) f(x) g(x) dx$$

defines an inner product in the vector space of the continuous functions from $[a, b]$ to $\mathbb{R}$. We get the degree-n L^2 approximation p^* to some function f by projecting f (using the projection associated with the above-defined inner product) on the subspace $\mathcal{P}_n$ constituted by the polynomials of degree less than or equal to n. This is illustrated by Figure 10.3. Computing p^* is easily done once we have precomputed an orthogonal basis of $\mathcal{P}_n$. The basis we will use is a family $(B_k)_{0 \le k \le n}$ of polynomials, called *orthogonal polynomials*, such that:

- B_k is of degree k;
- $\langle B_i, B_j \rangle = 0$ if $i \neq j$.

Once the B_k are computed,[5] p^* is obtained as follows:

- compute, for $0 \le k \le n$,

$$a_k = \frac{\langle f, B_k \rangle}{\langle B_k, B_k \rangle},$$

[5]For the usual weight functions $w(x)$ and the interval $[-1, 1]$, the orthogonal polynomials have been known for decades, so there is no need to recompute them.

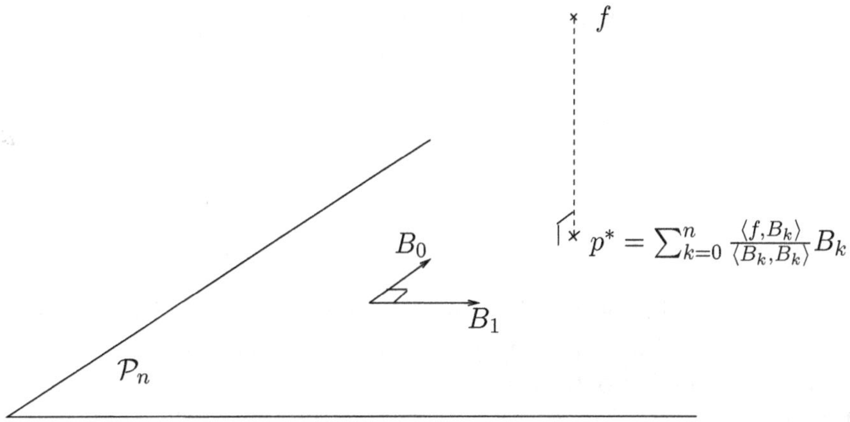

Figure 10.3: The L^2 approximation p^* is obtained by projecting f on the subspace generated by B_0, B_1, …, B_n.

- then get

$$p^* = a_0 B_0 + a_1 B_1 + \cdots + a_n B_n.$$

A very useful example of a family of orthogonal polynomials is the *Chebyshev polynomials*. The Chebyshev polynomials can be defined either by the recurrence relation

$$\begin{cases} T_0(x) &= 1, \\ T_1(x) &= x, \\ T_n(x) &= 2xT_{n-1}(x) - T_{n-2}(x); \end{cases}$$

or by

$$T_n(x) = \begin{cases} \cos\left(n \cos^{-1} x\right) \text{ for } |x| \leq 1, \\ \cosh\left(n \cosh^{-1} x\right) \text{ for } x > 1 \\ (-1)^n \cosh\left(n \cosh^{-1}(-x)\right) \text{ for } x < -1. \end{cases}$$

These are orthogonal polynomials for the inner product associated with the weight function $w(x) = 1/\sqrt{1-x^2}$ and the interval $[a, b] = [-1, 1]$. Detailed presentations of Chebyshev polynomials can be found in [59] and [510]. These polynomials play a central role in approximation theory.

10.3.2 L^∞, or minimax, case

As previously, define $\mathcal{P}_n$ as the set of the polynomials of degree less than or equal to n. The central result in minimax approximation theory is the following theorem, attributed in general to Chebyshev (according to Trefethen [606], this result was known to Chebyshev but the first proof was given by Kirchberger in his PhD dissertation [338]).

Theorem 10.1 (Chebyshev). *p^* is the minimax degree-n approximation to f on $[a, b]$ if and only if there exist at least $n + 2$ values*

$$a \leq x_0 < x_1 < x_2 < \cdots < x_{n+1} \leq b$$

such that:

$$p^*(x_i) - f(x_i) = (-1)^i \left[p^*(x_0) - f(x_0)\right] = \pm||f - p^*||_\infty.$$

Chebyshev's theorem shows that if p^* is the minimax degree-n approximation to f, then the largest approximation error is reached at least $n + 2$ times, and that the sign of the error *alternates*. This is illustrated by Figure 10.4. That property is used by an algorithm, due to Remez [246, 505], that computes the minimax degree-n approximation to a continuous function iteratively.

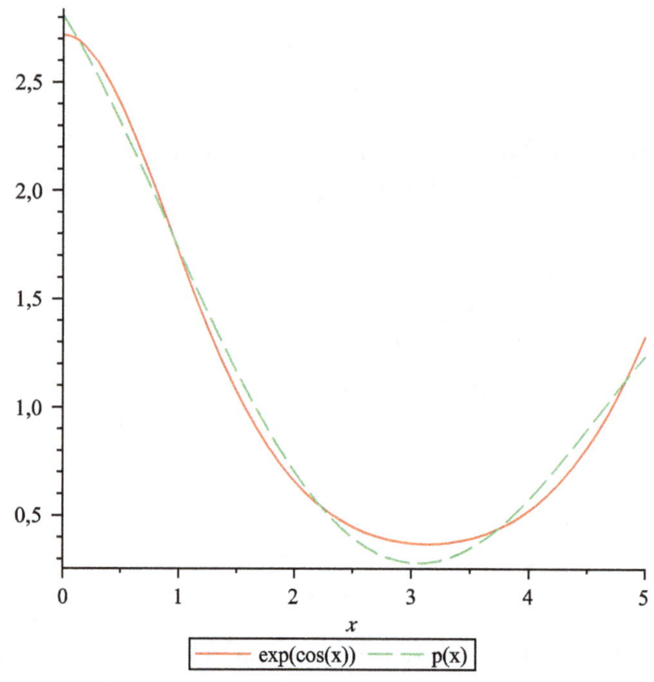

Figure 10.4: The $\exp(\cos(x))$ *function and its degree-4 minimax approximation on* $[0, 5]$*, denoted by* $p(x)$*. There are six values where the maximum approximation error is attained with alternate signs.*

There is a similar result concerning minimax *rational* approximations to functions [442].

A good implementation of the Remez algorithm that works even in slightly degenerate cases is fairly complex. Here, we just give a rough sketch of the algorithm. It consists in iteratively determining the set of points x_0, x_1, ..., x_{n+1} of Theorem 10.1. This is done as follows.

1. First, we start from an initial set of points $x_0, x_1, \ldots, x_{n+1}$ in $[a, b]$. There are some heuristics for choosing a good starting set of points. For instance
$$x_i = \frac{a+b}{2} + \frac{(b-a)}{2}\cos\left(\frac{i\pi}{n+1}\right), 0 \le i \le n+1$$
is in general a good choice.

2. We consider the linear system (whose unknowns are p_0, p_1, ..., p_n, and ϵ):
$$\begin{cases} p_0 + p_1x_0 + p_2x_0^2 + \cdots + p_nx_0^n - f(x_0) & = & +\epsilon \\ p_0 + p_1x_1 + p_2x_1^2 + \cdots + p_nx_1^n - f(x_1) & = & -\epsilon \\ p_0 + p_1x_2 + p_2x_2^2 + \cdots + p_nx_2^n - f(x_2) & = & +\epsilon \\ \cdots & & \cdots \\ p_0 + p_1x_{n+1} + p_2x_{n+1}^2 + \cdots + p_nx_{n+1}^n - f(x_{n+1}) & = & (-1)^{n+1}\epsilon. \end{cases}$$
That system is never singular (the corresponding matrix is a Vandermonde matrix [401]). It will therefore have a unique solution $(p_0, p_1, \ldots, p_n, \epsilon)$. Solving this system gives us a polynomial $P(x) = p_0 + p_1x + \cdots + p_nx^n$.

3. We compute a set of points y_i in $[a, b]$ where $P - f$ has its local extrema, and we start again (step 2), replacing the $x_i's$ by the y_i's.

It can be shown [206] that this is a convergent process and that, under certain conditions, the convergence is very fast: the rate of convergence is quadratic [615].

What we have done here (approximating a continuous function by a degree-n polynomial, i.e., a linear combination of the monomials $1, x, x^2, \ldots, x^n$) can be generalized to the approximation of a continuous function by a linear combination of continuous functions $g_0, g_1, \ldots, g_n$ that satisfy the *Haar condition*: for any subset of distinct points $x_0, x_1, \ldots, x_n$, the matrix
$$\begin{pmatrix} g_0(x_0) & g_1(x_0) & g_2(x_0) & \cdots & g_n(x_0) \\ g_0(x_1) & g_1(x_1) & g_2(x_1) & \cdots & g_n(x_1) \\ g_0(x_2) & g_1(x_2) & g_2(x_2) & \cdots & g_n(x_2) \\ \vdots & \vdots & \vdots & \cdots & \vdots \\ g_0(x_n) & g_1(x_n) & g_2(x_n) & \cdots & g_n(x_n) \end{pmatrix}$$
is nonsingular. An important special case is $g_i(x) = x^i/f(x)$: the coefficients of the best minimax approximation to the constant function 1 by a sum of the form $\sum_{i=0}^{n} p_ig_i(x)$ are the coefficients of the degree-n polynomial p that minimizes the maximum value of the relative error $|p(x) - f(x)|/|f(x)|$.

10.3.3 "Truncated" approximations

One frequently wishes to have polynomial approximations with special properties:

- approximations whose coefficients are exactly representable in floating-point arithmetic (or, for some coefficients that have much influence on the final accuracy—in general the low-order ones, that are equal to the sum of two or three floating-point numbers);
- approximations whose first coefficients coincide with those of the Taylor series (in general, to improve the behavior of the approximation near zero);
- approximations that have some coefficients equal to zero (for instance, of the form $x + x^3p(x^2)$ for the sine function: this preserves symmetry).

Efficient algorithms for generating approximations satisfying such constraints have been developed by Brisebarre, Chevillard, and Hanrot [68, 70]. These algorithms will not be detailed here. However, the following presents the Sollya package, which implements them.

Of course, one can first compute a conventional approximation (e.g., using the Remez algorithm), and then round its coefficients to satisfy the desired requirements. In most cases, the approximation obtained will be significantly worse than the best approximation that satisfies the requirements. As an illustration, consider the following example, drawn from CRlibm. The function arcsin, evaluated near 1 in the binary64 format, is reduced to the evaluation of

$$g(z) = \frac{\arcsin(1-(z+m)) - \frac{\pi}{2}}{\sqrt{2\cdot(z+m)}},$$

where $-0.110 \le z \le 0.110$ and $m \simeq 0.110$.

For correct rounding, an accuracy of $2^{-118} \approx 10^{35.5}$ is required. The Sollya tool (see below) is able to find an approximation polynomial of degree 21 whose coefficients are binary64 floating-point numbers (or, for some of them, double-binary64, see Section 14.1), with a maximum error of around 8×10^{-37}. If we round to nearest the coefficients of the degree-21 Remez polynomial, we get a maximum error of around 8×10^{-32}: this naive rounded Remez polynomial is almost $100,000$ times less accurate than the one found by Sollya.

10.3.4 In practice: using the Sollya tool to compute constrained approximations and certified error bounds

The Sollya package, designed by Lauter, Chevillard, and Joldeş [97], is available at http://sollya.gforge.inria.fr. It computes minimax polynomial approximations under constraints, using the method published in [68]. It also offers a *certified* supremum norm bound of the difference between a polynomial and a function obtained by using an algorithm published in [98, 96].

Let us give an example of a calculation that can be done using Sollya. The following commands compute an approximation of degree 5 to the exponential function on the interval $[0, 1/32]$, under the constraint that coefficients should be representable as binary32 numbers.

```
> P1 = fpminimax(exp(x),5,[|1,24...|],[0;1/32],relative);
> display=powers;
> P1;
1 + x * (1 + x * (1 * 2^(-1) + x * (2796203 * 2^(-24)
+ x * (2796033 * 2^(-26) + x * (4529809 * 2^(-29))))))
```

The "5" indicates that we are looking for a degree-5 polynomial approximation. The "`[|1,24...|]`" string indicates that we wish the degree-0 coefficient to be a 1-bit number (which is a simple way of requiring the coefficient to be 1), and that the other coefficients must be of precision 24 (i.e., they must be binary32/single precision numbers). The keyword "`relative`" indicates that we aim at minimizing the relative error.

The instruction "`display=powers`" indicates that we wish each number to be displayed as an integer multiplied by a power of two.

The following instructions return an interval that contains the maximum approximation error of the previous approximation.

```
> display=decimal;
Display mode is decimal numbers.
> supnorm(P1,exp(x),[0;1/32],relative,2^(-40));
[3.68934784163839051342476891962830035610293781682 99e-15;
3.68934784164164109816739368957045888810705176253 42e-15]
```

The string "`2^(-40)`" specifies the desired width of this interval, in other words the accuracy to which the approximation error should be computed. Sollya does not provide an *estimate* of the approximation error: we are certain that the approximation error is less than the upper bound of the interval. This is of utmost importance if we wish to be able to guarantee error bounds.

Now, assume that we wish to approximate the $\cos(x)$ function in $[-\pi/256, +\pi/256]$ by a degree-4 polynomial whose degree-0 coefficient is 1, whose coefficients of degrees 2 and 4 are binary64 floating-point numbers, and whose coefficients of degrees 1 and 3 are zero. This can be done with the following Sollya command line:

```
> P := fpminimax(cos(x),[|0,2,4|],[|1,53,53|],
                 [-0.0122719;0.0122719],relative);
```

and the polynomial returned by Sollya is

```
> 1 + x^2 * (-4503599627246363 * 2^(-53)
                    + x^2 * (1501189325363493 * 2^(-55)))
```

Again, using the command line

```
> supnorm(P,cos(x),[-0.0122719;0.0122719],relative,2^(-40));
```

we obtain an interval that encloses the approximation error bound:

```
[1.87401076984625593302749431903448641635845260093737e-16;
1.87401076984790707330416487722624988732787437790056e-16]
```

10.4 Evaluating Polynomials

10.4.1 Evaluation strategies

Once we have found a convenient polynomial approximation p, the problem is reduced to evaluating $p(x)$, where x lies in some small domain.

The polynomial $p(x) = a_0 + a_1x + a_2x^2 + \cdots + a_nx^n$ has been obtained using one of the methods presented in the previous section: its coefficients, and its variable x, are floating-point numbers.[6]

Evaluating $p(x)$ quickly and accurately on a modern pipelined floating-point unit is a nontrivial task. Of course, one should never evaluate $p(x)$ using the straightforward sequence of operations

$$a_0 + (a_1 \times x) + (a_2 \times x \times x) + (a_3 \times x \times x \times x) + \cdots + (a_n \times \underbrace{x \times x \times \cdots \times x}_{n-1 \text{ multiplications}}).$$

Evaluating a degree-n polynomial using this method would require $n(n+3)/2$ floating-point operations: $n(n+1)/2$ multiplications and n additions.

All recent evaluation schemes are combinations or generalizations of the well-known *Horner's rule* (Algorithm 5.3), and *Estrin's* method.

Horner's rule computes $p(x) = \sum_{i=0}^{n} a_ix^i$ by nested multiplications as

$$p(x) = \big(\ldots((a_nx + a_{n-1})x + a_{n-2})x + \cdots\big)x + a_0,$$

[6]When high accuracy is needed, some of the coefficients may be double-words or triple-words, *i.e.* unevaluated sums of floating-point numbers (see Section 14.1). The same holds for x, in particular when it comes from an inexact range reduction. For instance, a range reduction involving the constant π is necessarily inexact. An example will be detailed in Section 10.6.4.1.

while Estrin's method uses a binary evaluation tree implied by splitting $p(x)$ as

$$(\sum_{0\le i<h} a_{h+i}x^i)x^h + (\sum_{0\le i<h} a_i x^i),$$

where $h = (n+1)/2$ is assumed to be a power of 2.

Choosing a good evaluation scheme requires considering the target architecture (availability of an FMA instruction, depth of the pipelines, etc.) and performing some error analysis. A good rule of thumb is that Estrin's method wins for latency, and Horner's rule for throughput. Such studies have been published for the IA-64 [245, 225, 118] and ST-200 [433] architectures.

There are some generalizations of Horner's algorithm, such as Dorn's method [177]. See [442] for a recent account of the various evaluation schemes.

10.4.2 Evaluation error

To get a bound on the final error committed when evaluating a function approximated by a polynomial, we need to add two errors: the bound on the distance between the polynomial and the exact function, and the error committed when evaluating the polynomial. A bound on the error committed when evaluating a polynomial using Horner's rule in floating-point arithmetic is given by Equation (5.7). A similar bound can be derived for Estrin's method. These bounds are sometimes large overestimates. Much tighter bounds on the evaluation error can be obtained using the Gappa tool, designed by Melquiond [415] and presented in Section 13.3. The paper [147] explains how Gappa has been used to certify functions. Other examples can be found in Chapter 13 and in [442].

10.5 The Table Maker's Dilemma

What is the Table maker's dilemma (TMD) exactly? Assume we wish to implement function f, and let us call *rounding breakpoints* (*breakpoints* for short) the numbers where the value of $\circ$ changes:

- for "directed" rounding (i.e., toward $+\infty$, $-\infty$ or 0), the breakpoints are the (finite) floating-point numbers;
- for rounding to nearest, they are the "midpoints," i.e., the exact middle of two consecutive floating-point numbers (with the infinity replaced by $\beta^{e_{\max}+1}$).

The real number $f(x)$ cannot, in general, be represented with a finite number of digits. Furthermore, in most cases, the function f cannot be exactly reduced to a finite number of arithmetic operations. As we have already seen, the best we can do is to compute an *approximation* $\hat{f}(x)$ to $f(x)$, possibly using an intermediate precision higher than the target precision. It is possible, as per Section 10.4.2, to compute a (preferably tight) bound on the evaluation error $|\hat{f}(x) - f(x)|$. This defines a confidence interval $I_{f(x)}$ around $\hat{f}(x)$. The exact value $f(x)$ lies somewhere within this interval.

The TMD occurs when, for a given x, the interval $I_{f(x)}$ contains a breakpoint, i.e., when $\hat{f}(x)$ is so close to a breakpoint that, taking into account the error of the approximation, it is impossible to decide whether $f(x)$ is above or below the breakpoint. In such a case, one does not know which result should be returned: it could be the floating-point number right above the breakpoint or the one right below it. Figures 10.5 and 10.6 illustrate the two possible situations: when we can return a correctly rounded result, and when we cannot.

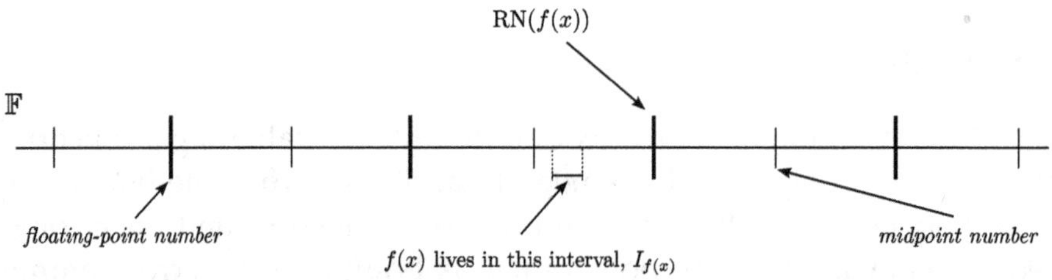

Figure 10.5: A situation where we can return $f(x)$ correctly rounded, assuming round-to-nearest is the rounding function. The interval $I_{f(x)}$ does not contain a breakpoint, so all elements of the interval round to the same value.

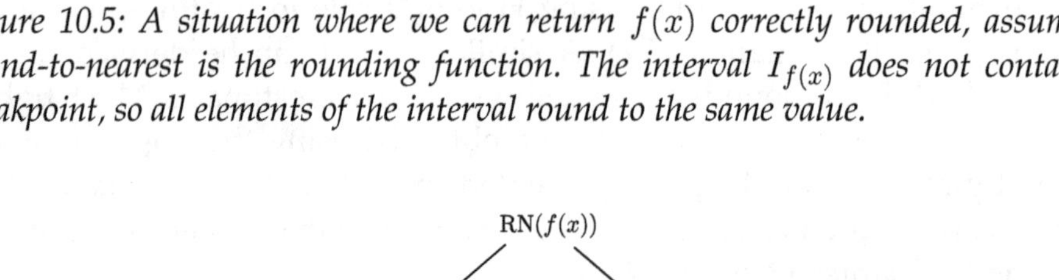

Figure 10.6: A situation where we cannot return $f(x)$ correctly rounded, assuming round-to-nearest is the rounding function. The interval $I_{f(x)}$ contains a breakpoint: hence for some elements of the interval, one should return the floating-point number below the breakpoint, and for some other elements, one should return the floating-point number above the breakpoint.

When this happens, a possible solution is to redo the computation again and again, each time with a higher accuracy (hence a smaller interval $I_{f(x)}$), until the approximation is accurate enough to determine whether $f(x)$ is above or below the breakpoint. This is Ziv's "multilevel strategy" [646]. We will call each of these individual evaluations of f with a given precision a *Ziv step*. That process does not terminate if $f(x)$ is exactly equal to a breakpoint: if we really aim at providing a correctly rounded implementation to function f, it is important to know beforehand what are the floating-point numbers x such that $f(x)$ is a breakpoint. Ziv's strategy was implemented in the `libultim` library.[7]

If we know that there are no floating-point numbers x such that $f(x)$ is a breakpoint (or if we know what are the input numbers whose image is a breakpoint, and only consider the other numbers), it is important to know how close $f(x)$ can be to a breakpoint: this will tell us what the accuracy of the approximation $\hat{f}$ to f must be to be certain that rounding $\hat{f}$ is equivalent to rounding f.

As we will see in this section, for the most common functions and a given, not too large, floating-point format (with our current knowledge and computing power, solutions for the binary128 and decimal128 formats seem out of reach), one can find satisfactory solutions to this problem. This is why the IEEE 754-2008 standard recommends (yet does not mandate), correct rounding of some functions (this is a novelty with respect to IEEE 754-1985). These functions are:

$$\begin{gathered}
e^x, e^x-1, 2^x, 2^x-1, 10^x, 10^x-1,\\
\ln(x), \log_2(x), \log_{10}(x), \ln(1+x), \log_2(1+x), \log_{10}(1+x),\\
\sqrt{x^2+y^2}, 1/\sqrt{x}, (1+x)^n, x^n, x^{1/n} (n \text{ is an integer}), x^y,\\
\sin(\pi x), \cos(\pi x), \arctan(x)/\pi, \arctan(y/x)/\pi,\\
\sin(x), \cos(x), \tan(x), \arcsin(x), \arccos(x), \arctan(x), \arctan(y/x),\\
\sinh(x), \cosh(x), \tanh(x), \sinh^{-1}(x), \cosh^{-1}(x), \tanh^{-1}(x).
\end{gathered}$$

Solving the TMD for a function f, a given floating-point format and a given rounding function consists either in computing its *hardest to round* points, or, if this is not possible, in finding an upper bound to its *hardness to round*. Let us now define these notions.

Definition 10.1 (Hardness to round). *The* **hardness to round** *for function f in interval $[a, b]$ is the smallest integer m such that for all floating-point numbers $x \in [a, b]$, either $f(x)$ is a breakpoint or the infinitely precise significand of $f(x)$ is not within 2^{-m} of a breakpoint.*

Definition 10.2 (Hardest to round point). *The floating-point number x is a hardest-to-round point for function f in interval $[a, b]$ if there is an integer m such that*

[7] `libultim` was released by IBM. An updated version is now part of the GNU `glibc` and available under the GNU General Public License.

- *the infinitely precise significand of* $f(x)$ *is within* 2^{-m+1} *of a breakpoint;*
- *for all floating-point numbers* y *in* $[a, b]$, *if* $f(y)$ *is not a breakpoint then* $f(y)$ *is not within* 2^{-m} *of a breakpoint.*

10.5.1 When there is no need to solve the TMD

For many functions, if the input argument is small enough, reasoning based on the Taylor expansion of the function being considered is enough to return correctly rounded results without having to actually find worst cases or compute the hardness to round.

For instance, in a radix-β, precision-p system, if one wishes to evaluate the exponential of x with $0 \leq x < \beta^{-p}$ (which implies $x \leq \beta^{-p} - \beta^{-2p}$), then

$$e^x \leq 1 + (\beta^{-p} - \beta^{-2p}) + \frac{\beta^{-2p}}{2} + \frac{\beta^{-3p}}{6} + \frac{\beta^{-4p}}{24} + \cdots,$$

which implies

$$e^x < 1 + \beta^{-p} = 1 + \frac{1}{\beta}\,\mathrm{ulp}(1) \leq 1 + \frac{1}{2}\,\mathrm{ulp}(1).$$

Therefore, in such a case, one can safely return 1 as the correctly rounded to the nearest value of e^x.

Similarly, in round-to-nearest mode and binary64 format,

- if $|x| \leq \mathrm{RN}\left(3^{1/3}\right) \cdot 2^{-26} = 1.4422\cdots \times 2^{-26}$, then $\sin(x)$ can be replaced by x;
- if $\mathrm{RN}\left(3^{1/3}\right) \cdot 2^{-26} < x \leq 2^{-25}$, then $\sin(x)$ can be replaced by $x^- = x - 2^{-78}$; the case $-2^{-25} \leq x < -\mathrm{RN}\left(3^{1/3}\right) \cdot 2^{-26}$ is obviously symmetrical.

Tables 10.3, 10.4, and 10.5 give results derived from similar reasoning for some functions, assuming the binary32 format (for Table 10.3) and the binary64 format (for Tables 10.4 and 10.5) of the IEEE 754 standard. For some functions, similar reasoning also applies to very large input arguments. For instance, if x is large enough, $\arctan(x)$ can be replaced by $\pi/2$ adequately rounded.

10.5.2 On breakpoints

Finding the breakpoints of a given function is in general a difficult task. The problem will be very differently handled depending on whether f is an *algebraic* or a *transcendental* function. Function f is algebraic if there exists a nonzero bivariate polynomial P with integer coefficients such that for all x in the domain of f we have $P(x, f(x)) = 0$. Examples of algebraic functions are $x^2 + 4$, $\sqrt{x}$, and $x^{5/4}$. A function is transcendental if it is not algebraic. Examples of transcendental functions are $\cos(x)$, 3^x, and $\sin(3x) + \exp(5x)$.

This expression	can be replaced by	when
$\exp(x)$	1	$-2^{-25} \leq x < 2^{-24}$
$\exp(x) - 1$	x	$\lvert x\rvert \leq \epsilon_1 = \mathrm{RN}(\sqrt{2}) \cdot 2^{-24}$ $\epsilon_1 =$ `0x1.6a09e6p-24`
$\mathrm{log1p}(x) = \ln(1+x)$	x	$-\epsilon_1 < x \leq \epsilon_1$
2^x	1	$-\epsilon_2/2 \leq x < \epsilon_2 = \mathrm{RN}(1/\ln 2) \cdot 2^{-24}$ $\epsilon_2 =$ `0x1.715476p-24`
10^x	1	$-\epsilon_3/2 < x < \epsilon_3 = \mathrm{RN}(1/\ln 10) \cdot 2^{-24}$ $\epsilon_3 =$ `0x1.bcb7b2p-26`
$\sin(x), \sinh(x), \sinh^{-1}(x)$	x	$\lvert x\rvert \leq \epsilon_4 = \mathrm{RN}((3/2)^{1/3}) \cdot 2^{-11}$ $\epsilon_4 =$ `0x1.250bfep-11`
$\arcsin(x)$	x	$\lvert x\rvert < \epsilon_4$
$\cos(x)$	1	$\lvert x\rvert \leq 2^{-12}$
$\cosh(x)$	1	$\lvert x\rvert \leq \epsilon_5 = \mathrm{RN}(\sqrt{2}) \cdot 2^{-12}$ $\epsilon_5 =$ `0x1.6a09e6p-12`
$\tan(x), \tanh(x), \arctan(x)$	x	$\lvert x\rvert \leq \epsilon_6 = \mathrm{RN}(3^{1/3}) \cdot 2^{-12}$ $\epsilon_6 =$ `0x1.713744p-12`
$\tanh^{-1}(x)$	x	$\lvert x\rvert < \epsilon_6$

Table 10.3: Some results for small values in the binary32 (single-precision) format, assuming ***rounding to nearest.***

10.5.2.1 Breakpoints of some algebraic functions

It has been known for years that the result of a reciprocal, division or square root of binary floating-point inputs cannot be a midpoint in the same format (except possibly in the subnormal domain for division) [406]. Jeannerod et al. [297] investigated some simple algebraic functions in radices 2 and 10. They gave for instance the following properties, which hold as long as the value of the function is not in the subnormal domain:

- if x is a binary or decimal floating-point number, then $\sqrt{x}$ and $x/\sqrt{x^2+y^2}$ cannot be a midpoint in the same format;
- if x is a binary floating-point number, then $1/\sqrt{x}$ cannot be a midpoint in the same format, and it can be a floating-point number in the same format only if x is 2^{2k} for some k;

This expression	can be replaced by	when
$\exp(x)$	1	$-2^{-54} \leq x < 2^{-53}$
$\exp(x) - 1$	x	$\lvert x\rvert < \epsilon_1 = \mathrm{RN}(\sqrt{2}) \cdot 2^{-53}$ $\epsilon_1 =$ `0x1.6a09e667f3bcdp-53`
	x^+	$\epsilon_1 \leq x < \epsilon_2 = \mathrm{RN}(\sqrt{3}) \cdot 2^{-52}$ $\epsilon_2 =$ `0x1.bb67ae8584caap-52`
$\mathrm{log1p}(x) = \ln(1+x)$	x	$\lvert x\rvert < \epsilon_1$
2^x	1	$-\epsilon_3/2 \leq x < \epsilon_3 = \mathrm{RN}(1/\ln 2) \cdot 2^{-53}$ $\epsilon_3 =$ `0x1.71547652b82fep-53`
10^x	1	$-\epsilon_4/2 \leq x < \epsilon_4 = \mathrm{RN}(1/\ln 10) \cdot 2^{-53}$ $\epsilon_4 =$ `0x1.bcb7b1526e50ep-55`
$\sin(x), \sinh(x), \sinh^{-1}(x)$	x	$\lvert x\rvert \leq \epsilon_5 = \mathrm{RN}(3^{1/3}) \cdot 2^{-26}$ $\epsilon_5 =$ `0x1.7137449123ef6p-26`
$\sin(x), \sinh^{-1}(x)$	$x^- = x - 2^{-78}$	$\epsilon_5 < x \leq 2^{-25}$
$\sinh(x)$	$x^+ = x + 2^{-78}$	$\epsilon_5 < x < 2^{-25}$
$\arcsin(x)$	x	$\lvert x\rvert < \epsilon_5$
	$x^+ = x + 2^{-78}$	$\epsilon_5 \leq x < 2^{-25}$
$\cos(x)$	1	$\lvert x\rvert < \epsilon_6 = \mathrm{RN}(\sqrt{2}) \cdot 2^{-27}$ $\epsilon_6 =$ `0x1.6a09e667f3bcdp-27`
	$1^- = 1 - 2^{-53}$	$\epsilon_6 \leq \lvert x\rvert \leq \epsilon_7 = \mathrm{RN}(\sqrt{3/2}) \cdot 2^{-26}$ $\epsilon_7 =$ `0x1.3988e1409212ep-26`
$\cosh(x)$	1	$\lvert x\rvert < 2^{-26}$
	$1^+ = 1 + 2^{-52}$	$2^{-26} \leq \lvert x\rvert \leq \epsilon_8 = \mathrm{RN}(\sqrt{3}) \cdot 2^{-26}$ $\epsilon_8 =$ `0x1.bb67ae8584caap-26`
	$1^{++} = 1 + 2^{-51}$	$\epsilon_8 < \lvert x\rvert < \epsilon_9 = \mathrm{RN}(\sqrt{5/4}) \cdot 2^{-25}$ $\epsilon_9 =$ `0x1.1e3779b97f4a8p-25`
$\tan(x), \tanh^{-1}(x)$	x	$\lvert x\rvert < \epsilon_{10} = \mathrm{RN}((3/2)^{1/3}) \cdot 2^{-26}$ $\epsilon_{10} =$ `0x1.250bfe1b082f5p-26`
	$x^+ = x + 2^{-78}$	$\epsilon_{10} \leq x \leq 1.650 \cdot 2^{-26}$
$\tanh(x), \arctan(x)$	x	$\lvert x\rvert \leq \epsilon_{10}$
	$x^- = x - 2^{-78}$	$\epsilon_{10} < x \leq 1.650 \cdot 2^{-26}$

Table 10.4: Some results for small values in the binary64 (double-precision) format, assuming ***rounding to nearest*** *(some of these results are extracted from [442]).*

This expression	can be replaced by	when
$\exp(x)$	1	$0 \leq x < 2^{-52}$
	$1^{-} = 1 - 2^{-53}$	$-2^{-53} \leq x < 0$
$\exp(x) - 1$	x	$\lvert x \rvert < \epsilon_1 = \mathrm{RN}(\sqrt{2}) \cdot 2^{-52}$ $\epsilon_1 =$ `0x1.6a09e667f3bcdp-52`
$\ln(1 + x)$	x^{-}	$-2^{-52} < x \leq \epsilon_1$ and $x \neq 0$
2^x	1	$0 \leq x < \epsilon_2 = \mathrm{RN}(1/\ln 2) \cdot 2^{-52}$ $\epsilon_2 =$ `0x1.71547652b82fep-52`
	$1^{-} = 1 - 2^{-53}$	$-\epsilon_2/2 \leq x < 0$
10^x	1	$0 \leq x < \epsilon_3 = \mathrm{RN}(1/\ln 10) \cdot 2^{-52}$ $\epsilon_3 =$ `0x1.bcb7b1526e50ep-54`
	$1^{-} = 1 - 2^{-53}$	$-\epsilon_3/2 \leq x < 0$
$\sin(x), \sinh^{-1}(x)$	x^{-}	$0 < x \leq \epsilon_4 = \mathrm{RN}(6^{1/3}) \cdot 2^{-26}$ $\epsilon_4 =$ `0x1.d12ed0af1a27fp-26`
	x^{--}	$\epsilon_4 < x \leq 2^{-25}$
	x	$-\epsilon_4 \leq x \leq 0$
	x^{+}	$-2^{-25} \leq x < -\epsilon_4$
$\arcsin(x), \sinh(x)$	x	$0 \leq x < \epsilon_4$
	$x^{+} = x + 2^{-78}$	$\epsilon_4 \leq x < 2^{-25}$
	x^{-}	$-\epsilon_4 < x < 0$
	$x^{--} = x - 2^{-77}$	$-2^{-25} < x \leq -\epsilon_4$
$\cos(x)$	$1^{-} = 1 - 2^{-53}$	$0 < \lvert x \rvert \leq 2^{-26}$
$\cosh(x)$	1	$\lvert x \rvert < \epsilon_5 = \mathrm{RN}(\sqrt{2}) \cdot 2^{-26}$ $\epsilon_5 =$ `0x1.6a09e667f3bcdp-26`
	$1^{+} = 1 + 2^{-52}$	$\epsilon_5 \leq \lvert x \rvert < 2^{-25}$
$\tan(x), \tanh^{-1}(x)$	x	$0 \leq x < \epsilon_6 = \mathrm{RN}(3^{1/3}) \cdot 2^{-26}$ $\epsilon_6 =$ `0x1.7137449123ef6p-26`
	x^{-}	$-\epsilon_6 < x < 0$
$\tanh(x), \arctan(x)$	x^{-}	$0 < x \leq \epsilon_6$
	x	$-\epsilon_6 \leq x \leq 0$

Table 10.5: Some results for small values in the binary64 format, assuming ***rounding toward*** $-\infty$ *(some of these results are extracted from [442]).*

- in decimal64 arithmetic, the only midpoints of function $1/\sqrt{x}$ are such that the integral significand of x is either 7036874417766400 (with an odd exponent), or 2814749767106560 (with an even exponent).

10.5.2.2 Breakpoints of transcendental functions

In 1882, Lindemann proved a famous theorem[8] stating that e^z is transcendental for every nonzero algebraic complex number z. Since floating-point numbers and breakpoints are algebraic numbers, we easily deduce that, if we except straightforward cases such as $\ln(1) = 0$ or $\cos(0) = 1$, the (base e) exponential or logarithm, and the sine, cosine, tangent, arctangent, arcsine, and arccosine of a floating-point number cannot be a breakpoint.

For more complicated functions (such as the Gamma and Bessel functions, etc.) we know (almost) nothing.

10.5.3 Finding the hardest-to-round points

10.5.3.1 Hardest-to-round points of transcendental functions in binary32 arithmetic

In binary32 arithmetic, the total number of possible input floating-point numbers is not so large: it is possible to find the hardest to round points by means of an exhaustive search. For instance, around one day of calculation and a very simple loop written in Maple suffices to show that a hardest-to-round point for function $\log_2(x)$ in the full binary32 domain for round-to-nearest is $x = 10517177/2^{22}$, whose base-2 logarithm is

$$\underbrace{1.01010011100001001011110}_{24 \text{ bits}}0\underbrace{11111111111111111111111111}_{26 \text{ ones}}010\cdots_2.$$

Similarly, the hardest-to-round point for function $\sin(x)$ in the full binary32 domain for directed roundings is[9]

$x = 77291789194529019661184401408 = 16367173 \times 2^{72}$, whose sine is

$$0.\underbrace{111111111111111111111111}_{24 \text{ bits}}\underbrace{11111111111111111111111111111111111}_{35 \text{ ones}}001_2.$$

In Tables 10.6 to 10.17, we give the hardest-to-round points in the full binary32 domain for functions e^x, $e^x - 1$, 2^x, 10^x, $\ln(x)$, $\ln(1+x)$, $\log_2(x)$, $\log_{10}(x)$, $\sinh(x)$, $\cosh(x)$, $\tanh(x)$, $\sin(x)$, $\cos(x)$, $\tan(x)$, $\mathrm{asin}(x)$, $\mathrm{acos}(x)$, $\mathrm{atan}(x)$, $\mathrm{erf}(x)$, $\mathrm{erfc}(x)$, $\Gamma(x)$, $\ln(|\Gamma(x)|)$, and the Bessel functions $J_0(x)$, $J_1(x)$,

---

[8] It was that result that showed the transcendence of π, which implies that squaring the circle is impossible.

[9] We already met this number (see Table 10.1). It is very close to a multiple of $\pi/2$, which makes it a worst case for range reduction for the trigonometric functions.

$Y_0(x)$, and $Y_1(x)$. These tables have been obtained with an exhaustive search by using the GNU MPFR library [204], presented in Chapter 14.

In these tables, each hardest-to-round point has the following format: the argument (binary32 number) and the truncated result (i.e., the result rounded toward zero in the binary32 format), both in C99's hexadecimal format (described below), and then the bits after the significand of the truncated result. The first bit after the significand is the rounding bit. Then, as these are hard-to-round points, the rounding bit is followed by a long run of zeros or ones; the value in exponent (superscript) is the length of this run. Finally, we give the next four bits.

Each datum in hexadecimal format in these tables consists of:

- a significand, written as a possible "−" if the number is negative, followed by a "1" or "0" (for subnormals), followed by a 6-hexadecimal-digit number. The last digit is necessarily even, since in fact it corresponds to a 3-bit number followed by a zero (hence the $1+4\times 5+3 = 24$ bits of precision). The digits are denoted `0`, `1`, `2`, `3`, `4`, `5`, `6`, `7`, `8`, `9`, `A`, `B`, `C`, `D`, `E`, and `F`;
- the letter `P` (as a separator);
- a binary exponent E, written in decimal, i.e., the significand is to be multiplied by 2^E to get the value.

10.5.3.2 Hardest-to-round points of transcendental functions in binary64 arithmetic

There are 2^{64} possible binary64 numbers: an exhaustive search of the hardest-to-round points, although not totally impossible, would be extremely costly (tens of thousands of CPU years, with current technology). Two algorithms (or, rather, two families of algorithms) have been suggested for accelerating this search:

- the L-algorithm, designed by Lefèvre [376, 378], uses piecewise linear approximations to function f and massive parallelism. A fast implementation of the L-Algorithm on GPUs is presented in [203].
- the SLZ algorithm, introduced by Stehlé, Lefèvre, and Zimmermann [570, 568, 571], consists of approximating the function being considered by piecewise degree-d polynomials and using a technique due to Coppersmith [113, 114] for finding the hardest-to-round points of each polynomial.

We have run the L-Algorithm to find hardest-to-round points in binary64 for the most common functions and domains. The results obtained so far, first partly published in [380], are given in Tables 10.18 to 10.24. In these

Function	Domain	Argument	Truncated result	Trailing bits
exp	$(-2^7, -2^{-50}]$	-1.000002P-22	1.FFFFF8P-1	$0\,0^{42}\,1010\cdots$
	$[2^{-49}, 2^7)$	1.FFFFFEP-24	1.000000P0	$1\,1^{46}\,0101\cdots$
		1.FFFFFCP-23	1.000002P0	$1\,1^{43}\,0101\cdots$
	$(-2^7, -2^{-16}]$	-1.659EC8P-9	1.FE9ADCP-1	$1\,1^{27}\,0010\cdots$
		-1.D2259AP3	1.FA6634P-22	$1\,0^{27}\,1010\cdots$
	$[2^{-16}, 2^7)$	1.627A9EP-10	1.0058AEP0	$0\,0^{27}\,1000\cdots$
$\exp(x)-1$	$(-2^5, -2^{-48}]$	-1.800006P-20	-1.7FFFF4P-20	$0\,0^{38}\,1000\cdots$
	$[2^{-48}, 2^7)$	1.7FFFFAP-20	1.80000AP-20	$1\,1^{38}\,0110\cdots$
	$(-2^5, -2^{-15}]$	-1.835330P-14	-1.834E9CP-14	$0\,0^{26}\,1000\cdots$
		-1.C93542P-13	-1.C9287EP-13	$1\,1^{29}\,0010\cdots$
		-1.F676D8P-9	-1.F5809AP-9	$1\,0^{26}\,1100\cdots$
		-1.2E3B12P1	-1.CFB6B8P-1	$0\,0^{26}\,1110\cdots$
	$[2^{-15}, 2^7)$	1.4A9CF4P-11	1.4AB7A4P-11	$1\,0^{27}\,1000\cdots$
		1.84A5BAP-4	1.97AED6P-4	$0\,1^{28}\,0001\cdots$

Table 10.6: Hardest-to-round points for functions e^x and $e^x - 1$ in binary32 arithmetic. Exponentials of numbers less than -2^7 are less than half the smallest positive machine number. Exponentials of numbers larger than 2^7 are overflows. Input numbers of absolute value less than 2^{-15} are easily handled using results such as those presented in Table 10.3.

Function	Domain	Argument	Truncated result	Trailing bits
2^x	$(-2^7, -2^{-49}]$	-1.5A3F34P-21	1.FFFFF0P-1	$0\,1^{33}\,0111\cdots$
		-1.48EF5EP-18	1.FFFF8EP-1	$0\,0^{27}\,1110\cdots$
		-1.A7A04CP-14	1.FFF6D2P-1	$0\,1^{28}\,0100\cdots$
		-1.D85680P-10	1.FF5C66P-1	$1\,0^{28}\,1111\cdots$
		-1.E7526EP-6	1.F58D60P-1	$1\,0^{31}\,1000\cdots$
	$[2^{-48}, 2^7)$	1.0F0AF8P-11	1.00177CP0	$0\,1^{27}\,0101\cdots$
		1.853A6EP-9	1.008708P0	$1\,0^{29}\,1101\cdots$
10^x	$(-2^6, -2^{-51}]$	-1.4D89C6P-25	1.FFFFFCP-1	$0\,1^{27}\,0110\cdots$
		-1.898CB8P-10	1.FE3BB0P-1	$0\,1^{28}\,0111\cdots$
		-1.E5B5CCP-5	1.BEA982P-1	$1\,1^{27}\,0100\cdots$
		-1.46110CP-1	1.D89618P-3	$0\,1^{27}\,0010\cdots$
	$[2^{-50}, 2^6)$	1.E0B45AP-7	1.08CB88P0	$1\,1^{27}\,0101\cdots$
		1.29B2ACP-5	1.1657C2P0	$1\,1^{28}\,0100\cdots$
		1.FAFEC0P3	1.8C880AP52	$1\,0^{28}\,1110\cdots$

Table 10.7: Hardest-to-round points for functions 2^x and 10^x in binary32 arithmetic. Tiny input numbers are easily handled using results such as those presented in Table 10.3.

Function	Domain	Argument	Truncated result	Trailing bits
ln	$[2^{-149}, 2^{128})$	1.108A5AP-66	-1.6D7B16P5	$1\,1^{32}\,0000\cdots$
		1.BACB4AP25	1.1E0694P4	$1\,0^{32}\,1011\cdots$
		1.C09D7CP27	1.346A56P4	$1\,0^{31}\,1001\cdots$
		1.B121A6P76	1.A9A3F0P5	$1\,0^{33}\,1111\cdots$
		1.2FE614P117	1.451436P6	$0\,0^{31}\,1001\cdots$
$\ln(1+x)$	$(-2^{0}, -2^{-48}]$	-1.7FFFFAP-21	-1.800002P-21	$0\,1^{41}\,0110\cdots$
		-1.7FFFF4P-20	-1.800004P-20	$1\,1^{38}\,0110\cdots$
		-1.7FFFE8P-19	-1.80000AP-19	$1\,1^{35}\,0110\cdots$
		-1.7FFFD0P-18	-1.800016P-18	$1\,1^{32}\,0110\cdots$
		-1.1FFFCAP-17	-1.20001AP-17	$0\,1^{31}\,0100\cdots$
		-1.1D9188P-9	-1.1DE148P-9	$1\,0^{30}\,1111\cdots$
	$[2^{-48}, 2^{128})$	1.800006P-21	1.7FFFFCP-21	$1\,0^{41}\,1000\cdots$
		1.80000CP-20	1.7FFFFAP-20	$0\,0^{38}\,1000\cdots$
		1.800018P-19	1.7FFFF4P-19	$0\,0^{35}\,1000\cdots$
		1.800030P-18	1.7FFFE8P-18	$0\,0^{32}\,1000\cdots$
		1.200036P-17	1.1FFFE4P-17	$1\,0^{31}\,1011\cdots$
		1.FB035AP-2	1.9BDDC2P-2	$0\,1^{32}\,0000\cdots$
		1.0F1FD6P3	1.1FCBCEP1	$0\,1^{30}\,0100\cdots$
		1.30BF04P43	1.DFAC90P4	$0\,1^{30}\,0110\cdots$
		1.B121A6P76	1.A9A3F0P5	$1\,0^{33}\,1111\cdots$
		1.5190C0P78	1.B2BC8AP5	$1\,0^{30}\,1110\cdots$
		1.2FE614P117	1.451436P6	$0\,0^{31}\,1001\cdots$

Table 10.8: Hardest-to-round points for functions $\ln(x)$ and $\ln(1+x)$ in binary32 arithmetic. For function $\ln(1+x)$, tiny input numbers are easily handled using results such as those presented in Table 10.3.

Function	Domain	Argument	Truncated result	Trailing bits
$\log_2$	$[2^{-67}, 2^{68})$	1.229520P-67	-1.0B44CEP6	$1\,0^{26}\,1111\ldots$
		1.229520P-66	-1.0744CEP6	$1\,0^{26}\,1111\ldots$
		1.229520P-65	-1.0344CEP6	$1\,0^{26}\,1111\ldots$
		1.40F572P-2	-1.AC7B42P0	$1\,0^{26}\,1010\ldots$
		1.40F572P1	1.5384BCP0	$0\,1^{26}\,0101\ldots$
		1.229520P64	1.00BB30P6	$0\,1^{26}\,0000\ldots$
		1.229520P65	1.04BB30P6	$0\,1^{26}\,0000\ldots$
		1.229520P66	1.08BB30P6	$0\,1^{26}\,0000\ldots$
		1.229520P67	1.0CBB30P6	$0\,1^{26}\,0000\ldots$
$\log_{10}$	$[2^{-149}, 2^{128})$	1.5D46ACP-110	-1.07D3B2P5	$1\,1^{29}\,0101\ldots$
		1.FDDCF4P-98	-1.D33A44P4	$1\,0^{29}\,1101\ldots$
		1.84DA26P-57	-1.0FA278P4	$0\,0^{30}\,1000\ldots$
		1.7BDB90P12	1.E4519EP1	$1\,0^{29}\,1010\ldots$
		1.0ACFC8P67	1.42FDD8P4	$0\,1^{31}\,0100\ldots$
		1.4D83BAP70	1.52FDD8P4	$0\,1^{31}\,0100\ldots$
		1.AD74BCP115	1.16BEBAP5	$0\,0^{33}\,1110\ldots$

Table 10.9: Hardest-to-round points for functions $\log_2(x)$ and $\log_{10}(x)$ in binary32 arithmetic. The hardest-to-round points for $\log_2(x)$ with $x \in [2^{65}, 2^{128})$ can be deduced from those with $x \in [2^{64}, 2^{65})$; only those with $x < 2^{68}$ are given in the table.

Function	Domain	Argument	Truncated result	Trailing bits
sinh	$[2^{-23}, 2^{7})$	1.250BFEP-11	1.250BFEP-11	$0\,1^{30}\,0000\cdots$
		1.6D543EP-9	1.6D545CP-9	$1\,0^{26}\,1101\cdots$
		1.AC6FE0P-9	1.AC7010P-9	$1\,1^{26}\,0000\cdots$
		1.ECFEB6P-4	1.EE2FA4P-4	$1\,1^{26}\,0101\cdots$
cosh	$[2^{-24}, 2^{7})$	1.DEEEA0P-11	1.000006P0	$1\,0^{29}\,1100\cdots$
		1.0F876CP-10	1.000008P0	$0\,1^{26}\,0110\cdots$
		1.1E3776P-9	1.000028P0	$0\,0^{26}\,1001\cdots$
		1.7FFFDCP-8	1.00011EP0	$1\,1^{28}\,0010\cdots$
		1.AAEEB8P-8	1.000162P0	$1\,1^{26}\,0111\cdots$
		1.C12A50P-5	1.006288P0	$1\,0^{28}\,1010\cdots$
tanh	$[2^{-23}, 2^{4})$	1.F51A66P-11	1.F51A5CP-11	$0\,0^{26}\,1101\cdots$
		1.B1C146P1	1.FED5B2P-1	$1\,1^{26}\,0111\cdots$

Table 10.10: Hardest-to-round points for functions $\sinh(x)$, $\cosh(x)$ *and* $\tanh(x)$ *in binary32 arithmetic. Tiny input numbers are easily handled using results such as those presented in Table 10.3.*

Function	Domain	Argument	Truncated result	Trailing bits
asinh	$[2^{-23}, 2^{128})$	1.BACB4AP24	1.1E0694P4	$1\,0^{30}\,1000\cdots$
		1.C09D7CP26	1.346A56P4	$1\,0^{31}\,1010\cdots$
		1.B121A6P75	1.A9A3F0P5	$1\,0^{33}\,1111\cdots$
		1.5190C0P77	1.B2BC8AP5	$1\,0^{30}\,1110\cdots$
		1.2FE614P116	1.451436P6	$0\,0^{31}\,1001\cdots$
acosh	$[2^{0}, 2^{128})$	1.BACB4AP24	1.1E0694P4	$0\,1^{32}\,0101\cdots$
		1.C09D7CP26	1.346A56P4	$1\,0^{31}\,1001\cdots$
		1.B121A6P75	1.A9A3F0P5	$1\,0^{33}\,1111\cdots$
		1.5190C0P77	1.B2BC8AP5	$1\,0^{30}\,1110\cdots$
		1.2FE614P116	1.451436P6	$0\,0^{31}\,1001\cdots$
atanh	$[2^{-23}, 2^{0})$	1.713744P-12	1.713744P-12	$1\,0^{27}\,1000\cdots$
		1.E3CF42P-11	1.E3CF4AP-11	$1\,0^{27}\,1000\cdots$
		1.F51A5CP-11	1.F51A64P-11	$1\,1^{26}\,0010\cdots$
		1.AC6FD6P-10	1.AC6FEEP-10	$0\,1^{27}\,0000\cdots$

Table 10.11: Hardest-to-round points for the inverse hyperbolic functions in binary32 arithmetic. Tiny input numbers are easily handled using results such as those presented in Table 10.3.

Function	Domain	Argument	Truncated result	Trailing bits
sin	$[2^{-23}, 2^{128})$	1.F9CBE2P7	1.FFFFFEP-1	$1\,1^{31}\,0101\cdots$
		1.333330P13	-1.63F4BAP-2	$0\,1^{29}\,0100\cdots$
		1.FBD9C8P22	-1.FF6DC0P-1	$1\,1^{31}\,0001\cdots$
		1.47D0FEP34	1.FFFFFEP-1	$1\,1^{33}\,0110\cdots$
		1.628D4CP40	-1.FFFFFEP-1	$1\,1^{30}\,0010\cdots$
		1.130930P76	1.FFFFFEP-1	$1\,1^{30}\,0000\cdots$
		1.32EDE2P85	1.FFFFFEP-1	$1\,1^{30}\,0011\cdots$
		1.F37C8AP95	1.FFFFFEP-1	$1\,1^{34}\,0011\cdots$
		1.487E0CP103	1.287506P-2	$1\,0^{29}\,1000\cdots$
		1.B08C4AP111	-1.FFFFFEP-1	$1\,1^{29}\,0000\cdots$
cos	$[2^{-24}, 2^{128})$	1.F9CBE2P8	-1.FFFFFEP-1	$1\,1^{29}\,0101\cdots$
		1.344860P19	-1.EDFE2EP-1	$1\,1^{29}\,0101\cdots$
		1.47D0FEP35	-1.FFFFFEP-1	$1\,1^{31}\,0110\cdots$
		1.47D0FEP36	1.FFFFFEP-1	$1\,1^{29}\,0110\cdots$
		1.64A032P43	1.9D4BA4P-1	$0\,0^{30}\,1110\cdots$
		1.887814P51	1.84BEC4P-1	$0\,1^{30}\,0111\cdots$
		1.48A858P54	1.F48148P-2	$0\,0^{29}\,1111\cdots$
		1.3170F0P63	1.FE2976P-1	$0\,1^{30}\,0101\cdots$
		1.2B9622P67	1.F0285CP-1	$1\,0^{30}\,1000\cdots$
		1.F37C8AP96	-1.FFFFFEP-1	$1\,1^{32}\,0011\cdots$
		1.F37C8AP97	1.FFFFFEP-1	$1\,1^{30}\,0011\cdots$
		1.96344EP117	1.F8A9B4P-1	$0\,1^{29}\,0101\cdots$

Table 10.12: Hardest-to-round points for functions $\sin(x)$ *and* $\cos(x)$ *in binary32 arithmetic. Tiny input numbers are easily handled using results such as those presented in Table 10.3.*

Function	Domain	Argument	Truncated result	Trailing bits
tan	$[2^{-23}, 2^{128})$	1.ADA6AAP27	1.E80304P-3	$0\,0^{30}\,1010\cdots$
		1.AF61DAP48	1.60D1C6P-2	$1\,1^{28}\,0110\cdots$
		1.0088BCP52	1.CA1ED0P0	$0\,0^{28}\,1111\cdots$
		1.FA6748P64	1.A0D916P0	$1\,0^{29}\,1001\cdots$
asin	$[2^{-23}, 2^{1})$	1.AC6FB8P-9	1.AC6FE8P-9	$1\,1^{26}\,0000\cdots$
		1.0F2B38P-5	1.0F37E6P-5	$0\,1^{26}\,0010\cdots$
		1.CBF43CP-4	1.CCED1CP-4	$0\,1^{27}\,0101\cdots$
		1.107434P-1	1.1F4B64P-1	$0\,1^{29}\,0100\cdots$
		1.55688AP-1	1.75B8A2P-1	$0\,0^{27}\,1110\cdots$
		1.EE836CP-1	1.4F0654P0	$0\,0^{27}\,1011\cdots$
acos	$[2^{-32}, 2^{1})$	1.110B46P-26	1.921FB4P0	$1\,0^{29}\,1000\cdots$
		1.04C444P-12	1.920F68P0	$1\,0^{32}\,1011\cdots$
		1.ECFCFEP-10	1.91A474P0	$1\,1^{26}\,0110\cdots$
		1.FA5036P-9	1.91228CP0	$1\,0^{26}\,1100\cdots$
		1.145F36P-6	1.8DCE2AP0	$1\,0^{26}\,1000\cdots$
		1.4D04F2P-6	1.8CEB88P0	$1\,1^{27}\,0011\cdots$
atan	$[2^{-23}, 2^{28})$	1.1AD646P-4	1.1A6384P-4	$1\,0^{30}\,1000\cdots$
		1.DDF9F6P0	1.143EC4P0	$0\,0^{29}\,1010\cdots$
		1.98C252P12	1.9215AEP0	$1\,1^{28}\,0111\cdots$
		1.71B3F4P16	1.921F02P0	$1\,1^{28}\,0101\cdots$

Table 10.13: Hardest-to-round points for functions $\tan(x)$, $\operatorname{asin}(x)$, $\operatorname{acos}(x)$ *and* $\operatorname{atan}(x)$ *in binary32 arithmetic. Tiny input numbers are easily handled using results such as those presented in Table 10.3.*

Function	Domain	Argument	Truncated result	Trailing bits
erf	$[2^{-149}, 2^2)$	1.4FF252P-24	1.7B133AP-24	$0\,0^{26}\,1001\cdots$
		1.81D5ACP-13	1.B35E26P-13	$1\,0^{31}\,1010\cdots$
		1.800AFAP-12	1.B15890P-12	$1\,0^{26}\,1010\cdots$
		1.CB2452P-1	1.972EA8P-1	$0\,0^{27}\,1011\cdots$
erfc	$(-2^2, -2^{-49}]$	-1.8D0798P-22	1.000006P0	$1\,0^{26}\,1001\cdots$
		-1.D93EC4P-17	1.00010AP0	$0\,1^{31}\,0110\cdots$
		-1.9A365EP-11	1.0039DAP0	$1\,1^{26}\,0000\cdots$
		-1.D9366EP-9	1.010AFAP0	$0\,1^{27}\,0101\cdots$
		-1.0DD174P-6	1.04C1B6P0	$1\,1^{26}\,0110\cdots$
		-1.CB2452P-1	1.CB9754P0	$0\,0^{28}\,1011\cdots$
	$[2^{-49}, 2^{128})$	1.8D0798P-23	1.FFFFF8P-1	$0\,1^{26}\,0110\cdots$
		1.D93EC4P-18	1.FFFEF4P-1	$0\,1^{26}\,0010\cdots$
		1.D93EC4P-17	1.FFFDEAP-1	$0\,0^{30}\,1001\cdots$
		1.949006P-16	1.FFFC6EP-1	$0\,1^{26}\,0011\cdots$
		1.D9366EP-9	1.FDEA0AP-1	$0\,0^{26}\,1010\cdots$
		1.E51B90P0	1.E2AEBAP-8	$1\,1^{26}\,0010\cdots$

Table 10.14: Hardest-to-round points for functions $\mathrm{erf}(x)$ *and* $\mathrm{erfc}(x)$ *in binary32 arithmetic.*

Function	Domain	Argument	Truncated result	Trailing bits
Γ	$(-2^{128}, -2^{-149}]$	`-0.400000P-126`	`-1.000000P128`	$0\,0^{104}\,1001\cdots$
		`-0.800000P-126`	`-1.000000P127`	$0\,0^{103}\,1001\cdots$
		`-1.000000P-126`	`-1.000000P126`	$0\,0^{102}\,1001\cdots$
		`-1.000000P-125`	`-1.000000P125`	$0\,0^{101}\,1001\cdots$
		`-1.000000P-124`	`-1.000000P124`	$0\,0^{100}\,1001\cdots$
	$[2^{-149}, 2^{128})$	`0.400000P-126`	`1.FFFFFEP127`	$1\,1^{103}\,0110\cdots$
		`0.800000P-126`	`1.FFFFFEP126`	$1\,1^{102}\,0110\cdots$
		`1.000000P-126`	`1.FFFFFEP125`	$1\,1^{101}\,0110\cdots$
		`1.000000P-125`	`1.FFFFFEP124`	$1\,1^{100}\,0110\cdots$
	$(-2^{128}, -2^{-54}]$	`-1.000000P-54`	`-1.000000P54`	$0\,0^{30}\,1001\cdots$
		`-1.000000P-53`	`-1.000000P53`	$0\,0^{29}\,1001\cdots$
		`-1.000000P-52`	`-1.000000P52`	$0\,0^{28}\,1001\cdots$
		`-1.77DF66P-19`	`-1.5CB6E2P18`	$1\,1^{28}\,0111\cdots$
		`-1.4D1DC4P-11`	`-1.899DDEP10`	$1\,0^{28}\,1111\cdots$
	$[2^{-54}, 2^{128})$	`1.000000P-54`	`1.FFFFFEP53`	$1\,1^{29}\,0110\cdots$
		`1.000000P-53`	`1.FFFFFEP52`	$1\,1^{28}\,0110\cdots$
		`1.A87A86P-48`	`1.34C894P47`	$1\,0^{28}\,1101\cdots$
		`1.B847BAP-48`	`1.29B38AP47`	$1\,0^{30}\,1000\cdots$
		`1.BD0D52P-48`	`1.268266P47`	$0\,1^{31}\,0111\cdots$
		`1.C0A8EAP-48`	`1.242422P47`	$0\,1^{30}\,0010\cdots$
		`1.C26D16P-48`	`1.22FEDCP47`	$0\,1^{29}\,0110\cdots$
		`1.C4A8E6P-48`	`1.218F44P47`	$0\,1^{29}\,0100\cdots$
		`1.F76AE0P-7`	`1.021D90P6`	$0\,1^{30}\,0110\cdots$

Table 10.15: Hardest-to-round points for function $\Gamma(x)$ in binary32 arithmetic.

Function	Domain	Argument	Truncated result	Trailing bits
$\ln\|\Gamma(x)\|$	$(-2^{128}, -2^{-149}]$	-1.108A5AP-66	1.6D7B16P5	$1\,1^{32}\,0000\cdots$
		-1.22D570P-65	1.676A7AP5	$1\,0^{30}\,1110\cdots$
		-1.ADE594P-30	1.446AB0P4	$1\,0^{30}\,1100\cdots$
		-1.C2F040P-30	1.43A6F6P4	$0\,0^{32}\,1000\cdots$
		-1.627346P7	-1.73235EP9	$0\,0^{30}\,1100\cdots$
	$[2^{-149}, 2^{128})$	1.108A5AP-66	1.6D7B16P5	$1\,1^{32}\,0000\cdots$
		1.22D570P-65	1.676A7AP5	$1\,0^{30}\,1110\cdots$
		1.F8A754P-9	1.63ACC2P2	$0\,1^{30}\,0010\cdots$
		1.F9413EP76	1.9D5AB2P82	$1\,0^{31}\,1000\cdots$

Table 10.16: Hardest-to-round points for function $\ln(|\Gamma(x)|)$ *in binary32 arithmetic.*

Function	Domain	Argument	Truncated result	Trailing bits
J_0	$[2^{-24}, 2^{128})$	1.000002P-9	1.FFFFE0P-1	$0\,0^{38}\,1010\cdots$
		1.000008P-8	1.FFFF80P-1	$0\,0^{32}\,1010\cdots$
		1.22081EP24	-1.497B48P-14	$1\,0^{30}\,1101\cdots$
		1.326C7EP24	1.728EC6P-13	$0\,0^{30}\,1001\cdots$
		1.FF10E4P88	-1.302FAEP-47	$0\,0^{30}\,1000\cdots$
		1.F14740P124	1.E7455AP-66	$0\,1^{33}\,0100\cdots$
J_1	$[2^{-23}, 2^{128})$	1.80000CP-8	1.7FFFA0P-9	$0\,0^{32}\,1100\cdots$
		1.B0DBC6P8	-1.386F80P-5	$0\,1^{30}\,0010\cdots$
		1.70A7A0P78	1.524326P-40	$1\,0^{30}\,1010\cdots$
		1.9D8DC6P97	-1.551388P-50	$1\,1^{32}\,0111\cdots$
Y_0	$[2^{-149}, 2^{128})$	1.6C7D4EP-55	-1.81E690P4	$0\,0^{32}\,1110\cdots$
		1.CABA12P-22	-1.2D225CP3	$0\,1^{30}\,0010\cdots$
		1.DB5DE8P56	1.777C22P-30	$0\,0^{30}\,1110\cdots$
		1.70A7A0P78	1.524326P-40	$1\,0^{30}\,1010\cdots$
		1.9D8DC6P97	-1.551388P-50	$1\,1^{32}\,0111\cdots$
Y_1	$[2^{-149}, 2^{128})$	1.F621C0P25	-1.7BE614P-14	$1\,1^{31}\,0101\cdots$
		1.B30EF0P40	1.1F8CACP-21	$0\,0^{30}\,1001\cdots$
		1.67615CP42	1.742D98P-23	$1\,0^{30}\,1100\cdots$
		1.FF10E4P88	1.302FAEP-47	$0\,0^{30}\,1000\cdots$
		1.F14740P124	-1.E7455AP-66	$0\,1^{33}\,0100\cdots$

Table 10.17: Hardest-to-round points for Bessel functions of the first kind (J_0 and J_1) and the second kind (Y_0 and Y_1) in binary32 arithmetic.

tables, each hardest-to-round point has the following format: the argument (binary64 number) and the truncated result (i.e., the result rounded toward zero in the binary64 format), both in C99's hexadecimal format (described below), and then the bits after the significand of the truncated result. The first bit after the significand is the rounding bit. Then, as these are hard-to-round points, the rounding bit is followed by a long run of zeros or ones; the value in the exponent (superscript) is the length of this run. Finally, we give the next four bits.

Each datum in hexadecimal format in these tables consists of:

- a significand, written as a possible "−" if the number is negative, followed by a "1," followed by a 13-hexadecimal-digit number (hence the $1 + 4 \times 13 = 53$ bits of precision), the digits being denoted `0`, `1`, `2`, `3`, `4`, `5`, `6`, `7`, `8`, `9`, `A`, `B`, `C`, `D`, `E`, and `F`;
- the letter `P` (as a separator);
- a binary exponent E, written in decimal, i.e., the significand is to be multiplied by 2^E to get the value.

For instance, the sixth line of Table 10.18 contains the argument

1.83D4BCDEBB3F4P2.

That number is equal to

$$\begin{aligned}\left(1 + \frac{\text{83D4BCDEBB3F4}_{16}}{2^{52}}\right) \cdot 2^2 &= \left(1 + \frac{2319195600303092_{10}}{2^{52}}\right) \cdot 2^2 \\ &= \frac{1705698806918397}{2^{48}},\end{aligned}$$

whose exponential is equal to $y \cdot 2^8$, with (in binary)

$$\begin{aligned} y \;=\; & 1.1010110001010000101101000000100111001000101011 \\ & 1011100000000000000000000000000000000000000000000 \\ & 000000000001000101100111000110011 \cdots \end{aligned}$$

If a and b belong to the same "binade" (they have the same sign and satisfy $2^p \le |a|, |b| < 2^{p+1}$, where p is an integer), let us call their *significand distance* the distance

$$\frac{|a-b|}{2^p}.$$

For instance, the significand distance between 14 and 15 is $(15 - 14)/8 = 1/8$. Tables 10.18 to 10.24 allow one to deduce properties such as the following [380].

Theorem 10.2 (Computation of exponentials). *Let y be the exponential of a binary64 number x. Let y^* be an approximation to y such that the significand distance*[10] *between y and y^* is bounded by ϵ.*

- *for $|x| \geq 2^{-30}$, if $\epsilon \leq 2^{-53-59-1} = 2^{-113}$, then for any of the four rounding modes, rounding y^* is equivalent to rounding y;*
- *for $2^{-54} \leq |x| < 2^{-30}$, if $\epsilon \leq 2^{-53-104-1} = 2^{-158}$ then rounding y^* is equivalent to rounding y;*

The case $|x| < 2^{-54}$ is easily dealt with using Tables 10.4 and 10.5.

Theorem 10.3 (Computation of logarithms). *Let y be the natural (base-e) logarithm of a binary64 number x. Let y^* be an approximation to y such that the significand distance between y and y^* is bounded by ϵ. If $\epsilon \leq 2^{-53-64-1} = 2^{-118}$ then for any of the 4 rounding modes, rounding y^* is equivalent to rounding y.*

10.5.3.3 Beyond binary64?

For the sake of simplicity, let us assume a *binary*, precision-p, floating-point format. To summarize very quickly, the L-algorithm consists in replacing the function f under study by a piecewise linear approximation, and then finding the worst cases of each of the linear approximations on its domain of definition. By taking into account the approximation error, we see that all bad rounding cases of f are (possibly slightly less) bad rounding cases for the linear approximations. The bad rounding cases of the linear approximations can be found using a slightly modified version of the well-known subtractive *Euclidean algorithm* for computing GCDs. If we are trying to find the worst cases of a regular elementary function f over a given binade, e.g., $[1, 2)$, then one can show that we need around $2^{2p/3}$ subintervals of length around $2^{-2p/3}$. Each subinterval requires a number of operations that is polynomial in p. Roughly speaking, the overall cost of the computation is around $2^{2p/3}$. Given the fact that with $p = 53$ (which corresponds to the binary64 format) we already need days of computation and massive parallelism to completely deal with a function,[11] we have little hope of being able to tackle formats wider than binary64 (or possibly, the Intel "double extended" format) with the L-algorithm.

The bottleneck of the L-algorithm lies in the number of linear approximations that are required to approximate the function with the accuracy that is necessary for the tests. In order to decrease the number of subintervals to be

[10]If one prefers to think in terms of relative error, one can use the following well-known properties: in radix-2 floating-point arithmetic, if the significand distance between y and y^* is less than ϵ, then their relative distance $|y-y^*|/|y|$ is less than ϵ. If the relative distance between y and y^* is less than ϵ_r, then their significand distance is less than $2\epsilon_r$.

[11]The time needed depends much on the function.

Function	Domain	Argument	Truncated result	Trailing bits
exp	$[\log(2^{-1074}), -2^{-36}]$	-1.12D31A20FB38BP5	1.5B0BF3244820AP-50	$0\,1^{58}\,0010\cdots$
		-1.A2FEFEFD580DFP-13	1.FFE5D0BB7EABFP-1	$0\,0^{57}\,1100\cdots$
		-1.ED318EFB627EAP-27	1.FFFFFF84B39C4P-1	$1\,1^{59}\,0001\cdots$
		-1.3475AC05CEAD7P-29	1.FFFFFFECB8A54P-1	$0\,0^{57}\,1001\cdots$
	$[2^{-31}, \log(2^{1025}))$	1.9E9CBBFD6080BP-31	1.000000033D397P0	$1\,0^{57}\,1010\cdots$
		1.83D4BCDEBB3F4P2	1.AC50B409C8AEEP8	$0\,0^{57}\,1000\cdots$
	$(\log(2^{-1074}), -2^{-54}]$	-1.0000000000001P-51	1.FFFFFFFFFFFFCP-1	$0\,0^{100}\,1010\cdots$
	$[2^{-53}, \log(2^{1025}))$	1.FFFFFFFFFFFFFP-53	1.0000000000000P0	$1\,1^{104}\,0101\cdots$
$\exp(x)-1$	$[2^{-35}, \log(2^{1024}))$	1.274BBF1EFB1A2P-10	1.2776572C25129P-10	$1\,0^{58}\,1000\cdots$
	$(-\infty, -2^{-34}]$	-1.19E53FCD490D0P-23	-1.19E53E96DFFA8P-23	$1\,0^{56}\,1110\cdots$
	$[2^{-51}, \log(2^{1024}))$	1.7FFFFFFFFFFFDP-49	1.8000000000005P-49	$1\,1^{96}\,0110\cdots$
	$(-\infty, -2^{-51}]$	-1.8000000000003P-49	-1.7FFFFFFFFFFFAP-49	$0\,0^{96}\,1000\cdots$
2^x	$(-\infty, +\infty)$	1.12B14A318F904P-27	1.00000017CCE02P0	$0\,0^{58}\,1101\cdots$
		1.BFBBDE44EDFC5P-25	1.0000009B2C385P0	$0\,0^{59}\,1011\cdots$
		1.E4596526BF94DP-10	1.0053FC2EC2B53P0	$0\,1^{59}\,0100\cdots$
10^x	$(0, \log10(2^{1024}))$	1.DF760B2CDEED3P-49	1.0000000000022P0	$1\,0^{58}\,1110\cdots$
		1.A83B1CF779890P-26	1.000000F434FAAP0	$0\,1^{60}\,0101\cdots$
		1.7C3DDD23AC8CAP-10	1.00DB40291E4F5P0	$0\,1^{58}\,0010\cdots$
		1.AA6E0810A7C29P-2	1.4DEC173D50B3EP1	$0\,1^{58}\,0001\cdots$
		1.D7D271AB4EEB4P-2	1.71CE472EB84C7P1	$1\,1^{64}\,0111\cdots$
		1.75F49C6AD3BADP0	1.CE41D8FA665F9P4	$1\,1^{64}\,0101\cdots$
	$(\log10(2^{-1074}), 0)$	-1.1416C72A588A6P-1	1.27D838F22D09FP-2	$1\,1^{65}\,0010\cdots$
		-1.F28E0E25574A5P-32	1.FFFFFFF70811EP-1	$0\,0^{59}\,1011\cdots$

Table 10.18: Hardest-to-round points for functions e^x, e^x-1, 2^x, and 10^x in binary64 arithmetic. The values given here and the results given in Tables 10.4 and 10.5 suffice to round functions e^x, 2^x, and 10^x correctly in the full binary64 range (for function e^x the input values between -2^{-53} and 2^{-52} are so small that the results given in Tables 10.4 and 10.5 can be applied, so they are omitted here) [380]. Base-β exponentials of numbers less than $\log_\beta(2^{-1074})$ are less than the smallest positive machine number. Base-β exponentials of numbers larger than $\log_\beta(2^{1024})$ are overflows.

Function	Domain	Argument	Truncated result	Trailing bits
ln	$[2^{-1074}, 2^{-1})$	1.EA71D85CEE020P-509	-1.60296A66B42FFP8	$1\,1^{60}\,0000\cdots$
		1.9476E304CD7C7P-384	-1.09B60CAF47B35P8	$1\,0^{60}\,1010\cdots$
		1.26E9C4D327960P-232	-1.4156584BCD084P7	$0\,0^{60}\,1001\cdots$
		1.613955DC802F8P-35	-1.7F02F9BAF6035P4	$0\,1^{60}\,0011\cdots$
	$[2^{-1}, 2^{1})$	1.BADED30CBF1C4P-1	-1.290EA09E36478P-3	$1\,1^{54}\,0110\cdots$
	$[2^{1}, 2^{1025})$	1.C90810D354618P245	1.54CD1FEA76639P7	$1\,1^{63}\,0101\cdots$
		1.62A88613629B6P678	1.D6479EBA7C971P8	$0\,0^{64}\,1110\cdots$
$\ln(1+x)$	$[2^{-35}, 2^{98}]$	1.AB50B409C8AEEP8	1.83D4BCDEBB3F3P2	$1\,1^{60}\,0101\cdots$
		1.8AA92BC84FF91P54	1.2EE70220FB1C4P5	$1\,1^{60}\,0011\cdots$
		1.0410C95B580B9P71	1.89D56A0C38E6FP5	$0\,0^{62}\,1011\cdots$
	$[2^{-35}, 2^{1024})$	1.C90810D354618P245	1.54CD1FEA76639P7	$1\,1^{63}\,0101\cdots$
		1.62A88613629B6P678	1.D6479EBA7C971P8	$0\,0^{64}\,1110\cdots$
	$(-1, -2^{-35}]$	-1.7FFFF3FCFFD03P-30	-1.7FFFF4017FCFEP-30	$1\,0^{58}\,1001\cdots$
	$(2^{-51}, 2^{1024})$	1.8000000000003P-50	1.7FFFFFFFFFFFEP-50	$1\,0^{99}\,1000\cdots$
	$(-1, -2^{-51}]$	-1.7FFFFFFFFFFFDP-50	-1.8000000000001P-50	$0\,1^{99}\,0110\cdots$

Table 10.19: Hardest-to-round points for functions $\ln(x)$ *and* $\ln(1+x)$ *in binary64 arithmetic. The values given here suffice to round functions* $\ln(x)$ *and* $\ln(1+x)$ *correctly in the full binary64 range.*

Function	Domain	Argument	Truncated result	Trailing bits
$\log_2$	$[2^{-1}, 2^{1024})$	1.B4EBE40C95A01P0	1.8ADEAC981E00DP-1	$1\,0^{53}\,1011\cdots$
		1.1BA39FF28E3EAP2	1.12EECF76D63CDP1	$0\,0^{53}\,1001\cdots$
		1.1BA39FF28E3EAP4	1.097767BB6B1E6P2	$1\,0^{54}\,1001\cdots$
		1.61555F75885B4P128	1.00EE05A07A6E7P7	$1\,1^{53}\,0011\cdots$
		1.D30A43773DD1BP256	1.00DE0E189B724P8	$1\,0^{53}\,1100\cdots$
		1.61555F75885B4P256	1.007702D03D373P8	$1\,1^{54}\,0011\cdots$
		1.61555F75885B4P512	1.003B81681E9B9P9	$1\,1^{55}\,0011\cdots$
$\log_{10}$	$[2^{-1074}, 2^{-1})$	1.365116686B078P-765	-1.CC68A4AEE240DP7	$0\,1^{61}\,0110\cdots$
		1.83E55C0285C96P-762	-1.CA68A4AEE240DP7	$0\,1^{61}\,0110\cdots$
		1.A8639E89F5E46P-625	-1.77D933C1A88E0P7	$1\,1^{61}\,0101\cdots$
		1.ED8C87C3BF5CFP-49	-1.CEE46399392D6P3	$0\,1^{62}\,0000\cdots$
		1.27D838F22D0A0P-2	-1.1416C72A588A5P-1	$1\,1^{65}\,0101\cdots$
	$[2^{-1}, 2^{1})$	1.B0CF736F1AE1DP-1	-1.2AE5057CD8C44P-4	$0\,1^{54}\,0110\cdots$
		1.89825F74AA6B7P0	1.7E646F3FAB0D0P-3	$1\,0^{57}\,1001\cdots$
	$[2^{1}, 2^{1024})$	1.71CE472EB84C8P1	1.D7D271AB4EEB4P-2	$0\,0^{64}\,1010\cdots$
		1.CE41D8FA665FAP4	1.75F49C6AD3BADP0	$0\,0^{66}\,1010\cdots$
		1.E12D66744FF81P429	1.02D4F53729E44P7	$1\,0^{68}\,1001\cdots$

Table 10.20: Hardest-to-round points for functions $\log_2(x)$ *and* $\log_{10}(x)$ *in binary64 arithmetic. The values given here suffice to round functions* $\log_2(x)$ *and* $\log_{10}(x)$ *correctly in the full binary64 range.*

Function	Domain	Argument	Truncated result	Trailing bits
sinh	$[2^{-25}, \text{asinh}(2^{1024}))$	1.DFFFFFFFFFE3EP-20	1.E000000000FD1P-20	$1\,1^{72}\,0001\cdots$
		1.DFFFFFFFFF8F8P-19	1.E000000003F47P-19	$1\,1^{66}\,0001\cdots$
		1.DFFFFFFFFE3E0P-18	1.E00000000FD1FP-18	$1\,1^{60}\,0001\cdots$
		1.67FFFFFFFD08AP-17	1.680000001AB25P-17	$1\,1^{57}\,0000\cdots$
		1.897374D74DE2AP-13	1.897374FE073E1P-13	$1\,0^{56}\,1011\cdots$
cosh	$[2^{-25}, 2^{6})$	1.465655F122FF5P-24	1.000000000000CP0	$1\,1^{61}\,0001\cdots$
		1.7FFFFFFFFFFF7P-23	1.0000000000047P0	$1\,1^{89}\,0010\cdots$
		1.7FFFFFFFFFFDCP-22	1.000000000011FP0	$1\,1^{83}\,0010\cdots$
		1.7FFFFFFFFFF70P-21	1.000000000047FP0	$1\,1^{77}\,0010\cdots$
		1.7FFFFFFFFFDC0P-20	1.00000000011FFP0	$1\,1^{71}\,0010\cdots$
		1.1FFFFFFFFFF0DP-20	1.0000000000A1FP0	$1\,1^{73}\,0110\cdots$
		1.DFFFFFFFFFB9BP-20	1.0000000001C1FP0	$1\,1^{69}\,0010\cdots$
		1.1FFFFFFFFFC34P-19	1.000000000287FP0	$1\,1^{67}\,0110\cdots$
		1.7FFFFFFFFF700P-19	1.00000000047FFP0	$1\,1^{65}\,0010\cdots$
		1.DFFFFFFFFEE6CP-19	1.000000000707FP0	$1\,1^{63}\,0010\cdots$
		1.1FFFFFFFFF0D0P-18	1.000000000A1FFP0	$1\,1^{61}\,0110\cdots$
		1.4FFFFFFFFE7E2P-18	1.000000000DC7FP0	$1\,1^{60}\,0011\cdots$
		1.7FFFFFFFFDC00P-18	1.0000000011FFFP0	$1\,1^{59}\,0010\cdots$
		1.AFFFFFFFFCCBEP-18	1.0000000016C7FP0	$1\,1^{58}\,0010\cdots$
		1.DFFFFFFFFB9B0P-18	1.000000001C1FFP0	$1\,1^{57}\,0010\cdots$
		1.EA5F2F2E4B0C5P1	1.710DB0CD0FED5P4	$1\,0^{57}\,1110\cdots$

Table 10.21: Hardest-to-round points for functions $\sinh(x)$ *and* $\cosh(x)$ *in binary64 arithmetic. The values given here suffice to round these functions correctly in the full binary64 range. If* x *is small enough, the results given in Tables 10.4 and 10.5 can be applied. If* x *is large enough, we can use the results obtained for the exponential function.*

Function	Domain	Argument	Truncated result	Trailing bits
asinh	$[2^{-25}, 2^{1024})$	1.E000000000FD2P-20	1.DFFFFFFFFFE3EP-20	$0\,0^{72}\,1110\cdots$
		1.E000000003F48P-19	1.DFFFFFFFFF8F8P-19	$0\,0^{66}\,1110\cdots$
		1.C90810D354618P244	1.54CD1FEA76639P7	$1\,1^{63}\,0101\cdots$
		1.8670DE0B68CADP655	1.C7206C1B753E4P8	$0\,0^{62}\,1111\cdots$
		1.62A88613629B6P677	1.D6479EBA7C971P8	$0\,0^{64}\,1110\cdots$
acosh	$[1, 2^{91}]$	1.297DE35D02E90P13	1.3B562D2651A5DP3	$0\,1^{61}\,0001\cdots$
		1.91EC4412C344FP85	1.E07E71BFCF06EP5	$1\,1^{61}\,0101\cdots$
	$[1, 2^{1024})$	1.C90810D354618P244	1.54CD1FEA76639P7	$1\,1^{63}\,0101\cdots$
		1.62A88613629B6P677	1.D6479EBA7C971P8	$0\,0^{64}\,1110\cdots$

Table 10.22: Hardest-to-round points for the inverse hyperbolic functions in binary64 arithmetic. Concerning function $\sinh^{-1}$*, if the input values are small enough, there is no need to compute the Hardest-to-round points : the results given in Tables 10.4 and 10.5 can be applied.*

Function	Domain	Argument	Truncated result	Trailing bits
sin	$[2^{-25}, u)$	1.E0000000001C2P-20	1.DFFFFFFFFF02EP-20	$0\,0^{72}\,1110\cdots$
		1.E000000000708P-19	1.DFFFFFFFFC0B8P-19	$0\,0^{66}\,1110\cdots$
		1.E000000001C20P-18	1.DFFFFFFFF02E0P-18	$0\,0^{60}\,1110\cdots$
		1.598BAE9E632F6P-7	1.598A0AEA48996P-7	$0\,1^{59}\,0000\cdots$
		1.FE767739D0F6DP-2	1.E9950730C4695P-2	$1\,1^{65}\,0000\cdots$
cos	$[2^{-17}, \mathrm{acos}(2^{-26}))$ $\cup\ [\mathrm{acos}(-2^{-27}), 2^2)$	1.06B505550E6B2P-9	1.FFFFBC9A3FBFEP-1	$0\,0^{58}\,1100\cdots$
		1.34EC2F9FC9C00P1	-1.7E2A5C30E1D6DP-1	$0\,1^{58}\,0110\cdots$
	$[0, \mathrm{acos}(2^{-26}))$ $\cup\ [\mathrm{acos}(-2^{-27}), 2^2)$	1.8000000000009P-23	1.FFFFFFFFFFF70P-1	$0\,0^{88}\,1101\cdots$
tan	$[2^{-25}, \pi/2)$	1.DFFFFFFFFFF1FP-22	1.E000000000151P-22	$0\,1^{78}\,0100\cdots$
		1.DFFFFFFFFFC7CP-21	1.E000000000545P-21	$1\,1^{72}\,0100\cdots$
		1.DFFFFFFFFF1F0P-20	1.E000000001517P-20	$1\,1^{66}\,0100\cdots$
		1.67FFFFFFFE845P-19	1.6800000002398P-19	$0\,1^{63}\,0100\cdots$
		1.DFFFFFFFFC7C0P-19	1.E00000000545FP-19	$1\,1^{60}\,0100\cdots$
		1.67FFFFFFA114P-18	1.6800000008E61P-18	$1\,1^{57}\,0100\cdots$
		1.50486B2F87014P-5	1.5078CEBFF9C72P-5	$1\,0^{57}\,1001\cdots$

Table 10.23: Hardest-to-round points for the trigonometric functions in binary64 arithmetic. So far, we only have hardest-to-round points in the following domains: $[2^{-25}, u)$ where $u = 1.100100100001 1_2 \times 2^1$ for the sine function ($u = 3.141357421875_{10}$ is slightly less than π); $[0, \arccos(2^{-26})) \cup [\arccos(-2^{-27}), 2^2)$ for the cosine function; and $[2^{-25}, \pi/2]$ for the tangent function. Sines of numbers of absolute value less than 2^{-25} are easily handled using the results given in Tables 10.4 and 10.5.

Function	Domain	Argument	Truncated result	Trailing bits
asin	$[2^{-25}, 1]$	1.DFFFFFFFFF02EP-20	1.E0000000001C1P-20	$1\,1^{72}\,0001\cdots$
		1.DFFFFFFFFC0B8P-19	1.E000000000707P-19	$1\,1^{66}\,0001\cdots$
		1.DFFFFFFFF02E0P-18	1.E000000001C1FP-18	$1\,1^{60}\,0001\cdots$
		1.67FFFFFFE54DAP-17	1.6800000002F75P-17	$1\,1^{57}\,0000\cdots$
		1.C373FF4AAD79BP-14	1.C373FF594D65AP-14	$1\,0^{57}\,1010\cdots$
		1.E9950730C4696P-2	1.FE767739D0F6DP-2	$0\,0^{64}\,1000\cdots$
acos	$[2^{-26}, 1]$	1.7283D529A146EP-19	1.921F86F3C82C5P0	$0\,0^{58}\,1000\cdots$
		1.FD737BE914578P-11	1.91E006D41D8D8P0	$1\,1^{62}\,0010\cdots$
		1.1CDCD1EA2AD3BP-9	1.919146D3F492EP0	$1\,1^{57}\,0010\cdots$
		1.60CB9769D9218P-8	1.90BEE93D2D09CP0	$0\,0^{57}\,1011\cdots$
		1.53EA6C7255E88P-4	1.7CDACB6BBE707P0	$0\,1^{57}\,0101\cdots$
	$[-1, -2^{-27}]$	-1.52F06359672CDP-2	1.E87CCC94BA418P0	$1\,0^{56}\,1101\cdots$
		-1.124411A0EC32EP-5	1.9AB23ECDD436AP0	$1\,0^{56}\,1101\cdots$
atan	$(2^{-25}, +\infty)$	1.E000000000546P-21	1.DFFFFFFFFFC7CP-21	$0\,0^{72}\,1011\cdots$
		1.E000000001518P-20	1.DFFFFFFFFF1F0P-20	$0\,0^{66}\,1011\cdots$
		1.E000000005460P-19	1.DFFFFFFFFC7C0P-19	$0\,0^{60}\,1011\cdots$
		1.6800000008E62P-18	1.67FFFFFFFA114P-18	$0\,0^{57}\,1011\cdots$
		1.22E8D75E2BC7FP-11	1.22E8D5694AD2BP-11	$1\,0^{59}\,1101\cdots$
		1.6298B5896ED3CP1	1.3970E827504C6P0	$1\,0^{63}\,1101\cdots$
		1.C721FD48F4418P19	1.921FA3447AF55P0	$1\,0^{58}\,1011\cdots$
		1.EB19A7B5C3292P29	1.921FB540173D6P0	$1\,1^{59}\,0011\cdots$
		1.CCDA26AD0CD1CP47	1.921FB54442D06P0	$0\,1^{57}\,0111\cdots$

Table 10.24: Hardest-to-round points for the inverse trigonometric functions in binary64 arithmetic. Concerning the arcsine function, the results given in Tables 10.4 and 10.5 and in this table make it possible to correctly round the function in its whole domain of definition.

considered, and thus the overall cost of the search for the hardest-to-round points, it is tempting to consider better approximations; namely, polynomial approximations of higher degree. To achieve that goal, Lefèvre, Stehlé, and Zimmermann [570, 571] suggested computing a piecewise, constant-degree, polynomial approximation to the function f considered. Over an interval of width $\tau < 1$, we can expect a good degree-d polynomial approximation to f to have a maximal error around τ^{d+1}: hence, by choosing a higher degree, we can get approximations that work in larger intervals.

It thus seems that by increasing the degree sufficiently, we could find all hardest-to-round points by studying a much smaller number of subintervals. Unfortunately, finding the bad rounding cases of a polynomial of degree $d > 1$ seems to be significantly more complicated than finding the bad rounding cases of a linear function. In addition to the quality of the approximation, one has to take into account the feasibility of finding the bad rounding cases of the approximating polynomials. Suppose we consider a precision p. In [570], the authors showed how to use degree-2 approximations on subintervals of length around $2^{-3p/5}$ to find the hardest-to-round points over a given binade in time $\approx 2^{3p/5}$. They refined their analysis in [571] to obtain a cost of around $2^{4p/7}$, still using degree-2 approximations. Later on, Stehlé [568, 569] strengthened the study further and showed that by using degree-3 polynomials the cost can be decreased to $2^{p/2}$. For the moment, higher degree polynomials seem useless for finding the hardest-to-round points.

However, variants of the SLZ algorithm can be used to find an upper bound on the hardness-to-round. If we do not require the bound to be tight, this can be very significantly faster than trying to find the hardest-to-round points. This is work-in-progress (see [604] for a recent account). Given the current state of knowledge and the currently available computing power, this is our only hope of getting useful results for binary128 (and decimal128) arithmetic.

10.6 Some Implementation Tricks Used in the CRlibm Library

Much effort has been devoted during the last 15 years to computing the actual worst-case accuracies needed to guarantee correct rounding of the main elementary functions in binary64 arithmetic,[12] using techniques discussed in the previous section.

[12]Some results have been obtained in decimal64 arithmetic: for example, the hardest-to-round case for the exponential function in that format is known. Getting hardest-to-round cases for the binary32 or decimal32 formats only requires a few hours of computation: we need to examine 2^{32} cases for each function, which is easily done with current computers. As explained above, finding hardest-to-round cases in formats significantly larger than binary64 or decimal64 seem to be a difficult challenge: precisions such as binary128 or decimal128 seem out of reach with current techniques.

Because of such results, it is now possible to obtain correct rounding of the most common functions in two Ziv steps only, which we may then optimize separately:

- the first *quick* step is as fast as a current `libm`, and provides an accuracy between 2^{-60} and 2^{-80} (depending on the function), which is sufficient to round correctly to the 53 bits of binary64 in most cases;
- the second *accurate* step is dedicated to challenging cases. It is slower but has a reasonably bounded execution time, being tightly targeted at the hardest-to-round cases given in Section 10.5. In particular, there is no need for arbitrary multiple precision.

This was the approach used in the CRlibm library. Let us now detail some tricks used in that library.

10.6.1 Rounding test

Let $\hat{y}_1$ be the approximation to $y = f(x)$ obtained at the end of the fast step. The test on $\hat{y}_1$, which either returns a correctly rounded value or launches the second step, is called a *rounding test*. The property that a rounding test must ensure is the following: a value will be returned only if it can be proven to be the correctly rounded value of y; otherwise (in doubt), the second step will be launched. To get good performance, it is essential that the rounding test be fast.

A rounding test depends on a bound $\overline{\epsilon}_1$ on the overall relative error of the first step. This bound is usually computed statically, although in some cases it can be refined at runtime.

The implementation of a rounding test depends on the rounding mode and the nature of $\hat{y}_1$, which may be a 80-bit double-extended number, or a double-binary64 number (see Section 14.1 for an introduction to double-word arithmetic). Besides, in each case, there are several sequences which are acceptable as rounding tests.

Some rounding tests are conceptually simple. If $\hat{y}_1$ is a double-extended number (with 64 bits of significand), it suffices to extract the significand of $\hat{y}_1$, then perform bit mask operations on the bits after bit 53, looking for a string of zeros in the case of directed rounding mode, or for a string of the form 10^k or 01^k for round to nearest. Here, k is deduced from $\overline{\epsilon}_1$. Examples can be found in Itanium-specific code inside the CRlibm distribution, because this architecture offers both native 64-bit arithmetic and efficient instructions for extracting the significand of a floating-point number.

If $\hat{y}_1$ is computed as a double-word number (y_h, y_ℓ), a better option is to use only binary64 floating-point operations. Listing 10.1 describes a rounding test for round to nearest, probably due to Ziv.

C listing 10.1 Floating-point-based rounding test.

```
if (yh == yh + yl * e)
   return yh;
else
   /* more accuracy is needed, launch accurate phase */
```

The constant e is slightly larger than 1, and its relationship to $\overline{\epsilon}_1$ is given by Theorem 10.4 below. An earlier version of this theorem appears in the CRlibm documentation [129], and a most recent version is proven in [149] (this test is already present in Ziv's `libultim` library, but neither the test itself nor the way the constants e have been obtained is documented).

Theorem 10.4. *Assume binary, precision-p, floating-point arithmetic. Assume that y_h is a floating-point number such that $\frac{1}{4}\,\mathrm{ulp}(y_h)$ is in the normal range. Also assume that $y_h = \mathrm{RN}(y_h + y_\ell)$ and $|(y_h + y_\ell) - y| < \overline{\epsilon}_1 \cdot |y|$, with $\overline{\epsilon}_1 < 1/(2^{p+1}+1)$. If*

$$e \geq \frac{1+2^{-p}}{1-\overline{\epsilon}_1 - 2^{p+1}\overline{\epsilon}_1}$$

then $y_h = \mathrm{RN}\left(y_h + \mathrm{RN}(y_\ell \cdot e)\right)$ implies $y_h = \mathrm{RN}(y)$.

10.6.2 Accurate second step

For the second step, correct rounding requires an accuracy of 2^{-120} to 2^{-150}, depending on the function. Several approaches are possible (some of them are presented in Chapter 14). The important point here is that this accuracy is known statically, so the overhead due to arbitrary multiple precision is avoided.

The result of the accurate step must finally be rounded to a binary64 number in the selected rounding mode. For some multiple-precision representations, this is a nontrivial task. For instance, in the CRlibm library, the result of the accurate step is represented by a triple binary64 number (see Section 14.1). Converting this result to a binary64 number is equivalent to computing the correctly rounded-to-nearest sum of three binary64 numbers. This can be done using techniques presented in Section 5.3.4.

10.6.3 Error analysis and the accuracy/performance tradeoff

The probability p_2 of launching the accurate step is the probability that the interval $[\hat{y}_1(1-\overline{\epsilon}_1), \hat{y}_1(1+\overline{\epsilon}_1)]$ contains the midpoint between two consecutive floating-point numbers (or a floating-point number in directed rounding modes). Therefore, it is expected to be proportional to the error bound $\overline{\epsilon}_1$ computed for the first step. Note that proven theoretical bounds of this probability were recently given in [71] for some common elementary functions.

This defines the main performance tradeoff one has to manage when designing a correctly rounded function: the average evaluation time will be

$$T_{\text{avg}} = T_1 + p_2 T_2, \tag{10.14}$$

where T_1 and T_2 are the execution time of the first and second phase respectively, and p_2 is the probability of launching the second phase.

For illustration, $T_2 \approx 10T_1$ in CRlibm, using triple-binary64 numbers. Recently, a correctly rounded logarithm implementation using only *integer* 64-bit and 128-bit arithmetic for both steps has been reported with $T_2 \approx 2T_1$ [372].

Typically, we aim at choosing (T_1, p_2, T_2) such that the average cost of the second step is negligible. Indeed, in this case the performance price to pay for correct rounding will be almost limited to the overhead of the rounding test, which is a few cycles only.

The second step is built to minimize T_2; there is no tradeoff there. Then, as p_2 is almost proportional to $\overline{\epsilon}_1$, to minimize the average time, we have to

- balance T_1 and p_2: this is a performance/precision tradeoff (the more accurate the first step, the slower),
- and compute a tight bound on the overall error $\overline{\epsilon}_1$.

Computing this tight bound is the most time-consuming part in the design of a correctly rounded elementary function. The proof of the correct rounding property only needs a proven bound, but a loose bound will mean a larger p_2 than strictly required, which directly impacts average performance. Compare $p_2 = 1/1000$ and $p_2 = 1/500$ for $T_2 = 100T_1$, for instance. As a consequence, when there are multiple computation paths in the algorithm, it may make sense to precompute different values of $\overline{\epsilon}_1$ on these different paths [151].

With the two-step approach, the proof that an implementation always returns the correctly rounded result reduces to two tasks:

- computing a bound on the overall error of the second step, and checking that this bound is less than the bound deduced from the hardest-to-round cases (e.g., 2^{-118} for natural logarithms);
- proving that the first step returns a value only if this value is correctly rounded, which also requires a proven (and tight) bound on the evaluation error of the first step.

10.6.4 The point with efficient code

Efficient code is especially difficult to analyze and prove because of all the techniques and tricks used by expert programmers.

For instance, many floating-point operations are exact, and the experienced developer of floating-point code will try to use them. Examples include multiplication by a power of the radix of the floating-point system, subtraction of numbers of similar magnitude due to Sterbenz's lemma (Lemma 4.1), exact addition and exact multiplication algorithms such as Fast2Sum (Algorithm 4.3) and 2MultFMA (Algorithm 4.8), multiplication of a small integer by a floating-point number whose significand ends with enough zeros, etc.

The expert programmer will also do his or her best to avoid computing more accurately than strictly needed. He or she will remove from the computation some operations that are not expected to improve the accuracy of the result by much. This can be expressed as an additional approximation. However, it soon becomes difficult to know what is an approximation to what, especially as the computations are re-parenthesized to maximize floating-point accuracy.

The resulting code obfuscation is best illustrated by an example.

10.6.4.1 Example: a double-binary64 polynomial evaluation

Listing 10.2 is an extract of the code of a sine function in CRlibm. The "target" arithmetic is binary64 (double-precision), and we sometimes need to represent big precision numbers as the unevaluated sum of two binary64 numbers ("double-binary64" numbers).

These three lines compute the value of an odd polynomial,

$$p(y) = y + s_3 \times y^3 + s_5 \times y^5 + s_7 \times y^7,$$

close to the Taylor approximation of the sine function (its degree-1 coefficient is equal to 1). In our algorithm, the reduced argument y is ideally obtained by subtracting from the floating-point input x an integral multiple of $\pi/256$. As a consequence, $y \in [-\pi/512, \pi/512] \subset [-2^{-7}, 2^{-7}]$.

However, as y is an irrational number, the implementation of this range reduction has to return a number more accurate than a binary64 number; otherwise, there is no hope of achieving an accuracy of the sine that allows for correct rounding in double precision. In our implementation, the range reduction step therefore returns a double-double number $\mathtt{yh} + \mathtt{yl}$.

To minimize the number of operations, Horner's rule is used for the polynomial evaluation:

$$p(y) = y + y^3 \times (s_3 + y^2 \times (s_5 + y^2 \times s_7)).$$

For a double-double input $y = \mathtt{yh} + \mathtt{yl}$, the expression to compute is thus

$$(\mathtt{yh} + \mathtt{yl}) + (\mathtt{yh} + \mathtt{yl})^3 \times (s_3 + (\mathtt{yh} + \mathtt{yl})^2 \times (s_5 + (\mathtt{yh} + \mathtt{yl})^2 \times s_7)).$$

The actual code uses an approximation to this expression: the computation is accurate enough if all the Horner steps but the last one are computed in binary64 arithmetic. Thus, y_l will be neglected for these iterations, and coefficients s_3 to s_7 will be stored as binary64 numbers noted s3, s5, and s7. The previous expression becomes:

$$(\mathtt{yh} + \mathtt{yl}) + \mathtt{yh}^3 \times (\mathtt{s3} + \mathtt{yh}^2 \times (\mathtt{s5} + \mathtt{yh}^2 \times \mathtt{s7})).$$

However, if this expression is computed as parenthesized above, it has poor accuracy. Specifically, the floating-point addition $\mathtt{yh} + \mathtt{yl}$ (by definition of a double-double number) returns yh, so the information held by yl is completely lost. Fortunately, the other part of the Horner evaluation also has a much smaller magnitude than yh—this is deduced from $|y| \leq 2^{-7}$, which gives $|y^3| \leq 2^{-21}$. The following parenthesizing leads therefore to a much more accurate algorithm:

$$\mathtt{yh} + \left(\mathtt{yl} + \mathtt{yh} \times \mathtt{yh}^2 \times (\mathtt{s3} + \mathtt{yh}^2 \times (\mathtt{s5} + \mathtt{yh}^2 \times \mathtt{s7}))\right).$$

In this last version of the expression, only the leftmost addition has to be accurate. It is therefore computed by a Fast2Sum (Algorithm 4.3) which, as seen in Chapter 4, computes the exact addition of two binary64 numbers, returning a double-binary64 number. The other operations use the native (and therefore fast) binary64 arithmetic. We obtain the code of Listing 10.2.

C listing 10.2 Three lines of C.

```
yh2 = yh * yh;
ts = yh2 * (s3 + yh2 * (s5 + yh2 * s7));
Fast2Sum(sh, sl, yh, yl + yh * ts);
```

To summarize, this code implements the evaluation of a polynomial with many layers of approximation. For instance, variable yh2 approximates y^2 through the following layers:

- y was approximated by $\mathtt{yh} + \mathtt{yl}$ with the relative accuracy ϵ_{argred};
- $\mathtt{yh} + \mathtt{yl}$ is approximated by yh in most of the computation;
- $\mathtt{yh}^2$ is approximated by yh2, with a floating-point rounding error.

In addition, the polynomial is an approximation to the sine function, with a relative error bound of ϵ_{approx} (which can be computed by Sollya as shown in Section 10.3.4).

Thus, the difficulty of evaluating a tight bound on an elementary function implementation is to combine all these errors without forgetting any of them, and without using overly pessimistic bounds when combining several sources of errors. The typical tradeoff here will be that a tight bound requires considerably more work than a loose bound (and its proof, since it is much longer and more complex, might inspire considerably less confidence: this illustrates the strong need for automation and formal proof). Some readers may get an idea of this tradeoff by relating each intermediate value with its error to confidence intervals, and propagating these errors using interval arithmetic. In many cases, a tighter error will be obtained by splitting confidence intervals into several cases, and treating them separately, at the expense of an explosion of the number of cases. This is one of the tasks that a tool such as Gappa (see Section 13.3) was designed to automate.

Part IV

Extensions

Chapter 11

Complex Numbers

11.1 Introduction

COMPLEX NUMBERS naturally appear in many domains (such as electromagnetism, quantum mechanics, and relativity). It is of course always possible to express the various calculations that use complex numbers in terms of real numbers only. However, this will frequently result in programs that are larger and less clear. A good complex arithmetic would make numerical programs devoted to these problems easier to design, understand, and debug.

If a and b are two floating-point numbers, the pair (a, b) can be used to represent the complex number $z = a + ib$. This is the *Cartesian representation* of z. Another possible representation of a complex number is its *polar* form $z = r \cdot e^{i\phi}$. Here we will concentrate on the Cartesian representation. Note that accurate conversions can be performed between the two representations once very accurate trigonometric functions and their inverses are implemented. We refer to Chapter 10 for these issues.

The sum of two complex numbers $a + ib$ and $c + id$ is $(a + c) + i \cdot (b + d)$. Hence, complex addition is easily performed, just by separately adding the real and imaginary parts of the operands. Things become more complicated with multiplication (Section 11.4), division (Section 11.5), and square root (Section 11.7). Also, the absolute value (which is useful for square root) already raises some interesting questions (Section 11.6).

The product of $a + ib$ and $c + id$ is

$$(ac - bd) + i \cdot (ad + bc), \tag{11.1}$$

and, if $c + id$ is nonzero, their ratio is given by

$$\frac{a + ib}{c + id} = \frac{(ac + bd) + i \cdot (bc - ad)}{c^2 + d^2}. \tag{11.2}$$

J.-M. Muller et al., *Handbook of Floating-Point Arithmetic*,
https://doi.org/10.1007/978-3-319-76526-6_11

The complex absolute value of $a + ib$ is

$$|a + ib| = \sqrt{a^2 + b^2}, \tag{11.3}$$

and the square root of $a + ib$ can be defined as the number $x + iy$ such that

$$\begin{aligned} x &= \sqrt{\frac{\sqrt{a^2+b^2}+a}{2}}, \\ y &= \text{sign}(b) \cdot \sqrt{\frac{\sqrt{a^2+b^2}-a}{2}}. \end{aligned} \tag{11.4}$$

A naive use of Equations (11.1) to (11.4) may lead to various problems.

- *Spurious underflows or overflows* may occur, i.e., we may encounter underflows or overflows in the intermediate calculations, although the exact final result is well within the normal number domain. As a consequence, the returned result may be infinity, NaN, or very inaccurate. For example, in binary64 rounded-to-nearest arithmetic, a naive implementation of (11.3) will return an infinite result for $a = 2^{600}$ and $b = 1$ (because the computation of a^2 will overflow), whereas the floating-point number nearest the exact result is 2^{600}.

- Even in the absence of overflow/underflow, *catastrophic cancellations* may lead to very poor results (at least from the point of view of *componentwise* relative error, see below for a definition). For example, in radix-2, precision-p floating-point arithmetic, assuming rounding to nearest ties-to-even, if the real part of (11.1) is evaluated as $\text{RN}(\text{RN}(ac) - \text{RN}(bd))$, then with the input values

$$\begin{aligned} a &= 2^{p-1} + 2^{p-2} - 1, \\ b &= 2^{p-1} + 2^{p-2}, \\ c &= a, \\ d &= 2^{p-1} + 2^{p-2} - 2, \end{aligned} \tag{11.5}$$

  the computed value of the real part is 2^p, whereas the exact value is 1.

These issues are well known. The C99 standard says that *the usual mathematical formulas for complex multiply, divide, and absolute value are problematic because of their treatment of infinities and because of undue overflow and underflow. The* `CX_LIMITED_RANGE` *pragma can be used to inform the implementation that (where the state is "on") the usual mathematical formulas are acceptable.*

However, reliable and accurate implementation of complex arithmetic is possible. Kahan gives good advice in [317], and *scaling* techniques that make it possible to (at least partly) avoid spurious underflows or overflows have been developed. As we are going to see, the availability of a fused multiply-add (FMA) instruction makes accurate complex arithmetic easier to implement.

11.2 Componentwise and Normwise Errors

Two notions of relative error are often considered while performing approximate computations on complex numbers. Suppose we have a floating-point computation producing a complex number $\hat{z} = \hat{a} + i\hat{b}$ which is meant to approximate a complex number $z = a + ib$. One sometimes requires a *componentwise* relative error bound, which corresponds to bounding both

$$\frac{|\hat{a} - a|}{|a|} \quad \text{and} \quad \frac{|\hat{b} - b|}{|b|}.$$

This is intrinsically linked with the Cartesian interpretation of complex numbers. Sometimes one wants a bound on the quantity

$$\frac{|\hat{z} - z|}{|z|}.$$

The latter is usually referred to as a *normwise* error bound, as for $z = a+ib$ the complex absolute value $|z|$ is also the Euclidean norm of the two-dimensional vector (a, b) of its coordinates. Having a bound of the first kind implies having a bound of the second kind, but the converse is not true, as the next lemma recalls.

Lemma 11.1. *Let $z = a + ib$ and $z' = a' + ib'$ be two complex numbers. Suppose that $|a' - a| \leq \epsilon|a|$ and $|b' - b| \leq \epsilon|b|$ for some $\epsilon > 0$. Then*

$$|z' - z| \leq \epsilon|z|.$$

The converse does not hold: We may simultaneously have $|z' - z| \leq \epsilon|z|$ and an arbitrarily large quantity $\frac{|a'-a|}{|a|}$.

Proof. We have the following relations:

$$\begin{aligned} |z' - z|^2 &= (a' - a)^2 + (b' - b)^2 \\ &\leq \epsilon^2|a|^2 + \epsilon^2|b|^2 \\ &= \epsilon^2|z|^2. \end{aligned}$$

For the second part of the lemma, consider $z = t + i$ and $z' = t + \sqrt{t} + i$, with $t \in (0, 1)$. Then

$$\frac{|a' - a|}{|a|} = \frac{1}{\sqrt{t}}$$

can be made arbitrarily large by letting t tend to zero, while

$$\frac{|z' - z|}{|z|} = \sqrt{\frac{t}{1 + t^2}} \leq \sqrt{t} \leq 1.$$

□

In the following, assuming radix-β, precision-p, floating-point arithmetic, normwise and componentwise error bounds will be expressed as functions of $\mathbf{u} = \frac{1}{2}\beta^{1-p}$.

Before giving algorithms for manipulating complex numbers, let us deal with an important basic building block of complex arithmetic: the calculation of $ad \pm bc$, where a, b, c, and d are real (floating-point) numbers.

11.3 Computing $ad \pm bc$ with an FMA

Expressions of the form $ad \pm bc$ appear in many circumstances, such as when performing complex multiplication, division, and 2D rotations, and when computing the discriminant of a quadratic equation. Due to possible catastrophic cancellations, it is not advisable to evaluate $ad \pm bc$ using the naive way, that is, assuming round-to-nearest is the rounding function, by computing $\mathrm{RN}(\mathrm{RN}(ad) \pm \mathrm{RN}(bc))$ or, if a fused multiply-add (FMA) instruction is available, $\mathrm{RN}(ad \pm \mathrm{RN}(bc))$.

However, the availability of an FMA makes it possible to design accurate yet reasonably fast algorithms for evaluating such expressions. As an example, consider the following algorithm (attributed to Kahan in [258, p. 60]).

Algorithm 11.1 Kahan's algorithm for computing $x = ad - bc$ with an FMA.

$\hat{w} \leftarrow \mathrm{RN}(bc)$
$e \leftarrow \mathrm{RN}(\hat{w} - bc)$ // exact with an FMA: $e = \hat{w} - bc$.
$\hat{f} \leftarrow \mathrm{RN}(ad - \hat{w})$
$\hat{x} \leftarrow \mathrm{RN}(\hat{f} + e)$
return $\hat{x}$

One easily recognizes that the first two lines of this algorithm are identical to the 2MultFMA algorithm (Algorithm 4.8, presented in Section 4.4.1). We deduce that $e = \hat{w} - bc$ exactly. The approach implemented by Algorithm 11.1 has been well known to be very accurate in practice (and especially for evaluating discriminants in radix 2, as reported in [319]). More recently, a detailed accuracy analysis was provided by Jeannerod, Louvet, and Muller [295], leading to the following results.

Theorem 11.2. *If no underflow or overflow occurs, then the value $\hat{x}$ returned by Algorithm 11.1 satisfies*

$$\frac{|\hat{x} - x|}{|x|} \leq 2\mathbf{u}$$

if $x \neq 0$, and $\hat{x} = 0$ if $x = 0$. Furthermore, when the radix β is even, the relative error bound $2\mathbf{u}$ is asymptotically optimal.

Theorem 11.3. *If no underflow or overflow occurs, then $|\hat{x} - x| \leq \frac{\beta+1}{2}\,\mathrm{ulp}(x)$ and, when $\beta/2$ is odd and $p \geq 4$, this absolute error bound is optimal.*

The term "asymptotically optimal" means that the ratio between the largest possible error and the error bound tends to 1 as $\mathbf{u} \to 0$ (or, equivalently, as the precision p tends to infinity for a given value of the radix β). The term "optimal" refers to an error bound that is attained for some floating-point input values a, b, c, d.

Another algorithm, which also exploits the availability of an FMA instruction, was introduced by Cornea, Harrison, and Tang in [118].

Algorithm 11.2 Cornea, Harrison, and Tang's algorithm for computing $x = ad - bc$ with an FMA.

$\pi_1 \leftarrow \mathrm{RN}(ad)$
$e_1 \leftarrow \mathrm{RN}(ad - \pi_1)$ // exact with an FMA: $e_1 = ad - \pi_1$.
$\pi_2 \leftarrow \mathrm{RN}(bc)$
$e_2 \leftarrow \mathrm{RN}(bc - \pi_2)$ // exact with an FMA: $e_2 = bc - \pi_2$.
$\pi \leftarrow \mathrm{RN}(\pi_1 - \pi_2)$
$e \leftarrow \mathrm{RN}(e_1 - e_2)$
$\hat{x} \leftarrow \mathrm{RN}(\pi + e)$
return $\hat{x}$

That algorithm is analyzed in [441, 287]. In particular, Jeannerod shows the following result.

Theorem 11.4 ([287]). *Assume $\beta^{p-1} \geq 24$. Then, if no underflow or overflow occurs, the value $\hat{x}$ returned by Algorithm 11.2 satisfies the following: for $x \neq 0$,*

$$\frac{|\hat{x} - x|}{|x|} \leq \begin{cases} 2\mathbf{u} & \text{if } \beta \text{ is odd or } \mathrm{RN}(1+\mathbf{u}) = 1, \\ \frac{2\beta\mathbf{u}+2\mathbf{u}^2}{\beta-2\mathbf{u}^2} & \text{otherwise;} \end{cases}$$

for $x = 0$, $\hat{x} = 0$. Furthermore, these relative error bounds are asymptotically optimal.

Note that the condition $\mathrm{RN}(1 + \mathbf{u}) = 1$ in Theorem 11.4 implies that if RN rounds ties *to even*, then the relative error is always at most $2\mathbf{u}$, similar to Kahan's method (Algorithm 11.1). On the other hand, when the radix is even and ties are rounded *to away*, then the error bound has the form $2\mathbf{u} + \mathcal{O}(\mathbf{u}^2)$ as $\mathbf{u} \to 0$ and, as shown in[287], the term $\mathcal{O}(\mathbf{u}^2)$ can in general not be removed.

Both Algorithms 11.1 and 11.2 thus have essentially the same, asymptotically optimal, relative error bound. Since Algorithm 11.1 uses fewer arithmetic operations, it might be preferred in most practical situations. However, a possible exception is when one wants to implement a *commutative* complex multiplication: in this case, the symmetry of Algorithm 11.2 will guarantee the required property.

11.4 Complex Multiplication

11.4.1 Complex multiplication without an FMA instruction

Let us assume that we use a straightforward implementation of (11.1), i.e., that we approximate $z = (a + ib) \cdot (c + id)$ by

$$\hat{z} = \mathrm{RN}(\mathrm{RN}(ac) - \mathrm{RN}(bd)) + i \cdot \mathrm{RN}(\mathrm{RN}(ad) + \mathrm{RN}(bc)). \tag{11.6}$$

Note that the example given in Equation (11.5) shows that the resulting componentwise error can be huge. However, as we are going to see, the normwise error always remains very small in the absence of underflow and overflow. Roughly speaking, the reason is that:

- if the real part of the product is very inaccurate, then it is negligible in front of the imaginary part and the imaginary part is computed accurately;
- if the imaginary part of the product is very inaccurate, then it is negligible in front of the real part and the real part is computed accurately.

Brent, Percival, and Zimmermann [65] have shown that if $\beta^{p-1} \geq 16$, and assuming that underflows and overflows do not occur, then

$$\hat{z} = z(1 + \epsilon), \qquad |\epsilon| < \sqrt{5}\,\mathbf{u}.$$

This implies that the normwise relative error is less than $\sqrt{5}\,\mathbf{u} = 2.236\ldots\mathbf{u}$. The sharpness of this bound is also established in [65], via the explicit construction of worst-case floating-point inputs. For example, in binary32 arithmetic ($p = 24$) with rounding to nearest ties-to-even, applying (11.6) with the floating-point inputs

$$a = \frac{3}{4}, \qquad b = \frac{3}{4}(1 - 2^{-22}), \qquad c = \frac{2}{3}(1 + 11 \cdot 2^{-24}), \qquad d = \frac{2}{3}(1 + 5 \cdot 2^{-24})$$

leads to the relative error $|\hat{z} - z|/|z| = \sqrt{4.99998\ldots}\,\mathbf{u}$.

11.4.2 Complex multiplication with an FMA instruction

If an FMA instruction is available, there are several ways of performing a complex product $(a + ib) \cdot (c + id)$, and Jeannerod et al. [294] consider three of them. A first possibility consists in simply returning

$$\mathrm{RN}(ac - \mathrm{RN}(bd)) + i \cdot \mathrm{RN}(ad + \mathrm{RN}(bc)). \tag{11.7}$$

Two other methods consist in using either Algorithm 11.1 or Algorithm 11.2, to separately evaluate with high relative accuracy the real and imaginary parts of the product.

If one is interested in minimizing the componentwise error, Equation (11.7) is to be avoided. Indeed, it can yield large componentwise relative errors, whereas with Kahan's method or the one by Cornea, Harrison, and Tang, the componentwise relative error is known to be at most $2\mathbf{u}$ or $2\mathbf{u} + \mathcal{O}(\mathbf{u}^2)$ and those bounds are asymptotically optimal (see Section 11.3).

However, if normwise error is at stake, the situation is quite different: with the three methods above, the normwise relative error is bounded by $2\mathbf{u}$ or $2\mathbf{u} + \mathcal{O}(\mathbf{u}^2)$, and specific inputs are described in [294] in order to show that these bounds are asymptotically optimal. Hence, if one is only interested in minimizing the normwise error, the first, cheaper algorithm defined by (11.7) may be a sensible choice.

11.5 Complex Division

Complex division is not as well understood as complex multiplication. It seems harder to obtain tight normwise error bounds. Moreover, the influence of underflows and overflows is more complicated to analyze.

11.5.1 Error bounds for complex division

Let us first give error bounds, assuming that no spurious overflows or underflows occur. We begin with methods that do not require the availability of an FMA instruction. Jeannerod et al. [298] first consider the *inversion* of a complex number $c + id$ in binary floating-point arithmetic, using the "naive" algorithm below (Algorithm 11.3), which is a direct implementation of the inversion formula

$$\frac{1}{c + id} = \frac{c - id}{c^2 + d^2}.$$

Algorithm 11.3 Inversion of a nonzero complex floating-point number $c + id$.

$\hat{\delta} \leftarrow \mathrm{RN}(\mathrm{RN}(c^2) + \mathrm{RN}(d^2))$
$\hat{R} \leftarrow \mathrm{RN}(c/\hat{\delta})$
$\hat{I} \leftarrow \mathrm{RN}(-d/\hat{\delta})$
return $\hat{R} + i\hat{I}$

It is shown in [298] that the componentwise relative error is at most $3\mathbf{u}$, and that this bound is asymptotically optimal when p is even, and sharp when p is odd. Furthermore, if $p \geq 24$, then the normwise relative error is at most $2.707131\mathbf{u}$. Although almost certainly not asymptotically optimal, that bound is reasonably sharp: for example, a normwise relative error of $2.70679\ldots \cdot \mathbf{u}$ is obtained in binary64 arithmetic ($p = 53$) for $c = 4503599709991314$ and $d = 6369051770002436 \cdot 2^{26}$.

Let us now turn to the problem of performing the complex division $(a + ib)/(c + id)$. From a componentwise point of view, a direct adaptation of Equation (11.2), i.e., the calculation of

$$\frac{\mathrm{RN}\left(\mathrm{RN}(ac) + \mathrm{RN}(bd)\right)}{\mathrm{RN}\left(\mathrm{RN}(c^2) + \mathrm{RN}(d^2)\right)} + i \cdot \frac{\mathrm{RN}\left(\mathrm{RN}(bc) - \mathrm{RN}(ad)\right)}{\mathrm{RN}\left(\mathrm{RN}(c^2) + \mathrm{RN}(d^2)\right)}, \tag{11.8}$$

results in arbitrarily large relative errors when catastrophic cancellations occur. This is not the case if we consider normwise errors. To our knowledge, assuming binary floating-point arithmetic, the best-known normwise relative error bound for the algorithm obtained using (11.8) is $(3 + \sqrt{5})\mathbf{u} + \mathcal{O}(\mathbf{u}^2) \approx 5.236\mathbf{u}$ (see [32]). However, a slightly smaller bound is obtained if we chose to first compute the inverse $c + id$ using Algorithm 11.3, and then multiply the obtained result by $a + ib$ using (11.6). In this case, the resulting error bound (for $p \geq 24$) is $(2.707131 + \sqrt{5})\mathbf{u} + \mathcal{O}(\mathbf{u}^2) \approx 4.943\mathbf{u}$.

Now, if an FMA instruction is available, one can use Kahan's algorithm (Algorithm 11.1) for evaluating the numerators accurately in (11.2). Several slightly different variants are possible. Consider for example the following algorithm.

Algorithm 11.4 Computation of $(a + ib)/(c + id)$ assuming that an FMA instruction is available.

$\hat{\delta} \leftarrow \mathrm{RN}(c^2 + \mathrm{RN}(d^2))$
$\hat{\nu}_{re} \leftarrow$ evaluation of $ac + bd$ with Algorithm 11.1
$\hat{\nu}_{im} \leftarrow$ evaluation of $bc - ad$ with Algorithm 11.1
$\hat{R} \leftarrow \mathrm{RN}\left(\hat{\nu}_{re}/\hat{\delta}\right)$
$\hat{I} \leftarrow \mathrm{RN}\left(\hat{\nu}_{im}/\hat{\delta}\right)$
return $\hat{R} + i\hat{I}$

Jeannerod, Louvet, and Muller [296] show that in binary floating-point arithmetic the componentwise relative error of Algorithm 11.4 is bounded by $5\mathbf{u} + 13\mathbf{u}^2$, and that this bound is asymptotically optimal when p is even, and sharp when p is odd. A consequence is that the normwise error too is bounded by $5\mathbf{u} + 13\mathbf{u}^2$ (however, this bound can probably be improved).

11.5.2 Scaling methods for avoiding over-/underflow in complex division

Overflows and underflows may occur while performing a complex division even if the result lies well within the limits. Consider for example, in the binary64 format of the IEEE 754-2008 standard, the division of $a + ib = 1 + i \cdot 2^{600}$ by $c + id = 2^{600} + i$. If we perform the calculations as indicated in (11.8), then the computations of $\mathrm{RN}(c^2)$ and $\mathrm{RN}(bc)$ will return $+\infty$, so that the final result will be $0 + i \cdot \mathrm{NaN}$, whereas the exact result is very close to $2^{-600} + i$.

The most famous method to work around such harmful intermediate underflows/overflows is due to Smith [561]. It consists in first comparing $|c|$ and $|d|$. After that,

- if $|c| \geq |d|$, one considers the formula

$$z = \frac{a + b(d/c)}{c + d(d/c)} + i \cdot \frac{b - a(d/c)}{c + d(d/c)},$$

and

- if $|c| \leq |d|$, one considers the formula

$$z = \frac{a(c/d) + b}{c(c/d) + d} + i \cdot \frac{b(c/d) - a}{c(c/d) + d}.$$

Doing this requires 1 comparison, 3 floating-point divisions, and 3 floating-point multiplications, instead of 2 divisions and 6 multiplications. Note that if dividing is much more expensive than multiplying, then one may consider (at the cost of a possible small loss of accuracy) inverting the common denominator of both the real and imaginary parts, and then multiplying the result by the numerators. This leads to 1 comparison, 2 divisions, and 5 multiplications for Smith's algorithm, and 1 division and 8 multiplications for the naive algorithm. In any case, Smith's algorithm can thus be significantly slower if (as it can happen) comparing and dividing are more expensive than multiplying. Note also that the error bounds obtained in Section 11.5.1 do not apply anymore, since one now has to take into account the additional rounding error of the division c/d or d/c.

Stewart [573] improves the accuracy of Smith's algorithm by performing a few additional comparisons. More precisely, he suggests computing the products adc^{-1}, bdc^{-1}, acd^{-1}, and bcd^{-1} in a way that prevents harmful underflows and overflows during that particular step: when computing $x \cdot y \cdot z$, if the result does not underflow or overflow, then it is safe to first compute the product of the two numbers of extremal magnitudes and then multiply the obtained result by the remaining term. However, Stewart's variant of Smith's algorithm can still suffer from harmful intermediate overflows and underflows. For instance, suppose that $0 < d < c$ and that both are close to the overflow limit. Then the denominator $c + d(dc^{-1})$ will overflow. Suppose furthermore that $a \approx c$ and $b \approx d$. Then the result of Stewart's algorithm will be NaN, although the result of the exact division is close to 1. Li et al. [384, Appendix B] show how to avoid harmful underflows and overflows in Smith's algorithm, by scaling the variables a, b, c, and d beforehand. We do not describe their method, since Priest's algorithm, given below, achieves the same result at a smaller cost.

Priest [497] uses a two-step scaling of the variables to prevent harmful overflows and underflows in the naive complex division algorithm in binary floating-point arithmetic. The scalings are particularly efficient because the common scaling factor s is a power of 2 that can be quickly determined from the input operands. Since, in radix-2 floating-point arithmetic, a multiplication by a power of 2 is exact (unless underflow or overflow occurs), the normwise error bound $(3+\sqrt{5})\mathbf{u}+\mathcal{O}(\mathbf{u}^2) \approx 5.236\mathbf{u}$ given in Section 11.5.1 applies to Algorithm 11.5 below in the absence of underflow and overflow. Note that if an FMA instruction is available, then the same scaling technique can be applied to Algorithm 11.4.

Algorithm 11.5 Priest's algorithm [497] for computing $(a+ib)/(c+id)$ using a scaling factor s, slightly modified: to avoid divisions, Priest computes $\mathrm{RN}(1/k)$ and obtains $\hat{R}$ and $\hat{I}$ by multiplying by this (approximate) inverse.

$c' \leftarrow \mathrm{RN}(sc)$ // exact operation: $c' = sc$
$d' \leftarrow \mathrm{RN}(sd)$ // exact operation: $d' = sd$
$k \leftarrow \mathrm{RN}(\mathrm{RN}(c'^2)+\mathrm{RN}(d'^2))$
$c'' \leftarrow \mathrm{RN}(sc')$ // exact operation: $c'' = sc'$
$d'' \leftarrow \mathrm{RN}(sd')$ // exact operation: $d'' = sd'$
$\hat{R} \leftarrow \mathrm{RN}\left(\mathrm{RN}(\mathrm{RN}(ac'')+\mathrm{RN}(bd''))/k\right)$
$\hat{I} \leftarrow \mathrm{RN}\left(\mathrm{RN}(\mathrm{RN}(bc'')-\mathrm{RN}(ad''))/k\right)$
return $\hat{R}+i\hat{I}$

In Algorithm 11.5, the scaling factor s is first applied to c and d to allow for a safe computation of (a scaling of) the denominator. It is applied a second time to allow for a safe computation of the numerators. Priest proves that one can always choose a scaling factor s that depends on the inputs (taking s close to $|c+id|^{-3/4}$ suffices for most values of $a+ib$), such that the following properties hold:

- If an overflow occurs in the computation, at least one of the two components of the exact value of
$$\frac{a+ib}{c+id}$$
is above the overflow threshold or within a few units in the last place of it.

- Any underflow occurring in the first five steps of Algorithm 11.5 or in the computations of $\mathrm{RN}(\mathrm{RN}(ac'')+\mathrm{RN}(bd''))$ and $\mathrm{RN}(\mathrm{RN}(bc'')-\mathrm{RN}(ad''))$ is harmless, i.e., its contribution to the relative error is small.

- Any underflow occurring in the divisions by k in the last two steps of Algorithm 11.5 incurs an error equal to at most a small integer times $\mathrm{ulp}(2^{e_{\min}}) = 2^{e_{\min}-p+1}$.

Furthermore, the restriction that no underflow nor overflow should occur may then be replaced by the weaker restriction that none of the components of the exact value of

$$\frac{a+ib}{c+id}$$

should lie above the overflow threshold nor within a few units in the last place of it.

In [497], Priest also provides a very efficient way of computing a scaling parameter s that satisfies the constraints for the above properties to hold, for the particular situation of IEEE binary64 arithmetic. In [317], Kahan describes a similar scaling method, but it requires possibly slow library functions to find and apply the scaling.

11.6 Complex Absolute Value

The complex absolute value of $|a+ib|$ is $\sqrt{a^2+b^2}$. As these absolute values are frequently used for "normalizing" a vector, accurately computing expressions of the form $c/\sqrt{a^2+b^2}$ is of interest too.

11.6.1 Error bounds for complex absolute value

Exactly as we did for division, let us first give error bounds for the computation of $\sqrt{a^2+b^2}$ and $c/\sqrt{a^2+b^2}$ assuming that no under-/overflow occurs. One easily shows that if these expressions are evaluated as

$$\mathrm{RN}\left(\sqrt{\mathrm{RN}(\mathrm{RN}(a^2)+\mathrm{RN}(b^2))}\right)$$

and

$$\mathrm{RN}\left(\frac{c}{\mathrm{RN}\left(\sqrt{\mathrm{RN}(\mathrm{RN}(a^2)+\mathrm{RN}(b^2))}\right)}\right),$$

then the relative error is bounded by $2\mathbf{u}+\mathcal{O}(\mathbf{u}^2)$ for the first function, and by $3\mathbf{u}+\mathcal{O}(\mathbf{u}^2)$ for the second one. With some care, it is possible to get rid of the $\mathcal{O}(\mathbf{u}^2)$ terms and get relative error bounds $2\mathbf{u}$ (always) and $3\mathbf{u}$ (in radix 2), respectively. This was shown in [303] for the first function, and in [299] for the second one. Furthermore, in binary floating-point arithmetic, these bounds are asymptotically optimal, and the same results hold if an FMA instruction is available and we replace $\mathrm{RN}(\mathrm{RN}(a^2)+\mathrm{RN}(b^2))$ by $\mathrm{RN}(a^2+\mathrm{RN}(b^2))$.

11.6.2 Scaling for the computation of complex absolute value

To prevent spurious overflows or underflows, a first possibility is to compare $|a|$ and $|b|$ and then to compute $|a|\cdot\sqrt{1+(b/a)^2}$ if $|a|\geq|b|$, and $|b|\cdot\sqrt{1+(a/b)^2}$

otherwise. The main drawback of this approach is that the division and the multiplication by a or b introduce additional errors, so that the relative error bound $2\mathbf{u}$ is to be replaced by $3.25\mathbf{u} + \mathcal{O}(\mathbf{u}^2)$.

A possible workaround is, as for division, to scale the operands by a power of the radix β of the floating-point format being used. Let s be the largest integral power of β less than or equal to $\max\{|a|, |b|\}$, that is, $\mathrm{ufp}\,(\max\{|a|, |b|\})$ with ufp denoting the unit in the first place; see [531] and Chapter 2. Then one can use the following algorithm.

Algorithm 11.6 Computation of a complex absolute value with operand scaling using $s = \mathrm{ufp}\,(\max\{|a|, |b|\})$.

$a' \leftarrow \mathrm{RN}(a/s)$ // exact operation: $a' = a/s$
$b' \leftarrow \mathrm{RN}(b/s)$ // exact operation: $b' = b/s$
$\rho \leftarrow \mathrm{RN}\left(\mathrm{RN}(a'^2) + \mathrm{RN}(b'^2)\right)$
$r' \leftarrow \mathrm{RN}\left(\sqrt{\rho}\right)$
$r \leftarrow \mathrm{RN}(r' \cdot s)$ // exact operation: $r = r's$
return r

To compute s, one can use, if they are implemented efficiently on one's system, the functions **logB** and **scaleB** specified by the IEEE 754-2008 standard: the largest power of β less than or equal to $|x|$, namely $\mathrm{ufp}(x)$, is $\mathbf{scaleB}(1, \mathbf{logB}(x))$. If these functions are either unavailable or slow, to compute $\mathrm{ufp}(x)$, one can directly modify the significand field in the binary representation of x or, in radix-2 arithmetic, to get a more portable program, use the following algorithm, due to Rump [522] (with the drawback that it does not work if $x \cdot (2^{p-1} + 1)$ overflows).

Algorithm 11.7 Computation of $\mathrm{sign}(x) \cdot \mathrm{ufp}(x)$ in precision-p, radix-2, floating-point arithmetic. It is Algorithm 3.5 in [521].

Require: $\varphi = 2^{p-1} + 1$ and $\psi = 1 - 2^{-p}$
$q \leftarrow \mathrm{RN}(\varphi x)$
$r \leftarrow \mathrm{RN}(\psi q)$
$\delta \leftarrow \mathrm{RN}(q - r)$
return δ

In Algorithm 11.6, a possible underflow in the computation of $\mathrm{RN}(b'^2)$ (if $|b| < |a|$) or $\mathrm{RN}(a'^2)$ (if $|a| < |b|$) is harmless, and the relative error bound $2\mathbf{u} + \mathcal{O}(\mathbf{u}^2)$ remains valid.

Kahan [317] gives another algorithm, significantly more complex than Algorithm 11.6, which is probably more accurate (although no tight error bound seems to be known). For computing $\sqrt{a^2 + b^2}$, assuming $a \geq b > 0$

with b not negligible, it uses the following formula

$$\sqrt{a^2+b^2} = a + \frac{b}{\frac{a}{b} + \sqrt{1+\left(\frac{a}{b}\right)^2}}. \tag{11.9}$$

Equation (11.9) is used in the straightforward manner if $a > 2b$. If $a \leq 2b$, then Kahan first computes

$$\theta = \left(\frac{a}{b}\right)^2 - 1$$

as $w(w+2)$ with $w \doteq (a-b)/b$. Here, note that $a-b$ is computed exactly thanks to Sterbenz' lemma (Lemma 4.1). Then, he computes the denominator $\frac{a}{b} + \sqrt{1+\left(\frac{a}{b}\right)^2}$ of the fraction in (11.9) as

$$(1+\sqrt{2}) + \frac{\theta}{\sqrt{2}+\sqrt{2+\theta}} + \left(\frac{a}{b} - 1\right).$$

11.7 Complex Square Root

11.7.1 Error bounds for complex square root

Using (11.4) to evaluate complex square roots may lead to spurious underflow or overflow and, even if all the intermediate variables are within the normal number range, may lead to poor accuracy if catastrophic cancellations occur.

A solution to avoid catastrophic cancellations is to evaluate $x + iy = \sqrt{a+ib}$ as follows:

- if $a \geq 0$, then compute

$$\begin{aligned} x &= \sqrt{\frac{\sqrt{a^2+b^2}+a}{2}}, \\ y &= \frac{b}{2x}; \end{aligned} \tag{11.10}$$

- if $a < 0$, then compute

$$\begin{aligned} y &= \operatorname{sign}(b) \cdot \sqrt{\frac{\sqrt{a^2+b^2}-a}{2}}, \\ x &= \frac{b}{2y}. \end{aligned} \tag{11.11}$$

The following algorithm can be traced back at least to Friedland [207]. It is presented assuming that $a \geq 0$, but the adaptation to the case $a < 0$ is straightforward using (11.11). We also assume that the case $a = b = 0$ has been handled in a preliminary step.

Algorithm 11.8 Calculation of $x+iy = \sqrt{a+ib}$ in binary, precision-p, floating-point arithmetic [207]. Here, we assume $a \geq 0$ and $a + ib \neq 0$.

$s \leftarrow \mathrm{RN}(\mathrm{RN}(a^2) + \mathrm{RN}(b^2))$
$\rho \leftarrow \mathrm{RN}(\sqrt{s})$
$\nu \leftarrow \mathrm{RN}(\rho + a)$
$\hat{x} \leftarrow \mathrm{RN}(\sqrt{\nu/2})$
$\hat{y} \leftarrow \mathrm{RN}(b/(2\hat{x}))$

It is easily shown that the componentwise relative error of Algorithm 11.8 is bounded by $\frac{7}{2}\mathbf{u}+\mathcal{O}(\mathbf{u}^2)$. This bound of course also applies to the normwise relative error. However, as shown by Hull, Fairgrieve, and Tang [264], it can be decreased to $\frac{\sqrt{37}}{2}\mathbf{u} + \mathcal{O}(\mathbf{u}^2)$, where $\frac{\sqrt{37}}{2} = 3.041\ldots$. Furthermore, note that this improved bound is rather tight: for example, in binary64 arithmetic ($p = 53$) and with

$$a = 650824205667/2^{52}$$

and

$$b = 4507997673885435/2^{51},$$

the normwise relative error of Algorithm 11.8 is $3.023\ldots\mathbf{u}$.

11.7.2 Scaling techniques for complex square root

A solution presented by Flannery et al. [201] is to first compute

$$w = \begin{cases} 0 & \text{if} \quad a = b = 0, \\ \sqrt{|a|}\sqrt{\dfrac{1+\sqrt{1+(b/a)^2}}{2}} & \text{if} \quad |a| \geq |b|, \\ \sqrt{|b|}\sqrt{\dfrac{|a/b|+\sqrt{1+(a/b)^2}}{2}} & \text{if} \quad |a| < |b|, \end{cases} \tag{11.12}$$

and then obtain

$$u + iv = \sqrt{a+ib} = \begin{cases} 0 & \text{if} \quad w = 0, \\ w + i\dfrac{b}{2w} & \text{if} \quad w \neq 0 \text{ and } a \geq 0, \\ \dfrac{|b|}{2w} + iw & \text{if} \quad w \neq 0 \text{ and } a < 0 \text{ and } b \geq 0, \\ \dfrac{|b|}{2w} - iw & \text{if} \quad w \neq 0 \text{ and } a < 0 \text{ and } b < 0. \end{cases} \tag{11.13}$$

This allows one to avoid intermediate overflows at the cost of more computation including several tests, divisions, and real square roots, which make the complex square root evaluation quite slow compared to a single arithmetic instruction. Also, compared to Algorithm 11.8, the additional multiplication and division by a (or b) lead to a slightly larger error bound. As in the case of the computation of complex absolute values, it is probably preferable, assuming binary arithmetic, to scale a and b by a power of two and then to apply Algorithm 11.8. Let $t = \text{ufp}(\max\{a, b\})$. We can define s as t if t is an even power of 2, and $2t$ otherwise. Then $\sqrt{a+ib}$ is computed as $\sqrt{s} \cdot \sqrt{(a/s) + i \cdot (b/s)}$. Here is a trick for easily checking if a number t that is known to be a power of 2 is an *even* power of 2: in binary32, binary64, Intel "double extended," and binary128 arithmetics, this is the case if and only if

$$\text{RN}\left[\left(\text{RN}\left(\sqrt{t}\right)\right)^2\right] = t.$$

In other words, one just has to square the square root of t. The reason behind this is that in radix-2 arithmetic, the only values between 3 and 114 of the precision p for which $\text{RN}\left[\left(\text{RN}\left(\sqrt{2}\right)\right)^2\right] = 2$ are 3, 8, 9, 10, 11, 16, 17, 18, 22, 26, 30, 31, 32, 33, 34, 35, 39, 40, 41, 43, 44, 45, 49, 51, 54, 57, 58, 59, 62, 63, 65, 67, 71, 73, 74, 75, 76, 77, 79, 81, 82, 87, 88, 90, 94, 95, 97, 98, 99, 101, 106, 108, 110, 112, and 114, and one can note that 24, 53, 64, and 113 do not belong to this list.

Kahan [317] gives a better solution than (11.12) and (11.13) that also correctly handles all special cases (infinities, zeros, NaNs, etc.), also at the cost of much more computations than the naive method.

11.8 An Alternative Solution: Exception Handling

Demmel and Li [163, 164], and Hull, Fairgrieve, and Tang [264, 265] suggest a very elegant solution for dealing with spurious underflows and overflows, using the exception-handling mechanisms specified by the IEEE standard. The idea is to begin with a direct evaluation of the simple formula. For instance, the absolute value of $a + ib$ is evaluated as $\sqrt{a^2 + b^2}$. In most cases, no intermediate underflow or overflow will occur, and we will obtain an accurate result. If an exception occurs then a significantly more complex program can be used. That program can be slow: it will seldom be used.

Chapter 12

Interval Arithmetic

THE AUTOMATION of the a posteriori analysis of floating-point error cannot be done in a perfect way (except possibly in straightforward or specific cases), yielding exactly the roundoff error. However, an approach based on interval arithmetic can provide results with a more or less satisfactory quality and with more or less efforts to obtain them. This is a historical reason for introducing interval arithmetic, as stated in the preface of R. Moore's PhD dissertation [427]: *"In a hour's time a modern high-speed stored-program digital computer can perform arithmetic computations which would take a "hand-computer" equipped with a desk calculator five years to do. In setting out a five year computing project, a hand computer would be justifiably (and very likely gravely) concerned over the extent to which errors were going to accumulate—not mistakes, which he will catch by various checks on his work—but errors due to rounding"* and discretization and truncation errors.

Very rapidly, interval computations have also been used to account for every source of error, such as measurement errors on physical data, truncation errors due to mathematical approximation, discretization errors due to finite representations on computer's memory, as well as roundoff errors. These errors are handled in a uniform way, oblivious to their origin.

The history of interval arithmetic is not developed here. Let us only cite a few pieces of work, to illustrate that interval arithmetic has been developed independently in Japan in 1956 by Sunaga in his Master's thesis [580], in the former USSR in 1962 by Kantorovich [327], in the USA by Moore in his above-cited PhD dissertation and in his subsequent books [428, 429], and in the UK as early as 1946, as reported by Wilkinson in his Turing award lecture in 1970 [635]: *"It is perhaps salutary to be reminded that as early as 1946 Turing had considered the possibility of working with both interval and significant digit arithmetic."* For more details about the history of interval arithmetic, see *Early papers* at http://www.cs.utep.edu/interval-comp/. In the following, references will be restricted to those published in English.

J.-M. Muller et al., *Handbook of Floating-Point Arithmetic*,
https://doi.org/10.1007/978-3-319-76526-6_12

12.1 Introduction to Interval Arithmetic

12.1.1 Definitions and the inclusion property

Before going into any details about the relations between interval arithmetic and floating-point arithmetic, let us define, from a mathematical point of view, interval arithmetic.

Definition 12.1 (Interval). *An interval is a closed and connected subset of* $\mathbb{R}$.

According to this definition, $\emptyset$, $[-1,3]$, $(-\infty,2]$, $[-7,+\infty)$ and $(-\infty,+\infty)$ are intervals. On the contrary, $(1,5]$, $(0,+\infty)$ and $[-1,3]\cup[7,9]$ are not intervals: the first two are not closed and the third one is not connected.

In the following, intervals are denoted in bold face, as recommended in [332]: $\mathbf{x} = \{x \in \mathbf{x}\}$. The *endpoints* of $\mathbf{x}$, also called *bounds* in [268], are denoted as $\underline{x}$ and $\overline{x}$ so that $\mathbf{x} = [\underline{x}, \overline{x}] = \{x \ :\ \underline{x} \le x \le \overline{x}\}$.

Definition 12.2 (Interval operation). *Using set theory, a n-ary operation*[1] op : $\mathbb{R}^n \to \mathbb{R}$ *can be extended to a n-ary* interval operation *as follows:*

$$\begin{aligned}\mathrm{op}(\mathbf{x}_1,\ldots,\mathbf{x}_n) \;&=\; \mathrm{Hull}\{\mathrm{op}(x_1,\ldots,x_n),\ x_i \in \mathbf{x}_i \textit{ for } 1 \le i \le n \\ &\qquad \textit{and } (x_1,\ldots,x_n) \in \mathrm{Dom}(\mathrm{op})\}\end{aligned}$$

where

- Dom(op) *is the domain of* op: $(x_1,\ldots,x_n) \in \mathrm{Dom}(\mathrm{op})$ *iff* $\mathrm{op}(x_1,\ldots,x_n)$ *is defined;*
- Hull *stands for the convex hull of the set: it is not mandatory in the set theory but it is required for interval arithmetic. Indeed, taking the convex hull of the range of* op *on* $\mathbf{x}_1,\ldots,\mathbf{x}_n$ *ensures that the result is an interval.*

This mathematical definition can be translated into formulas for each specific operation. These formulas are obtained by using the monotonicity of the operation:

$$\begin{aligned}
\mathbf{x}+\mathbf{y} &= [\underline{x}+\underline{y}, \overline{x}+\overline{y}] \\
-\mathbf{y} &= [-\overline{y}, -\underline{y}] \\
\mathbf{x}-\mathbf{y} &= [\underline{x}-\overline{y}, \overline{x}-\underline{y}] \\
\mathbf{x}\cdot\mathbf{y} &= [\min(\underline{x}\cdot\underline{y}, \underline{x}\cdot\overline{y}, \overline{x}\cdot\underline{y}, \overline{x}\cdot\overline{y}), \\
&\qquad \max(\underline{x}\cdot\underline{y}, \underline{x}\cdot\overline{y}, \overline{x}\cdot\underline{y}, \overline{x}\cdot\overline{y})] \\
\mathbf{x}/\mathbf{y} &= [\min(\underline{x}/\underline{y}, \underline{x}/\overline{y}, \overline{x}/\underline{y}, \overline{x}/\overline{y}), \\
&\qquad \max(\underline{x}/\underline{y}, \underline{x}/\overline{y}, \overline{x}/\underline{y}, \overline{x}/\overline{y})] \text{ if } 0 \notin \mathbf{y} \\
\sqrt{\mathbf{x}} &= \begin{cases} [\sqrt{\underline{x}}, \sqrt{\overline{x}}] & \text{if } \ \underline{x} \ge 0, \\ [0, \sqrt{\overline{x}}] & \text{if } \ \underline{x} \le 0 \le \overline{x}, \\ \emptyset & \text{if } \ \overline{x} < 0. \end{cases}
\end{aligned} \tag{12.1}$$

[1]It must be noted that in this chapter, the word "operation" denotes either an arithmetic operation or one of the elementary functions developed earlier. This denomination follows the IEEE 1788-2015 standard for interval arithmetic [268].

The formulas for multiplication and division are obtained by splitting $\mathbf{x}$ and $\mathbf{y}$ according to the sign of their elements and by using the monotonicity of the operation on each part $\mathbf{x} \cap \mathbb{R}^-$, $\mathbf{x} \cap \mathbb{R}^+$, $\mathbf{y} \cap \mathbb{R}^-$, and $\mathbf{y} \cap \mathbb{R}^+$. Partial monotonicity is also used to derive formulas for the exponential, the logarithm, and the trigonometric functions. Naturally, these formulas are simpler to express when the function is monotonic.

Once intervals and operations are defined, they can be combined to build more complex objects: following Neumaier [455, Chapter 1], let us define *arithmetic expressions*.

Definition 12.3 (Arithmetic expression in the variables $\xi_1, \ldots, \xi_n$). *An arithmetic expression in the formal variables $\xi_1, \ldots, \xi_n$ is defined using the following grammar:*

$$\begin{array}{lll} & c \in \mathbb{R} & \\ \mid & \xi_j & \textit{for } j \in \{1, \ldots, n\} \\ \mid & \mathrm{op}(e_1, \ldots, e_k) & \textit{if } \mathrm{op} \textit{ is a } k\textit{-ary operation and} \\ & & e_1, \ldots, e_k \textit{ are arithmetic expressions.} \end{array}$$

The operations belong to a predefined set including $+$, $-$, $\times$, $/$, $\sqrt{\cdot}$, $\square^2$, exp, log, sin... *An arithmetic expression can also be defined as a DAG (directed acyclic graph) with constants and variables as leaves and with operations and functions from a given set as internal nodes.*

From now on, it is assumed that every function under consideration is defined by an arithmetic expression. If every variable ξ_i in an arithmetic expression defining a function f is replaced by an interval $\mathbf{x}_i$, the expression can be evaluated using Formulas (12.1). The result is an interval $\mathbf{y}$ that includes, or encloses, the range of f over $\mathbf{x}_1, \ldots, \mathbf{x}_n$. This inclusion property is the main property of interval arithmetic [456, Chapter 1, Theorem 1.5.6]. It is often referred to as the *Fundamental Theorem of Interval Arithmetic*, and sometimes as the *Thou Shalt Not Lie* commandment of interval arithmetic. It is the most valuable feature of interval arithmetic.

Theorem 12.1 (Inclusion Property, or Fundamental Theorem of Interval Arithmetic). *Suppose that an arithmetic expression defining a mathematical function f on formal variables $\xi_1, \ldots, \xi_n$ can be evaluated at $\mathbf{x}_1, \ldots, \mathbf{x}_n$ where the $\mathbf{x}_i$ are intervals, and let denote by $\mathbf{f}(\mathbf{x}_1, \ldots, \mathbf{x}_n)$ the result of this evaluation, then*

$$\{f(x_1, \ldots, x_n), x_i \in \mathbf{x}_i\} \subseteq \mathbf{f}(\mathbf{x}_1, \ldots, \mathbf{x}_n).$$

The range of f over $\mathbf{x}_1, \ldots, \mathbf{x}_n$ is included in the result of the evaluation using interval arithmetic.

The result of the evaluation using interval arithmetic "does not lie" in the sense that it says where the values $f(x_1, \ldots, x_n)$ are. Indeed, no value

$f(x_1, \ldots, x_n)$ can lie outside. The inclusion property means that interval arithmetic provides guaranteed computations: the exact result of a real computation belongs to the computed interval. It also offers set computing, as it is based on set theory.

12.1.2 Loss of algebraic properties

Note that, while the bounds computed by naive interval arithmetic are guaranteed, they are not necessarily tight. The phenomenon of computing bounds which are too large is known as *overestimation*. Indeed, interval arithmetic does not offer the usual properties of real arithmetic. Let us detail these issues.

Consider a real number x belonging to the interval $[0, 1]$. By interval arithmetic, the expression $x - x$ is known to be contained in $[0, 1] - [0, 1] = [0 - 1, 1 - 0] = [-1, 1]$, which is much wider than the tightest enclosure $x - x \in [0, 0]$. To explain this result, let us recall the formula for the addition of two intervals $\mathbf{x}$ and $\mathbf{y}$:

$$\mathbf{x} + \mathbf{y} = [\underline{x}, \overline{x}] + [\underline{y}, \overline{y}] = [\underline{x} + \underline{y}, \overline{x} + \overline{y}].$$

A consequence of this formula is that an interval $\mathbf{x}$ has usually no opposite: there is no interval $\mathbf{y} = [\underline{y}, \overline{y}]$ such that $\mathbf{x} + \mathbf{y} = [0, 0]$. A simple proof consists in studying the widths of the intervals: if $\underline{x} \neq \overline{x}$ then the width of $\mathbf{x}$: $\overline{x} - \underline{x}$, is positive. As the width of $\mathbf{x} + \mathbf{y}$ is the sum of the width of $\mathbf{x}$ and the width of $\mathbf{y}$: $\mathrm{width}(\mathbf{x} + \mathbf{y}) = (\overline{x} + \overline{y}) - (\underline{x} + \underline{y}) = (\overline{x} - \underline{x}) + (\overline{y} - \underline{y})$, if $\underline{x} \neq \overline{x}$ then $\mathrm{width}(\mathbf{x} + \mathbf{y}) > \mathrm{width}(\mathbf{y}) \geq 0 = \mathrm{width}([0, 0])$. In other words, it is impossible to determine an interval $\mathbf{y}$ such that $\mathbf{x} + \mathbf{y} = [0, 0]$. More generally, every addition or subtraction increases the widths of the intervals. This means that a formula that uses the cancellation $x - x = 0$ cannot be employed when it is evaluated using interval arithmetic. Strassen's formulas for the fast product of matrices [577] are based on this rewriting; so are finite difference formulas for the approximation of derivatives. These formulas are therefore useless in interval arithmetic: they return much too large intervals.

A similar conclusion holds for multiplication. An interval $\mathbf{x} = [\underline{x}, \overline{x}] \not\ni 0$ has usually no inverse: if $\underline{x} < \overline{x}$, there is no interval $\mathbf{y} = [\underline{y}, \overline{y}]$ such that $\mathbf{x} \cdot \mathbf{y} = [\min(\underline{x} \cdot \underline{y}, \underline{x} \cdot \overline{y}, \overline{x} \cdot \underline{y}, \overline{x} \cdot \overline{y}), \max(\underline{x} \cdot \underline{y}, \underline{x} \cdot \overline{y}, \overline{x} \cdot \underline{y}, \overline{x} \cdot \overline{y})] = [1, 1]$. Let us prove it. As $\mathbf{x} \not\ni 0$, $\underline{x} \cdot \overline{x} > 0$. Let us assume that $\underline{x} > 0$, the case $\overline{x} < 0$ is similar. If $\mathbf{y}$ exists, then one also expects that $\underline{y} > 0$. The formula for the product $\mathbf{x} \cdot \mathbf{y}$ can then be simplified into $\mathbf{x} \cdot \mathbf{y} = [\underline{x} \cdot \underline{y}, \overline{x} \cdot \overline{y}]$. For $\mathbf{x} \cdot \mathbf{y}$ to be equal to $[1, 1]$, the following equalities must hold: $\underline{y} = 1/\underline{x}$ and $\overline{y} = 1/\overline{x}$. This means that $\mathbf{y} = [1/\underline{x}, 1/\overline{x}]$ with $\underline{x} \leq \overline{x}$ and $1/\underline{x} \leq 1/\overline{x}$: this is possible iff $\underline{x} = \overline{x}$ and thus $\underline{y} = \overline{y}$. In such a situation, $\mathbf{x}$ contains a single real number. Otherwise, as for the addition, if the width of $\mathbf{x}$ is positive, then $\mathbf{x}$ has no inverse.

Another important algebraic property that is lost in interval arithmetic is the distributivity of multiplication over addition. Only the inclusion $\mathbf{x} \cdot (\mathbf{y} +$

$\mathbf{z}) \subseteq \mathbf{x} \cdot \mathbf{y} + \mathbf{x} \cdot \mathbf{z}$ (called *subdistributivity*) holds. Let us check the following example for confirmation.

$$[1,2] \cdot ([-5,-3] + [4,7]) = [1,2] \cdot [-1,4] = [-2,8]$$

and

$$[1,2] \cdot [-5,-3] + [1,2] \cdot [4,7] = [-10,-3] + [4,14] = [-6,11] \supsetneq [-2,8].$$

The phenomenon explaining the loss of these properties is called *variable dependency*, a more accurate name should be *variable decorrelation*. Let us go back to the definition of an operation, for instance $\mathbf{x} - \mathbf{y} = \{x - y,\ x \in \mathbf{x},\ y \in \mathbf{y}\}$. If $\mathbf{y} = \mathbf{x}$, this formula becomes $\mathbf{x} - \mathbf{x} = \{x - y,\ x \in \mathbf{x},\ y \in \mathbf{x}\}$: this set contains 0, obtained when $x = y$, but also other elements, obtained when $x \neq y$. If the programmer had in mind $x - x = 0$, that is, the two occurrences of x are identical, then this information is lost in the formula for $\mathbf{x} - \mathbf{x}$: the dependency—or the correlation—between the two occurrences of x is lost.

The absence of these algebraic properties explains why the development of algorithms computing tight interval enclosures is an active research field. Indeed, the development of efficient expressions and algorithms using interval arithmetic is delicate. It requires to use $x \cdot (y + z)$ rather than $x \cdot y + x \cdot z$, or to avoid the rewriting of 0 into $x - x$. Developers should stick to the formulas acting on real numbers as long as possible, and should introduce intervals as late as possible, after every needed algebraic manipulation or simplification has taken place. They should also keep, as much as possible, real numbers rather than intervals. Several variants of interval arithmetic have been developed to cure the variable dependency problem, such as affine arithmetic or Taylor models. They will be briefly explained in Section 12.4.3.

12.2 The IEEE 1788-2015 Standard for Interval Arithmetic

Interval arithmetic is standardized, under the auspices of the IEEE, as the IEEE 1788-2015 standard for interval arithmetic [268]. One of the motivations for this standardization effort was that the libraries that implemented interval arithmetic were based on different paradigms, which were not compatible. Thus comparisons were difficult or even impossible.

Let us introduce the main features of this standard.

12.2.1 Structuration into levels

The first level is the *mathematical level*. It is the starting point of the standard: notions are defined at the mathematical level and the implementation issues derive from these notions. Level 1 defines intervals on real numbers and operations on intervals. Level 2 addresses *discretization* issues, prior to implementation questions that are the object of Level 3. Level 2 details how one goes from Level 1 to a discrete and finite world while preserving the inclusion property: from real values to approximations, from operations on real numbers to operations on Level 2 entities that return Level 2 entities. Topics at Level 2 are: rounding functions, handling of overflows (values too large to be represented), input and output. Level 3 is the *implementation level.* It specifies requirements about the type of the discrete approximations detailed at Level 2. Level 4 is the *encoding level*: it specifies which bit strings should represent objects of Level 3.

12.2.2 Flavors

The definition of an interval has been a controversial issue. The common basis, upon which everybody agrees, deals only with nonempty bounded closed connected subsets of $\mathbb{R}$: everybody agrees on the meaning of $[1,3]$, but unfortunately no more on the meaning of $[5,+\infty)$, $[5,2]$ or $[1,2]/[0,1]$. Each of these examples has a meaning in set theory, Kaucher arithmetic [330], modal intervals [218, 219] or Cset theory [499], but not in all of them and not necessarily the same. The standard is designed to accommodate several theories. It provides definitions and requirements for the intervals that have a common definition; these intervals are called *common intervals*. A theory can be incorporated in the standard if it complies with the definitions and requirements for common intervals and if it extends them to non-common intervals. In the phrasing of IEEE 1788-2015, such a theory is called a *flavor*. An implementation is IEEE 1788-2015 compliant if it implements at least one flavor.

12.2.2.1 Common intervals and operations

As stated above, a common interval is a nonempty bounded closed connected subset of $\mathbb{R}$. Operations on common intervals are defined when their operands are included in their domain. Required operations are the usual arithmetic operations $+$, $-$, $\times$, $/$, $\square^2$, $\sqrt{\cdot}$, fma, and several common functions: exponential, logarithm, and trigonometric functions [268, Table 9.1]. Other common—but probably less usual—operations include the reverse operations for the addition and the subtraction, `cancelPlus` and `cancelMinus`. For common intervals $\mathbf{x} = [\underline{x}, \overline{x}]$ and $\mathbf{y} = [\underline{y}, \overline{y}]$, the operation `cancelMinus` is defined if and only if the width of $\mathbf{x}$ is larger or equal to the width of $\mathbf{y}$, that is, $\overline{x} - \underline{x} \geq \overline{y} - \underline{y}$; it returns the unique interval $\mathbf{z}$ such that $\mathbf{y} + \mathbf{z} = \mathbf{x}$. The result $\mathbf{z}$ is given by the formula $\mathbf{z} = [\underline{x} - \underline{y}, \overline{x} - \overline{y}]$. The condition on the widths of $\mathbf{x}$ and $\mathbf{y}$ ensures that $\underline{z} \leq \overline{z}$. The operation `cancelPlus` is defined as `cancelMinus`$(\mathbf{x}, -\mathbf{y})$.

Other required operations that are specific to sets or to intervals are the intersection, the convex hull of the union of two intervals, and some numeric functions: if $\mathbf{x} = [\underline{x}, \overline{x}]$, then $\inf(\mathbf{x}) = \underline{x}$, $\sup(\mathbf{x}) = \overline{x}$, $\mathrm{mid}(\mathbf{x}) = (\underline{x} + \overline{x})/2$, $\mathrm{wid}(\mathbf{x}) = \overline{x} - \underline{x}$, $\mathrm{rad}(\mathbf{x}) = (\overline{x} - \underline{x})/2$, $\mathrm{mag}(\mathbf{x}) = \sup\{|x|, x \in \mathbf{x}\} = \max(|\underline{x}|, |\overline{x}|)$, $\mathrm{mig}(\mathbf{x}) = \inf\{|x|, x \in \mathbf{x}\} = 0$ if $0 \in \mathbf{x}$ and $\min(|\underline{x}|, |\overline{x}|)$ otherwise.

12.2.2.2 Set-based flavor

The definition of intervals and operations given in Section 12.1 includes the common intervals and operations. However, the empty set and unbounded intervals are also allowed in Definition 12.1. Operations acting on intervals not necessarily included in their domain are allowed in Definition 12.2. These definitions, based on set theory, correspond to the so-called *set-based flavor* [268, §10 to 14]. This flavor requires the availability of several reverse operations in addition to `cancelMinus` and `cancelPlus`, namely the reverse of the absolute value, of the square and more generally of the power by a natural integer, of the sine, cosine, tangent, and hyperbolic cosine, as well as the multiplication, power, and $\mathrm{atan}(x/y)$. These reverse functions are needed to solve constraints [541], which are sets of equalities and inequalities for which *every* solution is sought. For instance, solving the constraint $x^2 = 4$ yields the two values $x = 2$ and $x = -2$; this illustrates that the reverse of the square function is not the square root function. One of the strengths of interval arithmetic is to offer such functions and to enable to solve numeric constraints.

Another specific strength of interval arithmetic is that it allows one to determine all the zeros of a mathematical function f on a given interval x, even when the derivative of f contains 0. In particular, this happens when f has several zeros: its derivative vanishes between the zeros by Rolle's theorem. The interval version of the Newton–Raphson iteration [235] is the following:

$$\begin{cases} \mathbf{x}_0 \text{ is given,} \\ \mathbf{x}_{n+1} = \left(\mathbf{x}_n - \frac{\mathbf{f}(\{x_n\})}{\mathbf{f}'(\mathbf{x}_n)}\right) \bigcap \mathbf{x}_n \text{ where } x_n \in \mathbf{x}_n. \end{cases}$$

This iteration even takes benefit from the case where the derivative f' on $\mathbf{x}_0$ vanishes (and thus the denominator $\mathbf{f}'(\mathbf{x}_0)$ contains 0) because the function f has several zeros on $\mathbf{x}_0$: it is able to separate the zeros, as long as a specific division is provided. Let us illustrate this division on an example. If $\mathbf{x} = [4, 6]$ and $\mathbf{y} = [-1, 2]$, in the set theory, the division of $\mathbf{x}$ by $\mathbf{y}$ yields $(-\infty, -4] \cup [2, +\infty)$. The division from Definition 12.2 yields $(-\infty, +\infty)$ which is the convex hull of the result. However, in the framework of the Newton–Raphson iteration, if this gap $(-4, 2)$ were produced by the division in the formula above, it would have the following meaning: no zero of f lies in this gap.

To be able to preserve the gap, the set-based flavor mandates a division that can return either the empty set, or one, or two intervals. This function is called `mulRevToPair`. It cannot belong to the common part of the standard, as it can return the empty set or unbounded intervals, which are not common intervals.

Let us illustrate a last difference between the common part of the standard and the set-based flavor. It concerns the numeric functions specific to interval arithmetic: inf, sup, mid, wid, rad, mag, mig. Their return value, when given either the empty set or an unbounded interval, must be specified. At Level 1, values are dictated by the Set theory, such as $\mathrm{mag}([-3, +\infty)) = +\infty$ or $\mathrm{wid}(\emptyset)$ has no value. However, the behavior must be specified at Level 2, as an operation always returns a value or triggers an exception. When the Level 1 value does not exist, the operation at Level 2 returns either a special value (NaN for most functions) or a value deemed reasonable in practice; for instance, it was decided that returning 0 for $\mathrm{mid}((-\infty, +\infty))$ was most useful. For the whole specification of operations required by the set-based flavor, see Section 10.5 of the standard.

The set-based flavor is currently the only flavor of the IEEE 1788-2015 standard. During the development of the standard, other flavors have been discussed, namely Kaucher arithmetic [330], modal intervals [218, 219], Cset theory [499] and Rump's proposal for handling overflows [524]. However, none has been elaborated far enough to become a flavor. Any person proposing a new flavor should submit it as a revision of the standard.

Any conforming implementation shall provide at least one flavor. For the time being, this means that a conforming implementation must provide the set-based flavor.

12.2.3 Decorations

The standard aims at preserving the inclusion property of interval arithmetic. The set-based flavor is designed so as to preserve the unique capabilities of interval arithmetic for solving numeric constraints and for the determination of all zeros of a function by the Newton–Raphson iteration. Another valuable capability of interval arithmetic is that it can prove the existence of a fixed point of a function, or equivalently the existence of a zero.

What makes it possible is Brouwer's theorem. That theorem states that a continuous function f that maps a compact set K onto itself has a fixed point in K. This means that if the evaluation, using interval arithmetic, of an arithmetic expression of f on a bounded (and thus compact) interval $\mathbf{x}$ yields $\mathbf{f}(\mathbf{x})$ such that $\mathbf{f}(\mathbf{x}) \subseteq \mathbf{x}$, then, as $f(\mathbf{x}) \subseteq \mathbf{f}(\mathbf{x})$, $f(\mathbf{x}) \subseteq \mathbf{x}$ holds and thus Brouwer's theorem applies.

However, such a result must be used with caution, as shown by the following example.

Consider $\mathbf{x} = [-4, 9]$ and $f : x \mapsto \sqrt{x} - 2$. We have $\mathrm{f}(\mathbf{x}) = [0, 3] - 2 = [-2, 1] \subset [-4, 9]$. However, f has no real fixed point and in particular no fixed point in $[-4, 9]$. What prevents us from applying Brouwer's theorem here is the fact that f is not continuous on $\mathbf{x}$, as f is not everywhere defined on $\mathbf{x}$. The standard must provide a means to detect this case, otherwise false conclusions could be drawn, which are not compatible with the "Thou shalt not lie" commandment. This is the motivation for the decoration system of the IEEE 1788-2015 standard. It was decided to avoid the use of global information, such as the status flags of IEEE 754. Instead, it was decided to attach this piece of information to the resulting interval. This piece of information is called *decoration* and an interval bearing a decoration is called a *decorated interval*.

The decoration `com` is the only decoration for common intervals. Each flavor must define its own set of decorations, with the rules used to compute them when evaluating an arithmetic expression. For any flavor, *a decoration primarily describes a property, not of the interval it is attached to, but of the function defined by some code that produced the interval by evaluating over some input box* [268, §8.1]. The decorations that were deemed relevant for the set-based flavor are listed below.

- `com` corresponds to a common interval and every flavor must provide it. An interval is decorated as `com` if it is a common interval and it is the result of the evaluation of an arithmetic expression, where every interval is a nonempty bounded interval and where every operation is defined and continuous on its arguments.
- `dac` stands for defined and continuous. An interval is decorated as `dac` if it is the result of the evaluation of an arithmetic expression, where every interval is a nonempty interval and where every operation is defined and continuous on its arguments.
- `def` stands for defined. An interval is decorated as `def` if, in the previous definition, the continuity cannot be established.
- `trv` stands for trivial. This decoration conveys no information.
- `ill` stands for ill-formed. An interval is decorated `ill` if it is an invalid interval, typically the result of a constructor with invalid arguments such as "$[\text{incorrect}, 2]$."

Decorations `com`, `dac`, `def`, and `trv` are given here in decreasing order of information they convey. More precisely, the set of intervals decorated `com` is a subset of intervals decorated `dac`, which is a subset of intervals decorated `def`, which is a subset of intervals decorated `trv`.

It has been mentioned that the decoration is a property, not of an interval, but of the computation that led to this interval, because, in the set-based flavor, a decoration is computed as follows. Let $\mathbf{x}_{dx}$ be a vector of decorated intervals $((\mathbf{x}_1, dx_1), \ldots, (\mathbf{x}_n, dx_n))$ and let op be a n-ary operation. The *decorated interval extension* of op is defined as follows:

$$(\mathbf{y}, dy) = \mathrm{op}((\mathbf{x}_1, dx_1), \ldots, (\mathbf{x}_n, dx_n))$$

if and only if

- $\mathbf{y} = \mathrm{op}(\mathbf{x}_1, \ldots, \mathbf{x}_n)$: $\mathbf{y}$ does not depend on the decorations dx_i;
- some decoration d states whether $\mathbf{y}$ is a common interval and whether op is defined and continuous on $\mathbf{x}_1, \ldots, \mathbf{x}_n$, or simply defined, or none of the above; this is a local information;
- $dy = \min(d, dx_1, \ldots, dx_n)$ where the minimum is taken with respect to this propagation order: `com` > `dac` > `def` > `trv` > `ill`. This takes into account both the local decoration and the full history of the computation that leads to $\mathbf{y}$. The resulting decoration dy is a property of this computation.

The system of decorations enables the expert user to check whether the conditions of Brouwer's theorem hold before jumping into its conclusion. As the interval part is computed independently of the decoration system, the user who does not need the decorations can ignore them safely.

The standard does not mandate that the tightest decoration is determined, as it may be too difficult. However, the standard states, for the set-based flavor, a decorated version of the fundamental theorem of interval arithmetic and its proof [268, Annex B].

Theorem 12.2 (Fundamental Theorem of Decorated Interval Arithmetic). *Let* $\mathbf{f}$ *be an arithmetic expression defining a mathematical function* f. *Let* $(\mathbf{x}, dx) = ((\mathbf{x}_1, dx_1), \ldots, (\mathbf{x}_n, dx_n))$ *be a vector of decorated intervals, that is,* $\mathbf{x}_i$ *is an interval for* $1 \le i \le n$ *and* dx_i *is the decoration of interval* $\mathbf{x}_i$, *and denote by* $(\mathbf{y}, dy)$ *the result of the evaluation* $\mathbf{f}((\mathbf{x}, dx))$.
If any component of $(\mathbf{x}, dx)$ *is decorated* `ill`, *then* $dy = \texttt{ill}$.
If no component of $(\mathbf{x}, dx)$ *is decorated* `ill`, *and if none of the operations in* $\mathbf{f}$ *is an everywhere undefined function, then*

$$dy \neq \texttt{ill},$$
$$\mathbf{y} \supseteq \mathrm{range}(f, \mathbf{x}),$$
$$dy \textit{ holds for the evaluation of } f \textit{ on } \mathbf{x}.$$

Every flavor must define its set of decorations with their propagation rules and must establish the corresponding Fundamental Theorem of Decorated Interval Arithmetic.

Other issues related to decorations are detailed in the standard, such as getting access to the decoration, or setting a specified or a default value for a decoration (in a constructor in particular). An optional part regards compressed intervals, where only intervals with a decoration above a given threshold are relevant; other ones are discarded and only the decoration is kept.

12.2.4 Level 2: discretization issues

It has been stated, at the beginning of this description of IEEE 1788-2015, that the starting point for the elaboration of the standard was the mathematical level, or Level 1. Going from Level 1 to Level 2, the discretization level, implies to solve issues related to the use of a finite set of intervals. This finite set is called an *interval type* in what follows. The choice of another finite set corresponds to another interval type.

Any Level 1 interval $\mathbf{z}$ satisfies the inclusion property. At Level 2, $\mathbf{z}$ must be represented by an interval $\widehat{\mathbf{z}}$ that belongs to the given interval type and that also satisfies the inclusion property, i.e., $\widehat{\mathbf{z}} \supseteq \mathbf{z}$. The interval $\widehat{\mathbf{z}}$ can be much larger than $\mathbf{z}$; for instance, in case of an overflow, $\widehat{\mathbf{z}}$ can be unbounded whereas $\mathbf{z}$ is bounded.

Generally speaking, the relation between a Level 1 operation and its Level 2 counterpart is similar to the notion of correct rounding in the IEEE 754-2008 standard. For most operations, it acts as if the operation is first performed at Level 1.

If at Level 1 the result is defined, then that result is converted to a Level 2 result of the appropriate interval type. This ensures that, for most operations, the result at Level 2 will be the tightest possible, that is, the smallest Level 2 interval for inclusion.

Another issue when going from Level 1 to Level 2 is that, at Level 2, every operation must return a result. However, at Level 1, for some operations and some particular inputs, there might be no valid result. When the Level 1 result does not exist, the operation at Level 2 returns either a special value indicating this event (e.g., NaN for most of the numeric functions) or a value considered reasonable in practice, as exemplified in Section 12.2.2.2 and detailed in [268, §7.5.3].

The representation of intervals at Level 2 is related to the number formats. Two representations are mentioned in the standard. The most common one is the representation of an interval by its bounds: $\widehat{\mathbf{z}} = [\underline{\widehat{z}}, \overline{\widehat{z}}]$ where $\underline{\widehat{z}}$ and $\overline{\widehat{z}}$ are floating-point numbers, together with the empty set. The other one is the representation of an interval by its midpoint and its radius: $\widehat{\mathbf{z}} = \langle \widehat{z}_m, \widehat{z}_r \rangle = [\widehat{z}_m - \widehat{z}_r, \widehat{z}_m + \widehat{z}_r]$ with $\widehat{z}_m$ and $\widehat{z}_r$ finite, $\widehat{z}_m$ and $\widehat{z}_r$ are floating-point numbers, $\widehat{z}_r \geq 0$, along with the representation of the empty set, of $\mathbb{R}$, and possibly of semi-bounded intervals.

The relationship with the IEEE 754-2008 standard is given in more details: a representation is *754-conforming* if the intervals are represented by their bounds $\widehat{\mathbf{z}} = [\underline{\widehat{z}}, \overline{\widehat{z}}]$ and if $\underline{\widehat{z}}$ and $\overline{\widehat{z}}$ belong to one of the IEEE 754-2008 basic formats (see Section 3.1).

A last point that is specified by the standard, but is not really relevant for this book, is the description of literals for intervals. For instance, `[]` stands for the empty set, `[entire]` for $\mathbb{R}$, and `[3.560?2]` for $[3.558, 3.562]$ as the quantity following the question mark corresponds to the uncertainty on the last digit.

12.2.5 Exact dot product

Another topic, related exclusively to Level 2, is the recommendation that, for each supported 754-conforming interval type, the four operations `sum`, `dot`, `sumSquare`, and `sumAbs` [267, §9.4] are provided with correct rounding. The rationale is that IEEE 1788-2015 addresses the numerical quality of a computation and this recommendation concurs with this concern.

12.2.6 Levels 3 and 4: implementation issues

Numbers that are used to handle intervals, such as the bounds or the width, are dealt with at Level 2: they are either exact values, or floating-point values. Level 3 and Level 4 issues for these numbers are thus specified elsewhere.

The main point concerns decorations, as they are not defined anywhere else. The bit string representation of the decorations for the set-based flavor is given in Clause 14.4 of [268]; the compressed intervals are also detailed.

12.2.7 Libraries implementing IEEE 1788–2015

To our knowledge, only two libraries are compliant with the IEEE 1788-2015 standard. One of them is `libieee1788`, developed by Nehmeier [452]. It is a C++ library that is fully compliant with the standard. It has been developed as a proof-of-concept of the standard. The other one is the Gnu Octave package `interval`, developed by Heimlich [252]. As its name suggests, it is called through Gnu Octave, which makes it very easy to use and thus allows one to test and check ideas and algorithms very rapidly. To keep the development simple, the only numerical type allowed for the endpoints in particular is the binary64 format (see Section 3.1.1.1). Unfortunately, neither author is still in academy and it is not clear how these libraries will be maintained over time.

An issue raised by the developers (of these libraries and of other attempts) is the difficulty to deal with intervals represented by bounds of any type. It is arduous to implement conversions and comparisons, to ensure for instance that the left bound is not greater than the right bound. These difficulties have led to the development of a simplified version of IEEE 1788-2015, namely the *IEEE 1788.1-2017 Standard for Interval Arithmetic (Simplified)*, that considers only binary64 as the numerical type for the representation of the endpoints and that contains some other minor simplifications.

The goal of the working group that elaborated the IEEE 1788-2015 standard, and that now works on a simplified version of it, is really to help developers and experimenters to have and share a common basis. This common basis is mandatory for portability, or even simpler, for comparison purposes. The working group hopes to see the fruits of its efforts and of its numerous and lively discussions. Needless to say, the introduction of flavors and of decorations were subject to hot debates. It is hard to tell which subject led to the most vigorous discussions.

12.3 Intervals with Floating-Point Bounds

12.3.1 Implementation using floating-point arithmetic

Formulas (12.1) depend on the arithmetic on real numbers. However, it is possible to design formulas that use floating-point numbers, and still satisfy the inclusion property: the computed result $\widehat{\mathbf{z}} = [\widehat{\underline{z}}, \widehat{\overline{z}}]$ encloses the exact result $\mathbf{z} = [\underline{z}, \overline{z}]$, that is, both inequalities $\widehat{\underline{z}} \le \underline{z}$ and $\overline{z} \le \widehat{\overline{z}}$ hold. Indeed, directed rounding functions in floating-point arithmetic (see Section 2.2) provide an efficient implementation of interval arithmetic. For instance,

$$\begin{aligned}
\sqrt{[\underline{x}, \overline{x}]} &:= [\mathrm{RD}(\sqrt{\max(0, \underline{x})}), \mathrm{RU}(\sqrt{\overline{x}})] \quad \text{if } \overline{x} \ge 0, \\
[\underline{x}, \overline{x}] + [\underline{y}, \overline{y}] &:= [\mathrm{RD}(\underline{x} + \underline{y}), \mathrm{RU}(\overline{x} + \overline{y})], \\
[\underline{x}, \overline{x}] - [\underline{y}, \overline{y}] &:= [\mathrm{RD}(\underline{x} - \overline{y}), \mathrm{RU}(\overline{x} - \underline{y})].
\end{aligned}$$

Thanks to the properties of RD and RU (see Section 2.2), if the input intervals have finite bounds, the interval result is guaranteed to contain the exact mathematical value, even if the operations on the bounds are inexact.

Moreover, the lower bound can be replaced by $-\infty$ in order to represent an interval unbounded on the negative numbers. Similarly, $+\infty$ can be used for the upper bound.[2] Then, thanks to the properties of the IEEE 754-2008 arithmetic with respect to infinite values, the inclusion property is still valid on the final result even when overflows to infinities occur during intermediate computations. Notice that the operations $\infty - \infty$ can never happen in these formulas, as long as the inputs are valid.

Floating-point arithmetic also makes it possible to handle the empty set. For instance, it can be represented by the pair $[\mathrm{NaN}, \mathrm{NaN}]$, which will properly propagate when given to the preceding formulas.

For multiplication, the situation is slightly more complicated. Consider the product of the intervals $[0, 7]$ and $[10, +\infty]$ (The $+\infty$ bound may have been obtained by overflow, if the real bound happens to be too large to be representable by a finite floating-point number). Assume that, as in the case of real bounds, the lower bound is computed by taking the minimum of the four products $\mathrm{RD}(0 \times 10) = 0$, $\mathrm{RD}(7 \times 10) = 70$, $\mathrm{RD}(0 \times \infty) = \mathrm{NaN}$, and $\mathrm{RD}(7 \times \infty) = +\infty$. A NaN datum is obtained for a product whose result would be zero if there had been no overflow. Therefore, when the underlying floating-point arithmetic is compliant with IEEE 754-2008, some special care is needed to prevent propagating incorrect bounds [253].

12.3.2 Difficulties

The presence of directed rounding functions in the IEEE 754-1985 standard for floating-point arithmetic was advocated by Kahan, and his motivation

[2] An interval with floating-point bounds $[x, +\infty]$ contains all of the real numbers greater than or equal to x.

was to enable the implementation of interval arithmetic [314]. However, even if these directed rounding functions were already mandatory in IEEE 754-1985, it took a long time until high level programming languages gave easy access to these functions. The only way was to call routines in assembly and it was neither easy nor portable. Nowadays, the situation has improved thanks to the standard C11 header `<fenv.h>`, as mentioned in Section 6.2.1; access to the rounding functions is achieved through calls to `fegetround` and `fesetround`.

Another difficulty with the use of directed rounding functions is the loss of performance in terms of execution time: observed slowdowns are multiplicative factors from 10 to 50 and sometimes even in the thousands between programs computing with interval arithmetic, compared to the same programs computing with floating-point arithmetic. A first reason is the fact that each interval operation involves several floating-point operations, see Formulas (12.1), and consequently an interval operation is slower than its floating-point counterpart. For instance, an interval addition requires two floating-point additions. For multiplication, the ratio goes as high as 8, or even higher if the minimum and maximum operations are counted: indeed, in the multiplication $\mathbf{z} = \mathbf{x} \cdot \mathbf{y}$ with $\mathbf{x} = [\underline{x}, \overline{x}]$ and $\mathbf{y} = [\underline{y}, \overline{y}]$, the left bound $\underline{\widehat{z}} = \min(\mathrm{RD}(\underline{x} \cdot \underline{y}), \mathrm{RD}(\underline{x} \cdot \overline{y}), \mathrm{RD}(\overline{x} \cdot \underline{y}), \mathrm{RD}(\overline{x} \cdot \overline{y}))$ requires 4 floating-point multiplications with rounding toward $-\infty$, whereas the right bound $\overline{\widehat{z}} = \max(\mathrm{RU}(\underline{x} \cdot \underline{y}), \mathrm{RU}(\underline{x} \cdot \overline{y}), \mathrm{RU}(\overline{x} \cdot \underline{y}), \mathrm{RU}(\overline{x} \cdot \overline{y}))$ requires 4 floating-point multiplications with rounding toward $+\infty$. In comparison, the formula on real numbers required only 4 real multiplications.

But this is not the only reason. The observed slowdown is largely due to the dynamic handling of rounding through a status flag on architectures such as x86. When the rounding mode is modified in the code, this status flag is modified; the consequence is that the pipeline is flushed, every operation is stopped and restarted according to the new rounding mode. If the chosen formulas for interval arithmetic require a call to a different rounding function for each endpoint, no computation can be vectorized anymore. In the context of parallel computations, other issues related to the programming environment and language arise: they are detailed in [508] and are not developed here.

In addition to the inherent algorithmic complexity of interval computations and to the execution cost due to hardware architecture, there is a third reason for the inefficiency of interval computations. Indeed, while numerical libraries such as BLAS or Lapack are usually highly optimized with respect to cache use, this is not (yet) the case for interval libraries.

Let us end this section on an optimistic note. On less classical architectures, such as GPUs, or on more recent architectures, such as Intel's Xeon Phi Knights Landing, the rounding mode is handled statically: it is specified in the opcode of the floating-point operation. (For a more detailed explanation, see Section 3.4.4.) There is hope for better performance of interval computations on these architectures. For instance, CUDA implementations on GPU exhibit speed-ups from 2 to 10, for Newton iteration in [33], compared to interval computations on CPU, and up to 100 for an AXPY in [110].

12.3.3 Optimized rounding

Solutions to obviate these issues have been proposed: relying on the symmetry property of the rounding modes can be of help. Indeed, the identity $\forall x\ \mathrm{RD}(-x) = -\mathrm{RU}(x)$ makes it possible to use one single rounding direction for most of the arithmetic operations.

For instance, let us assume that interval operations should only use RU. Addition and subtraction can be rewritten as

$$\begin{aligned}[\underline{x}, \overline{x}] + [\underline{y}, \overline{y}] &:= [-\mathrm{RU}((-\underline{x}) - \underline{y}), \mathrm{RU}(\overline{x} + \overline{y})] \\ [\underline{x}, \overline{x}] - [\underline{y}, \overline{y}] &:= [-\mathrm{RU}(\overline{y} - \underline{x}), \mathrm{RU}(\overline{x} - \underline{y})]\end{aligned}$$

The computation of the upper bound is left unchanged, but the lower bound now requires some (cheap) sign flipping. In order to avoid these extra operations, we can store the lower bound with its sign bit already flipped. Let us denote $\mathbf{x}^*$ an interval stored using this convention; addition and subtraction then become

$$\begin{aligned}[\underline{x}, \overline{x}]^* + [\underline{y}, \overline{y}]^* &:= [\mathrm{RU}(\underline{x} + \underline{y}), \mathrm{RU}(\overline{x} + \overline{y})]^* \\ [\underline{x}, \overline{x}]^* - [\underline{y}, \overline{y}]^* &:= [\mathrm{RU}(\underline{x} + \overline{y}), \mathrm{RU}(\overline{x} + \underline{y})]^*\end{aligned}$$

Notice that if intervals are now considered as length-2 floating-point vectors, interval addition is just a vector addition with rounding toward $+\infty$. Interval subtraction is also a vector addition, but the components of the second interval have to be exchanged first. Similar optimizations can be applied to other arithmetic operations; they lead to efficient implementations of floating-point interval arithmetic on SIMD architectures [358, 221].

Some floating-point environments may also have issues with subnormal results (see Section 2.1.2). The hardware may not support them directly; gradual underflow is then handled either by microcode or by software trap, which may incur some slowdowns. Abrupt underflow alleviates this issue at the expense of some properties mandated by IEEE 754. As long as this abrupt underflow is not implemented by a flush to zero,[3] the inclusion property is still satisfied. Therefore, even though abrupt underflow may cause the enclosures to be a bit wider, they are still guaranteed.

[3]That is, a positive subnormal result is rounded to the smallest positive normal floating-point when rounding toward $+\infty$, and to zero when rounding toward $-\infty$.

12.4 Interval Arithmetic and Roundoff Error Analysis

Interval arithmetic incorporates gracefully roundoff errors of floating-point arithmetic. A step further consists in using interval arithmetic to bound roundoff errors in floating-point numerical computations.

Let us start with a short comment regarding infinite bounds. Every definition, unless stated otherwise, applies to intervals with infinite bound(s). Regarding the implementation: as the IEEE 754-2008 standard for floating-point arithmetic defines $-\infty$ and $+\infty$, intervals can be implemented using floating-point arithmetic, as far as the handling of infinities is concerned. However, roundoff analyses do not hold when infinite bounds occur; for this reason, it is assumed in this section that every value is finite.

In this section, we consider an arithmetic expression on real inputs, that is, inputs that are real numbers and not intervals, and we assume that interval arithmetic is implemented using floating-point arithmetic. The evaluation of the arithmetic expression using this implementation of interval arithmetic results in an interval $\widehat{\mathbf{z}}$ with floating-point bounds $\underline{\widehat{z}}$ and $\overline{\widehat{z}}$. This interval $\widehat{\mathbf{z}}$ contains the exact real result z and roundoff errors that occurred during the computation. In other words, $\mathrm{width}(\widehat{\mathbf{z}})$ is an upper bound on the absolute error on z, and we are interested in this value.

As roundoff errors on the outward rounded bounds, for each intermediate operation, are also accounted for, $\widehat{\mathbf{z}}$ does not only enclose roundoff errors on the computation of z, but all these extra roundoff errors on the computation of the bounds. As a consequence, the width of $\widehat{\mathbf{z}}$ is often pessimistic: overestimation occurs again.

The use of interval arithmetic to bound roundoff errors can be classified as

- a *certain bound*, as opposed to an estimated one: it is guaranteed that the width of $\widehat{\mathbf{z}}$ is an upper bound on the absolute error on z;
- *a posteriori*: the computation of $\widehat{\mathbf{z}}$ depends on the actual values of the inputs;
- *running*: the error bound is computed along with the computation of the result, in particular with the mid-rad representation described below, as the midpoint usually corresponds to the computation of an approximation of z using floating-point arithmetic with rounding to nearest. With the representation of intervals by their bounds, these bounds are computed instead of the approximate result.

12.4.1 Influence of the computing precision

Let us study the width of $\widehat{\mathbf{z}}$ in relation with the computing precision p defined in Section 2.1.1. First, let us establish the width of the smallest, for inclusion, interval $\widehat{\mathbf{z}}$ with floating-point bounds that contains a real number $z \in \mathbb{R}$. As seen in Section 2.2.1, $\mathrm{RD}(z)$ and $\mathrm{RU}(z)$ define the unique pair of consecutive floating-point numbers (or equal numbers if z is exactly representable as a floating-point number) such that $z \in [\mathrm{RD}(z), \mathrm{RU}(z)]$, and this interval is $\widehat{\mathbf{z}}$. Using $\mathbf{u} = \frac{1}{2}\beta^{1-p}$ (it is worth specifying the value of $\mathbf{u}$ here, as several rounding modes will be in use) and the inequalities of Equation (2.6) in Section 2.3.1, one can deduce a bound on the width of $\widehat{\mathbf{z}}$:

$$\mathrm{width}(\widehat{\mathbf{z}}) = \mathrm{RU}(z) - \mathrm{RD}(z) \leq 2\mathbf{u}|z|.$$

The width of an interval used as input is thus, at most, linear in $\mathbf{u}$ and in the absolute value of the real input.

Let us now consider the result of an operation $\widehat{\mathbf{z}} = \widehat{\mathrm{op}}(\mathbf{x}_1, \ldots, \mathbf{x}_n)$: $\widehat{\mathbf{z}} = [\mathrm{RD}(z_1), \mathrm{RU}(z_2)]$ with z_1 and z_2 given by the formulas for the operation, as in Equation (12.1). The quantity that is added to the width of the exact result $[z_1, z_2]$ can be bounded using Equation (2.6). For instance, if $\mathbf{x} = [\underline{x}, \overline{x}] \geq 0$ and $\mathbf{y} = [\underline{y}, \overline{y}] \geq 0$, that is, $\underline{x} \geq 0$ and $\underline{y} \geq 0$, and if $\mathbf{z} = \mathbf{z} + \mathbf{y}$ and $\widehat{\mathbf{z}}$ is the addition of $\mathbf{x}$ and $\mathbf{y}$ in interval arithmetic implemented with floating-point arithmetic:

$$\widehat{\mathbf{z}} = [\mathrm{RD}(\underline{x} + \underline{y}), \mathrm{RU}(\overline{x} + \overline{y})],$$

then

$$\begin{aligned}
\mathrm{width}(\widehat{\mathbf{z}}) &= \mathrm{RU}(\overline{x} + \overline{y}) - \mathrm{RD}(\underline{x} + \underline{y}) \\
&\leq (1 + 2\mathbf{u})(\overline{x} + \overline{y}) - (1 - 2\mathbf{u})(\underline{x} + \underline{y}) \\
&\leq (\overline{x} + \overline{y}) - (\underline{x} + \underline{y}) + 2\mathbf{u}((\underline{x} + \underline{y}) + (\overline{x} + \overline{y})) \\
&\leq \mathrm{width}(\mathbf{x} + \mathbf{y}) + 4\mathbf{u}\mathrm{mid}(\mathbf{x} + \mathbf{y}) \\
&\leq \mathrm{width}(\mathbf{z}) + 4\mathbf{u}\mathrm{mid}(\mathbf{z}).
\end{aligned}$$

If $\mathrm{mid}(\mathbf{z}) > 0$, this means that the relative width of $\mathbf{z}$: $\mathrm{width}(\mathbf{z})/\mathrm{mid}(\mathbf{z})$, is increased by at most $4\mathbf{u}$ to produce the result $\widehat{\mathbf{z}}$ computed in interval arithmetic implemented with floating-point arithmetic. Similar properties hold for other operations. These properties can be seen as an analogous, for interval arithmetic, of the first standard model (see Section 5.1.1) for floating-point arithmetic: the relative width of the computed interval exceeds the relative width of the exact interval by a factor at most $4\mathbf{u}$.

More generally, if the expression that defines $\mathbf{z}$ and each of its subexpressions are Lipschitz in their variables, the overestimation of the result $\widehat{\mathbf{z}}$ computed in interval arithmetic implemented using floating-point arithmetic, compared to the result $\mathbf{z}$ computed in interval arithmetic relying on exact, real arithmetic, is at most linear in $\mathbf{u}$. The proof of this result is an induction on the DAG that defines $\mathbf{z}$. It is completely developed in [455, Theorem 2.1.5]. This bound on the overestimation means that the overestimation is reduced when a larger computing precision p is employed.

Let us illustrate this result on two examples. We use the MPFI library [507] for the computations. The MPFI library implements interval arithmetic, using the arbitrary precision floating-point arithmetic offered by the MPFR library [204] presented in Chapter 14. The MPFI library uses the radix 2. The first example is a straightforward illustration: let us numerically "check" Machin's formula

$$\frac{\pi}{4} = 4\arctan\left(\frac{1}{5}\right) - \arctan\left(\frac{1}{239}\right).$$

Table 12.1 shows the difference between the left-hand side and the right-hand side for varying (doubling) values of the computing precision p. Each time the computing precision doubles, the absolute value of the exponent of the difference doubles as well, as predicted.

p	Difference
20	$[-1.9073487 \cdot 10^{-6}, 1.9073487 \cdot 10^{-6}]$
40	$[-1.8189895 \cdot 10^{-12}, 1.8189895 \cdot 10^{-12}]$
80	$[-1.6543613 \cdot 10^{-24}, 1.6543613 \cdot 10^{-24}]$
160	$[-1.3684556 \cdot 10^{-48}, 1.3684556 \cdot 10^{-48}]$
320	$[-9.3633528 \cdot 10^{-97}, 9.3633528 \cdot 10^{-97}]$
640	$[-2.1918094 \cdot 10^{-193}, 6.5754281 \cdot 10^{-193}]$
1280	$[-4.8040283 \cdot 10^{-386}, 1.4412085 \cdot 10^{-385}]$
2560	$[-2.3078688 \cdot 10^{-771}, 6.9236062 \cdot 10^{-771}]$
5120	$[-5.3262581 \cdot 10^{-1542}, 1.5978775 \cdot 10^{-1541}]$
10240	$[-5.6738049 \cdot 10^{-3083}, 5.6738049 \cdot 10^{-3083}]$

Table 12.1: Difference of the interval evaluation of Machin's formula for various precisions.

The second example, which was already mentioned in Chapter 1, is Rump's formula, Equation (1.2):

$$\left(333 + \frac{3}{4}\right) b^6 + a^2 \left(11a^2b^2 - b^6 - 121b^4 - 2\right) + \frac{11}{2}b^8 + \frac{a}{2b}$$

with

$$a = 77617.0 \text{ and } b = 33096.0.$$

The evaluation of this formula causes a series of catastrophic cancellations as long as precisions less than 122 bits are used for the computations. In the interval evaluations below, one observes large intervals when the computing precision is small and very tight ones when the computing precision is large enough.

```
Rump's formula:
(333+3/4)*b^6 + a^2*(11*a^2*b^2-b^6-121*b^4-2) + 11/2*b^8 + a/2b
with a = 77617.0 and b = 33096.0
Floating-point evaluation:
   binary32: 1366525287113805923940668623252619264.000000
   binary64: 1366524611239166270608683216459005952.000000

Computing precision: 24
    Eval. =  [-8.23972891e30,8.23972951e30]
Computing precision: 53
    Eval. =  [-5.9029581035870566e21,4.7223664828696463e21]
Computing precision: 121
    Eval. =  [-2.8273960599468213681411650954798162924,
               1.17260394005317863185883490452018370884]
Computing precision: 122
    Eval. =  [-8.2739605994682136814116509547981629201e-1,
              -8.27396059946821368141165095479816291629e-1]
```

As advocated by Kahan in [320], using interval arithmetic and varying the computing precision is a rather mindless approach to assess the numerical quality of a code, in the sense that this approach requires neither expertise in numerical analysis nor lengthy development time (but the execution time can be long), that nevertheless yields useful results regarding the numerical quality of a computed result. In other words, it is an automatic enough approach that is worth trying, at least as a first step.

Gappa, the tool presented in Section 13.3, uses interval arithmetic, either in a naive way when it suffices to establish the required bounds, or, more frequently, in a more elaborated way when needed. Gappa provides an additional level of guarantee by providing a formal proof on the bounds it establishes.

12.4.2 A more efficient approach: the mid-rad representation

In the previous numerical experiments, the last interval was displayed by its two very close bounds. It would be easier to see the difference between the two bounds by displaying $-8.2739605994682136814116509547981629 16 \cdot 10^{-1} \pm 2 \cdot 10^{-37}$, that is by displaying the midpoint of the interval and its radius. Such a representation is called a *mid-rad representation*. Several (non IEEE 1788-2015 compliant) libraries use the mid-rad representation, such as INTLAB [520] which stores both the midpoint and the radius as floating-point numbers, or Arb [304] and Mathemagix [373] which store an interval using a arbitrarily high precision for the midpoint and a fixed precision for the radius. *Staggered* interval arithmetic represents an interval as a floating-point expansion (see Section 14.2), that is, with a high but limited precision floating-point number, and an interval containing the error term. It is implemented in the C-XSC language [39].

Formulas adapted to the mid-rad representation can be derived. Let us denote $\mathbf{x} = \langle x_m, x_r \rangle$ the interval $\mathbf{x}$ represented by its midpoint x_m and its radius $x_r \geq 0$, $\mathbf{y} = \langle y_m, y_r \rangle$ with $y_r \geq 0$, and $\mathbf{z} = \langle z_m, z_r \rangle = \mathbf{x} \,\mathrm{op}\, \mathbf{y}$ with $z_r \geq 0$. Let us consider addition and subtraction:

$$\begin{aligned} \mathbf{x} + \mathbf{y} &= \langle x_m + y_m, x_r + y_r \rangle, \\ \mathbf{x} - \mathbf{y} &= \langle x_m - y_m, x_r + y_r \rangle. \end{aligned}$$

The implementation of these formulas using floating-point arithmetic uses the rounding to nearest function to compute the midpoint. The radius is computed using the rounding to $+\infty$ function, to preserve the inclusion property. Furthermore, the radius incorporates a bound on the roundoff error on the computation of the midpoint. For instance, Arb [304] and Mathemagix [373] add $\mathbf{u}|z_m|$ to the formula for the radius.

When the operation is a multiplication, the exact formula can be found in [455, Property 1.6.5]. It is a rather complicated formula and it involves 9 multiplications of real numbers. Quite often, an approximate, simpler formula is used. As early as 1999, Rump suggested in [519] the following formulas for z_m and z_r:

$$\left\{ \begin{aligned} z_m &= x_m \cdot y_m, \\ z_r &= (|x_m| + x_r) \cdot y_r + x_r \cdot |y_m|. \end{aligned} \right. \tag{12.2}$$

Nguyen [461] in 2011, Rump [523] in 2012, Théveny [600] in 2014, in particular, gave other formulas that involve from 2 to 7 multiplications of real numbers. Of course, the overestimation of the resulting intervals, compared to the exact product, is larger when the number of multiplications is reduced. For the implementation using floating-point arithmetic, roundoff errors are accounted for in the formula for the radius. For instance, the previous formulas become

$$\left\{ \begin{aligned} z_m &= \mathrm{RN}(x_m \cdot y_m), \\ z_r &= \mathrm{RU}((|x_m| + x_r) \cdot y_r + x_r \cdot |y_m| + \mathbf{u} \cdot |z_m|). \end{aligned} \right.$$

12.4.2.1 Why is the mid-rad representation of intervals interesting?

It has already been mentioned that, for tight intervals, the mid-rad representation is more "natural" because the midpoint can be seen as an approximation to the considered real value and the radius conveys the error. Another advantage of the mid-rad representation is exemplified on matrix computations. Let us consider the product of two interval matrices (that is, matrices with interval coefficients) $\mathbf{A}$ and $\mathbf{B}$: $\mathbf{A} \cdot \mathbf{B} = \langle A_m, A_r \rangle \cdot \langle B_m, B_r \rangle$. The formulas for the multiplication using the mid-rad representation apply, if $|A|$ is understood as the matrix of the absolute values of each element of the matrix A. Each multiplication appearing in the formulas is now a matrix-matrix multiplication. When interval arithmetic is implemented using floating-point arithmetic, optimized routines for the multiplication of matrices can be called and the corresponding interval matrix multiplication is fast: the slowdown factor in [600] is 3, which is a drastic improvement compared to the 1 to 3 orders of magnitude mentioned in Section 12.3.2.

Furthermore, the mid-rad representation is particularly interesting for parallel computations. First, if radii are stored using binary32 floating-point numbers rather than binary64 ones, each operation on the radii is usually very fast and each data move is more efficient, both in time and energy, than the use of binary64 radii. Second, each numeric value, which is made of a midpoint and a radius, incurs several floating-point arithmetic operations every time it is involved in a computation: the numerical intensity of the computation is higher for interval computations than for the analogous floating-point computations.

Guaranteed computations could benefit from the mid-rad representation and from the improvements described above to become faster, more comparable with floating-point computations, and thus more attractive, if issues regarding data storage, data moves, and numerical intensity are addressed.

12.4.2.2 Influence of roundoff errors

First, note that the different formulas for the multiplication, from the "exact" formula with 9 multiplications of numbers to the formula with only 2 multiplications of numbers, have been studied to determine the quality of the approximation they provide. For instance, the ratio of the radius obtained by Equation (12.2) to the radius of the exact product does not exceed $3/2$, and that value can be attained. Not surprisingly, that bound increases, i.e., the quality decreases, when the number of multiplications decreases.

When floating-point arithmetic is used to implement interval arithmetic, roundoff errors must also be taken into account. The radius of the computed result is a bound both on the roundoff errors due to the computation of the midpoint and on the width of the interval. Let us study the part due to roundoff error and the part due to the width of the input intervals. Results presented here are taken from Théveny's PhD thesis [600]. Two interval matrices $\mathbf{A} = \langle A_m, A_r \rangle$ and $\mathbf{B} = \langle B_m, B_r \rangle$ of dimension 128×128 are chosen randomly such that the maximal relative radius equals a prescribed value e:

$$\max_{i,j} \frac{A_{r_{i,j}}}{|A_{m_{i,j}}|} = e, \quad \max_{i,j} \frac{B_{r_{i,j}}}{|B_{m_{i,j}}|} = e.$$

The value of e is given on the x-axis of Figure 12.1, which is Figure 4.14 of [600].

$\mathbf{C} = \langle C_m, C_r \rangle$ is the result of one of the formulas for interval matrix multiplication, either one with 3 multiplications, called `MMul3`, or one involving 5 multiplications, called `MMul5`. Again, the focus is on the relative radius of $\mathbf{C}$: $\max_{i,j} C_{r_{i,j}}/|C_{m_{i,j}}|$, which is given on the y-axis.

It is expected that `MMul5` is more accurate than `MMul3`, and thus that the curve for `MMul5` (solid line) lies below the curve for `MMul3` (dashed line). However, when the underlying arithmetic is the floating-point arithmetic on binary64 numbers, and thus when roundoff errors enter the scene, the opposite happens when $\mathbf{A}$ and $\mathbf{B}$ are very tight, as shown on the left of Figure 12.1. In this case, the radii of the elements of $\mathbf{C}$ merely account for roundoff errors. As `MMul5` implies more operations than `MMul3`, it incurs more roundoff errors and is thus less accurate, i.e., higher on the figure. As e increases, the relative importance of roundoff errors decreases and the width of input intervals prevails. On the right of Figure 12.1, the curve of `MMul5` is below the curve for `MMul3`, which corresponds to the fact that the overestimation due to `MMul5` is smaller than the one due to `MMul3`.

Other experiments have been performed in [600] to study further the left part of this figure: the radii of the elements of $\mathbf{A}$ and $\mathbf{B}$ are now one ulp of the corresponding midpoints. In $\mathbf{C}$, computed using `MMul3`, most of the radii are close to one ulp of the corresponding midpoint: the initial uncertainty is not amplified.The remaining ones are close to the theoretical, a priori, bounds

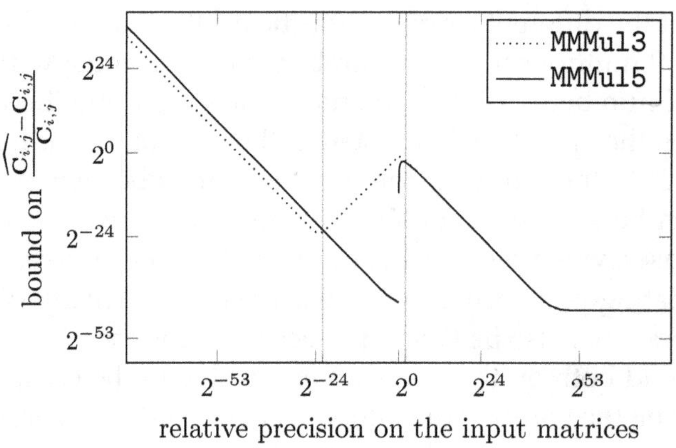

Figure 12.1: Comparison of the relative widths of matrices computed using `MMul3` and `MMul5`.

and they correspond to ill-conditioned problems for which a large amplification of initial errors is predicted. There is no intermediate behavior.

The change of representation, from endpoints to mid-rad, opened the path for practical, present and future, improvements. Let us now broaden the possibilities of representations.

12.4.3 Variants: affine arithmetic, Taylor models

The variants mentioned in this section have been developed to reduce the problem of variable dependency (see Section 12.1.2). With these variants, inputs and intermediate values are not expressed as values carrying an uncertainty, but as simple functions in some variables. As they depart from the strict framework of representing a numerical value by a small set of numbers, we just briefly sketch them. However, they deserve to be mentioned because of their practical usage.

With these variants, the execution time of one arithmetic operation is 100 to 1000 slower than the execution time of the corresponding floating-point operation. However, at the level of a complete application, the benefit in terms of accuracy, and thus the reduction of the number of operations required to get a tight result, may compensate for this overhead.

12.4.3.1 Affine arithmetic

Affine arithmetic was introduced by Comba and Stolfi in 1993 [111] and fully developed in [575]. The main principle consists in replacing every variable x (either an input variable or an intermediate variable) by an affine combination $x_0 + \sum_{i=1}^{n} x_i \varepsilon_i$, where each x_i is a real number and each ε_i is a "noise" symbol. The noise symbols ε_i are assumed to belong to $[-1, 1]$ and to be independent of the other noise symbols $\varepsilon_j, j \neq i$.

The formula for the addition or subtraction of $x = x_0 + \sum_{i=1}^{n} x_i \varepsilon_i$ and $y = y_0 + \sum_{i=1}^{n} y_i \varepsilon_i$ is

$$z = x \pm y = z_0 + \sum_{i=1}^{n} z_i \varepsilon_i \text{ with } \begin{cases} z_0 &= x_0 \pm y_0, \\ z_i &= x_i \pm y_i \text{ for } i = 1 \dots n. \end{cases}$$

Keeping a symbolic expression solves the problem of variable dependency when the terms are linear: $x - x$ is now exactly 0.

The multiplication of x and y is more delicate: terms of the form $\varepsilon_i \varepsilon_j$ are created. To keep the expression affine, a new noise symbol ε_{n+1} is created and a new term $z_{n+1}\varepsilon_{n+1}$ appears in the expression for z. The approach is similar for the division and any nonlinear function: $\sqrt{\cdot}$, exp, tan...

In practice, several variants have been developed to handle these new noise variables, to limit their number and to account for roundoff errors when affine arithmetic is implemented using floating-point arithmetic to compute the coefficients z_i. Implementations exist in MATLAB [529] or in C++ in IBEX.[4]

A flagship application of affine arithmetic (and other more elaborate versions of it, using zonotopes) is the Fluctuat software, which allows one to test numerical codes. Good descriptions of Fluctuat are [407] and [500]. Let us

[4] http://www.ibex-lib.org/.

quote [500]: *Fluctuat is a static analyzer by abstract interpretation, relying on the general-purpose zonotopic abstract domains, and dedicated to the analysis of numerical properties of programs.* Affine arithmetic is used to bound results and roundoff errors, and, via the use of symbolic noises, to keep track of the operations that produced these errors. It is then possible to analyze these errors and to identify when they become troublesome.

12.4.3.2 Polynomial models, Taylor models

These models use polynomial (and no more affine) expansions of every computed value with respect to the input variables. They track the input variables but not the uncertainty attached to each of them. Even if polynomials are used, their maximal possible degree is fixed beforehand. As for affine arithmetic, terms of higher degree must be handled: they are bounded and added to a last and specific term of the expansion, sometimes called the remainder term, which is an interval accounting for truncation, but also for roundoff errors. It is possible to bound these quantities and to drop them in the remainder term by assuming (modulo prior translations and homotheties) that every input variable belongs to $[-1, 1]$.

Very few implementations of these polynomial models are available. The oldest one dates back from Eckman et al. in 1991 [182]. Let us also mention COSY [402, 506], Zumkeller's PhD thesis [647] and FPTaylor by Solovyev et al. [562]. Both use Taylor expansions and the canonical basis for polynomials. In the literature, other bases can be found, such as Bernstein in [449, 447] which allows one to easily obtain bounds on the range of a function, or Chebyshev in [307, Chap. 4] as they offer rather uniformly tight enclosures on the whole considered interval.

12.5 Further Readings

The literature on interval arithmetic vastly exceeds the topics covered in this chapter. To begin with, further readings naturally contain the text of the IEEE 1788-2015 standard for interval arithmetic [268]. A whole branch of numerical analysis, devoted to interval computations, concerns algorithms that yield results not attainable with usual arithmetic on real numbers or on their approximations. Newton–Raphson iteration has already been evoked for the determination of every zero of a function on a given interval. The method is able to conclude that no zero exists, or to determine all the zeros, often with a proof of their existence and uniqueness in the enclosing interval, by applying Brouwer's theorem. Global optimization, that is, the determination of all extrema of a function over an interval, along with the guarantee that these extrema are global ones and not local, is made possible via the use of interval arithmetic. Indeed, a key ingredient is the possibility to compute with sets.

Good introduction to global optimization with interval arithmetic are [331, 236]. Interval arithmetic is also successfully employed for the guaranteed integration of ordinary differential equations, for instance in the proof that Lorenz' system has an attractor [608]. In this case, many variants of interval arithmetic are at stake, for instance affine arithmetic as in [8]. For an introduction to interval analysis, recommended readings are [429, 518, 355, 430, 609].

Chapter 13

Verifying Floating-Point Algorithms

WHILE THE PREVIOUS CHAPTERS have made clear that it is common practice to verify floating-point algorithms with pen-and-paper proofs, this practice can lead to subtle bugs. Indeed, floating-point arithmetic introduces numerous special cases, and examining all the details would be tedious. As a consequence, the verification process tends to focus on the main parts of the correctness proof, so that it does not grow out of reach.

For instance, the proof and even the algorithm may no longer be correct when underflow occurs, or when some value is equal to or near a power of the radix, as being a discontinuity point of the ulp function. Moreover pen-and-paper proofs may be ambiguous, e.g., by being unclear on whether the exact value or its approximation is considered for the ulp.

Unfortunately, experience has shown that simulation and testing may not be able to catch the corner cases this process has ignored. By providing a stricter framework, formal methods provide a means for ensuring that algorithms always follow their specifications.

13.1 Formalizing Floating-Point Arithmetic

In order to perform an in-depth proof of the correctness of an algorithm, its specification must be precisely described and formalized. For floating-point algorithms, this formalization has to encompass the arithmetic: number formats, operators, exceptional behaviors, undefined behaviors, and so on. A new formalization may be needed for any variation in the floating-point environment.

J.-M. Muller et al., *Handbook of Floating-Point Arithmetic*,
https://doi.org/10.1007/978-3-319-76526-6_13

Fortunately, the IEEE 754 standard precisely defines some formats and how the arithmetic functions behave on these formats: "Each operation shall be performed as if it first produced an intermediate result correct to infinite precision and with unbounded range, and then rounded that result [...]." This definition makes a formalization of floating-point arithmetic both feasible and practical, as long as the implementation (language, compiler, etc.) strictly follows the IEEE requirements, which may not be the case in practice (see Chapter 6 and Section 3.1.2.5). However, a formalization can still take into account the specificity of the implementation; for the sake of simplicity and because it is not possible to be exhaustive, such a specificity will be ignored in the following.

Moreover, this single formalization can be used for describing the specification of any algorithm whose implementation relies on this standardized arithmetic.

13.1.1 Defining floating-point numbers

The first stage of a formalization lies in a proper definition of the set of floating-point numbers. This definition can be done at several levels. First of all, one should define the set itself and the values that parameterize it, e.g., the radix and the precision. Then comes the actual representation of the numbers: how they translate from and to streams of bits. Finally, a semantics of the numbers is needed. It is generally provided by their interpretation as a subset of the real numbers.

13.1.1.1 Structural definition

The IEEE 754 standard describes five categories of floating-point data: signed zeros, subnormal numbers, normal numbers, signed infinities, and Not a Number (NaN) data. These categories can be used to define the set of data as a disjoint union of sets of data. Any given floating-point datum is described by one and only one of the following branches. For each category, some additional pieces of information represent the datum, e.g., its significand.

Floating-point data ::=
| **Zero**: sign
| **Subnormal**: sign, significand
| **Normal**: sign, significand, exponent
| **Infinity**: sign
| **NaN**: sign, payload

While the same disjoint union could be used to define both binary and decimal floating-point numbers, the formalization may be simpler if the radix β is encoded as a parameter of the whole type. The type of the "significand" fields has to be parameterized by the precision p, while the type of the "exponent" is parameterized by $e_{\min}$ and $e_{\max}$. The type of the "payload" could also be parameterized by p; but for clarity, we will assume it is not. The type of "sign" is simply the set $\{+, -\}$, possibly encoded by a Boolean. The fully featured disjoint union has now become:

Floating-point data $(\beta, p, e_{\min}, e_{\max})$ ::=
| **Zero**: sign
| **Subnormal**: sign, significand $\in M_s(\beta, p)$
| **Normal**: sign, significand $\in M_n(\beta, p)$, exponent $\in E(e_{\min}, e_{\max})$
| **Infinity**: sign
| **NaN**: sign, payload

Notice that the parameters do not necessarily have to be restricted to the set of values mandated by the standard. A generic formalization can be written for any radix at least 2, any precision at least 1, and any extremal exponents. A specific standard-compliant instantiation of the generic formalization can then be used when verifying an algorithm.

Some other parameters can be added to represent floating-point numbers that do not follow the scheme set by the IEEE 754 standard, e.g., the use of two's complement signed significands. The "Subnormal" branch could also be removed for implementations that do not support such numbers.

13.1.1.2 Binary representation

This part of the formalization describes how the numbers are actually stored. It mainly provides two functions for converting the structural definition of a number from/to its physical representation as a vector of bits. These functions are indispensable when working at the bit level (What does happen if the 35th bit of a number gets set?), for example, in hardware or software implementation of floating-point operators.

Note that the floating-point operations are better specified as functions operating on the structural definition of numbers, which is a more expressive representation than vectors of bits. As a consequence, the conversion functions can be ignored if the algorithm does not perform any bit twiddling. Indeed, the algorithm can then be verified on the structural definitions only. Such a verification is equivalent to the one made on vectors of bits, but it allows for the use of a simplified formalism, as shown in Section 13.1.2.

13.1.1.3 Semantic interpretation

Floating-point arithmetic is designed as a replacement for the arithmetic on real numbers in most algorithms. The specification then has to relate the floating-point results with the corresponding real numbers. This can be achieved by providing functions between the two sets. Unfortunately, neither are these sets isomorphic, nor is one a subset of the other.

First, normal numbers, subnormal numbers, and signed zeros can easily be converted into real numbers.[1] When the specification focuses on the behavior of the algorithm with respect to real numbers, this conversion function r is usually sufficient, and its inverse is not needed. For example, expressing that a nonzero real number x has to be close to a floating-point number $\tilde{x}$ can be achieved by the following inequality on real numbers: $|r(\tilde{x})/x - 1| \leq \varepsilon$.

As for floating-point infinities and NaNs, the set of real numbers could be extended to include corresponding elements. The function r would then be defined on the whole set of floating-point data. In order for this extended real set to be usable, a consistent arithmetic has to be defined on it. This may prove difficult and, more importantly, unneeded. Indeed, as these exceptional values are usually avoided or handled with special care in algorithms, they do not warrant a special arithmetic. As a consequence, the conversion function can be restricted to a partial domain of floating-point data. When employed in a verification, this formalism will then require the use of this function to be accompanied with a proof that the floating-point datum being converted is a finite number.

This conversion function is usually hidden: finite floating-point numbers are implicitly coerced to real numbers in mathematical formulas.

13.1.2 Simplifying the definition

The more complicated the arithmetic formalization is, the less efficient its use will be when verifying an algorithm, as it requires taking into account many more corner cases. So, a simplified structural definition with less branches may ease the verification process. However, care should be taken that the formalization is still meaningful for verifying a given algorithm.

First of all, some branches of the disjoint union can be merged: instead of splitting zeros, subnormal, and normal numbers apart, they can all be expressed as a triple (s, m, e) which maps to the real number $(-1)^s \cdot m \cdot \beta^{e-p+1}$. The exponent e is still between $e_{\min}$ and $e_{\max}$, but a normal floating-point number may no longer have a normalized representation since the branch of the disjoint union should also deal with other numbers. In particular, all the numbers with $m = 0$ would now represent a floating-point zero. Note that using an integral significand (Section 2.1.1) can greatly improve the ability of proof assistants to automatically discharge some proof obligations, hence reducing the amount of work left to the user.

Depending on its usage, the formalization can be simplified further. For most algorithms, the actual bit pattern of the computed values does not matter much. Indeed, higher-level properties are usually specified: the accuracy of a result, whether it is in a given domain, and so on. Therefore, the

[1]Both floating-point signed zeros are mapped to the same real zero, so the conversion function is not injective.

formalization can be simplified, as long as it does not change the truth of the properties described in the specification of the algorithm. For instance, if the algorithm never accesses the sign and payload of a NaN, then they can be removed from the definition. Indeed, from the point of view of the algorithm, NaN data will be indistinguishable from each other.

The following formalization may therefore be sufficient to prove the exact same properties as the full formalization on a given algorithm:

Floating-point numbers $(\beta, p, e_{\min}, e_{\max})$ ::=
| **Finite**: sign, significand $\in M(\beta, p)$, exponent $\in E(e_{\min}, e_{\max})$
| **Infinity**: sign
| **NaN**:

Let us now consider the case of signed zeros. Their sign mostly matters when a floating-point computation has to return a signed zero. So, if the sign of zero were to be always ignored, it would not have any impact on the interpretation of these computations. However, there are some other cases where the sign has an influence, e.g., a division by zero. In this case, the sign of zero is involved in the choice of the sign of the nonzero (infinite) result. Yet, if the verification is meant to include proofs that no divisions by zero occur during computations, then the sign of zero can be safely discarded from the definition. In particular, it means that the sign can be embedded into the significand: any zero, subnormal, or normal number would therefore be represented by a pair (m, e) with m a signed value.

In order to simplify the definition even further, the exponent bounds $e_{\min}$ and $e_{\max}$ could be removed. It would, however, introduce values that cannot be represented as floating-point data. Section 13.1.4 details this approach.

13.1.3 Defining rounding operators

Thanks to the framework that the IEEE 754 standard provides, floating-point operators do not have to be formalized to great lengths. They can be described as the composition of a mathematical operation on real numbers ("infinite precision" and "unbounded range") and a rounding operator that converts the result to a floating-point number. As a consequence, assuming arithmetic on real numbers has already been properly formalized, most of the work involved in formalizing floating-point arithmetic operators will actually focus on defining rounding operators on real numbers.

13.1.3.1 Range and precision

In our preceding structural definition, the exponent range $[e_{\min}, e_{\max}]$ and the precision p are part of the datum type: they restrict the ranges of the available significands and exponents. As a consequence and by definition, a finite floating-point number is bounded. Moreover, a formal object representing a floating-point number cannot be created unless one proves that its significand and exponent, if any, are properly bounded.

Another approach would be to remove these bounds from the datum type and use them to parameterize the rounding operators only. Then a finite floating-point number would only be a multiple of a power of the radix β, and hence unbounded *a priori*. The property that it is bounded would instead come from the fact that such a number was obtained by a rounding operation.

This makes it much easier to manipulate floating-point numbers in a formal setting, but it comes at the expense of having to carry extraneous hypotheses around. For example, stating that an algorithm takes a floating-point input might no longer provide sufficient information to prove its correctness, one might also have to state that this number has a bounded range and a fixed precision (assuming it is finite). So, in practice, one often ends up juggling between the two representations, in order to benefit from their respective strengths.

13.1.3.2 Relational and functional definitions

There are two approaches to defining rounding operators. The first one considers rounding operators as relations between the set of floating-point numbers and the set of real numbers. Two such numbers are related if the first one is the rounded value of the second one.

In addition to standard rounding modes, this approach makes it possible to define nondeterministic rounding modes. For example, one could imagine that, when rounding to nearest, the tie breaking is performed in a random direction, so several floating-point values could be obtained when rounding a real number. Such a property can be expressed with a relation. However, nondeterministic rounding modes are rarely used in practice.

More interestingly, the relational approach can deal with underspecified rounding operators. This makes it possible to verify more generic properties about an algorithm, without having to change the way the algorithm is described. For instance, an algorithm may compute the expected value, even if some intermediate results are only faithfully rounded. Or some languages may allow a multiplication followed by an addition to be contracted to an FMA on some processors (e.g., see Section 6.2.3.2), and algorithms may still compute the expected value if this happens.

The other approach applies to deterministic rounding modes only. It describes them by a function from the set of real numbers to the set of floating-point numbers. This approach is especially interesting when the image of the function is just the subset of finite floating-point numbers; this property arises naturally when the formalism has been extended by removing the upper bound $e_{\max}$ (see Section 13.1.4). This functional definition makes it possible to manipulate floating-point expressions as if they were expressions on real numbers, as long as none of the floating-point operations of the expressions cause an exceptional behavior.

13.1.3.3 Monotonicity

Powerful theorems can be applied on rounded computations as long as the rounding operators satisfy the following two properties. First, any floating-point number is its own and only rounded value. Second, if $\tilde{u}$ and $\tilde{v}$ are finite rounded values of the real numbers u and v, the ordering $u \leq v$ implies $\tilde{u} \leq \tilde{v}$. As a consequence, the functional version of a rounding operator is useful if it is the identity function on the set of floating-point numbers and a locally constant yet increasing function on the whole set of real numbers—or at least on the subset that rounds to finite floating-point numbers. As an example of using these two properties, let us consider the proof of the following simple lemma.

Lemma 13.1 (Approximate midpoint of two floating-point numbers in radix-2 arithmetic). *Consider the formalism of a binary floating-point arithmetic with no upper bound $e_{\max}$ on the exponents of rounded numbers. When evaluated with any useful rounding operator $\circ$, the expression* `(a + b)/2` *returns a floating-point value no less than* $\min(a, b)$ *and no greater than* $\max(a, b)$, *assuming a and b are finite floating-point numbers.*

Proof. We start from the mathematical inequality

$$2 \cdot \min(a, b) \leq a + b \leq 2 \cdot \max(a, b).$$

The rounding operator is a nondecreasing function, so we get

$$\circ(2 \cdot \min(a, b)) \leq \circ(a + b) \leq \circ(2 \cdot \max(a, b)).$$

Since the floating-point radix is 2, the values $2 \cdot \min(a, b)$ and $2 \cdot \max(a, b)$ are representable floating-point numbers. As a consequence,

$$2 \cdot \min(a, b) \leq \circ(a + b) \leq 2 \cdot \max(a, b).$$

Halving the terms and using once again the monotonicity of the rounding operator, we get

$$\circ(\min(a, b)) \leq \circ\left(\frac{\circ(a + b)}{2}\right) \leq \circ(\max(a, b)).$$

As the rounding operator is the identity on $\min(a, b)$ and $\max(a, b)$, we obtain the final inequality:

$$\min(a, b) \leq \circ\left(\frac{\circ(a + b)}{2}\right) \leq \max(a, b). \qquad \square$$

Note that the identity and monotonicity properties of the rounding operator are independent of the constraint $e_{\min}$. As a consequence, the lemma holds even when subnormal numbers are computed. The correctness of the proof, however, depends on the numbers $2 \cdot \min(a, b)$ and $2 \cdot \max(a, b)$ being in the range of finite floating-point numbers. Otherwise, the rounding operation would no longer be the identity for these two numbers, hence invalidating the proof. Therefore, a hypothesis was added in order to remove the upper bound $e_{\max}$. This means that the lemma no longer applies to a practical floating-point arithmetic. Section 13.1.4 will consider ways to use such a formalization so that it can be applied to a bounded $e_{\max}$.

13.1.4 Extending the set of numbers

As mentioned before, a formalization is sufficient for verifying an algorithm, if it does not change the truth value of the properties one wants to prove on this algorithm. Therefore, if one can prove that no infinities or NaNs can be produced by the algorithm,[2] a formalization without infinities and NaNs is sufficient.

Moreover, the simplified formalization does not need a bound $e_{\max}$. Indeed, in order to prove that this formalization is sufficient, one has already proved that all the floating-point numbers produced while executing the algorithm are finite. Therefore, embedding $e_{\max}$ in the formalization does not bring any additional confidence in the correctness of the algorithm.

Such a simplification eases the verification process, but it also requires some extra discipline. Formal tools will no longer systematically require proofs of the exponents not overflowing. Therefore, the user has to express these properties explicitly in the specification, in order to prove that the simplified formalization is valid, and hence to obtain a correct certificate.

Let us consider Lemma 13.1 anew. This lemma assumes there is no $e_{\max}$, so that a value like $2 \cdot \max(a, b)$ is rounded to a finite number. Notice that this value is not even computed by the algorithm, so whether the product overflows or not does not matter. Only the computation of `(a + b)/2` is important. In other words, if one can prove that no overflows can occur in the algorithm, then the formalization used in Lemma 13.1 is a valid simplification for the problem at hand. Therefore, an immediate corollary of Lemma 13.1 states: for a full-fledged binary floating-point arithmetic, `(a + b)/2` is between a and b if the sum $a + b$ does not overflow.

Similarly, if subnormal numbers are proved not to occur, the lower bound $e_{\min}$ can be removed from the description of the floating-point datum

[2]When arithmetic operations produce a trap in case of exception, the proof is straightforward; but the way the traps are handled must also be verified (the absence of correct trap handling was one of the causes of the explosion of Ariane 5 in 1996). Another way is to prove that all the infinitely precise values are outside the domain of exceptional behaviors.

and finite floating-point numbers will be left with zero and normal numbers only. These modifications are extensions of the floating-point model, as exponent bounds have been removed. The main consequence is that some theorems are then valid on the whole domain of floating-point finite numbers. For instance, one can assume that $|\,\mathrm{RN}(x) - x|/x$ is bounded by $\frac{1}{2}\beta^{1-p}$ for any (nonzero) real x, without having to first prove that x is in the normal range. In particular, x does not have to be a value computed by the studied program; it can be a ghost value that appears in the proof only, in order to simplify it.

13.2 Formalisms for Verifying Algorithms

Once there is a mathematical setting for floating-point arithmetic, one can start to formally verify algorithms that depend on it. Among the early attempts, one can cite Holm's work that combined a floating-point formalism with Hoare's logic in order to check numerical algorithms [261]. Barrett later used the Z notation to specify the IEEE 754 standard and refine it to a hardware implementation [31]. Priest's work is an example of a generic formalism for designing guaranteed floating-point algorithms [495]. All of these works were based on detailed pen-and-paper mathematical proofs.

Nowadays, computer-aided proof systems make it possible, not only to state formalisms on floating-point arithmetic, but also to mechanically check the proofs built on these formalisms. This considerably increases the confidence in the correctness of floating-point hardware or software. The following paragraphs present a survey of some of these formalisms. The choice of a given formalism for verifying an algorithm should depend on the scope of the algorithm (low level or high level?) and on the features of the corresponding proof assistant (proof automation, support for real numbers, support for programming constructs, and so on).

13.2.1 Hardware units

Processor designers have been early users of formal verification. If a bug goes unnoticed at design time, it may incur a tremendous cost later, as it may cause its maker to recall and replace the faulty processor. Unfortunately, extensive testing is not practical for floating-point units, as their highly combinatorial nature makes it time consuming, especially at pre-silicon stages: hence the need for formal methods, in order to extend the coverage of these units.

At this level, floating-point arithmetic does not play an important role. Verified theorems mostly state that "given these arrays of bits representing the inputs, and assuming that these logical formulas describe the behavior of the unit, the computed output is the correct result." Floating-point arithmetic is only meaningful here for defining what a "correct result" is, but none of its properties will usually appear in the formal proof.

As an example of such formal verifications, one can cite the various works around the floating-point units embedded in AMD-K5 processors. Moore, Lynch, and Kaufmann were interested in the correctness of the division algorithm [426], while Russinoff tackled the arithmetic operators at the RTL level [534, 536] and the square root at the microcode level [535]. All these proofs are based on a formalism written for the ACL2 first-order proof assistant.[3]

One specificity of this formalism is the use of rational numbers only. Since inputs, outputs, and ideal results are all rational numbers, this restriction does not hinder the verification process of addition, multiplication, and division. But it is an issue for the square root, as the ideal results often happen to be irrational numbers. It implies that the correctness theorem for the square root cannot be stated as follows:

$$\forall x \in \text{Float}, \quad \texttt{sqrt}(x) = \circ(\sqrt{x}).$$

It has to be slightly modified so that the exact square root can be avoided:

$$\begin{aligned}\forall x \in \text{Float}, \forall a, b \in \text{Rational}, \quad 0 \le a \le b \quad &\Longrightarrow \\ a^2 \le x \le b^2 \quad \Longrightarrow \quad &\circ(a) \le \texttt{sqrt}(x) \le \circ(b).\end{aligned}$$

From this formally verified theorem, one can then conclude with a short pen-and-paper proof that the floating-point square root is correctly computed. First of all, rounding operators are monotonic, so $a \le \sqrt{x} \le b$ implies $\circ(a) \le \circ(\sqrt{x}) \le \circ(b)$. If $\sqrt{x}$ is rational, taking $a = b = \sqrt{x}$ forces $\texttt{sqrt}(x) = \circ(\sqrt{x})$. If $\sqrt{x}$ is not rational, it is not the rounding breakpoint between two floating-point values either.[4] Therefore, there is a neighborhood of $\sqrt{x}$ in which all the real values are rounded to the same floating-point number. Since rational numbers are a dense subset of real numbers, choosing a and b in this neighborhood concludes the proof.

ACL2 is not the only system that was used to verify a floating-point unit. One can also cite the Forte system,[5] which was used by O'Leary et al. to verify the Intel Pentium Pro floating-point unit [474]. It was also used by Kaivola and Kohatsu to verify the Intel Pentium 4 divider [323]. As with ACL2, the Forte system is only able to reason about integer properties, which makes some specifications a bit awkward.

---

[3]http://www.cs.utexas.edu/users/moore/acl2/.

[4]One could certainly devise a rounding operator that has irrational breakpoints. Fortunately, none of the standard modes is that vicious.

[5]https://www.cs.ox.ac.uk/tom.melham/res/forte.html.

Another instance of a formal system used to verify hardware units is PVS.[6] It was used by Jacobi and Berg to verify the floating-point unit of the VAMP processor [285]. There is also HOL,[7] which was used by Akbarpour, Abdel-Hamid, and Tahar to transpose Harrison's verification of an algorithm for exponential (see below) to an actual circuit [6].

13.2.2 Floating-point algorithms

The verification process of some hardware units mentioned above is actually much closer to the one of software implementations. This is for example the case for the division and square root circuits of the AMD-K5 processor, as they are in fact microcoded. These operators are computed by small algorithms performing a sequence of additions and multiplications, on floating-point numbers with a 64-bit significand and a 17-bit exponent, with various rounding operators. The verification of these algorithms then relies on properties of floating-point arithmetic, e.g., the notion of rounding error.

ACL2 is not the only proof assistant with a formalism rich enough to prove such algorithms. Harrison devised a framework for HOL Light[8] [240, 243] and used it for proving the correctness of the division operator on Intel Itanium processors [242]. Fused multiply-add (FMA) is the only floating-point operator provided by these processors, so division has to be implemented in software by a sequence of FMA. As such, the approach is similar to AMD-K5's microcoded division, and so is the verification. Harrison also proved the correctness of a floating-point implementation of the exponential function [238] and of some trigonometric functions [241].

Later, Jacobsen, Solovyev, and Gopalakrishnan, devised a new formalism for HOL Light. It was much simpler than Harrison's one, since its main purpose was to define and prove an error bound for the rounding operation rather than to verify whole algorithms [286].

There are some other formalisms for proving floating-point algorithms, one of which is available for two proof assistants: Coq[9] and PVS. This effort was started by Daumas, Rideau, and Théry in Coq [136] and was ported to PVS by Boldo. Later, Boldo extended the Coq version in order to prove numerous properties of floating-point arithmetic, e.g., Dekker's error-free multiplication for various radices and precisions [42], the conditional faithfulness of Horner's polynomial evaluation [56], the use of an exotic rounding operator for performing correct rounding [52], Kahan's algorithm for computing the discriminant of a quadratic polynomial [43], algorithms for computing the rounding error of an FMA [55].

---

[6] http://pvs.csl.sri.com/.

[7] https://hol-theorem-prover.org/.

[8] http://www.cl.cam.ac.uk/~jrh13/hol-light/.

[9] http://coq.inria.fr/.

The Coq proof assistant had two other formalisms for floating-point arithmetic devised by Melquiond: one for the Gappa tool [134] and the other for the CoqInterval library [416, 417]. Boldo and Melquiond encompassed these three competing formalisms into the Flocq[10] library [53, 54]. Among the floating-point properties proved by Boldo using Flocq, one can note the accurate computation of the area of a triangle [44] and of the average [45]. Figueroa's results about double rounding [200] were formalized by Roux and extended to cover corner cases such as radices other than 2 and subnormal numbers [516]. A development by Martin-Dorel, Melquiond, and Muller analyzed the impact of double rounding on basic blocks such as the 2Sum algorithm [409].

The Flocq library was also used by Boldo et al. to formally verify the correctness of the floating-point passes of the CompCert C compiler [51], including the sequences of assembly instructions it generates when compiling conversions between floating-point numbers and integers (see Section 4.10).

13.2.3 Automating proofs

When verifying more complicated algorithms, e.g., elementary functions, some automation can be introduced to ease the formal proof. Indeed, for such functions, verification is mainly geared toward proving that the global roundoff error between an infinitely precise expression g and its floating-point evaluation $\tilde{g}$ is small enough. This proof is usually done by separately bounding the error between $\tilde{g}$ and its infinitely precise version $\hat{g}$, and the error between $\hat{g}$ and g. Let us see how to formally bound the latter error. While $\hat{g}$ and g might be equal, it is generally not the case as it might not even be possible to perform the evaluation of g using only the basic arithmetic operators on real numbers. For instance, an elementary function g may be replaced by a polynomial $\hat{g}$ that approximates it. So one has to compute a bound on the error function $\varepsilon(x) = \hat{g}(x) - g(x)$ for any x in the input domain X.

As an illustration, let us consider Harrison's HOL Light proof of Tang's binary32 implementation of the exponential function [238]. In this case, X is a small domain around zero, $\hat{g}$ is a degree-3 polynomial, and g is the exponential. Since Harrison did not have any tool to directly verify a bound on ε, he first rewrote it as $\varepsilon(x) = (\hat{g}(x) - P(x)) + (P(x) - g(x))$ with P a large-degree polynomial approximating g. For this proof, he used a degree-6 truncation of the Taylor series of exponential, which makes it simple to formally bound $P(x) - g(x)$ using Taylor-Lagrange inequality. The other part $Q = \hat{g} - P$ is simply a polynomial, so it can be automatically bounded by formal methods. Let us see several approaches that have been employed to obtain and prove bounds on such a polynomial.

---

[10] http://flocq.gforge.inria.fr/.

Harrison originally computed small intervals enclosing the roots of the derivative Q' of Q using its Sturm sequence, so as to locate the extrema of Q [239]. Harrison also considered using sum-of-square decompositions for proving properties of polynomial systems [244], which is a problem much more general than simply bounding Q. The decomposition $\sum R_i^2$ of a polynomial $R = Q + a$ is particularly well suited for formal proofs. Indeed, while the decomposition may be difficult to obtain, the algorithm does not have to be formally proved. Only the equality $R = \sum R_i^2$ has to, which is easy and fast. A straightforward corollary of this equality is the property $Q \geq -a$.

However, both approaches seem to be less efficient than doing a variation analysis of Q in HOL Light [241]. Again, this analysis looks for the roots of the derivative Q', but instead of using Sturm sequences to obtain the enclosures of these roots, Harrison recursively enclosed all the roots of its derivatives. Indeed, a polynomial is monotonic between two consecutive roots of its derivative. The recursion stops on linear polynomials, which have trivial roots.

Another approach is the representation of Q' in the Bernstein polynomial basis. Indeed, if the number of sign changes in the sequence of coefficients is zero or one, then it is also the number of roots (similar to Descartes' Law of Signs). If it is bigger, then De Casteljau's algorithm can be used to efficiently compute the Bernstein representations of Q on two subintervals. A bisection can therefore isolate all the extrema of Q. Zumkeller implemented and proved this method in Coq [647]. Muñoz and Narkawicz implemented a similar approach in PVS [447].

Instead of expressing the approximation error ε as $(\hat{g} - P) + (P - g)$ with P a single polynomial with a larger degree than $\hat{g}$, one can also split the input interval into numerous subintervals X_i and compute a different polynomial P_i on each of them. This is the approach followed by Sawada and Gamboa when verifying the square root algorithm of the IBM Power 4 processor [540]. They split the input interval into 128 subintervals; bounding each of the $\hat{g} - P_i$ polynomials was then performed using a naive method akin to interval arithmetic. This was done using the ACL2(r) experimental extension of ACL2 for supporting real numbers and nonstandard analysis. (Quantifier support in ACL2 is too limited for statements of standard analysis.)

In the context of verifying approximation bounds, the main issue with the approaches above is that they require that P be a polynomial (i.e., no use of the division and square root operators) and that $P(x) - g(x)$ be easy to bound formally. So other approaches for directly obtaining the extrema of $\varepsilon(x) = \hat{g}(x) - g(x)$ were investigated. One such approach is interval arithmetic and variation analysis to obtain extrema. For this purpose, Melquiond formalized and implemented in Coq an automatic differentiation and an interval arithmetic with floating-point bounds [416]. A similar method was first experimented in PVS [135], but it proved inefficient due to the use of rational numbers as interval bounds.

While this approach based on interval arithmetic was sufficient for verifying bounds on low-accuracy approximations, e.g., binary32 implementations, it proved to be inadequate for more accurate ones. Indeed, a first-order automatic differentiation requires to subdivide the input interval into too many subintervals for the verification to complete quickly. Therefore, Martin-Dorel and Melquiond ended up extending the CoqInterval library [408],[11] so that it could compute large-degree Taylor models of ε using ideas and proofs from [96] and [72], thus considerably reducing the number of subdivisions. In a way, this is similar to Harrison's original approach, except that it is not g but ε which gets approximated by a large-degree polynomial and that this polynomial approximation is found automatically.

Quality of approximations is not the only part that can be automated, as tools can also help to formally prove bounds on roundoff errors $\tilde{g}-\hat{g}$. The next section will be devoted to the Gappa tool, which is able to produce proofs for Coq. For the HOL Light system, the FPTaylor tool,[12] designed by Solovyev et al., is able to compute and bound an affine term representing the roundoff error of a floating-point straight-line code [563]. This affine term is expressed with respect to the rounding errors generated by each floating-point operator. In a well-behaved program, only first-order errors will have a noticeable effect, so other errors can be coarsely overestimated without degrading too much the bound on the final roundoff error. A peculiarity of FPTaylor is that the coefficients of the affine term are computed symbolically rather than numerically; actual bounds on these coefficients are evaluated as late as possible using global optimization.

Let us consider a small example: the computation of $g = x - x^2$ using binary64 floating-point multiplication and subtraction rounded to nearest. By introducing error terms, we know that the computed value $\tilde{g}$ satisfies

$$\tilde{g} = (x - (x^2 \cdot (1 + \varepsilon_1) + \mu_1)) \cdot (1 + \varepsilon_2)$$

with $|\varepsilon_1|, |\varepsilon_2| \le 2^{-53}$ and $|\mu_1| \le 2^{-1075}$. The first step is to express $\tilde{g} - g$ as an affine term of ε_1 and ε_2:

$$\tilde{g} - g = -x^2 \cdot \varepsilon_1 + (x - x^2) \cdot \varepsilon_2 + \nu$$

with ν a term that encompasses all the errors that are ignored at first order. The bound on these ignored errors is obtained along the computation of the affine form by coarse interval arithmetic. Here, for an input domain $X = [0, 2]$, one would obtain a bound on ν slightly larger than 2^{-104}. The next step

[11] http://coq-interval.gforge.inria.fr/.

[12] https://github.com/soarlab/FPTaylor.

is to bound each of the coefficients of the affine form, e.g., by using one of the above methods for bounding polynomials. This gives the inequality

$$|\tilde{g} - g| \leq 4 \cdot |\varepsilon_1| + 2 \cdot |\varepsilon_2| + |\nu| \leq 3 \cdot 2^{-52} + \cdots .$$

Note that the optimal error bound is about 2^{-52}. One cause for this discrepancy comes from the approach being unable to capture the fact that the subtraction is exact when $x \in [0.5, 2]$, as x and $\mathrm{RN}(x^2)$ then satisfy Sterbenz's conditions (see Section 4.2).

13.3 Roundoff Errors and the Gappa Tool

When verifying that some floating-point algorithm satisfies a specification expressed using real arithmetic (e.g., to express roundoff errors), the theorems one has to prove end up in an undecidable logic fragment. So there cannot be an automatic process able to verify any valid floating-point algorithm. Some parts of the verification may have to be manually performed, possibly in a formal environment. Nevertheless, most proofs on floating-point algorithms involve some repetitive, tedious, and error-prone tasks, e.g., verifying that no overflows occur. Another common task is a forward error analysis in order to prove that the distance between the computed result and the ideal result is bounded by a specified constant.

The Gappa[13] tool is meant to automate most of these tasks. Given a high-level description of a binary floating-point (or fixed-point) algorithm, it tries to prove or exhibit some properties of this algorithm by performing interval arithmetic and forward error analysis. When successful, the tool also generates a formal proof that can be embedded into a bigger development,[14] hence greatly reducing the amount of manual verification needed. The methods used in Gappa depend on the ability to perform numerical computations during the proofs, so there are two main constraints on the floating-point algorithms: their inputs have to be bounded somehow, and the precision of the arithmetic has to be specified. In practice, this is not an issue. For example, the inputs of algorithms evaluating elementary functions are naturally bounded, due to the argument reduction step (see Section 10.1.2). Moreover, most algorithms are written with a specific floating-point format in mind, e.g., binary64.

While Gappa is meant to help verify floating-point algorithms, it manipulates expressions on real numbers only and proves properties of these

[13]http://gappa.gforge.inria.fr/.

[14]Verifying the numerical accuracy of an implementation is often only a small part of a verification effort. One may also have to prove that there are no infinite loops, no accesses out of the bounds of an array, and so on. So the user has to perform and combine various proofs in order to get a complete verification.

expressions. Floating-point arithmetic is expressed using the functional rounding operators presented in Section 13.1.3. As a consequence, infinities, NaNs, and signed zeros are not first-class citizens in this approach. Gappa is unable to propagate them through computations, but it is nonetheless able to prove they do not occur during computations.

In developing the tool, several concepts related to the automatic verification of the usual floating-point algorithms were brought to light. The following sections describe them in the context of Gappa.

13.3.1 Computing on bounds

Gappa proves floating-point properties by exhibiting facts on expressions of real numbers. These facts are of several types, characterized by some predicates mixing numerical values with expressions. The main predicate expresses numerical bounds on expressions. These bounds are obtained and can then be verified with interval arithmetic (see Chapter 12). For example, if the real x is enclosed in the interval $[-3, 7]$, one can easily prove that the real $\sqrt{1+x^2}$ is well specified and enclosed in the interval $[1, 8]$.

13.3.1.1 Exceptional behaviors

Enclosures are sufficient for expressing properties usually encountered while verifying numerical algorithms. The first kind of property deals with exceptional behaviors. An exceptional behavior occurs when the input of a function is out of its definition domain. It also occurs when the evaluation causes a trap, e.g., in case of overflow or underflow.

Consider the computation $\mathrm{RN}\left(\sqrt{\mathrm{RD}(1+\mathrm{RU}(x^2))}\right)$. Let us assume that the software will not be able to cope with an overflow or an invalid operation if one is caused by this formula. Verifying that no exceptional behavior occurs then amounts to checking that neither $\mathrm{RU}(x^2)$ nor $\mathrm{RD}(1+\mathrm{RU}(x^2))$ nor $\mathrm{RN}\left(\sqrt{\mathrm{RD}(\ldots)}\right)$ overflows, and that $\mathrm{RD}(1+\mathrm{RU}(x^2))$ is nonnegative. This last property is easy enough to express with an enclosure: $\mathrm{RD}(1+\mathrm{RU}(x^2)) \in [0, +\infty)$. What about the ones on overflow?

Let Ω be the largest representable floating-point number and Ω^+ be the first power of the radix too large to be representable.[15] Expressing that $\mathrm{RU}(x^2)$, or more generally $\mathrm{RU}(y)$ does not overflow may be achieved with the property $y \in (-\Omega^+, +\Omega]$. Negative values outside this half-open interval would be rounded to $-\infty$, and positive values to $+\infty$. This approach has a drawback: it depends on the rounding function. The enclosure $y \in (-\Omega^+, +\Omega]$ is used for $\mathrm{RU}(y)$, $y \in [-\Omega, +\Omega^+)$ for $\mathrm{RD}(y)$, $y \in (-\Omega^+, \Omega^+)$ for $\mathrm{RZ}(y)$, and $y \in (-\frac{1}{2}(\Omega+\Omega^+), +\frac{1}{2}(\Omega+\Omega^+))$ for $\mathrm{RN}(y)$.

[15] So Ω^+ would be the successor of Ω if the exponent range was unbounded.

Let us consider a set of floating-point numbers without an upper bound on the exponent instead, as described in Section 13.1.4. In other words, while e_{max} still characterizes numbers small enough to be stored in a given format, the rounding operators ignore it and may return real numbers of magnitude larger than Ω, e.g., $\mathrm{RN}(\beta \cdot \Omega) = \beta \cdot \Omega$. This extended set, therefore, makes it simpler to characterize an operation that does not overflow: one only has to prove that $\circ(y) \in [-\Omega, +\Omega]$, whatever the rounding function $\circ(\cdot)$.

13.3.1.2 Quantifying the errors

Proving that no exceptional behaviors occur is only a part of the verification process. If the algorithm is not meant (or unable) to produce exact values at each step due to rounding errors, one wants to bound the distance between the computed value $\tilde{v}$ and the infinitely precise value $\hat{v}$ at the end of the algorithm. An expression computing $\hat{v}$ can be obtained by removing all the rounding operators from an expression computing $\tilde{v}$. But $\hat{v}$ may not even be the mathematical value v that the algorithm is meant to approximate. Indeed, some terms of v may have been neglected in order to speed up the algorithm ($1 + x^{-42} \rightsquigarrow 1$ for large x). Or they may have been simplified so that they can actually be implemented ($\cos x \rightsquigarrow 1 - \frac{x^2}{2}$ for small x).

There are two common distances between the computed value and the mathematical value: the absolute error and the relative error. In order to reduce the number of expressions to analyze, Gappa favors errors written as $\tilde{v} - v$ and $\frac{\tilde{v}-v}{v}$, but other forms could be handled as well.

Error expressions are bounded by employing methods from forward error analysis. For instance, the expression $(\tilde{u} \times \tilde{v}) - (u \times v)$ is bounded in two steps. First, the expressions $\tilde{u} - u$ and $\tilde{v} - v$, which are still error expressions, are bounded separately. The same is done for expressions u and v. Second, these four bounds are combined by applying interval arithmetic to the following formula:

$$(\tilde{u} \times \tilde{v}) - (u \times v) = (\tilde{u} - u) \times v + u \times (\tilde{v} - v) + (\tilde{u} - u) \times (\tilde{v} - v).$$

The previous example is the decomposition of the absolute error between two products. Similar decompositions exist for the two kinds of error and for all the arithmetic operators: $+, -, \times, \div, \sqrt{\cdot}$.

Another important class of operators is the rounding operators, as presented in Section 13.1.3. The absolute error $\circ(\tilde{u}) - u$ is decomposed into a simpler error $\tilde{u} - u$ and a rounding error $\circ(\tilde{u}) - \tilde{u}$. The rounding error is then bounded depending on the properties of the rounding operator. For floating-point arithmetic, this often requires a coarse bound on $\tilde{u}$.

Gappa therefore works by decomposing the target expression into smaller parts by relying on methods related to forward error analysis. Once all these parts can be bounded—either by hypotheses (e.g., preconditions of the studied function) or by properties of rounding operators—the tool performs interval arithmetic to compute an enclosure of the expression.

13.3.2 Counting digits

Representable numbers for a given format are a discrete subset of the real numbers. This is especially noticeable when considering fixed-point arithmetic. Indeed, since we do not consider overflow, a fixed-point format can be defined as the weight or position of the least significant bit (LSB) of the numbers. Therefore, given three fixed-point variables a, b, and c with the same number format, the addition $c := \circ(a + b)$ is exact. Such exact operations are inherently frequent when using fixed-point arithmetic, be it in software or in hardware design.

Thus, we would like the tool to be able to prove that $c - (a + b)$ is equal to zero, so that the roundoff error is not overestimated. As described above, the tool knows how to bound $\circ(u) - u$ for a real number u. For instance, if $\circ$ rounds toward $+\infty$ and the weight of the LSB is 2^k, then the property $\circ(u) - u \in [0, 2^k)$ holds for any $u \in \mathbb{R}$. Notice that $[0, 2^k)$ is the smallest possible subset, since u can be any real number. So this enclosure is not helpful in proving $\circ(a + b) - (a + b) \in [0, 0]$.

The issue comes from $a + b$ not being any real number. It is the sum of two numbers belonging to the fixed-point set $\{m \cdot 2^k \mid m \in \mathbb{Z}\}$, which happens to be closed under addition. Since $\circ$ is the identity on this set, the rounding error is zero. So manipulating enclosures only is not sufficient: the tool has to gather some information on the discreteness of the expressions.

13.3.2.1 Fixed-point arithmetic

As stated above, reasoning on discrete expressions could be achieved by considering the stability of fixed-point sets. However, it would be overly restrictive, as this approach would not cater to multiplication. Instead, Gappa keeps track of the positions of the LSB of all the expressions, if possible. In the previous example, a and b are multiples of 2^k, and so is their sum. Since the operator $\circ$ rounds at position k, it does not modify the result of the sum $a + b$. More generally, if an expression u is a multiple of 2^ℓ and if an operator $\circ$ rounds at position $h \le \ell$, then $\circ(u) = u$.

For directed rounding, this proposition can be extended to the opposite case $h > \ell$. Indeed the rounding error $\varepsilon = \circ(u) - u$ satisfies $|\varepsilon| < 2^h$ and is a multiple of 2^ℓ. Thus, $|\varepsilon| \le 2^h - 2^\ell$. As a consequence, when rounding toward zero at position h a number u multiple of 2^ℓ, the rounding error $\circ(u) - u$ is enclosed in $[-\delta, \delta]$ with $\delta = \max(0, 2^h - 2^\ell)$.

Let us denote $\mathrm{FIX}(u, k)$ the predicate stating that the expression u is a multiple of 2^k: $\exists m \in \mathbb{Z},\ u = m \cdot 2^k$. Assuming that the properties $\mathrm{FIX}(u, k)$

and $\mathrm{FIX}(v, \ell)$ hold for the expressions u and v, the following properties can be deduced:

$$\begin{aligned}
&\mathrm{FIX}(-u, k),\\
&\mathrm{FIX}(u + v, \min(k, \ell)),\\
&\mathrm{FIX}(u \times v, k + \ell).
\end{aligned}$$

Moreover, if the operator $\circ$ rounds at position k, then the property $\mathrm{FIX}(\circ(u), k)$ holds. The predicate is also monotonic:

$$\mathrm{FIX}(u, k) \wedge k \geq \ell \quad \Rightarrow \quad \mathrm{FIX}(u, \ell).$$

13.3.2.2 Floating-point arithmetic

The error bounds could be handled in a similar way for floating-point arithmetic, but the improvements are expected to be less dramatic. For instance, when rounding toward zero the product of two numbers in the binary64 format, the improved bound on the relative error (assuming a nonunderflowing result) is $2^{-52} - 2^{-105}$. This better bound is hardly more useful than the standard error bound 2^{-52} for proving the correctness of most floating-point algorithms.

As a consequence, Gappa does not try to improve relative error bounds, but it still tries to detect exact operations. Consider that rounding happens at precision p and that the smallest positive number is 2^k. A rounding operator $\circ_{p,k}$ will leave an expression u unmodified when its number w of significant digits is small enough ($w \leq p$) and its least significant digit has a weight 2^e large enough ($e \geq k$). The second property can be expressed with the FIX predicate: $\mathrm{FIX}(u, k)$. The first property requires a new predicate:

$$\mathrm{FLT}(u, p) \quad \equiv \quad \exists m, e \in \mathbb{Z},\ u = m \cdot 2^e \wedge |m| < 2^p.$$

Gappa combines both predicates into the following theorem:

$$\mathrm{FIX}(u, k) \wedge \mathrm{FLT}(u, p) \quad \Rightarrow \quad \circ_{p,k}(u) - u \in [0, 0].$$

Assuming that the properties $\mathrm{FLT}(u, p)$ and $\mathrm{FLT}(v, q)$ hold for the expressions u and v, the following properties can be deduced:

$$\begin{aligned}
&\mathrm{FLT}(-u, p),\\
&\mathrm{FLT}(u \times v, p + q).
\end{aligned}$$

Note that if p or q is equal to 1 (that is, u or v is a power of 2), then $p + q$ is an overestimation of the number of bits needed to represent $u \times v$. Thus, Gappa uses $p + q - 1$ in that case.

As with the fixed-point case, some properties can be deduced from a floating-point rounding operator:

$$\mathrm{FLT}(\circ_{p,k}(u), p) \wedge \mathrm{FIX}(\circ_{p,k}(u), k).$$

There is again a monotonicity property:

$$\mathrm{FLT}(u, p) \wedge p \le q \quad \Rightarrow \quad \mathrm{FLT}(u, q).$$

Finally, both predicates can be related as long as bounds are known on the expressions:

$$\begin{aligned} \mathrm{FIX}(u, k) \wedge |u| < 2^g \quad &\Rightarrow \quad \mathrm{FLT}(u, g - k), \\ \mathrm{FLT}(u, p) \wedge |u| \ge 2^g \quad &\Rightarrow \quad \mathrm{FIX}(u, g - p + 1). \end{aligned}$$

13.3.2.3 Application

Here is an example showing how the various predicates interact. Consider two expressions $u \in [3.2, 3.3]$ and $v \in [1.4, 1.8]$, both with p significant bits at most. How can we prove that the difference $u - v$ also has p significant bits at most? Notice that Lemma 4.1 (Sterbenz's lemma, see Section 4.2) does not apply, since $\frac{3.3}{1.4} > 2$.

First of all, by interval arithmetic, we can prove $|u - v| \le 1.9$. Moreover, since we have $|u| \ge 2^1$ and $\mathrm{FLT}(u, p)$, the property $\mathrm{FIX}(u, 2 - p)$ holds. Similarly, $\mathrm{FIX}(v, 1 - p)$ holds. Therefore, we can deduce a property on their difference: $\mathrm{FIX}(u - v, 1 - p)$. By combining this property with the bound $|u - v| < 2^1$, we get $\mathrm{FLT}(u - v, p)$. This concludes the proof that $u - v$ has at most p significant bits.

This reasoning could then be extended so as to prove that the floating-point subtraction is actually exact. Indeed, if the smallest positive floating-point number is 2^k, we have both $\mathrm{FIX}(u, k)$ and $\mathrm{FIX}(v, k)$. As a consequence, we also have $\mathrm{FIX}(u - v, k)$. By combining this property with $\mathrm{FLT}(u - v, p)$, we conclude that $u - v$ is representable as a floating-point number.

13.3.3 Manipulating expressions

13.3.3.1 Errors between structurally similar expressions

As presented in Section 13.3.1, Gappa encloses an expression representing an error, either $\tilde{x} - x$ or $\frac{\tilde{x} - x}{x}$, by first rewriting it in another form. This new expression is meant to make explicit the errors on the subterms of $\tilde{x}$ and x. For instance, if $\tilde{x}$ and x are the quotients $\tilde{u}/\tilde{v}$ and u/v, respectively, then Gappa relies on the following equality[16]:

$$\frac{\tilde{u}/\tilde{v} - u/v}{u/v} = \frac{\varepsilon_u - \varepsilon_v}{1 + \varepsilon_v} \quad \text{with} \quad \varepsilon_u = \frac{\tilde{u} - u}{u} \quad \text{and} \quad \varepsilon_v = \frac{\tilde{v} - v}{v}.$$

The left-hand side contains the correlated expressions $\tilde{u}$ and u, and $\tilde{v}$ and v, while the right-hand side does not. Therefore, the latter should have a better structure for bounding the error by interval arithmetic. It still has a bit of correlation, since the expression ε_v appears twice; however, this does not induce an overestimation of the bounds. Indeed, the whole expression is a homographic transformation with respect to ε_v, hence monotonic on the two parts of its domain. Gappa takes advantage of this property when computing its enclosure. Therefore, as long as the errors ε_u and ε_v are not correlated, tight bounds on the relative error between $\tilde{u}/\tilde{v}$ and u/v are obtained. In fact, unless the partial errors ε_u and ε_v are correlated intentionally, this correlation hardly matters when bounding the whole error.

For this approach to work properly, the expressions $\tilde{x}$ and x have to use the same operator—a division in the example above. Moreover, their subterms should match, so that the error terms ε_u and ε_v are meaningful. In other words, $\tilde{u}$ should somehow be an approximation to u, and $\tilde{v}$ to v.

For instance, Gappa's approach will fail if the user wants to analyze the error between the expressions $\tilde{x} = (a_1 \times \tilde{u})/(a_2 \times \tilde{v})$ and $x = u/v$. If a_1 and a_2 are close to each other, then computing $\tilde{x}$ may be a sensible way of getting an approximate value for x, hence the need to bound the error. Unfortunately, $a_1 \times \tilde{u}$ is not an approximation to u, only $\tilde{u}$ is. So the transformation above will fail to produce useful bounds on the error. Indeed, an error expression ε_u between $a_1 \times \tilde{u}$ and u does not present any interesting property.

In such a situation, the user should tell Gappa that $\tilde{x}$ and $x' = \tilde{u}/\tilde{v}$ are close, and how to bound the (relative) error between them. When asked about the error between $\tilde{x}$ and x, Gappa will separately analyze its two parts—the error between $\tilde{x}$ and x' and the error between x' and x—and combine them.

13.3.3.2 Using intermediate expressions

This process of using an intermediate expression in order to get errors between structurally similar expressions is already applied by Gappa for rounding operators. Indeed, while the user may sometimes want to bound the error between $\tilde{x} = \tilde{u}/\tilde{v}$ and $x = u/v$, a more usual case is the error between

[16]Making use of that equality requires to prove that u, v, and $\tilde{v}$ are nonzero. Section 13.3.4 details how Gappa eliminates these hypotheses by using a slightly modified definition of the relative error.

$\tilde{y} = \circ(\tilde{x})$ and x. The two expressions no longer have a similar structure, since the head symbol of $\tilde{y}$ is the rounding function $\circ$, while the head symbol of x is the division operator.

To avoid this issue, Gappa registers that $\tilde{y}$ is an approximation to $y = \tilde{x}$, by definition and purpose of rounding operators. Whenever it encounters an error expression involving $\tilde{y}$, e.g., $\frac{\circ(\tilde{x})-x}{x}$, Gappa will therefore try to decompose it by using y as an intermediate expression:

$$\frac{\tilde{y} - x}{x} = \varepsilon_1 + \varepsilon_2 + \varepsilon_1 \times \varepsilon_2$$

with

$$\varepsilon_1 = \frac{\tilde{y} - y}{y} \quad \text{and} \quad \varepsilon_2 = \frac{y - x}{x} = \frac{\tilde{x} - x}{x}.$$

Rounded values are not the only ones Gappa considers to be approximations to other values. Whenever the tool encounters a hypothesis of the form $\tilde{x} - x \in I$ or $\frac{\tilde{x}-x}{x} \in I$, it assumes that $\tilde{x}$ is an approximation to x, as the enclosed expression looks like an error between $\tilde{x}$ and x. This heuristic is useful in automatically handling problems where there are method errors in addition to rounding errors. In the example of Section 13.3.5, the user has a polynomial $p(x)$ that approximates $\sin x$. The goal is to get a bound on the error between the computed value $\tilde{p}(x)$ and the ideal value $\sin x$. Since the user provides a bound on the method error between p and $\sin$, Gappa deduces that the following equality may be useful:

$$\frac{\tilde{p}(x) - \sin x}{\sin x} = \varepsilon_p + \varepsilon_s + \varepsilon_p \times \varepsilon_s$$

with

$$\varepsilon_p = \frac{\tilde{p}(x) - p(x)}{p(x)} \quad \text{and} \quad \varepsilon_s = \frac{p(x) - \sin x}{\sin x}.$$

Gappa can then automatically compute a bound on ε_p, as this is an error between two expressions with the same structure once rounding operators have been removed from $\tilde{p}$.

13.3.3.3 Cases of user hints

When Gappa's heuristics are unable to reduce expressions to errors between terms with similar structures, the user can provide additional hints. There are two kinds of hints. The first kind tells Gappa that two expressions have the same bounds under some constraints. In particular, if one of the expressions is composed of errors between structurally similar terms, then Gappa will be able to deduce tight bounds on the other expression. The second kind of hint tells Gappa that a specific term approximates another one, so the tool will try to exhibit an error expression between the two of them. The following paragraphs detail these two kinds of hints.

Matching formula structures If the developer has decided some terms are negligible and can be ignored in the computations, then the formula of the computed value will have some missing terms with respect to the formula of the exact value. As a consequence, the structures of the two formulas may be different.

As an example, let us consider a double-double multiplication algorithm (see Algorithm 14.3). The inputs are the pairs of numbers (x_h, x_ℓ) and (y_h, y_ℓ) with x_ℓ and y_ℓ negligible with respect to x_h and y_h, respectively. The output of the algorithm is a pair (z_h, z_ℓ) that should approximate the exact product $p = (x_h + x_\ell) \times (y_h + y_\ell)$. The algorithm first computes the exact result $z_h + t$ of the product $x_h \times y_h$ (see Section 4.4). The products $x_h \times y_\ell$ and $x_\ell \times y_h$ are then added to the lower part t in order to get z_ℓ:

$$\begin{aligned} z_h + t &\leftarrow x_h \times y_h \\ z_\ell &\leftarrow \mathrm{RN}(t + \mathrm{RN}(\mathrm{RN}(x_h \times y_\ell) + \mathrm{RN}(x_\ell \times y_h))). \end{aligned}$$

The product $x_\ell \times y_\ell$ is supposedly negligible and was ignored when computing z_ℓ. The goal is to bound the error $(z_h + z_\ell) - p$, but the two subterms do not have the same structure: p is a product, $z_h + z_\ell$ is not. So we rewrite the error by distributing the multiplication and setting aside $x_\ell \times y_\ell$:

$$(z_h + z_\ell) - p = \underbrace{(z_h + t) - (x_h \times y_h)}_{\delta_1} + \underbrace{z_\ell - (t + (x_h \times y_\ell + x_\ell \times y_h))}_{\delta_2} - x_\ell \times y_\ell\,.$$

In the preceding formula, δ_1 is zero by definition of the exact product, while δ_2 is the difference between two expressions with the same structure once the rounding operators have been removed. An evaluation by interval of the right-hand-side expression will therefore give tight bounds on the left-hand-side expression.

Handling converging formulas Gappa does not recognize converging expressions. Handling them with the usual heuristics will fail, since their structure is generally completely unrelated to the limit of the sequence. The developer should therefore explain to the tool why a particular formula was used. Let us consider Newton's iteration for computing the multiplicative inverse of a number a (the innermost RN should be ignored in case of an FMA):

$$x_{n+1} = \mathrm{RN}(x_n \times \mathrm{RN}(2 - \mathrm{RN}(a \times x_n))).$$

This sequence quadratically converges toward a^{-1}, when there are no rounding operators. So the first step is to tell Gappa that the computed value x_{n+1} is an approximation to x'_{n+1}, which is defined as x_{n+1} without rounding operators:

$$x_{n+1} \simeq x'_{n+1} = x_n \times (2 - a \times x_n).$$

Notice that x'_{n+1} is not defined as depending on x'_n but on x_n. Indeed, the iteration is self-correcting, so using x_n will help Gappa to notice that rounding errors are partly compensated at each step. The developer can now tell Gappa that the error with the ideal value decreases quadratically:

$$\frac{x'_{n+1} - a^{-1}}{a^{-1}} = -\left(\frac{x_n - a^{-1}}{a^{-1}}\right)^2.$$

When evaluating the error $(x_{n+1} - a^{-1})/a^{-1}$ at step $n+1$, Gappa will split it into two parts: $(x_{n+1} - x'_{n+1})/x'_{n+1}$, which is the total roundoff error at this step, and $(x'_{n+1} - a^{-1})/a^{-1}$, which is given by the identity above and depends on the error at step n.

13.3.4 Handling the relative error

Expressing a relative error as $\frac{\tilde{u}-u}{u}$ implicitly puts a constraint $u \neq 0$. Therefore, if the enclosure $\frac{\tilde{u}-u}{u} \in I$ appears in a hypothesis or as a conclusion of a theorem, it is generally unavailable until one has also proved $u \neq 0$. This additional constraint is not an issue when dealing with the relative error of a product, that is, when $\tilde{u} = \circ(x \times y)$ and $u = x \times y$. Indeed, the relative error $\frac{\tilde{u}-u}{u}$ can usually be bounded only for u outside of the underflow range. As a consequence, the property $u \neq 0$ comes for free.

For floating-point addition, the situation is not that convenient. Indeed, the relative error between $\tilde{u} = \circ(x + y)$ and $u = x + y$ (with x and y floating-point numbers) is commonly admitted to be always bounded, even for subnormal results. So the relative error should be expressed in a way that does not require the ideal value u to be nonzero. Gappa achieves this property by introducing another predicate:

$$\mathrm{REL}(\tilde{u}, u, I) \quad \equiv \quad -1 < \mathrm{lower}(I) \wedge \exists \varepsilon_u \in I,\ \tilde{u} = u \times (1 + \varepsilon_u).$$

13.3.4.1 Always-bounded relative errors

The relative error of a sum $z = \mathrm{RN}(x) + \mathrm{RN}(y)$ can now be enclosed as follows, for binary64:

$$\mathrm{REL}(\mathrm{RN}(z), z, [-2^{-53}, 2^{-53}]).$$

Note that z is $\mathrm{RN}(x) + \mathrm{RN}(y)$ and not just $x + y$. This guarantees that z is not the sum of any two real numbers, but of two floating-point numbers represented with the same format as $\mathrm{RN}(z)$, i.e., same precision and $e_{\min}$. This constraint on z is too restrictive, though. The two subterms do not necessarily have to be rounded to nearest in binary64 format. For instance, the bound on the relative error would still hold for $z = \mathrm{RD}(x) + \mathrm{RZ}_{\mathrm{binary32}}(y)$. As a matter of fact, z does not even have to be a sum. So, which conditions are sufficient for the relative error to be bounded? For $\mathrm{RN}(z)$ with $z = \mathrm{RN}(x) + \mathrm{RN}(y)$, the bound is a consequence of the floating-point addition not losing any information with a magnitude smaller than the smallest positive floating-point number 2^k. This is generalized to the following theorem, which Gappa relies on to bound the relative error when rounding to nearest:

$$\mathrm{FIX}(u,k) \quad \Longrightarrow \quad \mathrm{REL}(\circ_{p,k}(u), u, [-2^{-p}, 2^{-p}]).$$

The original property can then be recovered with the following proof. Since both $\mathrm{RN}(x)$ and $\mathrm{RN}(y)$ are binary64 numbers, they satisfy the properties $\mathrm{FIX}(\mathrm{RN}(x), -1074)$ and $\mathrm{FIX}(\mathrm{RN}(y), -1074)$. As a consequence (Section 13.3.2), their sum does too: $\mathrm{FIX}(\mathrm{RN}(x) + \mathrm{RN}(y), -1074)$. Applying the theorem above hence gives $\mathrm{REL}(\mathrm{RN}(z), z, [-2^{-53}, 2^{-53}])$.

13.3.4.2 Propagation of relative errors

Rules for the REL predicate are straightforward. Given the two properties $\mathrm{REL}(\tilde{u}, u, I_u)$ and $\mathrm{REL}(\tilde{v}, v, I_v)$, multiplication and division are as follows:

$$\begin{array}{ll} \mathrm{REL}(\tilde{u} \times \tilde{v}, u \times v, J) & \text{with } J \supseteq \{\varepsilon_u + \varepsilon_v + \varepsilon_u \times \varepsilon_v \mid \varepsilon_u \in I_u, \varepsilon_v \in I_v\}, \\ \mathrm{REL}(\tilde{u}/\tilde{v}, u/v, J) & \text{with } J \supseteq \{\frac{\varepsilon_u - \varepsilon_v}{1+\varepsilon_v} \mid \varepsilon_u \in I_u, \varepsilon_v \in I_v\}. \end{array}$$

Assuming $u + v \neq 0$ and $u/(u+v) \in I$, the relative error for addition is given by $\mathrm{REL}(\tilde{u} + \tilde{v}, u + v, J)$ with

$$J \supseteq \{\varepsilon_u \times t + \varepsilon_v \times (1-t) \mid \varepsilon_u \in I_u, \varepsilon_v \in I_v, t \in I\}.$$

Composing relative errors is similar[17] to multiplication: Given the two properties $\mathrm{REL}(z, y, I_1)$ and $\mathrm{REL}(y, x, I_2)$, we have $\mathrm{REL}(z, x, J)$ with

$$J \supseteq \{\varepsilon_1 + \varepsilon_2 + \varepsilon_1 \times \varepsilon_2 \mid \varepsilon_1 \in I_1, \varepsilon_2 \in I_2\}.$$

13.3.5 Example: toy implementation of sine

Let us consider the C code shown in Listing 13.1. Note that the literal `0x28E9.p-16f` is a C hexadecimal notation for the floating-point constant of type `float` equal to 10473/65536. Since the input x is such that $|x| \leq 1$, the function computes a value $y \simeq x - x^3/6 \simeq \sin x$.

[17]Encountering the same formula is not unexpected: taking $z = \tilde{u} \times \tilde{v}$, $y = \tilde{u} \times v$, and $x = u \times v$ makes it appear.

C listing 13.1 Toy implementation of sine around zero.

```
float my_sine(float x)
{
  assert(fabsf(x) <= 1);
  float y = x * (1.f - x*x * 0x28E9.p-16f);
  return y;
}
```

Assuming all the computations are rounded to nearest in the binary32 format,[18] what is the relative error between the computed value y and the mathematical value $\sin x$? The Gappa tool will be used to find some bounds on this error. Note that, for the sake of this example, x will first be assumed not to be too small. So one of the hypotheses will be $|x| \in [2^{-100}, 1]$.

Gappa is designed to prove some logical propositions involving enclosures of real-valued expressions. These expressions can be built around the basic arithmetic operators only. In particular, Gappa does not know about the sine function. So the value $\sin x$ will be represented by a new variable sin_x. Consider the following script:

```
{ |x| in [1b-100,1] -> (y - sin_x) / sin_x in ? }
```

Since it contains a question mark, Gappa will try to find an interval I such that the following statement holds:

$$\forall x, y, sin\_x \in \mathbb{R}, \quad |x| \in [2^{-100}, 1] \Rightarrow \frac{y - sin\_x}{sin\_x} \in I.$$

Obviously, there is no such interval I, except $\mathbb{R}$, assuming sin_x is implicitly nonzero.[19] The proposition is missing a lot of data. In particular, y should not be a universally quantified variable: it should be an expression depending on x. So, we can either substitute y with its actual expression, or the definition of y can be added to the Gappa script before the logical proposition, as follows:

```
y = ...; # to be filled
{ |x| in [1b-100,1] -> (y - sin_x) / sin_x in ? }
```

The computations are performed in rounding to nearest and the numbers are stored in the binary32 format. In Gappa's formalism, this means that the rounding operator `float<24,-149,ne>` should be applied to each infinitely precise result. The parameters of the rounding operator express its

[18]With this C code, this is guaranteed by the ISO C standard under the condition that `FLT_EVAL_METHOD` is equal to 0 (see Section 6.2.3) and that floating-point expressions are not contracted (in particular because an FMA could be used here, see Section 6.2.3.2).

[19]Gappa never assumes that a divisor cannot be zero, unless running in *unconstrained* mode.

properties: target precision is 24 bits, the smallest positive representable number is 2^{-149}, and results are rounded to nearest even (ne). Note that the operator float<ieee_32,ne> uses a predefined synonym to avoid to remember these parameters. Still it would be cumbersome to type this operator for each floating-point operation, so a shorter alias rnd can be defined beforehand:

```
@rnd = float<ieee_32,ne>;
y = rnd(x * rnd(1 - rnd(rnd(x * x) * 0x28E9p-16)));
```

Even with the alias, the definition of y is still obfuscated. Since all the operations are using the same rounding operator, it can be prefixed to the equal symbol, so that the definition is almost identical to the original C code:

```
y rnd= x * (1 - x*x * 0x28E9p-16);
```

Although y is now defined, Gappa is still unable to tell anything interesting about the logical proposition, since sin_x is a universally quantified variable instead of being $\sin x$. It would be possible to define sin_x as the sum of a high-degree polynomial in x with an error term δ. Then a hypothesis stating some bounds on δ would be sufficient from a mathematical point of view.

However, this is not the usual approach with Gappa. Rather, we should look at the original C code again. The infinitely precise expression

$$My = x \times \left(1 - \frac{10473}{65536} \times x^2\right)$$

was chosen on purpose so that it is close to $\sin x$, since this is what we want to compute. Therefore, sin_x should be defined with respect to that infinitely precise expression. The relative error between My and $sin\_x = \sin x$ can be obtained (and proved) by means external to Gappa. For example, the CoqInterval library [408] can formally verify that this relative error is bounded by $155 \cdot 10^{-5}$. This property is added as a hypothesis to the proposition:

```
My = x * (1 - x*x * 0x28E9p-16);
{ |x| in [1b-100,1] /\ |(My - sin_x) / sin_x| <= 1.55e-3
  -> ((y - sin_x) / sin_x) in ? }
```

Note that constants in Gappa are never implicitly rounded. So the literal 1.55e-3 is the real number $155 \cdot 10^{-5}$ and the literal 1b-100 represents the real number $1 \cdot 2^{-100}$—it could also have been written 0x1.p-100. Similarly, the hexadecimal literal 0x28E9p-16 that appears in y and My is a compact way of expressing the real number $10473/65536$.

At this point, Gappa is still unable to compute an enclosure of the relative error, although all the information seems to be present. In order to understand why, the tool can be run in *unconstrained* mode with the -Munconstrained command-line option. This mode causes Gappa to assume any theorem hypothesis stating that an expression is nonzero or does not underflow. Thus, the bounds computed by Gappa are potentially incorrect and no formal proof is generated (since it would contain holes), but it might help uncovering where Gappa initially got stuck.

On the example above, Gappa now answers that the relative error is bounded by $1.551 \cdot 10^{-3}$. The tool also explains that it has assumed that sin_x (as well as a few other expressions) is nonzero when computing this bound. Indeed, without such an assumption, Gappa is unable to use the hypothesis on the relative error $(My - sin\_x)/sin\_x$. This assumption cannot be deduced from the current hypotheses, so we need to enrich the problem with a new hypothesis. Since $|\sin x|$ is known to be larger than $\frac{1}{2}|x|$ for $|x| \leq \frac{\pi}{2}$, we can tell the tool that $|sin\_x|$ is larger than 2^{-101}.

Gappa script 13.1 Relative error of a toy implementation of sine.

```
# Floating-point format is binary32; operations are rounded to nearest
@rnd = float<ieee_32,ne>;

# Value computed by the floating-point function
y rnd= x * (1 - x*x * 0x28E9p-16);
# Value computed with an infinitely-precise arithmetic (no rounding)
My   = x * (1 - x*x * 0x28E9p-16);

# Proposition to prove
{
  # Input x is smaller than 1 (but not too small)
  |x| in [1b-100,1] /\
  # My is close to sin x and the bound is known
  |(My - sin_x) / sin_x| <= 1.55e-3 /\
  # Helper hypothesis: sin x is not too small either
  |sin_x| in [1b-101,1]
->
  # Target expression to bound
  ((y - sin_x) / sin_x) in ?
}
```

Given Script 13.1, Gappa is able to compute and prove a bound on the relative error between y and sin_x. Since sin_x will appear as a universally quantified variable in a formal development, it can later be instantiated by $\sin x$ without much trouble. This instantiation will simply require the user to prove that both inequalities

$$|(My - \sin x)/\sin x| \leq 1.55 \cdot 10^{-3}$$

and

$$|\sin x| \in [2^{-100}, 1]$$

hold for $x \in [2^{-100}, 1]$.

In order to get rid of the extraneous hypotheses on x and $\sin x$ not being too small, we can express the relative errors directly with the predicate REL (Section 13.3.4). In Gappa syntax, the relative error between two expressions

u and v is written `u -/ v`. It amounts to saying that there exists $\varepsilon \in \mathbb{R}$ such that $u = v \times (1+\varepsilon)$, and that bounds on `u -/ v` are simply syntactic sugar for bounds on ε. This leads to Script 13.2.

Gappa script 13.2 Relative error of a toy implementation of sine (variant).

```
@rnd = float<ieee_32,ne>;

# x is a binary32 floating-point number
x = rnd(x_);

y rnd= x * (1 - x*x * 0x28E9p-16);
My   = x * (1 - x*x * 0x28E9p-16);

{
  |x| <= 1 /\
  |My -/ sin_x| <= 1.55e-3
->
  |y -/ sin_x| <= 1.551e-3
}
```

Notice that this script passes one extra piece of information to Gappa by stating that x is a binary32 number rather than an arbitrary real number. This property is expressed by defining x as the rounded value of a dummy number $x\_$. Therefore, x is no longer universally quantified (but $x\_$ is). As a consequence, the generated formal proof can no longer be used by instantiating x with the actual argument X of the `my_sine` function. Instantiating $x\_$ with X achieves the same effect. Indeed, by virtue of X being a representable floating-point number, we have $x = \mathrm{RN}(x\_) = \mathrm{RN}(X) = X$. Another way of expressing that x is a binary32 number would have been to use the predicates presented in Section 13.3.2 to replace the hypothesis about x by

```
|x| <= 1 /\ @FIX(x, -149) /\ @FLT(x, 23)
```

Note that, without the hypothesis that x is a binary32 number, there would be counterexamples to the property we are trying to prove. Indeed, if it was possible for x to be equal to 2^{-150}, then we would get $y = \mathrm{RN}(x) = 0$ and $My \simeq x = 2^{-150}$, which means that the relative error between y and My would be -1 while we expect it to be negligible for x close to zero. Thus, the hypothesis is not just meant to help Gappa find a proof.

Another important point about this example is that the proof depends on the value of x. If x is close to zero, then Gappa proves that the last multiplication of y is exact because its right-hand side is rounded to 1. If $|x|$ is close to 1, Gappa instead proves that y does not fall in the subnormal range. Thus, in both cases, the roundoff error caused by the last multiplication is negligible. This means that proving the property requires a case split.

Fortunately, Gappa guesses that x is a sensible expression on which to perform a case split. By trial and error, the tool eventually finds that the following three subcases make it possible to complete the proof: $x \in [-1, -2^{-12}]$, $x \in [-2^{-12}, 2^{-12}]$, and $x \in [2^{-12}, 1]$. If Gappa had not found such a case split for some reason, we could have indicated it at the end of the script using the following hint:

```
$ x in (-1b-12, 1b-12);
```

Note that x is not the best expression on which to perform a case split, since the proof does not depend on the sign of x. Thus, by using the following hint instead, Gappa would perform a case split on $|x|$ and generate a shorter formal proof:

```
$ |x| in (1b-12);
```

Both previous hints explicitly tell Gappa which split points it should use. But we could also ask the tool to automatically find them. For example, the following hint causes Gappa to split the interval of $|x|$ until it is able to prove that the relative error between y and My is less than 2^{-20} on each subinterval:

```
|y -/ My| <= 1b-20 $ |x|;
```

13.3.6 Example: integer division on Itanium

The second example is taken from the software algorithm for performing a 16-bit unsigned integer division using floating-point operations on Intel Itanium [119, 242]. It is described by Algorithm 13.1. A peculiarity of this algorithm is that all the operations are without rounding error and the correctness of the algorithm heavily relies on this property. Thus, Gappa's ability to keep count of the significant digits of computed expressions is important.

Algorithm 13.1 Quotient of two unsigned 16-bit integers using FMA [119].

Require: a and b are two 16-bit positive integers
Ensure: $q = \lfloor a/b \rfloor$
$y_0 \leftarrow 1/b$ approximately
$q_0 \leftarrow \mathrm{RN}(a \cdot y_0)$
$e_0 \leftarrow \mathrm{RN}(1 + 2^{-17} - b \cdot y_0)$
$q_1 \leftarrow \mathrm{RN}(e_0 \cdot q_0 + q_0)$
$q \leftarrow \lfloor q_1 \rfloor$

The inputs of the algorithm are two 16-bit positive integers a and b. The first step of the algorithm approximates the reciprocal b^{-1}. This approximation y_0 is computed by the Itanium assembly instruction `frcpa`, which returns an 11-bit value with a relative error ε_0 of at most $2^{-8.886}$. The Gappa script expresses these hypotheses as follows.

```
@FIX(a, 0) /\ a in [1,65535] /\
@FIX(b, 0) /\ b in [1,65535] /\
@FLT(y0, 11) /\ |eps0| <= 0.00211373
```

The algorithm then performs three FMA operations in 64-bit extended precision (operator rnd). These operations could be expressed as follows.

```
@rnd = float<x86_80,ne>;
q0 = rnd(a * y0);
e0 = rnd(1 + 1b-17 - b * y0);
q1 = rnd(q0 + e0 * q0);
```

Yet all these operations are exact by design of the algorithm. So we instead define q_0, e_0, and q_1 without any rounding operator and we ask Gappa to prove that rounding these values does not modify them. We also want to prove that the computed result $\lfloor q_1 \rfloor$ is equal to $\lfloor a/b \rfloor$, which leads to the following tentative script.

```
q0 = a * y0;
e0 = 1 + 1b-17 - b * y0;
q1 = q0 + e0 * q0;
{ ... # hypotheses on a, b, and y0
->
  int<dn>(q1) = int<dn>(a/b) /\
  rnd(q0) = q0 /\ rnd(e0) = e0 /\ rnd(q1) = q1
}
```

Unfortunately, Gappa is unable to prove such a property. Theoretically, the tool could succeed by verifying the algorithm one pair (a, b) after the other until all possible inputs have been considered, but this does not scale to more complicated algorithms. Therefore, we will instead prove a different property that implies $\lfloor q_1 \rfloor = \lfloor a/b \rfloor$.

Let ε_1 denote the relative error between q_1 and a/b. If the product $a \times \varepsilon_1 = q_1 \times b - a$ lies in the interval $[0, 1)$, then q_1 lies in the interval $[\frac{a}{b}, \frac{a+1}{b})$, which does not contain any integer except for a/b potentially. Thus, if Gappa succeeds in proving $a \times \varepsilon_1 \in [0, 1)$, then we have enough information to conclude that $\lfloor q_1 \rfloor = \lfloor a/b \rfloor$ holds.

The tool is still not able to prove the property, though. We have to give it a bit more information. In particular, we have to somehow tell Gappa that the algorithm is a variation on the Newton–Raphson iterations presented in Section 4.7. It relies on the fact that ε_1 is almost the square of ε_0. Thus, even though $q_0 = a \times y_0$ is a poor approximation to a/b, the next iterate q_1 is a good approximation to a/b. So we tell Gappa about this quadratic convergence by adding the following hypothesis to the proposition we want to prove:

```
eps1 = -(eps0 * eps0) + (1 + eps0) * 1b-17
```

Note that this equality being a hypothesis, it is not verified by Gappa. Thus, it has to be verified outside the tool, for instance using the Coq proof assistant. Fortunately, unfolding the definitions of q_1 and ε_0 and simplifying both sides of the equality are sufficient to prove that it holds.

Gappa script 13.3 Correctness of the 16-bit integer division on IA-64.

```
@rnd = float<x86_80,ne>;

# Values computed by FMA, with rounding operators omitted
q0 = a * y0;
e0 = 1 + 1b-17 - b * y0;
q1 = q0 * (1 + e0);

# Relative errors between computed and ideal values
eps0 = (y0 - 1 / b) / (1 / b);
eps1 = (q1 - a / b) / (a / b);

{
  # Inputs are unsigned 16-bit integers
  @FIX(a, 0) /\ a in [1,65535] /\
  @FIX(b, 0) /\ b in [1,65535] /\
  # Initial approximation is obtained by frcpa
  @FLT(y0, 11) /\ |eps0| <= 0.00211373 /\
  # Convergence is almost quadratic
  eps1 = -(eps0 * eps0) + (1 + eps0) * 1b-17
->
  # The quotient is correct
  a * eps1 in [0,0.99999] /\
  # The intermediate computations are exact
  rnd(q0) = q0 /\ rnd(e0) = e0 /\ rnd(q1) = q1
}

# Perform bisection along b; needed when q1 is defined as q0 + e0 * q0
# rnd(q1) = q1 $ b;
```

At this stage, the Gappa script is almost complete. The tool is able to prove all the properties, except for $\mathsf{rnd}(q_1) = q_1$. Indeed, it is unable to prove that 64 bits are sufficient for representing q_1. This is due to Gappa overestimating some enclosures. Indeed, naive interval arithmetic tends to overestimate bounds when there are dependencies between subexpressions, and the wider the input intervals are, the larger the overestimation is. So, a simple way to guide Gappa is to force it to bisect the enclosure of b until it proves $\mathsf{rnd}(q_1) = q_1$ on each subinterval:

```
rnd(q1) = q1 $ b;
```

This hint is sufficient for Gappa to prove all the properties. A look at the generated formal proof shows that the tool decided to separately study two subcases: $b \in [1, 8192]$ and $b \in [8193, 65535]$.

Another way of helping Gappa would be to search which enclosures are overestimated, and then add hints (or modify formulas) in order to tighten the bounds. Here, the overestimation comes from the variable q_0 appearing twice in the definition of q_1. Thus, changing its definition from $q_0 + e_0 \times q_0$ to $q_0 \times (1 + e_0)$ is sufficient to avoid the overestimation. But this also means that the definition of q_1 is now a bit farther from the usual definition of a floating-point FMA (yet provably equivalent). We could also have added $q_1 = q_0 \times (1 + e_0)$ as a new hypothesis. The complete example is Script 13.3.

Note that reducing overestimation is not meant to increase the confidence in the proved proposition; it only reduces the size of the generated formal proof and the time it takes to find it, since the tool no longer needs to study several cases. So, in practice, one would look for reducing overestimation only if Gappa failed to prove the proposition in a reasonable amount of time.

Chapter 14

Extending the Precision

THOUGH SATISFACTORY in most situations, the fixed-precision floating-point formats that are available in hardware or software in our computers may sometimes prove insufficient. There are reasonably rare cases when the binary64/decimal64 or binary128/decimal128 floating-point numbers of the IEEE 754 standard are too crude as approximations of the real numbers. Also, at the time of writing these lines, the binary128 and decimal128 formats are very seldom implemented in hardware.[1]

The need for higher precision occurs for example when dealing with ill-conditioned numerical problems: internal computations with very high precision may be needed to obtain a meaningful final result. For instance, "double-word" arithmetic has sometimes proven useful for linear algebra [385], for algorithmic geometry [555, 493], or for the determination of integer relations between real or complex numbers [27].

Another example is the evaluation of transcendental functions with correct rounding (see Chapter 10): if all intermediate calculations are done in the target precision, it is very difficult to guarantee last-bit accuracy in the final result. In this case, double-word or triple-word arithmetic may be needed. For instance, the CRlibm library of correctly rounded elementary functions uses "double-double" or "triple-double" operations in critical parts [142].

A situation where the usual basic precisions are obviously insufficient is when the result of the computation itself is required with very high precision. This frequently happens in "experimental mathematics" [58]; for example, when one wants billions of decimals of mathematical constants such as π [537, 25], the constant e or the zeros of Riemann's zeta function. For

[1]To our knowledge, the only commercially significant platform that has supported binary128 in hardware for the last decade has been the IBM z Systems [387].

J.-M. Muller et al., *Handbook of Floating-Point Arithmetic*,
https://doi.org/10.1007/978-3-319-76526-6_14

instance, a famous formula such as the Bailey–Borwein–Plouffe formula for π [28]:

$$\pi = \sum_{k=0}^{\infty} \frac{1}{16^k} \left(\frac{4}{8k+1} - \frac{2}{8k+4} - \frac{1}{8k+5} - \frac{1}{8k+6} \right),$$

which allows one to compute the zillionth bit of π without needing to compute the previous ones, was first found using "experimental mathematics" techniques. Other examples from several domains of science and engineering are given by Bailey, Barrio, and Borwein [26].

Of course, big precisions may be costly, both in terms of time and memory consumption, so one must find convenient tradeoffs between cost and precision. Precision is not the only problem; the dynamic range of usual formats may be a problem as well. The exponent range may be too restricted for the application under scope, giving rise to frequent overflows and underflows.

The scope of this chapter is to describe some usual ways of extending the precision and the exponent range. Multiple-precision arithmetic (i.e., algorithms that allow one to manipulate numbers represented with thousands of digits) is out of the scope of this book. The reader can for instance refer to [66]. However, since it has useful applications for people who design floating-point algorithms, we will quickly present a few basic methods, give some references, and list a few useful packages. We will mainly focus on methods that allow a programmer to roughly double, triple, or even quadruple the usual precision for critical parts of programs, without requiring the use of multiple-precision packages.

14.1 Double-Words, Triple-Words...

When the need for increased precision is limited to twice (or thrice, or maybe even four times) the largest precision of the available floating-point arithmetic, a possible solution is to resort to arithmetics that are usually called "double-double," "triple-double," or "quad-double" arithmetics in the literature. These clumsy names come from the fact that, at the time of writing this book, the format with the largest precision that is available on most platforms of commercial significance is the double-precision format of IEEE 754-1985 (now much better named binary64 in the IEEE 754-2008 standard). We will call them "double-word" and "triple-word" arithmetics, since there is no special reason for necessarily having the underlying floating-point format being double precision—it makes sense to choose, as the underlying format, the largest one available in hardware—and because the format that was once called "double precision" now has another name in the IEEE 754-2008 standard.

14.1.1 Double-word arithmetic

In Chapter 4, we have studied algorithms[2] that allow one to represent the sum or product of two floating-point numbers as the unevaluated sum of two floating-point numbers $x_h + x_\ell$, where x_h is nearest the exact result. A natural idea is to try to directly perform arithmetic operations on such unevaluated sums. This is the underlying principle of *double-word arithmetic* [158]. It consists in representing a number x as the unevaluated sum of two basic precision floating-point numbers[3]:

$$x = x_h + x_\ell,$$

such that the significands of x_h and x_ℓ do not overlap, which means here that

$$x_h = \mathrm{RN}(x). \tag{14.1}$$

If the basic precision floating-point numbers are of precision p, double-words are *not* equivalent to precision-$2p$ floating-point numbers. For instance, if $p = 53$ (from the binary64 format), the double-word that best approximates π is constituted by

$$p_1 = 11.001001000011111101101010100010001000010110100011000_2$$

and

$$p_2 = 1.0001101001100010011000110011000101000101110000000111_2 \times 2^{-53}.$$

Here, $p_1 + p_2$ is equivalent to a precision-107 binary floating-point approximation to π. Now, $\ln(12)$ will be approximated by

$$\ell_1 = 10.011111000010001011010111100110100111001111001111101_2$$

and

$$\ell_2 = -1.1001100011100100000011111000010110111101011110010111_2 \times 2^{-55}.$$

Here, $\ell_1 + \ell_2$ is equivalent to a precision-109 binary floating-point approximation to $\ln(12)$.

Due to this "wobbling precision" and the fact that the arithmetic algorithms will be either quite complex or slightly inaccurate, double-word arithmetic does not exhibit the "clean, predictable behavior" of a correctly rounded, precision-$2p$, floating-point arithmetic. It might sometimes be very useful (there are nice applications in computational geometry or linear algebra, for instance), but it must be used with caution. Li et al. [385] qualify

---

[2]Such as 2Sum (Algorithm 4.4, page 108), Fast2Sum (Algorithm 4.3, page 104), Dekker product (Algorithm 4.10, page 116), and 2MultFMA (Algorithm 4.8, page 112).

[3]Of course, when we write $x_h + x_\ell$, the addition symbol corresponds to the exact, mathematical addition.

double-double arithmetic as an "attractive nuisance except for the BLAS"[4] and even compare it to an unfenced backyard swimming pool. However, we now have reasonably fast and accurate algorithms for double-word arithmetic with proven and tight error bounds [308].

When Dekker introduced the algorithms that are now known as the "Dekker product" and "Fast2Sum" in his seminal paper [158], he also suggested ways of performing operations on double-word numbers (addition, multiplication, division, and square root), and he provided an analysis of the rounding error for these operations.

Libraries that offer double-word arithmetic (with binary64 as the underlying floating-point format) have been written by Bailey [255], Briggs [67], and Joldes et al. [309] (Briggs no longer maintains his library). Functions for double-word arithmetic are included in the QD[5] ("quad-double") library by Hida, Li, and Bailey [254, 255], as well as in the CAMPARY library.[6]

Let us now present some basic algorithms for double-word arithmetic. These algorithms will use several small algorithms from Chapter 4 as "building blocks," namely:

- Fast2Sum (Algorithm 4.3);
- 2Sum (Algorithm 4.4);
- 2MultFMA (Algorithm 4.8);
- DekkerProduct (Algorithm 4.10).

As the Fast2Sum algorithm does not work in radices higher than 3, *we need to assume in the following that the underlying floating-point arithmetic is of radix* 2.

It is important to understand that the arithmetic operations on double-word numbers implemented by the following algorithms are not "correctly rounded."[7] As stated above, in most cases, double-word arithmetics do not behave as well as a precision-$2p$ correctly rounded floating-point arithmetic. However, when designing algorithms for performing the arithmetic operations with double word numbers, we aim at obtaining a relative error not more than a small constant times 2^{-2p}.

[4]BLAS is an acronym for *Basic Linear Algebra Subroutines*.

[5]QD is available at http://crd-legacy.lbl.gov/~dhbailey/mpdist/.

[6]CAMPARY is available at http://homepages.laas.fr/mmjoldes/campary/.

[7]By the way, "correct rounding" is not clearly defined for double-words: Does this mean that we get the double-word closest to the exact value, or that we get the $2p$-digit number closest to the exact value (if the underlying arithmetic is of precision p)? This can be quite different. For instance, in radix-2, precision-p arithmetic, the double-word closest to $a = 2^p + 2^{-p} + 2^{-p-1}$ is a itself, whereas the precision-$2p$ number closest to a is $2^p + 2^{-p+1}$.

14.1.1.1 Addition of double-word numbers

Dekker's algorithm for adding two double-word numbers (x_h, x_ℓ) and (y_h, y_ℓ) is shown in Algorithm 14.1.

Algorithm 14.1 Dekker's algorithm for adding two double-word numbers (x_h, x_ℓ) and (y_h, y_ℓ) [158].

---

if $|x_h| \geq |y_h|$ **then**
  $(r_h, r_\ell) \leftarrow \text{Fast2Sum}(x_h, y_h)$
  $s \leftarrow \text{RN}(\text{RN}(r_\ell + y_\ell) + x_\ell)$
else
  $(r_h, r_\ell) \leftarrow \text{Fast2Sum}(y_h, x_h)$
  $s \leftarrow \text{RN}(\text{RN}(r_\ell + x_\ell) + y_\ell)$
end if
$(t_h, t_\ell) \leftarrow \text{Fast2Sum}(r_h, s)$
return (t_h, t_ℓ)

Algorithm 14.1 can sometimes be very inaccurate. The relative error

$$\left|\frac{(t_h + t_\ell) - (x + y)}{x + y}\right|$$

can be very large. Assuming binary, radix-p, floating-point arithmetic, Dekker showed an absolute error bound $(|x| + |y|) \cdot 2^{2-2p}$. This means that Algorithm 14.1 is to be used only when we know in advance that x and y have the same sign.

The following algorithm, presented by Li et al. [384, 385] and implemented in the QD library under the name of "IEEE addition," is much more accurate, yet more expensive.

Algorithm 14.2 – **AccurateDWPlusDW**$(x_h, x_\ell, y_h, y_\ell)$. Calculation of $(x_h, x_\ell) + (y_h, y_\ell)$ in binary, precision-p, floating-point arithmetic.

$(s_h, s_\ell) \leftarrow \text{2Sum}(x_h, y_h)$
$(t_h, t_\ell) \leftarrow \text{2Sum}(x_\ell, y_\ell)$
$c \leftarrow \text{RN}(s_\ell + t_h)$
$(v_h, v_\ell) \leftarrow \text{Fast2Sum}(s_h, c)$
$w \leftarrow \text{RN}(t_\ell + v_\ell)$
$(z_h, z_\ell) \leftarrow \text{Fast2Sum}(v_h, w)$
return (z_h, z_ℓ)

Joldes, Muller, and Popescu [308] analyse this algorithm and show that in radix-2, precision-p arithmetic, unless overflow occurs, the relative error of Algorithm 14.2 is upper bounded by

$$\frac{3 \cdot 2^{-2p}}{1 - 4 \cdot 2^{-p}} = 3 \cdot 2^{-2p} + 12 \cdot 2^{-3p} + 48 \cdot 2^{-4p} + \cdots, \tag{14.2}$$

which is less than $3 \cdot 2^{-2p} + 13 \cdot 2^{-3p}$ as soon as $p \geq 6$.

14.1.1.2 Multiplication of double-word numbers

Dekker's algorithm for multiplying two double-word numbers is shown in Algorithm 14.3.

Algorithm 14.3 – DekkerDWTimesDW$(x_h, x_\ell, y_h, y_\ell)$. Dekker's algorithm for multiplying two double-word numbers (x_h, x_ℓ) and (y_h, y_ℓ) [158]. If a fused multiply-add (FMA) instruction is available, it is advantageous to replace DekkerProduct(x_h, y_h) by 2MultFMA(x_h, y_h) (defined in Section 4.4.1).

$(c_h, c_\ell) \leftarrow$ DekkerProduct(x_h, y_h)
$p_1 \leftarrow \mathrm{RN}(x_h \cdot y_\ell)$
$p_2 \leftarrow \mathrm{RN}(x_\ell \cdot y_h)$
$c_\ell \leftarrow \mathrm{RN}(c_\ell + \mathrm{RN}(p_1 + p_2))$
$(t_h, t_\ell) \leftarrow$ Fast2Sum(c_h, c_ℓ)
return (t_h, t_ℓ)

Joldes, Muller, and Popescu [308] show that in radix-2, precision-p arithmetic, unless underflow or overflow occurs, the relative error of Algorithm 14.3 is bounded by $7 \cdot 2^{-2p}$. If an FMA instruction is available, we can make Algorithm 14.3 significantly faster and slightly more accurate by replacing the "Dekker Product" at Line 1 by the 2MultFMA algorithm (Section 4.4.1) and by using an FMA for approximating $x_h y_\ell + x_\ell y_h$ at lines 2 and 3. This gives the following algorithm.

Algorithm 14.4 – FastDWTimesDW$(x_h, x_\ell, y_h, y_\ell)$. Another algorithm for computing $(x_h, x_\ell) \times (y_h, y_\ell)$ in binary, precision-p, floating-point arithmetic, assuming an FMA instruction is available.

$(c_h, c_{\ell 1}) \leftarrow$ 2MultFMA(x_h, y_h)
$p_1 \leftarrow \mathrm{RN}(x_h \cdot y_\ell)$
$c_{\ell 2} \leftarrow \mathrm{RN}(p_1 + x_\ell y_h)$
$c_\ell \leftarrow \mathrm{RN}(c_{\ell 1} + c_{\ell 2})$
$(z_h, z_\ell) \leftarrow$ Fast2Sum(c_h, c_ℓ)
return (z_h, z_ℓ)

If $p \geq 5$ (which always holds in practice), the relative error of this algorithm is less than or equal to $6 \cdot 2^{-2p}$ [308]. Let us now give an example of use of Algorithm 14.4.

Kornerup et al. [346] use Algorithm 14.4 for evaluating very accurately integer powers. Consider Algorithm 14.5 below.

Algorithm 14.5 IteratedProductPower(x, n), Kornerup et al. [346].

```
Require: n integer ≥ 1
  i ← n
  (h, ℓ) ← (1, 0)
  (u, v) ← (x, 0)
  while i > 1 do
    if i is odd then
      (h, ℓ) ← FastDWTimesDW(h, ℓ, u, v)
    end if
    (u, v) ← FastDWTimesDW(u, v, u, v)
    i ← ⌊i/2⌋
  end while
  (h, ℓ) ← FastDWTimesDW(h, ℓ, u, v)
  return (h, ℓ)
```

Using the error bound on Algorithm 14.4, one can show that in radix-2, precision-p, floating-point arithmetic, the relative error of Algorithm 14.5 is bounded by $(1 + 6 \cdot 2^{-2p})^{n-1} - 1$. From this one can deduce that if algorithm IteratedProductPower is implemented in binary64 arithmetic, then $\mathrm{RN}(h + \ell)$ is a *faithful* result (see Section 2.2) for x^n, as long as $n \leq 2^{49}$ (see [362] for recent similar results). To guarantee a correctly rounded result in binary64 arithmetic, we must know how close x^n can be to the exact midpoint between two consecutive floating-point numbers. This problem is an instance of the *Table maker's dilemma*, which is the main topic of Section 10.5. For instance, in binary64 arithmetic, the hardest-to-round case for function x^{952} corresponds to

$$x = (1.0101110001101001001000000010110101000110100000100001)_2,$$

for which we have

$$\begin{aligned} x^{952} &= (\underbrace{1.0011101110011001001111100000100010101010110100100110}_{53 \text{ bits}}\ 1 \\ &\qquad \underbrace{00000000 \cdots 00000000}_{63 \text{ zeros}} 1001 \cdots)_2 \times 2^{423}. \end{aligned}$$

For this example, x^n is extremely close to the exact middle of two consecutive binary64 numbers. There is a run of 63 consecutive zeros after the rounding bit. This case is the worst case for all values of n between 3 and 1035. Using this result, Kornerup et al. have shown the following result.

Theorem 14.1. *[346] If algorithm IteratedProductPower is performed in the Intel double-extended precision format (x87 instruction set), and if* $3 \leq n \leq 1035$, *then* $\mathrm{RN}_{\mathrm{binary64}}(h+\ell) = \mathrm{RN}_{\mathrm{binary64}}(x^n)$. *Hence, by rounding* $h+\ell$ *to the nearest binary64 number, we obtain a correctly rounded result.*

14.1.1.3 Division of a double-word number by a floating-point number

Several algorithms have been suggested for dividing a double-word number by a floating-point number. The following [308] is a simplification of an algorithm suggested in [384] (it is as accurate, but requires less arithmetic operations).

Algorithm 14.6 Calculation of $(x_h, x_\ell) \div y$ in binary, precision-p, floating-point arithmetic.

$t_h \leftarrow \mathrm{RN}(x_h/y)$
$(\pi_h, \pi_\ell) \leftarrow$ 2MultFMA(t_h, y)
$\delta_h \leftarrow \mathrm{RN}(x_h - \pi_h)$
$\delta_\ell \leftarrow \mathrm{RN}(x_\ell - \pi_\ell)$
$\delta \leftarrow \mathrm{RN}(\delta_h + \delta_\ell)$
$t_\ell \leftarrow \mathrm{RN}(\delta/y)$
$(z_h, z_\ell) \leftarrow$ Fast2Sum(t_h, t_ℓ)
return (z_h, z_ℓ)

Assuming radix 2 and precision p, unless under/overflow occurs, the relative error of Algorithm 14.6 is bounded by $\frac{7}{2} \cdot 2^{-2p}$ [308].

14.1.2 Static triple-word arithmetic

For triple-word arithmetic, where numbers are represented as unevaluated sums of three floating-point numbers, the implementation by Lauter [370, 371] is worth mentioning. It is used for handling critical parts in the CRlibm library for correctly rounded elementary functions in double-precision/binary64 floating-point arithmetic. It was specifically designed for the implementation of such functions, which typically require about 120 bits of accuracy. For such accuracies, double-double (i.e., double-word arithmetic based on the binary64 format) is not enough, but triple-double, with $3 \times 53 = 159$ bits, is an overkill (if double-extended precision—$p = 64$—is available, double-word arithmetic based on this format is an interesting alternative). The originality of Lauter's implementation is therefore to allow some overlap in the significands of the three binary64 numbers. Here, two floating-point numbers are said to overlap if their exponent difference is smaller than their significand size.[8] Overlap in a triple-double means that

[8]Note that even if the general idea remains the same, the notion of overlap slightly changes depending on the authors!

it is not as accurate as it could be. This is acceptable in static code such as polynomial-based elementary function evaluation, for which one knows in advance the number of operations to perform and the accuracy required for each operation.

As the following will show, removing overlap (an operation called *renormalization*) is expensive. It requires several invocations of the Fast2Sum algorithm, which removes overlap from a double-double. The approach of Lauter is to provide a separate renormalization procedure, so renormalizations are invoked explicitly in the code, and as rarely as possible.

Each operation is provided with a theorem that expresses a bound on its relative accuracy, as a function of the overlaps of the inputs. For illustration, we give in Algorithm 14.7 and in Theorem 14.2 one example of an operation implemented by Lauter with its companion theorem.

Algorithm 14.7 Evaluating the sum of a triple-double and a double-double as a triple-double [370] (i.e., we assume an underlying binary64 arithmetic).

Require: $a_h + a_\ell$ is a double-double number and $b_h + b_m + b_\ell$ is a triple-double number such that

$$\begin{aligned} |b_h| &\leq 2^{-2} \cdot |a_h|, \\ |a_\ell| &\leq 2^{-53} \cdot |a_h|, \\ |b_m| &\leq 2^{-\beta_o} \cdot |b_h|, \\ |b_\ell| &\leq 2^{-\beta_u} \cdot |b_m|. \end{aligned}$$

Ensure: $r_h + r_m + r_\ell$ is a triple-double number approximating $a_h + a_\ell + b_h + b_m + b_\ell$ with a relative error given by Theorem 14.2.

$(r_h, t_1) \leftarrow \text{Fast2Sum}(a_h, b_h)$
$(t_2, t_3) \leftarrow \text{Fast2Sum}(a_\ell, b_m)$
$(t_4, t_5) \leftarrow \text{Fast2Sum}(t_1, t_2)$
$t_6 \leftarrow \text{RN}(t_3 + b_\ell)$
$t_7 \leftarrow \text{RN}(t_6 + t_5)$
$(r_m, r_\ell) \leftarrow \text{Fast2Sum}(t_4, t_7)$

Here, β_o and β_u are positive numbers that measure the possible overlap of the significands of the inputs.

Theorem 14.2 (Relative error of Algorithm 14.7). *If $a_h + a_\ell$ and $b_h + b_m + b_\ell$ are the values in the argument of Algorithm 14.7, such that the preconditions hold, then the values r_h, r_m, and r_ℓ returned by Algorithm 14.7 satisfy*

$$r_h + r_m + r_\ell = \left((a_h + a_\ell) + (b_h + b_m + b_\ell) \right) \cdot (1 + \epsilon),$$

where ϵ is bounded by

$$|\epsilon| \leq 2^{-\beta_o - \beta_u - 52} + 2^{-\beta_o - 104} + 2^{-153}.$$

The values r_m and r_ℓ will not overlap at all, and the overlap of r_h and r_m will be bounded by

$$|r_m| \leq 2^{-\gamma} \cdot |r_h|$$

with

$$\gamma \geq \min(45, \beta_o - 4, \beta_o + \beta_u - 2).$$

For instance, if $\beta_o = 47$ and $\beta_u = 50$ then the relative error ϵ is bounded by $21 \cdot 2^{-153}$, and γ is larger than or equal to 43. Lauter's triple-double operators are freely available as part of the CRlibm library.[9] There are about 30 such operators that turned out to be useful for the development of this library [148]. Some operators have inputs of different types, as in the previous example. Several implementations of the correct rounding of a triple-double to a double-precision number are also provided. The technical report that describes these operators [370] is available as part of the documentation of CRlibm [129].

14.2 Floating-Point Expansions

A natural extension of double-word representation is the notion of *floating-point expansion* i.e., the representation of real numbers as the unevaluated sum of several floating-point numbers. As in the previous section, we assume radix-2 floating-point arithmetic.

Definition 14.1. *A floating-point expansion x with n terms is the unevaluated sum of n floating-point numbers $x_0, \ldots, x_{n-1}$, in which all nonzero terms are ordered by magnitude (i.e., if y is the sequence obtained by removing all zeros in the sequence x, and if y contains m terms, then $|y_i| \geq |y_{i+1}|$ for all $0 \leq i < m-1$). Each x_i is called a component (or a term) of x.*

The arithmetic on floating-point expansions was first developed by Priest [495], and in a slightly different way by Shewchuk [555]. A specific implementation, quad-word arithmetic (each of the four terms being a binary64 number), has been implemented in the QD library by Hida, Li, and Bailey [254] with the requirement that

$$x_{i+1} \leq \frac{1}{2}\,\mathrm{ulp}(x_i). \tag{14.3}$$

This means that these numbers are *nonoverlapping* in a sense close to that used in (14.1) for defining double-word numbers, yet slightly stronger than what we will consider later in this section.

[9]CRlibm was developed by the Arénaire/AriC team of CNRS, INRIA, UCBL and ENS Lyon, France. It is available at https://gforge.inria.fr/scm/browser.php?group_id=5929&extra=crlibm.

This constraint on the components is imposed because in general, the notion of expansion is redundant since a nonzero number always has more than one representation as a floating-point expansion. There is no unicity of nonoverlapping representations of a number. A nonoverlapping expansion has better properties, such as "compacity": it takes less terms for achieving the same accuracy. There are several slightly different notions of nonoverlapping. Besides the one that was required for quad-words (Equation (14.3)), other slightly different definitions of interest are given below. An expansion may contain interleaving zeros, but the definitions that follow apply only to the nonzero terms of the expansion (i.e., the sequence y in Definition 14.1).

Definition 14.2 ($\mathcal{P}$-nonoverlapping floating-point numbers). *Assuming x and y are normal numbers with representations $M_x \cdot 2^{e_x-p+1}$ and $M_y \cdot 2^{e_y-p+1}$ (with $2^{p-1} \le |M_x|, |M_y| \le 2^p - 1$), they are $\mathcal{P}$-nonoverlapping (that is, nonoverlapping according to Priest's definition [496]) if $|e_y - e_x| \ge p$.*

Definition 14.3. *A floating-point expansion $x_0 + x_1 + \cdots + x_{n-1}$ is $\mathcal{P}$-nonoverlapping (that is, nonoverlapping according to Priest's definition) if all its nonzero terms are mutually $\mathcal{P}$-nonoverlapping that is $|x_i| < \mathrm{ulp}(x_{i-1})$, for all $1 \le i \le n-1$.*

Shewchuk [555] weakens this into *nonzero-overlapping* expansions, as it follows in Definition 14.4.

Definition 14.4. *A floating-point expansion $x_0 + x_1 + \cdots + x_{n-1}$ is $\mathcal{S}$-nonoverlapping (that is, nonoverlapping according to Shewchuk's definition) if for all $1 \le i \le n-1$, we have $e_{x_{i-1}} - e_{x_i} \ge p - z_{x_{i-1}}$, where $e_{x_{i-1}}$ and e_{x_i} are the exponents of x_{i-1} and x_i, respectively, and $z_{x_{i-1}}$ is the number of trailing zeros of x_{i-1}.*

Note that zero is $\mathcal{S}$-nonoverlapping with any nonzero floating-point number.

For example, in a binary floating-point system of precision $p = 4$, the numbers $1.100_2 \times 2^3$ and $1.010_2 \times 2^1$ are $\mathcal{S}$-nonoverlapping , whereas they are not $\mathcal{P}$-nonoverlapping .

In general, since Priest's condition is stronger than Shewchuk's, for the same number of terms, a $\mathcal{P}$-nonoverlapping expansion carries more information than an $\mathcal{S}$-nonoverlapping one. In extreme cases, in radix 2, an $\mathcal{S}$-nonoverlapping expansion with 53 components may not contain more information than one binary64 number (it suffices to put each bit of a floating-point number in a separate component). And yet, $\mathcal{S}$-nonoverlapping expansions are of interest due to the simplicity of the related arithmetic algorithms. Let us give an important example. First, following Priest [495], we define expansions whose terms "overlap by at most $0 \le d \le p-2$ bits," where p is the underlying precision.

Definition 14.5. *Consider n precision-p floating-point numbers: $x_0, x_1, \ldots, x_{n-1}$. They overlap by at most d binary digits ($0 \le d < p$) if and only if for all $i, 0 \le i \le n-2$, there exist integers k_i, δ_i such that*

$$2^{k_i} \le |x_i| < 2^{k_i+1}, \tag{14.4}$$

$$2^{k_i-\delta_i} \le |x_{i+1}| \le 2^{k_i-\delta_i+1}, \tag{14.5}$$

$$\delta_i \ge p - d, \tag{14.6}$$

$$\delta_i + \delta_{i+1} \ge p - z_{i-1}, \tag{14.7}$$

where z_{i-1} is the number of trailing zeros at the end of x_{i-1} and for $i = 0$, $z_{-1} = 0$.

Loosely speaking, this definition states that when written in positional notation, the binary digits of any two successive nonzero terms coincide in at most d positions, and no three terms mutually coincide in any digit position.

Joldes et al. [310] prove that, under mild assumptions, VecSum(x) (where VecSum is Algorithm 5.12) which is simply a chain of 2Sum, applied to an expansion x with n terms, makes it $\mathcal{S}$-nonoverlapping as soon as its terms *overlap by at most $0 \le d \le p-2$ bits*. They also prove that for a better performance, when $d \le p-2$, the 2Sum calls in the VecSum algorithm can be replaced by calls to Fast2Sum.

The fact that VecSum transforms an expansion whose terms overlap by at most $d \le p-2$ bits into an $\mathcal{S}$-nonoverlapping expansion is important, because, as we are going to see in Section 14.2.1, an $\mathcal{S}$-nonoverlapping expansion can easily be transformed into another expansion with a "stronger" nonoverlapping property, that we now define.

Definition 14.6. *[310] A floating-point expansion $x_0 + x_1 + \cdots + x_{n-1}$ is* ulp-*nonoverlapping if for all $1 \le i \le n-1$, $|x_i| \le \mathrm{ulp}(x_{i-1})$.*

In other words, the components are either $\mathcal{P}$-nonoverlapping or they overlap by one bit, in which case the second component is a power of two, equal to the ulp of the first one.

Depending on the nonoverlapping type of an expansion, when using standard floating-point formats as underlying arithmetic, the exponent range forces a constraint on the number of terms. The largest expansion can be obtained when the largest term is close to overflow and the smallest is close to underflow. This gives a maximum ulp-nonoverlapping expansion size of 40 for binary64 and 12 for binary32.

Concerning ulp-nonoverlapping expansions, we recall in what follows several algorithms implemented in the CAMPARY library, which are collected and proven in detail in Popescu's PhD dissertation [494]. They allow for a performance versus accuracy tradeoff and target specialized architectures such as GPUs.

14.2.1 Renormalization of floating-point expansions

To restore the nonoverlapping property after different manipulations with expansions, one needs to perform a so-called renormalization algorithm. We have seen that an expansion that "does not overlap too much" (its terms overlap by at most $d \leq p-2$ bits) can easily be transformed into an $\mathcal{S}$-nonoverlapping expansion. Let us now give an algorithm, VecSumErrBranch, that transforms an $\mathcal{S}$-nonoverlapping expansion into an ulp-nonoverlapping expansion.

14.2.1.1 The VecSumErrBranch algorithm

Algorithm 14.8 is a variation of the VecSum algorithm 5.12, which starts from the most significant term and instead of propagating the partial sums, propagates the errors. If however, the error after a 2Sum block is zero, the sum is propagated instead. It is formally proved by Boldo et al. [50] that this algorithm transforms an $\mathcal{S}$-nonoverlapping expansion into an ulp-nonoverlapping one and that, in all practical cases (in particular, the IEEE formats), 2Sum calls can be safely replaced by Fast2Sum.

Algorithm 14.8 – VecSumErrBranch$(e_0, \ldots, e_{n-1}, m)$. The parameter m is the number of terms of the result.

```
j ← 0
ε_0 = e_0
for i ← 0 to n − 2 do
    (r_j, ε_{i+1}) ← 2Sum(ε_i, e_{i+1})
    if ε_{i+1} ≠ 0 then
        if j ≥ m − 1 then
            return r_0, r_1, ..., r_{m−1}    //enough output terms
        end if
        j ← j + 1
    else
        ε_{i+1} ← r_j
    end if
end for
if ε_{n−1} ≠ 0 and j < m then
    r_j ← ε_{n−1}
end if
return r_0, r_1, ..., r_{m−1}
```

14.2.1.2 The renormalization algorithm

By successively using VecSum and VecSumErrBranch, we can convert an array of numbers overlapping by at most $p-2$ bits into an ulp-nonoverlapping expansion. This gives Algorithm 14.9 below. It is presented in [310, 494], and a formal proof of this algorithm is given by Boldo et al. [50].

Algorithm 14.9 – Renormalize$(x_0, \dots, x_{n-1}, m)$.

Input: a sequence $x_0, x_1, \dots, x_{n-1}$ of floating-point numbers overlapping by at most $p-2$ bits.
Output: an ulp-nonoverlapping expansion $r_0, r_1, \dots, r_{m-1}$.
$(e_0, e_1, \dots, e_{n-1}) \leftarrow \text{VecSum}(x_0, x_1, \dots, x_{n-1})$
$(r_0, r_1, \dots, r_{m-1}) \leftarrow \text{VecSumErrBranch}(e_0, e_1, \dots, e_{n-1}, m)$
return $r_0, r_1, \dots, r_{m-1}$

14.2.1.3 Renormalization of arbitrary numbers

We have seen how to "renormalize" a sequence that "does not overlap too much." If we are given an arbitrary sequence of floating-point numbers as input, renormalization is still possible, but at a much higher cost. For Algorithm 14.10, given below, we have the following theorem.

Theorem 14.3. *Let $x_0, x_1, \dots, x_{n-1}$ be a sequence of n floating-point numbers that may contain interleaving* 0s, *and let m be an integer such that $1 \le m \le n-1$. Provided that no underflow/overflow occurs during the calculations, Algorithm 14.10 returns the first m terms of an* ulp-*nonoverlapping floating-point expansion $r = r_0 + \cdots + r_{n-1}$ such that $x_0 + \cdots + x_{n-1} = r$.*

Algorithm 14.10 – Renormalize_arbitrary$(x_0, \dots, x_{n-1}, m)$.

Input: an arbitrary sequence $x_0, x_1, \dots, x_{n-1}$ of floating-point numbers.
Output: an ulp-nonoverlapping expansion $r_0, r_1, \dots, r_{m-1}$.
$e_0^{(0)} \leftarrow x_0$
for $i \leftarrow 1$ **to** $n-1$ **do**
    $(e_0^{(i)}, e_1^{(i)}, \dots, e_i^{(i)}) \leftarrow \text{VecSum}(e_0^{(i-1)}, e_1^{(i-1)}, \dots, e_{i-1}^{(i-1)}, x_i)$
end for
$(r_0, r_1, \dots, r_{m-1}) \leftarrow \text{VecSumErrBranch}(e_0^{(n-1)}, e_1^{(n-1)}, \dots, e_{n-1}^{(n-1)}, m)$
return $r_0, r_1, \dots, r_{m-1}$

14.2.2 Addition of floating-point expansions

The sum of two expansions x and y with n and m terms may contain up to $n+m$ terms. Hence, when successive computations are performed, in order to reduce the number of terms, both renormalization and truncation must be used. Based on this idea, many algorithms that compute the sum of two expansions have been proposed in literature [495, 555, 254, 531]. Priest's algorithm [495] is based on merge-sorting the components of the two expansions, which are assumed to be $\mathcal{P}$-nonoverlapping , by increasing magnitude, followed by applying a renormalization algorithm, in order to render the expansion $\mathcal{P}$-nonoverlapping.

14.2.2.1 Accurate addition algorithm

Similar to Priest's idea, Algorithm 14.11 from Popescu's thesis [494] consists in merging two ulp-nonoverlapping expansions in decreasing order of magnitude, and renormalizing the resulting array using Algorithm 14.9. The output expansion s, with r terms, is ulp-nonoverlapping and satisfies the error bound:

$$|x + y - s| < \frac{9}{2} \cdot 2^{-(p-1)r} \cdot (|x| + |y|), \tag{14.8}$$

as soon as the underlying precision p is at least 4, which always holds in practice.

Algorithm 14.11 – Addition_accurate$(x_0, \ldots, x_{n-1}, y_0, \ldots, y_{m-1}, r)$.

$n' \leftarrow \min(n, r)$; $m' \leftarrow \min(m, r)$
$(f_0, f_1, \ldots, f_{m'+n'-1}) \leftarrow$ Merge$((x_0, x_1, \ldots x_{n'}), (y_0, y_1, \ldots, y_{m'}))$
$(s_0, \ldots, s_{r-1}) \leftarrow$ Renormalize$(f_0, \ldots, f_{m'+n'-1}, r)$ //using Alg. 14.9
return $s_0, s_1, \ldots, s_{r-1}$

When adding a floating-point expansion and a floating-point number, the aforementioned algorithm can be simplified. In this case, Popescu [494] proved that a simple renormalization given in Algorithm 14.9 suffices (in the special case when only one number does not satisfy the condition of "overlapping by at most $d \leq p - 2$ bits") and allows for a slightly tighter error bound on the result.

14.2.2.2 QD-like addition algorithm

A generalization of the double-double and quad-double addition implemented in the QD library [254] was studied by Popescu [494] and is given in Algorithm 14.12 below. Given two ulp-nonoverlapping expansions, x and y, as above, the ulp-nonoverlapping output expansion satisfies a coarser error bound than (14.8):

$$|x + y - s| < 24 \cdot 2^{-(p-1)r} \cdot (|x| + |y|).$$

On the other hand, in practice, this algorithm proved to be faster than Algorithm 14.11 in the testings of the tuned implementation of CAMPARY library [494].

Note that, for all these algorithms, subtraction can be performed simply by negating the terms of y.

Algorithm 14.12 – Addition_QD-like$(x_0, \ldots, x_{r-1}, y_0, \ldots, y_{r-1}, r)$.

```
(f_0, e_0) ← 2Sum(x_0, y_0)
for n ← 1 to r − 1 do
    (s_n, e_n) ← 2Sum(x_n, y_n)
    (f_n, e_0, e_1, ..., e_{n−1}) ← VecSum(s_n, e_0, e_1, ..., e_{n−1})
end for
f_r ← 0
for i ← 0 to r − 1 do
    f_r ← f_r + e_i
end for
(s_0, ..., s_{r−1}) ← Renormalize_arbitrary(f_0, ..., f_r, r) //using Alg. 14.10
return s_0, s_1, ..., s_{r−1}
```

14.2.3 Multiplication of floating-point expansions

We focus on two multiplication algorithms for ulp-nonoverlapping expansions. Algorithm 14.13 generalizes the quad-double multiplication of the QD library, while Algorithm 14.14 proposes a different method for accumulating the partial products. First, similar to the double-word multiplication, both algorithms use as a building block an algorithm that expresses the product of two floating-point numbers as a double word. This algorithm can be either DekkerProduct (Algorithm 4.10), or, preferably, 2MultFMA (Algorithm 4.8) when an FMA instruction is available. We will name it "2Prod" in the following.

Second, the two expansion multiplication algorithms under study in this section rely on the so-called *on-the-fly* truncation, based on the following intuition: consider computing the first r terms of a product ulp-nonoverlapping expansion $\pi = x \cdot y$, where x and y have respectively n and m terms. Roughly speaking, if the product $x_0 y_0$ is of order Λ, then, for the product $(\pi', e) = \text{2Prod}(x_i, y_j)$, π' is of $\mathcal{O}(\varepsilon^k \Lambda)$ and e is $\mathcal{O}(\varepsilon^{k+1}\Lambda)$, where $k = i + j$ and $\varepsilon = 2^{-p}$. To gain performance, the partial products that have an order of magnitude less than $\mathcal{O}(\varepsilon^r \Lambda)$ are discarded, while the first $\sum_{k=0}^{r}(k+1)$ partial products are computed via the 2Prod Algorithm. Theorem 14.4, which holds for both algorithms, bounds the error caused by this truncation.

Theorem 14.4 ([494]). *Let x and y be two* ulp*-nonoverlapping floating-point expansions, with n and m terms, respectively. For a given r and underlying precision p, the truncation error of the partial product $\sum_{k=0}^{r} \sum_{i+j=k} x_i y_j$ satisfies:*

$$\left| xy - \sum_{k=0}^{r} \sum_{i+j=k} x_i y_j \right| \le |x_0 y_0| \cdot 2^{-(p-1)(r+1)} \left(\frac{-2^{-(p-1)}}{(1-2^{-(p-1)})^2} + \frac{m+n-r-2}{1-2^{-(p-1)}} \right).$$

14.2.3.1 QD-like multiplication algorithm

For computing an approximation to $\sum_{k=0}^{r}\sum_{i+j=k} x_i y_j$, Algorithm 14.13 sums up all the products π'_i with the same order of magnitude (as well as all the errors resulting from the previous step) with the VecSum algorithm. Specifically, for each k such that $1 \le k \le r-1$, one has $k+1$ products to add. Besides these, k^2 error terms result from the previous iteration. One accumulates all these terms using VecSum, which provides at each step an approximate term f_k. Finally, the products with the same order of magnitude as π_r are intended as an extra error correction term. Hence, f_r is obtained by standard multiplication of partial products $x_i y_{r-i}$, $1 \le i \le r-1$ followed by standard addition of all remaining errors. The array f is then renormalized in order to make the result ulp-nonoverlapping . Theorem 14.5, proved by Popescu in [494], bounds the overall error.

Algorithm 14.13 – QD-like Multiplication$(x_0, \ldots, x_{r-1}, y_0, \ldots, y_{r-1}, r)$.

$(f_0, e_0) \leftarrow$ 2Prod(x_0, y_0)
for $k \leftarrow 1$ **to** $r-1$ **do**
  for $i \leftarrow 0$ **to** k **do**
    $(\pi'_i, \hat{e}_i) \leftarrow$ 2Prod(x_i, y_{k-i})
  end for
  $(f_k, e_0, e_1, \ldots, e_{k^2+k-1}) \leftarrow$ VecSum$(\pi'_0, \pi'_1, \ldots, \pi'_k, e_0, e_1, \ldots, e_{k^2-1})$
  $(e_0, e_1, \ldots, e_{k^2+2k}) \leftarrow (e_0, e_1, \ldots, e_{k^2+k-1}, \hat{e}_0, \ldots, \hat{e}_k)$
end for
$f_r \leftarrow 0$
for $i \leftarrow 1$ **to** $r-1$ **do**
  $f_r \leftarrow f_r + x_i \cdot y_{r-i}$
end for
for $i \leftarrow 0$ **to** r^2-1 **do**
  $f_r \leftarrow f_r + e_i$
end for
$(\pi_0, \pi_1, \ldots, \pi_{r-1}) \leftarrow$ Renormalize_arbitrary$(f_0, f_1, \ldots, f_r, r)$
return $\pi_0, \pi_1, \ldots, \pi_{r-1}$

Theorem 14.5. *Let x and y be two* ulp-*nonoverlapping floating-point expansions, with n, and m terms, respectively. Assume $p \ge 8$. Provided that no underflow/overflow occurs during computations, the output π of Algorithm 14.13 is an* ulp-*nonoverlapping floating-point expansion with r terms, which satisfies*

$$|xy - \pi| \le C\,|x_0 y_0|\,2^{-(p-1)(r+1)},$$

where

$$C = \frac{128}{127}(m+n) - \frac{129}{254}r - \frac{385}{254} + 2^{p-1} + 2^{-p-r}(r^2+r)((r+1)!)^2.$$

14.2.3.2 Accurate multiplication algorithm

Similar to the previous algorithm, one discards the same partial products, but the remaining ones are summed up in a different manner by Algorithm 14.14, suggested by Muller, Popescu, and Tang [443], and inspired from Rump, Ogita and Oishi's summation algorithm [521, 531] presented in Section 5.3.3.1. The idea is to accumulate via Algorithm 14.15 numbers of size at most b in *bins* B_i, which are floating-point variables whose least significant bit (LSB) has a fixed weight. This allows for errorless accumulation provided that at most 2^c numbers are added in one bin, where $b + c = p - 1$. In [443], the following parameter choices are proposed:

- for binary64, $b = 45$, which allows for $c = 7$ bits of carry to happen; this means one can add at most 128 numbers for the result to be exact;
- for binary32, $b = 18$, which implies $c = 5$, allowing for summing up to 32 numbers into one bin.

The constants $\frac{3}{2} \cdot 2^{t-(i+1)b+p-1}$ that appear in Algorithm 14.14 are used for "initializing" the bins.

The number of allocated bins is chosen from the number of output terms r as $\lfloor r \cdot p/b \rfloor + 2$ and the LSB of each bin is set according to the starting exponent, $t = t_{x0} + t_{y0}$ at a distance of b bits, where t_{x0} and t_{y0} are the exponents of x_0 and y_0, respectively. Each bin B_i is initialized with a constant value depending on their LSB, that is, $\frac{3}{2} \cdot 2^{t-(i+1)b+p-1}$. These constant values are subtracted before the renormalization step.

For each partial product (π', e) computed by 2Prod, one determines its corresponding bins using the formula $\lfloor (t - t_{x_i} - t_{y_j})/b \rfloor$, where t_{x_i} and t_{y_j} are the exponents of x_i and y_j, respectively.

Besides the pair (π', e) and the array of bins B, Algorithm 14.15 also receives two integer parameters. The number sh represents the first corresponding bin for the pair and $\ell := t - t_{x_i} - t_{y_j} - sh \cdot b$ is the number of leading bits. This value gives the difference between the LSB of B_{sh-1} and the sum of the exponents of x_i and y_j of the corresponding pair (π', e). The renormalization step consists in a call to the VecSumErrBranch algorithm. Finally, one has the following error bound proved by Muller, Popescu, and Tang [443].

Theorem 14.6. *Let x and y be two* ulp*-nonoverlapping floating-point expansions, with n, and m terms, respectively. Provided that no underflow/overflow occurs during the calculations, when computing their product using Algorithm 14.14, the result π is an* ulp*-nonoverlapping floating-point expansion with r terms that satisfies:*

$$\begin{aligned} |xy - \pi| &\leq |x_0 y_0| \, 2^{-(p-1)r} \\ &\times \left[1 + (r+1)2^{-p} + 2^{-(p-1)} \left(\tfrac{-2^{-(p-1)}}{(1-2^{-(p-1)})^2} + \tfrac{m+n-r-2}{1-2^{-(p-1)}} \right) \right]. \end{aligned}$$

Algorithm 14.14 – Multiplication_accurate$(x_0, \dots, x_{n-1}, y_0, \dots, y_{m-1}, r)$.

```
t ← t_{x0} + t_{y0}
for i ← 0 to ⌊r · p/b⌋ + 1 do
  B_i ← 3/2 · 2^{t−(i+1)b+p−1}
end for
for i ← 0 to min(n − 1, r) do
  for j ← 0 to min(m − 1, r − 1 − i) do
    (π′, e) ← 2Prod(x_i, y_j)
    ℓ ← t − t_{x_i} − t_{y_j}
    sh ← ⌊ℓ/b⌋
    ℓ ← ℓ − sh · b
    B ← Accumulate(π′, e, B, sh, ℓ)    //using Algorithm 14.15
  end for
  if j < m − 1 then
    π′ ← x_i · y_j
    ℓ ← t − t_{x_i} − t_{y_j}
    sh ← ⌊ℓ/b⌋
    ℓ ← ℓ − sh · b
    B ← Accumulate(π′, 0., B, sh, ℓ)    //using Algorithm 14.15
  end if
end for
for i ← 0 to ⌊r · p/b⌋ + 1 do
  B_i ← B_i − 3/2 · 2^{t−(i+1)b+p−1}
end for
(π_0, π_1, ..., π_{r−1}) ← VecSumErrBranch(B_0, B_1, ..., B_{⌊r·p/b⌋+1}, r)    //using
Algorithm 14.8
return π_0, π_1, ..., π_{r−1}
```

14.2.4 Division of floating-point expansions

Most algorithms employed for dividing expansions are a generalization of the digit-recurrence algorithms [186], which themselves are a generalization of the *paper-and-pencil* method we learned at school. Examples are Priest's division algorithm [495] and the division in the QD library [254]. For instance, let $x = x_0 + x_1 + x_2 + x_3$ and $y = y_0 + y_1 + y_2 + y_3$ be quad-word numbers. First, one approximates the quotient by $q_0 = \mathrm{RN}(x_0/y_0)$, then one computes the remainder $r = x - q_0 y = r_0 + r_1 + r_2 + r_3$ in quad-word arithmetic.

The next correction term is $q_1 = \mathrm{RN}(r_0/y_0)$. Subsequent terms q_i are obtained by continuing this process. At each step when computing the remainder r, full quad-word multiplication and subtraction are performed since most of the bits will be cancelled out when computing q_3 and q_4. A renormalization step is performed only at the end, on $q_0 + q_1 + q_2 + \cdots$ in order to ensure the nonoverlapping requirement. While no error bound is given in [254],

Algorithm 14.15 – Accumulate(π', e, B, sh, ℓ).

```
if ℓ < b − 2c − 1 then
  (B_sh, B_sh+1) ← Deposit(π') // that is, (B_sh, π') ←Fast2Sum(B_sh, π'), and
                               // B_sh+1 ← B_sh+1 + π'
  (B_sh+1, B_sh+2) ← Deposit(e)
else if ℓ < b − c then
  (B_sh, B_sh+1) ← Deposit(π')
  (B_sh+1, e) ← Fast2Sum(B_sh+1, e)
  (B_sh+2, B_sh+3) ← Deposit(e)
else
  (B_sh, p) ← Fast2Sum(B_sh, π')
  (B_sh+1, B_sh+2) ← Deposit(π')
  (B_sh+2, B_sh+3) ← Deposit(e)
end if
return B
```

Priest proved in [495] the following error bound for a quotient q represented as a $\mathcal{P}$-nonoverlapping expansion with d terms in precision-p arithmetic:

$$\left|\frac{q - x/y}{x/y}\right| < 2^{1-\lfloor (p-4)d/p \rfloor}. \tag{14.9}$$

Daumas and Finot [133] modified Priest's division algorithm by using only estimates of r_0, the most significant component of the remainder, and storing the less significant components of the remainder and the terms $-q_i y$ unchanged in a set that is managed with a priority queue. While the asymptotic complexity of this algorithm is better, in practical simple cases Priest's algorithm is faster due to the control overhead of the priority queue [133]. The error bound obtained with the algorithm of Daumas and Finot is (using the same notation as above)

$$\left|\frac{q - x/y}{x/y}\right| < 2^{-d(p-1)} \prod_{i=0}^{d-1} (4i + 6). \tag{14.10}$$

A different approach based on the Newton-Raphson iteration (see also Section 4.7 and the references therein) was revived by Joldes et al. [310] for computing reciprocals and square roots of ulp-nonoverlapping expansions.

14.2.4.1 Newton-Raphson based reciprocal algorithm

As already explained in Section 4.7, the classical Newton-Raphson iteration for computing the reciprocal of x is

$$r_{n+1} = r_n(2 - x r_n), \tag{14.11}$$

Algorithm 14.16 below adapts it to the computation of floating-point expansions of length an integral power of 2. The intermediate addition and multiplication operations are *truncated* and an error analysis provides a reasonably tight error bound on the result, given in Theorem 14.7.

Algorithm 14.16 – Reciprocal$(x_0, x_1, \ldots, x_{2^k-1}, 2^q)$.

$r_0 = \mathrm{RN}(1/x_0)$
for $i \leftarrow 0$ **to** $q-1$ **do**
  $(\hat{v}_0, \ldots, \hat{v}_{2^{i+1}-1}) \leftarrow$ Multiplication_accurate$(r_0, \ldots, r_{2^i-1}, x_0, \ldots, x_{2^{i+1}-1}, 2^{i+1})$ //using Algorithm 14.14
  $(\hat{w}_0, \ldots, \hat{w}_{2^{i+1}-1}) \leftarrow$ Renormalize$(-(\hat{v}_0, \hat{v}_1, \ldots, \hat{v}_{2^{i+1}-1}), 2.0, 2^{i+1})$ //using Algorithm 14.9
  $(r_0, \ldots, r_{2^{i+1}-1}) \leftarrow$ Multiplication_accurate$(r_0, \ldots, r_{2^i-1}, \hat{w}_0, \ldots, \hat{w}_{2^{i+1}-1}, 2^{i+1})$
end for
return $r_0, r_1, \ldots, r_{2^q-1}$

Theorem 14.7. *Let x be an* ulp-*nonoverlapping floating-point expansion with $n = 2^k$ terms and let $q \geq 0$ be an integer parameter such that 2^q is the required number of output terms. Assume $p \geq 6$ and $q \leq \frac{p}{2} - 2$, which always hold in practice. Provided that no underflow/overflow occurs during the calculations, Algorithm 14.16 computes an approximation to $\frac{1}{x}$ given as an* ulp-*nonoverlapping floating-point expansion $r = r_0 + \cdots + r_{2^q-1}$, such that*

$$\left| r - \frac{1}{x} \right| \leq \frac{2^{-2^q(p-4)-2}}{(1-2^{-p+1})} \cdot \frac{1}{|x|}. \tag{14.12}$$

14.2.4.2 Newton-Raphson based division algorithm

In this setting the division of two floating-point expansions is simply performed with Algorithm 14.16 followed by a multiplication with the numerator expansion. The multiplication can be done with Algorithm 14.14 or Algorithm 14.13, as shown in Algorithm 14.17.

Algorithm 14.17 – Division$(x_0, \ldots, x_{n-1}, y_0, \ldots, y_{m-1}, r)$.

$(f_0, f_1, \ldots, f_{r-1}) \leftarrow$ Reciprocal$(y_0, y_1, \ldots, y_{m-1}, r)$ //using Algorithm 14.16
$(d_0, d_1, \ldots, d_{r-1}) \leftarrow$ Multiplication$(x_0, \ldots, x_{n-1}, f_0, f_1, \ldots, f_{r-1}, r)$ //using Algorithm 14.14 or Algorithm 14.13
return $d_0, d_1, \ldots, d_{r-1}$

14.3 Floating-Point Numbers with Batched Additional Exponent

A limitation of the basic-precision floating-point formats is their restricted exponent range. For instance, in binary64 arithmetic, the smallest representable number is

$$\alpha = 2^{-1074}$$

and the largest one is

$$\Omega = (2 - 2^{-52}) \cdot 2^{1023}.$$

For example, this restriction may prevent one from converting very large integers into floating-point numbers. If this is deemed necessary, there are two main ways to work around this restriction. The easiest way is to use an arbitrary-precision floating-point library, since for these libraries, the exponent of the considered floating-point number is most often stored in a 32- or 64-bit integer, which allows for much larger numbers. However, this is an expensive solution since arbitrary-precision numbers can be significantly more expensive to manipulate than basic-precision floating-point numbers, and in our case, a larger precision may not be needed.

Another usual approach is to batch each floating-point number f with an integer e. The represented number is then $f \cdot \beta^e$ (where β is the radix of the floating-point system), which we will denote by (f, e). A given number has several representations, due to the presence of two exponents, the one contained in f, and the additional one, i.e., e. For example, assuming $\beta = 2$, $(1.0, 0) = (0.5, 1)$. To make the representation unique, one may require f to be between two given consecutive powers of β (and thus make the internal exponent useless). Normalizing a pair (f, e) into an equivalent one (f', e') such that f' belongs to the prescribed domain can be performed by using a simple exponent manipulation; for example, with the C function `ldexp`. This normalization has the advantage of making all basic arithmetic operations simpler. Basic arithmetic operations are rather simple to program. Such numbers are available in the `dpe` library of Pélissier and Zimmermann [489], in the INTLAB[10] library of Rump, and internally in many libraries, including Magma[11] [60] and NTL[12] [557] (in which it corresponds to the XD class).

[10]INTLAB is the Matlab toolbox for self-validating algorithms; it is available at http://www.ti3.tu-harburg.de/rump/intlab/.

[11]Magma is available at http://magma.maths.usyd.edu.au/magma/.

[12]NTL is a library for performing number theory, available at http://www.shoup.net/ntl/.

14.4 Large Precision Based on a High-Radix Representation

The most common approach for building an arbitrary-precision arithmetic consists in generalizing the usual radix-β floating-point representation, with an arbitrary-precision significand represented as an array of machine words.

This approach has been used since the early days of programming. For instance, Brent [64] used it in the mid-1970s for his MP library written in FORTRAN. Bailey et al. [29] pursued MP in ARPREC.[13] Other examples are the mpf layer of GNU MP[14] (a.k.a. GMP) and the GNU MPFR[15] library [204], which is also based on the layers of GMP that implement arbitrary precision integers. The specificity of MPFR is to extend the spirit of the IEEE 754 standard to arbitrary-precision floating-point arithmetic: correct rounding for *all* supported mathematical functions (which is actually more than what IEEE 754-2008 requires) in various rounding modes, special data such as infinities and NaNs, and IEEE 754-like exception handling. Despite these requirements, MPFR is slightly faster than mpf. Moreover, many elementary functions and even several special functions are implemented in MPFR, while mpf provides the basic floating-point operations only. For all these reasons, MPFR is now recommended as a replacement for mpf. It will be presented with more details in Section 14.4.4.

Each word of the significand can be regarded as a digit in a high-radix system, where the radix B of this system is a power of the radix β chosen for the arbitrary-precision floating-point arithmetic being implemented. For instance, on a 64-bit machine, the integer registers can hold unsigned integers between 0 and $2^{64} - 1$, thus $B = 2^{64}$ will typically be chosen. This can be used to implement an arbitrary-precision system of radix β, where β can be 2 (such as in MPFR) or B (such as in mpf). More generally, B is typically chosen of the form 2^w, where w is the width of the integer registers of the processor because in general, the integer arithmetic of current CPUs is particularly suited for implementing such a multiple-precision arithmetic. But floating-point registers could be used as well if this turned out to be faster.

For its floating-point arithmetic, the computer algebra system Maple uses $\beta = 10$. Thus B is a power of 10. While the choice $\beta = 10$ has advantages for humans on some simple computations, as already discussed in Chapter 1, it implies a poor choice for B in terms of performance.

In general, the digits are integers between 0 and $B - 1$, but variants are possible, possibly yielding redundant representations of the significand: digits may be signed (in which case the sign field of the floating-point representation may be part of the significand) or digits may be larger than the base,

[13]ARPREC is available at http://crd-legacy.lbl.gov/~dhbailey/mpdist/.
[14]GNU MP is available at https://gmplib.org/.
[15]MPFR is available at http://www.mpfr.org/.

so carries can be accumulated in digits instead of having to be propagated. For instance, GMP has experimental support for "nails": a register of w bits (where $w = 32$ or 64) consists of n "number bits" (i.e., B equals 2^n), and $w-n$ "nail bits" for absorbing carries.

The precision is roughly equal to the number of words of the significand multiplied by the constant $\log_\beta B$ (it will actually be a bit higher in case of accumulated carries in the most significant word, or a bit smaller if the least significant word is partly used as in MPFR). The way the precision is specified depends on the software. For instance, in mpf and MPFR, the precision is a component of the number, i.e., each variable has its own precision.

In this context, arbitrary-precision integer arithmetic is used to implement the basic arithmetic operations, as the significands can be regarded as multiple-precision integers. Then these operations are used to implement other functions, for example, elementary or special functions (the algorithms are out of the scope of this book: see for instance [66]). After a brief discussion on the specifications, we describe how to use integer arithmetic to perform the basic floating-point arithmetic operations. Then we quickly review arbitrary-precision integer arithmetic. Finally, we present the GNU MPFR library and explain how to use it.

14.4.1 Specifications

Often, on recent packages, arbitrary-precision floating-point numbers have been implemented in such a way that they can be viewed as "smooth extensions" of the fixed precisions of the IEEE 754-2008 standard. In particular, if we require the precision to be the same as in one of the basic formats of the standard, one should essentially see the behavior described in the standard. For instance, the RR module of NTL implements multiple-precision real numbers with correctly rounded-to-nearest arithmetic operations. For the transcendental functions, correct rounding is not guaranteed, but the computed result has a relative error less than 2^{-p+1}, where p is the current precision. However, in NTL, there are no special values such as $\pm\infty$ or NaN. In addition to the support of special values, MPFR goes much further by offering correct rounding even for the transcendental functions (this is possible using Ziv's "multilevel strategy," see Chapter 10, Section 10.5), in all the supported rounding modes. Contrary to fixed, small precision, new Ziv iterations do not need specific code in arbitrary precision: it is generally sufficient to increase the internal precision and execute the generic code again.

As a basic-precision floating-point number, an arbitrary-precision floating-point number is made of a sign, a significand, and an exponent. Note that a clear difference from conventional formats is that underflows and overflows are less likely to occur since the exponent range is usually very large; this is done without any special effort because the exponent can be represented as a native integer. However, exceptional cases may occur anyway. GNU MPFR handles them in a way similar to what is specified by the IEEE 754 standard. And yet, it does not support subnormals since they are much less interesting in arbitrary precision and would make the code more complex (MPFR provides a function to emulate their support for applications that need them, for instance to use MPFR as a reference implementation of the IEEE 754 standard).

More precisely, concerning the exponent range, it is advised to choose a range significantly larger than the maximum precision that will be used in practice, for instance in order to allow the ulp of a number to be representable as a normal number in general. In particular, [267, §3.7] requires that for an extendable precision, when only the precision p is specified, $e_{\max}$ should be at least $1000 \cdot p$. However, a huge exponent range also means that some accurate computations will be infeasible in practice, e.g., $\sin(x)$ for a huge value of x (even in small precision) since π would need to be computed with a precision of the order of e_x bits, where e_x is the exponent of x.

Two strategies exist for handling the precision p. First, it may be a global variable. In this case, at a given time, all floating-point numbers have the same precision, which may be changed over time. This strategy is used for NTL's RR class [557]. The second strategy (available for example in MPFR [204]) consists in batching each floating-point number with a local precision. This can be significantly more efficient, because the user can allocate large precisions only for the variables for which it is deemed necessary, and low precisions for the other ones. However, this makes the semantics and the implementation more complicated: the precisions of the operands as well as the precision of the target must be considered, with different cases that must be handled carefully.

14.4.2 Using arbitrary-precision integer arithmetic for arbitrary-precision floating-point arithmetic

Suppose we are trying to perform a basic operation op $\in \{+, \times, \div\}$ on arbitrary-precision floating-point numbers. We first perform a related integer operation on the significands (with appropriate scaling), and then postprocess the integer result to obtain the floating-point result.

Let a and b be two radix-2 floating-point numbers of precisions p_a and p_b, respectively. Suppose we want to compute $c = \circ(a \text{ op } b)$ for some precision p_c, some rounding function $\circ$, and some basic operation op $\in \{+, \times, \div\}$. The numbers a, b, and c can be written

$$x = (-1)^{s_x} \cdot m_x \cdot 2^{e_x - p_x + 1}, \quad \text{for } x \in \{a, b, c\},$$

where s_x belongs to $\{0, 1\}$, m_x is an integer whose bit length is p_x, and e_x is a small integer.

Quite similar to what we have presented in Chapter 7, the addition of a and b reduces to the integer addition or subtraction of m_a with a shifted value of m_b, the amount of the shift being determined by the difference between e_a and e_b. Similarly, the multiplication of a and b reduces to the integer multiplication of m_a and m_b. Finally, the division of a by b can be reduced to the Euclidean division of a shifted value of m_a by m_b.

Note that, in all cases, only the first p_c bits of the result of the integer operation are actually needed to produce most of m_c. Additional bits may be needed to determine the correctly rounded result; most often, just a few bits are sufficient, but in the worst case, all the bits of the inputs may be necessary (this corresponds to some form of the Table maker's dilemma, presented in Section 10.5). In the case of floating-point multiplication, this remark can be used to save a possibly significant proportion of the overall cost. Suppose for example that the input and target precisions match. Then only the most significant half (plus a few) of the bits of the integer product of the two significands is actually needed. Krandick and Johnson [351] designed an efficient algorithm to compute that half of the bits. It was later improved by Mulders [435] and then by Hanrot and Zimmermann [234]. In the last two references, the authors concentrate on the lower half of the product of two polynomials, but their analyses extend to the upper half of the product of two integers (with complications due to the propagation of carries).

14.4.3 A brief introduction to arbitrary-precision integer arithmetic

Arbitrary-precision integer arithmetic has been extensively studied for more than 55 years [328, 603, 547, 545, 546, 209, 247]. A big integer is in general stored as a list or an array of processor integers. Big integer arithmetic is usually much slower than processor integer arithmetic, even when the "big" integers turn out to be small enough to be representable by processor integers. A way of using the best of both worlds is the following. Consider a machine integer. If its first (or last) bit is 1, the other bits encode the address of a multiple-precision integer; otherwise, they encode a small integer. This trick is used, e.g., for the integers of the Magma computational algebra system [60]. The overhead for small integers can thus be significantly decreased.

Adding or subtracting arbitrary-precision integers is relatively straightforward. The algorithmic aspects become significantly more involved when one considers multiplication and division. Here we will consider only multiplication. Note that any multiplication algorithm can be turned into a division algorithm via the Newton–Raphson iteration (see Section 4.7, page 124), with a constant factor overhead in the asymptotic complexity, thanks to the quadratic convergence (see [66] for an extensive discussion). There exists a full hierarchy of multiplication algorithms. By decreasing complexity upper bounds, with n being the maximum of the bit lengths of the two integers to be multiplied, we have:

- the "school-book" multiplication, with complexity $\mathcal{O}(n^2)$;

- Karatsuba's multiplication [328], with complexity $\mathcal{O}(n^{\log_2 3})$;
- the Toom-Cook-3 multiplication [603], with complexity $\mathcal{O}(n^{\log_3 5})$;
- its higher-degree generalizations, with complexities $\mathcal{O}(n^{\log_4 7})$, $\mathcal{O}(n^{\log_5 9}), \ldots$;
- and the fast multiplication of Schönhage and Strassen, which relies on the discrete Fourier transform [547], with complexity $\mathcal{O}(n \cdot \log n \cdot \log \log n)$.

Note that integer multiplication is still the focus of active research, and has been improved in the last decade by Fürer [209], and Harvey, van der Hoeven, and Lecerf [247]. By changing the ring to which the Fourier transform maps the integers to multiply, Fürer was able to obtain an $n \cdot \log n \cdot 2^{\mathcal{O}(\log^* n)}$ complexity bound, where $\log^* n$ is the number of times the log function has to be applied recursively to obtain a number less than 1. Harvey, van der Hoeven, and Lecerf present a modified algorithm with complexity $O\left(n \cdot \log n \cdot 8^{\log^* n}\right)$.

Several practical improvements over the Schönhage–Strassen algorithm have also been reported by Gaudry, Kruppa, and Zimmermann [210].

In Table 14.1, we give a list of some multiplication algorithms, with their asymptotic complexities. We also give the approximate numbers of machine words that the involved integers require for any given algorithm to become more efficient in practice than the algorithms above it in the table. These thresholds derive from the GNU MP library [224]. See also some experiments by Zuras [648].

Algorithm	Asymptotic complexity	GNU MP thresholds
School-book	$\mathcal{O}(n^2)$	-
Karatsuba	$\mathcal{O}(n^{\log_2 3})$	$[10, 30]$
Toom–Cook-3	$\mathcal{O}(n^{\log_3 5})$	$[40, 130]$
Toom–Cook-4	$\mathcal{O}(n^{\log_4 7})$	$[100, 300]$
Schönhage–Strassen	$\mathcal{O}(n \cdot \log n \cdot \log \log n)$	$[2000, 10000]$
Fürer	$n \cdot \log n \cdot 2^{\mathcal{O}(\log^* n)}$	unknown
Harvey et al.	$O\left(n \cdot \log n \cdot 8^{\log^* n}\right)$	unknown

Table 14.1: Asymptotic complexities of some multiplication algorithms and approximate practical thresholds in numbers of machine words on current machines.

The fast multiplication algorithms rely on the evaluation-interpolation paradigm. We explain it briefly with Karatsuba's multiplication, and refer to [38] for a detailed survey on integer multiplication algorithms. Suppose we want to multiply two positive integers a and b, that are both n bits long. Karatsuba's algorithm first splits a and b into almost equal sizes: a is written $a_1 2^{\lceil n/2 \rceil} + a_0$ and similarly b is written $b_1 2^{\lceil n/2 \rceil} + b_0$, where a_0 and b_0 are both in $[0, 2^{\lceil n/2 \rceil} - 1]$. The product $c = a \cdot b$ can be expressed as

$$c = c_2 \cdot 2^{2\lceil n/2 \rceil} + c_1 \cdot 2^{\lceil n/2 \rceil} + c_0,$$

where $c_2 = a_1 \cdot b_1, c_1 = a_1 \cdot b_0 + a_0 \cdot b_1$, and $c_0 = a_0 \cdot b_0$. Using these formulas as such leads to an algorithm for multiplying two n-bit-long integers that uses four multiplications of numbers of size around $n/2$. Applying this idea recursively, i.e., replacing (a, b) by (a_1, b_1), (a_1, b_0), (a_0, b_1), and (a_0, b_0) respectively, leads to a multiplication algorithm of complexity $\mathcal{O}(n^2)$, which is the complexity of the school-book algorithm. Indeed, if C_n is the cost of computing the product of two n-bit-long integers with this algorithm, then we just proved that $C_n = 4C_{n/2} + \mathcal{O}(n)$, which leads to $C_n = \mathcal{O}(n^2)$. Karatsuba's contribution was to notice that in order to obtain c_0, c_1, and c_2, we only need three multiplications of numbers of size around $n/2$ instead of four, because of the formulas

$$\begin{array}{lll} c_2' = a_1 \cdot b_1, & c_1' = (a_1 + a_0) \cdot (b_1 + b_0), & c_0' = a_0 \cdot b_0, \\ c_2 = c_2', & c_1 = c_1' - c_2' - c_0', & c_0 = c_0'. \end{array}$$

We therefore readily deduce that only three multiplications of numbers of size around $n/2$ are needed to perform the product of two numbers of size n (as well as a few additions, which turn out to be negligible in the cost). If we apply these formulas recursively (for the computation of the c_i'), then we obtain $C_n = 3C_{n/2} + \mathcal{O}(n)$, which leads to the asymptotic complexity bound $\mathcal{O}\left(n^{\log_2 3}\right)$, with $\log_2 3 = 1.584 \cdots$.

These formulas may seem surprising at first sight, but they can be interpreted in an elegant mathematical way. Replace $2^{\lceil n/2 \rceil}$ by an indeterminate x. Then the integers a, b, and c become three polynomials:

$$a(x) = a_1 x + a_0,$$

$$b(x) = b_1 x + b_0,$$

and

$$c(x) = c_2 x^2 + c_1 x + c_0.$$

If we evaluate $c(x)$ for $x = 0$, $x = 1$, and $x = +\infty$ (more precisely, at $+\infty$, we take the dominant coefficient), then we obtain

$$\begin{aligned} c(0) = c_0 &= c_0', \\ c(1) = c_2 + c_1 + c_0 &= c_1', \\ c(+\infty) = c_2 &= c_2'. \end{aligned}$$

The c_i' are the evaluations of $c(x)$ at the three points we selected. Furthermore, since $c = a \cdot b$ and $c(x) = a(x) \cdot b(x)$, we have

$$\begin{aligned} c_0' = a(0) \cdot b(0) &= a_0 \cdot b_0, \\ c_1' = a(1) \cdot b(1) &= (a_0 + a_1) \cdot (b_0 + b_1), \\ c_2' = a(+\infty) \cdot b(+\infty) &= a_1 \cdot b_1. \end{aligned}$$

Overall, Karatsuba's algorithm can be summarized as follows: evaluate a and b at points 0, 1, and $+\infty$; by taking the products, obtain the evaluations of c at these points; finally, interpolate the evaluations to recover c. The costly step in Karatsuba's multiplication is the computation of the evaluations of c from the evaluations of a and b.

This evaluation-interpolation principle was also generalized with more than three evaluation points. The Toom-Cook-3 multiplication algorithm consists in splitting a and b into three equal parts (instead of two as in the previous algorithm), and using five evaluation points, for example, 0, 1, -1, -2, and $+\infty$. The multiplication of n-bit-long operands can then be performed with 5 multiplications between operands of bit-size around $n/3$. This provides an asymptotic complexity bound $\mathcal{O}(n^{\log_3 5})$, with $\log_3 5 = 1.464\cdots$. If a and b have different sizes, it may also be interesting to split them in different numbers of components, so that all components have the same sizes approximately; this idea is used in GMP. By letting the number of evaluation points grow to infinity, one can obtain multiplication algorithms of complexity $\mathcal{O}(n^{1+\epsilon})$ for any $\epsilon > 0$. However, when n becomes large, it is more interesting to consider particular evaluation points, namely, roots of unity, which gives rise to the Fourier transform-based multiplication of Schönhage and Strassen [547].

14.4.4 GNU MPFR

GNU MPFR [204], which has briefly been presented at the beginning of this section, is a free library for efficient arbitrary-precision floating-point computation in radix 2, with well-defined semantics. We now describe this library with more details and show how to use it. This section applies to MPFR 3.1.5 [232] (the latest release at the time of writing this book), but also mentions changes and new features present in the development version, which will be in MPFR 4.

The parts of the following concerning the interface apply to the C interface (which is the only official interface to MPFR). There exist several other interfaces to MPFR, provided by third parties, for other programming languages.

14.4.4.1 MPFR data

Computations are done on MPFR floating-point objects, representing numbers or NaN (Not a Number). Each object has its own precision in bits, starting from 2; with MPFR 4, the minimum precision will be 1.[16] The maximum precision depends on several factors: a (large) limit imposed by MPFR, internal algorithmic limitations, the available memory (the actual limitation on 64-bit machines), and the address space. A MPFR datum is either a nonzero finite floating-point number in radix 2, called *regular* datum, or a *singular* datum: a signed zero (± 0), a signed infinity ($\pm\infty$), or a NaN (Not a Number).

The representation of MPFR data is a bit like the one of the binary interchange formats of IEEE 754 (see Section 3.1.1.1). In addition to the precision field (regarded mainly as a parameter), the three usual fields *sign*, *exponent*, and *significand* are used. The regular data are always normalized, and the leading bit of the significand is stored (i.e., there is no "hidden bit convention") as this is more practical and also since saving one bit is useless in multiple precision.

The exponent field is represented by a native integer of the same type as the precision (currently, the C type `long int`); thus its actual size depends on the ABI[17] (32 or 64 bits in practice). To avoid integer overflows in internal computations, the valid values of the exponent are allowed to range in $[1 - 2^{30}, 2^{30} - 1]$ and $[1 - 2^{62}, 2^{62} - 1]$ on 32-bit and 64-bit machines, respectively. At the same time, this leaves unused values of the exponent type, which are exploited to represent singular data.

The above choice of the exponent range can be explained by reasons given in Section 14.4.1. Thus the goal was to choose a maximum exponent significantly larger than the maximum precision used in practice. This goal is not necessarily fulfilled for all supported precisions, but this choice is regarded as a good compromise between performance and practical use, and on a 64-bit machine, the precision is limited by the available memory (or computation time) anyway.

The exponent range is $[1 - 2^{30}, 2^{30} - 1]$ by default on all machines, but the user can change it, for instance to increase it to its maximum on 64-bit machines, or to reduce it in order to emulate an IEEE 754 binary format.

Apart from the variable precision, the main differences with IEEE 754 about the floating-point data are:

- MPFR does not have subnormals: to perform computations, they are regarded as useless due to the huge exponent range. But MPFR provides

[16]The minimum precision was initially chosen as 2 because the ties-to-even rule was not defined for precision 1 in IEEE 754. But since this case can occur in conversions to character strings, a complete definition has been proposed for the 2018 revision (see Section 3.1.2.1), and the minimum precision could be changed to 1 in MPFR, even though the context is different (there are no formats with 1-bit precision in IEEE 754).

[17]Application Binary Interface, which specifies the sizes of the scalar types, in particular.

a function `mpfr_subnormalize` to emulate them for particular applications, such as emulating IEEE 754 binary formats.

- MPFR has a single kind of NaN, whose behavior is somehow between quiet NaN and signaling NaN. However, just like in IEEE 754, the sign of the NaN is taken into account for some particular operations.

- The convention on the exponent and interpretation of the significand are different from those of IEEE 754. In IEEE 754-2008, for radix 2, one uses either a real significand in the interval $[1,2)$ with the exponent e, or an integral significand with the exponent $q = e - p + 1$. In MPFR, the significand is regarded as a number in the interval $[1/2, 1)$, so the fractional point lies on a word boundary (just before the most significant word). However, for a given range, the set of the floating-point data and the rounding behavior do not depend on the convention. The user should just remember that, when using functions and macros dealing with exponent values, there is a shift by 1 compared to IEEE 754.

Each MPFR object must be explicitly initialized before use, which has the effect of assigning a precision and allocating the memory for the significand. The default value is NaN, which is the equivalent of an uninitialized value, in the sense that an initial value is not provided by the user.

14.4.4.2 Rounding modes

MPFR supports 5 rounding modes:

- `MPFR_RNDN`, for roundTiesToEven ("round to nearest even");

- `MPFR_RNDD`, for roundTowardNegative (toward $-\infty$);

- `MPFR_RNDU`, for roundTowardPositive (toward $+\infty$);

- `MPFR_RNDZ`, for roundTowardZero (toward zero);

- `MPFR_RNDA`, for rounding away from zero, which is not part of IEEE 754, but is provided as the opposite of `MPFR_RNDZ`.

MPFR does not support roundTiesToAway, but the development code currently provides an experimental macro to emulate it.

The development code also has limited support for faithful arithmetic; the goal is to be slightly faster in internal computations, in particular by avoiding the Table maker's dilemma (see Section 10.5). Indeed, one may just be interested in an approximation to the exact result and an error bound. Almost all algorithms used to implement correctly rounded mathematical functions just need this.

14.4.4.3 Exceptions

MPFR supports the 5 exceptions of IEEE 754. The invalid exception, which is also called NaN exception in MPFR, is raised each time a result is a NaN (even on NaN inputs, i.e., as if the NaN inputs were signaling NaNs); so, in particular, the use of an uninitialized value will generally raise a NaN exception, following the idea behind [267, §6.2] (or Section 3.1.7) related to uninitialized values. MPFR uses the *after-rounding* underflow definition (see Sections 2.1.3 and 3.1.6.4) for consistency with overflow. This is also more useful as it avoids raising some underflow exceptions that do not correspond to a loss of precision (in the sense that the usual error analysis still applies). It has also an interesting property: in a computation without overflows and underflows, if all intermediate results have exponents in some range $[e_1, e_2]$, then the exponent range of the system can be reduced to $[e_1, e_2]$ without affecting the floating-point computations.

MPFR defines a sixth kind of exception: range error, which more or less corresponds to the invalid exception, but for functions that do not return a floating-point result. For instance, in a comparison function, which returns the sign ($+1$, 0 or -1) of the difference of two MPFR numbers, this exception occurs when one of the operands is NaN, and in the functions that convert a MPFR number to a native integer, this exception also occurs when the operand is too big for the return type. In short, the invalid exception of IEEE 754 has been split into two different exceptions in MPFR.

Exceptions in MPFR are implemented by status flags, similar to the default exception handling of IEEE 754-2008.

14.4.4.4 The ternary value

Most MPFR functions that compute a floating-point result return an integer, called *ternary value*, that gives the sign of the rounding error:

- Zero: the rounded result is actually the exact result of the mathematical function.
- Positive: the rounded result is greater than the exact result.
- Negative: the rounded result is less than the exact result.

For instance, with the `MPFR_RNDU` rounding mode, the ternary value is usually positive, except when the result is exact, in which case it is zero.

An infinite result is regarded as inexact when it was obtained by an overflow, and exact otherwise. A NaN result is always regarded as exact.

In particular, the ternary value can be used to handle double rounding, as done internally in the following cases:

- As said earlier, the exponent range can be changed. For instance, near the beginning of the code of a mathematical function, the exponent range is usually extended to its maximum in order to avoid spurious underflows and overflows in the internal computations, and at the end, the exponent range is restored. MPFR provides a function `mpfr_check_range` to check whether a floating-point datum (e.g., a result computed with the extended exponent range) is in the current exponent range, and round it to this exponent range; so, this function is called after the exponent range is restored. But in the `MPFR_RNDN` rounding mode (roundTiesToEven), if the number (in absolute value) is the middle between zero and the minimum positive number, then one needs to consider the sign of the error to avoid the double-rounding problem. This is why this function takes a ternary value as an argument.

- The `mpfr_subnormalize` function also takes a ternary value as an argument in order to handle midpoint cases in the `MPFR_RNDN` rounding mode. It works as follows. The user chooses a floating-point system with subnormals. To emulate it with MPFR, he or she must choose the minimum exponent for MPFR in such a way that it is the largest one allowing all subnormals of his/her system to be represented in the MPFR floating-point system. This means that when a MPFR computational function is called, a result in the subnormal range may have more nonzero digits than possible in the emulated system. The `mpfr_subnormalize` function just rounds this result to the wanted precision. So, here one has two roundings: the one for the computation in the MPFR exponent range (with no subnormals), then the one from `mpfr_subnormalize`. The ternary value from the first rounding allows one to avoid the double-rounding problem.

  Let us give an example. For simplicity, assume that we want to emulate the binary16 format (as if it were used for computation). Table 3.1 gives for this format: $p = 11$ and $e_{\min} = -14$ (with the usual IEEE 754 convention). The smallest positive value, or smallest positive subnormal, is $2^{e_{\min}+1-p} = 2^{-24}$. We first need to determine the minimum exponent $E_{\min}$ to choose for MPFR, with the MPFR convention, in order to satisfy the requirements for `mpfr_subnormalize`, as given above. The smallest positive value in MPFR with the chosen exponent range is $(1/2)\cdot 2^{E_{\min}} = 2^{E_{\min}-1}$. We want the smallest positive subnormal of binary16 to be this smallest positive value in MPFR. Thus $E_{\min} = e_{\min} + 2 - p = -23$. Now, consider the binary16 value $x = 0.1000001 \cdot 2^{-9}$, and assume that we wish to compute $\mathrm{RN}(x^2)$ in binary16, which will give a subnormal. One has $x^2 = 0.\mathbf{10000}10000001 \cdot 2^{-19}$, where the 5 bold digits form the truncated significand in binary16, to be completed with some zero digits on the left. In MPFR, the user calls the `mpfr_sqr` function (square function) on x with the rounding mode `MPFR_RNDN`, which yields a value y equal to x^2 rounded on $p = 11$ digits (i.e., $y = \mathrm{RN}_{11}(x^2) = 0.\mathbf{10000}100000 \cdot 2^{-19}$) and the corresponding ternary value, which is negative here (because the rounded result is less than the exact result). Then the user calls `mpfr_subnormalize`. This function will round y on $(-19) - E_{\min} + 1 = 5$ digits, taking the ternary value into account if need be. Here, we have a halfway case in the round-to-nearest mode, so the ternary value will be used. Since the ternary value is negative, this means that the exact value (unknown to the `mpfr_subnormalize` function) is greater than y, thus `mpfr_subnormalize` must round y away from zero, yielding $0.10001 \cdot 2^{-19}$.

- The experimental macro to emulate roundTiesToAway works by first computing the rounded result in a precision equal to the target precision plus one, and then deciding the result for roundTiesToAway. Here, this is not exactly the double-rounding problem, but one also needs the ternary value of the first rounding for the same reason.

14.4.4.5 Types and calling convention

MPFR is inspired by the choices done in GMP, on which it is based. In the C interface, the type name for the MPFR floating-point data is `mpfr_t`. Like GMP types, this type is internally defined as a one-element array of a structure, so that one does not need an explicit memory allocation for the structure itself and one can pass variables to functions by reference (i.e., the pointer to the structure is actually passed). The type name for the pointer to the structure is `mpfr_ptr`.

For the MPFR floating-point operations (with one result), a function call has the following form:

```
mpfr_t a, b, c;
int r;
/* ... (initialization, etc.) */
r = mpfr_add (a, b, c, MPFR_RNDN);
```

The first argument is the destination; the object must have been initialized, so that the target precision is taken from it, but the floating-point datum it contains before the call will be lost. Then one has the input argument(s), and the rounding mode is the last argument. The return value (here, stored in `r`) is the ternary value. This function call computes $a \leftarrow \mathrm{RN}_{p(a)}(b+c)$, the result being rounded to the precision $p(a)$ of variable `a`. It is always possible to use the same MPFR object in several arguments, including for the destination. For example,

```
mpfr_mul (x, x, x, MPFR_RNDN);
```

has the effect of computing the square of `x` and storing it back to `x`.

14.4.4.6 Memory handling

MPFR uses some global variables, such as the current exponent range and the status flags. When MPFR is built as thread-safe (which is the default when threading is supported by the system), these variables are actually declared as thread-local ("thread local storage").

Some MPFR functions create and use caches for some usual constants. For instance, the trigonometric functions need the value of π. An approximation to this constant is cached, so π is not recomputed each time a trigonometric function is called. It is recomputed only if the precision of the cached value is not sufficient. A change has been done in the development version of MPFR in order to avoid too many recomputations of the constant in some applications: the new precision of the cache is chosen in such a way that it is at least 10% larger than the old precision. Caches are currently thread-local, but the development version of MPFR has experimental support for shared caches.

14.4.4.7 Implemented functions

MPFR implements the following operations and mathematical functions that give a datum as the result: negate; absolute value; minimum and maximum; addition; subtraction; multiplication; square; FMA and FMS[18]; division; various kinds of integer and remainder functions; square root; reciprocal square root; cubic root; power; Euclidean norm (hypot); logarithm and exponential in bases e, 2, and 10; log1p and expm1; trigonometric functions (sine, cosine, tangent, their inverse, and cotangent, secant, cosecant); hyperbolic functions (ditto); factorial; Gamma function and the logarithm of its absolute value (lgamma); zeta; error function and complementary error function; exponential integral; dilogarithm; Digamma; first and second kind Bessel functions; arithmetic-geometric mean (AGM); Airy function; various kinds of pseudo-random functions.

The behavior corresponds to the IEEE 754-2008 one when specified, NaN being regarded as a quiet NaN in this context. The only exception is the reciprocal square root on -0: In IEEE 754-2008, rSqrt(-0) gives $-\infty$, while `mpfr_rec_sqrt` gives $+\infty$ on -0; the choice done by MPFR comes from the fact that the mathematical function is defined only for positive arguments, and when $x \to 0$, one has $1/\sqrt{x} \to +\infty$ (this is the same rule used to define the logarithm on -0, for instance). Moreover, negation and absolute value are conventional functions, not sign-bit operations like in IEEE 754-2008; but since MPFR has a single NaN, this actually makes very little difference.

In IEEE 754, some operations are necessarily exact, but this is not always true in MPFR, due to the fact that the destination may have a lower precision

[18]Fused multiply-subtract.

than the input. Similarly, some operations cannot overflow in IEEE 754, but this may not be the case in MPFR.

MPFR also provides conversions, comparisons, input/output functions, formatted output functions, and other basic functions related to the floating-point system.

MPFR 3.1.5 provides a function `mpfr_sum` that computes the correctly rounded sum of the elements of an array of MPFR data, but the user should be aware of its limitations:

- the sign of an exact zero result is not specified;
- the return value is guaranteed to be the usual ternary value only if it is zero (meaning that the result is exact);
- when the Table maker's dilemma occurs, this implementation just recomputes the sum with more precision, so in extreme cases (easy to construct), it may take much time, exhaust the memory of the machine, or crash due to a lack of memory.

This function has completely been specified and rewritten for MPFR 4 to resolve the above limitations. [379]

14.4.4.8 An example of how to use MPFR

Listing 14.1 gives a small example of how MPFR can be used. It computes the first 200 terms of the sequence

$$\begin{cases} u_0 &= 2, \\ u_1 &= -4, \\ u_n &= 111 - \dfrac{1130}{u_{n-1}} + \dfrac{3000}{u_{n-1}u_{n-2}}, \qquad n \geq 2, \end{cases}$$

presented in Section 1.3.2.1. It can be compiled with a command line of the form:

```
cc program.c -o program -lmpfr -lgmp
```

The `-lmpfr` option tells the compiler that the program will be linked with the MPFR library. One also needs `-lgmp` because MPFR uses the GMP library. Note that in doubt, `-lgmp` must come after `-lmpfr`.[19]

The program works as follows. The `mpfr_inits2` function is used to initialize several variables in a single call to the same precision, here 600 bits (this is more practical than using `mpfr_init2`, which initializes a single variable); the argument `(mpfr_ptr) 0` terminates the list.[20] Functions whose

[19]Some linkers will include only unresolved symbols seen so far in the command line, and the requisite GMP symbols are known only after `-lmpfr` has been taken into account.

[20]This is how functions with a variable number of arguments are usually designed in C.

C listing 14.1 Simple example computing the following sequence with MPFR: $u_0 = 2$; $u_1 = -4$; $u_n = 111 - 1130/u_{n-1} + 3000/(u_{n-1}u_{n-2})$.

```c
#include <mpfr.h>

static void display (mpfr_t v, int k)
{
  mpfr_printf ("u[%d] = %.20Rg\n", k, v);
}

int main (void)
{
  mpfr_t u, v, x, y, z;
  int k;

  mpfr_inits2 (600, u, v, x, y, z, (mpfr_ptr) 0);

  mpfr_set_si (u, 2, MPFR_RNDN);    // u[0] = 2
  mpfr_set_si (v, -4, MPFR_RNDN);   // u[1] = -4

  display (u, 0);
  display (v, 1);

  for (k = 2; k < 200; k++)
    {
      // u = u[n-2], v = u[n-1]
      mpfr_ui_div (x, 1130, v, MPFR_RNDN);  // x = 1130 / v
      mpfr_ui_sub (y, 111, x, MPFR_RNDN);   // y = 111 - x
      mpfr_mul (z, v, u, MPFR_RNDN);        // z = v * u
      mpfr_ui_div (x, 3000, z, MPFR_RNDN);  // x = 3000 / z
      mpfr_set (u, v, MPFR_RNDN);           // u = v
      mpfr_add (v, y, x, MPFR_RNDN);        // v = y + x
      display (v, k);
    }

  mpfr_clears (u, v, x, y, z, (mpfr_ptr) 0);
  return 0;
}
```

name starts with `mpfr_set` such as `mpfr_set_si` are assignment functions (with rounding). The `si` component of the function name means that the argument is a native signed integer (type `long int`). Similarly, `ui` (as found in the loop) means that the argument is a native unsigned integer (type `unsigned long int`). For instance, the `mpfr_ui_div` in the loop means that an unsigned integer is divided by a MPFR datum (not to be confused with `mpfr_div_ui`, which means that a MPFR datum is divided by an unsigned integer). The `ui` and `si` components (or similar ones for some other C or GMP types) are supported for the most basic functions only; see the MPFR manual corresponding to your MPFR version for details. Back to the code, the performed operations (rounded to the target precision, always 600 here) are mentioned in comments. The `mpfr_printf` formatted-output function is an extension to GMP's `gmp_printf` function, itself an extension to C's `printf` function: `%.20Rg` is similar to C's `%.20g` for `double` or `%.20Lg` for `long double`, where `.20` means that 20 digits are output; the `R` length modifier is used for MPFR data (type `mpfr_ptr`).

With the provided precision 600, running this program shows that the sequence seems to converge to 6 until u_{140}, then seems to converge to 100 as explained in Section 1.3.2.1; note, however, that only 20 decimal digits are displayed here, so the output value 100 just means that the internal value is an approximation to 100 on 20 decimal digits (i.e., about 66 bits). One can replace `%.20Rg` by `%Ra` to get the exact value of the MPFR variable in the hexadecimal notation (i.e., on 600 bits), and increase the number of iterations to at least 300 to see that 100 is actually reached after 300 iterations.

This program does not allow one to prove anything, but observing the behavior in large precisions can give a better idea of what is going on. By using the directed rounding modes, one can also get lower and upper bounds on each term of the sequence, i.e., enclosing intervals. If one assumes that for each variable, both bounds have the same sign,[21] the program remains quite simple. Otherwise, one can do full interval arithmetic, as described in Chapter 12; Section 14.4.4.10 lists software providing interval arithmetic based on MPFR.

14.4.4.9 Caveats

Users should be aware of some common mistakes and limitations.

The most common issue is the use of MPFR functions with an argument of type `double` or `long double` (native C floating-point types), such as `mpfr_set_d`. Using such a function is acceptable when the floating-point value is exact in the C type, such as the constant `1.25`; otherwise the precision is limited to the one of the native C type (e.g., 53 bits for `double` on the usual platforms). For instance, if one writes

[21] This can easily be detected dynamically, and one can choose to stop the iterations when this is no longer true, as the enclosing interval gets too wide to provide any interesting information.

```
mpfr_init2 (x, 500);
mpfr_set_d (x, 0.1, MPFR_RNDN);
```

then x will be an approximation to 0.1 with a 53-bit accuracy only, even though it has a 500-bit precision. This is how the C language is designed, and 0.1 is rounded to a 53-bit precision before the function call, so MPFR cannot do anything about it. For such constants, it is recommended to use `mpfr_set_str` (but its return value is not a ternary value) or `mpfr_strtofr`. For example,

```
mpfr_set_str (x, "0.1", 10, MPFR_RNDN);
```

This will always work as expected because the period is accepted even when the decimal point is a different character in the current locale. MPFR also provides functions for some constants, such as `mpfr_const_pi` for π. With MPFR interfaces in other languages, the rules for decimal constants (literals) may be different.

Users should also remember that for algorithmic reasons, evaluating some functions in particular domains may be expensive (in terms of time and/or memory), even in small precision. For instance, this is the case with the trigonometric and Bessel functions for very large arguments, as mentioned in Section 14.4.1. For some functions, the memory usage may also depend on the precision of the input arguments (even when the target precision is small).

Just like with native floating point, algorithms generally need more precision for the intermediate calculations than the target precision in order to provide accurate results.[22] Thus using a target precision close to the maximum precision supported by MPFR is not guaranteed to work, even if there is enough memory and address space; an attempt to exceed this limitation will trigger an assertion failure. As most computers nowadays, especially those for calculations, have a 64-bit ABI (which is the default), this issue should not appear, but users may also reach a limitation from GMP, which currently uses a 32-bit type for the size of integers.

Modern computers have several cores, but the implementation of MPFR functions is sequential only, just like GMP. Thus, if the user wishes to exploit this level of parallelism, he or she needs to parallelize the code on his/her side. Anyway, coarse-grained parallelism is expected at this level, to avoid wasting too much time in communication (including implicit communication from shared memory).

[22]The results are actually correctly rounded, so that even more precision may be needed, but the issue mainly comes from the accuracy requirement, and the future support for faithful arithmetic will not avoid it.

14.4.4.10 Beyond GNU MPFR

The goal of GNU MPFR is to provide basic operations and the most common mathematical functions on real floating-point numbers, all correctly rounded and with a well-defined semantic (thanks to special values and exceptions), and various other support functions. It can be regarded as the equivalent of IEEE 754 to arbitrary precision. Still, partly because it is based on GMP, it is very efficient, even in small precision. MPFR 4 will even be faster in precisions smaller than twice the number of bits per word (e.g., in precisions up to 127 bits on 64-bit machines) thanks to specific code for these precisions [381]; in particular, emulating the binary64 and binary128 formats will be faster than with previous MPFR versions.

Users who wish more advanced features in arbitrary precision (such as interval arithmetic, see Chapter 12) may be interested in other software based on MPFR, such as:

- Arb[23] [304], a C library for arbitrary-precision interval arithmetic, where an interval is represented by a midpoint and a radius ("mid-rad interval arithmetic," a.k.a. "ball arithmetic"). Arb supports complex numbers, which are represented in rectangular form: the real and imaginary parts are represented by intervals (balls). It also supports polynomials, power series, and matrices, with real and complex coefficients. Although Arb uses MPFR, it has its own implementation of transcendental functions using ball arithmetic.

- GNU MPC,[24] a C library for arbitrary-precision arithmetic on complex numbers with correct rounding.

- MPFI,[25] for interval arithmetic, where an interval is represented by its endpoints, both MPFR numbers.

- iRRAM,[26] a C++ package for error-free real arithmetic based on the concept of a Real-RAM; iRRAM internally uses mid-rad interval arithmetic.

[23]Arb is available at http://arblib.org/.
[24]GNU MPC is available at http://www.multiprecision.org/mpc/.
[25]MPFI is available at https://gforge.inria.fr/projects/mpfi/.
[26]iRRAM is available at http://irram.uni-trier.de/.

Appendix A

Number Theory Tools for Floating-Point Arithmetic

A.1 Continued Fractions

Continued fractions make it possible to build very good (indeed, the best possible, in a sense that will be made explicit by Theorems A.1 and A.2) rational approximations to real numbers. As such, they naturally appear in many problems of number theory, discrete mathematics, and computer science. Since floating-point numbers are rational approximations to real numbers, it is not surprising that continued fractions play a role in some areas of floating-point arithmetic. A typical example is Table 10.1: that table gives, among other results, the floating-point numbers that are nearest to an integer multiple of $\pi/2$, which is a typical continued-fraction problem. It allows one to design accurate range-reduction algorithms for evaluating trigonometric functions. Another example is the continued-fraction-based proof of the algorithm for performing correctly rounded multiplication by an arbitrary-precision constant, presented in Section 4.11.

Excellent surveys can be found in [237, 565, 334]. Here, we will just present some general definitions, as well as the few results that are needed in this book, especially in Chapters 4 and 10.

Let α be a real number. From α, consider the two sequences (a_i) and (r_i) defined by

$$\left\{ \begin{array}{lcl} r_0 & = & \alpha, \\ a_i & = & \lfloor r_i \rfloor, \\ r_{i+1} & = & \dfrac{1}{r_i - a_i}, \end{array} \right. \tag{A.1}$$

J.-M. Muller et al., *Handbook of Floating-Point Arithmetic*,
https://doi.org/10.1007/978-3-319-76526-6

where "$\lfloor . \rfloor$" is the usual floor function. Note that the a_i's are integers and that the r_i's are real numbers.

If α is an irrational number, then these sequences are defined for any $i \geq 0$ (i.e., r_i is never equal to a_i), and the rational number

$$\frac{P_i}{Q_i} = a_0 + \cfrac{1}{a_1 + \cfrac{1}{a_2 + \cfrac{1}{a_3 + \cfrac{1}{\ddots + \cfrac{1}{a_i}}}}}$$

is called the i-th *convergent* of α. The a_i's constitute the *continued fraction expansion* of α. We write

$$\alpha = a_0 + \cfrac{1}{a_1 + \cfrac{1}{a_2 + \cfrac{1}{a_3 + \cdots}}}$$

or, to save space,

$$\alpha = [a_0; a_1, a_2, a_3, \ldots].$$

If α is rational, then these sequences terminate for some k, and $P_k/Q_k = \alpha$ exactly. The P_i's and the Q_i's can be deduced from the a_i's using the following recurrences:

$$\begin{cases} P_0 &=& a_0, \\ Q_0 &=& 1, \\ P_1 &=& a_1 a_0 + 1, \\ Q_1 &=& a_1, \\ P_k &=& P_{k-1} a_k + P_{k-2} \quad \text{for } k \geq 2, \\ Q_k &=& Q_{k-1} a_k + Q_{k-2} \quad \text{for } k \geq 2. \end{cases}$$

Note that these recurrences give irreducible fractions P_i/Q_i: the values P_i and Q_i that are deduced from them satisfy $\gcd(P_i, Q_i) = 1$.

The major interest in the continued fractions lies in the fact that P_i/Q_i is the best rational approximation to α among all rational numbers of denominator less than or equal to Q_i. More precisely, we have the following two results [237].

Theorem A.1. *Let $(P_j/Q_j)_{j \geq 0}$ be the convergents of α. If a rational number P/Q is a better approximation to α than P_k/Q_k (namely, if $|P/Q - \alpha| < |P_k/Q_k - \alpha|$), then $Q > Q_k$.*

Theorem A.2. *Let $(P_j/Q_j)_{j \geq 0}$ be the convergents of α. If Q_{k+1} exists, then for any $(P, Q) \in \mathbb{Z} \times \mathbb{N}^*$, with $Q < Q_{k+1}$, we have*

$$|P - \alpha Q| \geq |P_k - \alpha Q_k|.$$

If Q_{k+1} does not exist (which implies that α is rational), then the previous inequality holds for any $(P, Q) \in \mathbb{Z} \times \mathbb{N}^*$.

Interestingly enough, a kind of converse result exists: if a rational approximation to some number α is extremely good, then it must be a convergent of its continued fraction expansion.

Theorem A.3 ([237]). *Let P, Q be integers, $Q \neq 0$. If*

$$\left|\frac{P}{Q} - \alpha\right| < \frac{1}{2Q^2},$$

then P/Q is a convergent of α.

An example of continued fraction expansion of an irrational number is

$$e = \exp(1) = 2 + \cfrac{1}{1 + \cfrac{1}{2 + \cfrac{1}{1 + \cfrac{1}{1 + \cfrac{1}{4 + \cfrac{1}{\ddots}}}}}} = [2; 1, 2, 1, 1, 4, \ldots],$$

which gives the following rational approximations to e:

$$\frac{P_0}{Q_0} = 2, \quad \frac{P_1}{Q_1} = 3, \quad \frac{P_2}{Q_2} = \frac{8}{3}, \quad \frac{P_3}{Q_3} = \frac{11}{4}, \quad \frac{P_4}{Q_4} = \frac{19}{7}, \quad \frac{P_5}{Q_5} = \frac{87}{32}.$$

Other examples are

$$\pi = 3 + \cfrac{1}{7 + \cfrac{1}{15 + \cfrac{1}{1 + \cfrac{1}{292 + \cfrac{1}{1 + \cfrac{1}{\ddots}}}}}} = [3; 7, 15, 1, 292, 1, \ldots]$$

and

$$\sqrt{2} = 1 + \cfrac{1}{2 + \cfrac{1}{2 + \cfrac{1}{2 + \cfrac{1}{2 + \cfrac{1}{\ddots}}}}} = [1; 2, 2, 2, 2, \ldots] = [1; \overline{2}].$$

A.2 Euclidean Lattices

A Euclidean lattice is a set of points that are regularly spaced in the Euclidean space $\mathbb{R}^n$ (see Definition A.1). It is a discrete algebraic object that is encountered in several domains of various sciences, including mathematics, computer science, electrical engineering, and chemistry. It is a rich and powerful modeling tool, thanks to the deep and numerous theoretical results, algorithms, and implementations available (see [90, 112, 226, 397, 557, 599] for example).

Applications of Euclidean lattices to floating-point arithmetic include the calculation of polynomial approximations with coefficients that are floating-point numbers (or double-word numbers) [68, 97] and the search for hardest-to-round points of elementary functions (see Chapter 10 and [571]).

Let $x = (x_1, \ldots, x_n) \in \mathbb{R}^n$. We set

$$\|x\|_2 = (x|x)^{1/2} = \left(x_1^2 + \cdots + x_n^2\right)^{1/2} \text{ and } \|x\|_\infty = \max_{1 \le i \le n} |x_i|.$$

Definition A.1. *Let L be a nonempty subset of $\mathbb{R}^n$. The set L is a (Euclidean) lattice if there exists a set of $\mathbb{R}$-linearly independent vectors $b_1, \ldots, b_d$ such that*

$$L = \mathbb{Z} \cdot b_1 \oplus \cdots \oplus \mathbb{Z} \cdot b_d = \left\{ \sum_{i \le d} x_i \cdot b_i, x_i \in \mathbb{Z} \right\}.$$

The family $(b_1, \ldots, b_d)$ is a basis of the lattice L, and d is called the rank of the lattice L.

For example, the set $\mathbb{Z}^n$ and all of its additive subgroups are lattices. These lattices play a central role in computer science since they can be represented exactly. We say that a lattice L is integer (resp. rational) when $L \subseteq \mathbb{Z}^n$ (resp. $\mathbb{Q}^n$). An integer lattice of rank 2 is given in Figure A.1, as well as one of its bases.

A lattice is often given by one of its bases (in practice, a matrix whose rows or columns are the basis vectors). Unfortunately, as soon as the rank of the lattice is greater than 1, there are infinitely many such representations for any given lattice. In Figure A.2, we give another basis of the lattice of Figure A.1.

Proposition A.1. *If $(e_1, \ldots, e_k)$ and $(f_1, \ldots, f_j)$ are two families of $\mathbb{R}$-linearly independent (column) vectors that generate the same lattice, then $k = j$ (this is the rank of the lattice), and there exists a $(k \times k)$ matrix M with integer coefficients and determinant equal to ± 1 such that $(e_i) = (f_i) \cdot M$.*

Among the infinitely many bases of a specified lattice (if $k \geq 2$), some are more interesting than others. One can define various notions of what a

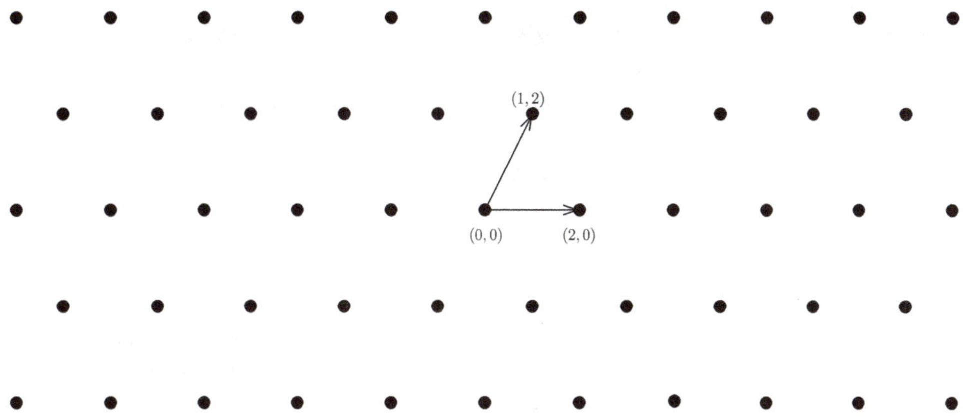

Figure A.1: The lattice $\mathbb{Z}(2,0) \oplus \mathbb{Z}(1,2)$.

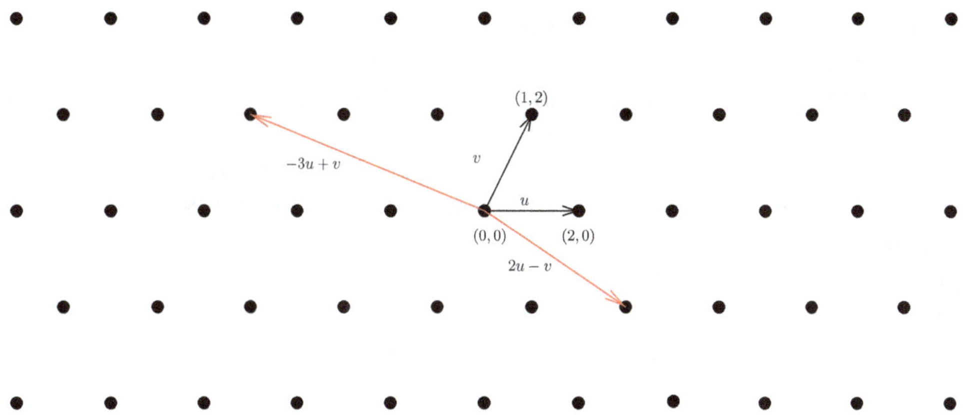

Figure A.2: Two bases of the lattice $\mathbb{Z}(2,0) \oplus \mathbb{Z}(1,2)$.

"good" basis is, but most of the time, it is required to consist of somewhat short vectors.

The two most famous computational problems related to lattices are the shortest and closest vector problems (SVP and CVP). Since a lattice is discrete, it contains a vector of smallest nonzero norm. That norm is denoted by λ and is called the *minimum* of the lattice. Note that the minimum is reached at least twice (a vector and its opposite), and may be reached more times. The discreteness also implies that, given an arbitrary vector of the space, there always exists a lattice vector closest to it (note that there can be several such vectors). We now state the search versions of SVP and CVP.

Problem A.1. Shortest vector problem (SVP). *Given a basis of a lattice $L \subseteq \mathbb{Q}^n$, find a shortest nonzero vector of L, i.e., a vector of norm $\lambda(L)$.*

SVP naturally leads to the following approximation problem, which we call γ-SVP, where γ is a function of the rank only: given a basis of a lattice $L \subseteq \mathbb{Q}^n$, find $b \in L$ such that

$$0 < \|b\| \leq \gamma \cdot \lambda(L).$$

Problem A.2. Closest vector problem (CVP). *Given a basis of a lattice $L \subseteq \mathbb{Q}^n$ and a target vector $t \in \mathbb{Q}^n$, find $b \in L$ such that $\|b - t\| = \operatorname{dist}(t, L)$.*

CVP naturally leads to the following approximation problem, which we call γ-CVP, where γ is a function of the rank only: given a basis of a lattice $L \subseteq \mathbb{Q}^n$ and a target vector $t \in \mathbb{Q}^n$, find $b \in L$ such that

$$\|b - t\| \leq \gamma \cdot \operatorname{dist}(t, L).$$

Note that SVP and CVP can be defined with respect to any norm of $\mathbb{R}^n$, and we will write SVP_2 and CVP_2 to explicitly refer to the Euclidean norm. These two computational problems have been studied extensively. We very briefly describe some of the results, and refer to [420] for more details.

Ajtai [4] showed in 1998 that the decisional version of SVP_2 (i.e., given a lattice L and a scalar x, compare $\lambda(L)$ and x) is NP-hard under randomized polynomial reductions. The NP-hardness had been conjectured in the early 1980s by van Emde Boas [612], who proved the result for the infinity norm instead of the Euclidean norm. Khot [335] showed that Ajtai's result still holds for the decisional version of the relaxed problem γ-SVP_2, where γ is an arbitrary constant. Goldreich and Goldwasser [216] proved that, under very plausible complexity theory assumptions, approximating SVP_2 within a factor $\gamma = \sqrt{d/\ln d}$ is not NP-hard, where d is the rank of the lattice. No polynomial-time algorithm is known for approximating SVP_2 within a factor $f(d)$ with f a polynomial in d. On the constructive side, Kannan [325] described an algorithm that solves SVP_2 in time

$$d^{\frac{d(1+o(1))}{2e}} \approx d^{0.184 \cdot d},$$

and in polynomial space (the complexity bound is proved in [233]). Ajtai, Kumar, and Sivakumar [5] gave an algorithm of complexity $2^{\mathcal{O}(d)}$ both in time and space that solves SVP with high probability. Becker et al. [34] described a heuristic variant of that algorithm that runs is time $\approx 2^{0.292 \cdot d}$. Similar results hold for the infinity norm instead of the Euclidean norm.

In 1981, van Emde Boas [612] proved that the decisional version of CVP_2 is NP-hard (see also [419]). On the other hand, Goldreich and Goldwasser [216] showed, under very plausible assumptions, that approximating CVP_2 within a factor $\sqrt{d/\ln d}$ is not NP-hard. Similar results also hold for the infinity norm (see [488]). Unfortunately, no polynomial-time algorithm is known for approximating CVP to a polynomial factor. On the constructive side, Kannan [325] described an algorithm that solves CVP in time

$$d^{\frac{d(1+o(1))}{2}}$$

for the Euclidean norm and

$$d^{d(1+o(1))}$$

for the infinity norm (see [233] for the proofs of the complexity bounds).

If we sufficiently relax the parameter γ, the situation becomes far better. In 1982, Lenstra, Lenstra, and Lovász [382] gave an algorithm that allows one to get relatively short vectors in polynomial time. Their algorithm is now commonly referred to by the acronym LLL.

Theorem A.4 (LLL [382]). *Given an arbitrary basis $(a_1, \ldots, a_d)$ of a lattice $L \subseteq \mathbb{Z}^n$, the LLL algorithm provides a basis $(b_1, \ldots, b_d)$ of L that is made of relatively short vectors. Among others, we have $\|b_1\| \leq 2^{(d-1)/2} \cdot \lambda(L)$. Furthermore, LLL terminates within $\mathcal{O}(d^5 n \ln^3 A)$ bit operations with $A = \max_i \|a_i\|$.*

More precisely, the LLL algorithm computes what is called a δ-LLL-reduced basis, where δ is a fixed parameter which belongs to $(1/4, 1)$ (if δ is omitted, then its value is $3/4$, which is the historical choice). To define what a LLL-reduced basis is, we need to recall the Gram–Schmidt orthogonalization of a basis. Consider a basis $(b_1, \ldots, b_d)$. Its Gram–Schmidt orthogonalization $(b_1^*, \ldots, b_d^*)$ is defined recursively as follows:

- the vector b_1^* is b_1;
- for $i > 1$, we set $b_i^* = b_i - \sum_{j<i} \mu_{i,j} b_j^*$, where $\mu_{i,j} = \frac{\langle b_i, b_j^* \rangle}{\|b_j^*\|^2}$.

Geometrically, the vector b_i^* is the projection of the vector b_i orthogonally to the span of the previous basis vectors $b_1, \ldots, b_{i-1}$. We say that the basis $(b_1, \ldots, b_d)$ is δ-LLL-reduced if the following two conditions are satisfied:

- for any $i > j$, the quantity $\mu_{i,j}$ has magnitude less than or equal to $1/2$. This condition is called the *size-reduction condition;*
- for any i, we have $\delta \|b_i^*\|^2 \leq \|b_{i+1}^*\|^2 + \mu_{i+1,i}^2 \|b_i^*\|^2$. This condition is called Lovász's condition. It means that orthogonally to the first $i-1$ vectors, the $(i+1)$-th vector cannot be arbitrarily small compared to the i-th vector.

LLL-reduced bases have many interesting properties. The most important one is probably that the first basis vector cannot be more than $2^{(d-1)/2}$ times longer than the lattice minimum. The LLL algorithm computes an LLL-reduced basis $(b_1, \ldots, b_d)$ of the lattice spanned by $(a_1, \ldots, a_d)$ by incrementally trying to make the LLL conditions satisfied. It uses an index k, which starts at 2 and eventually reaches $d+1$. At any moment, the first $k-1$ vectors satisfy the LLL conditions, and we are trying to make the first k vectors satisfy the conditions. To make the size-reduction condition satisfied for the k-th vector, one subtracts from it an adequate integer linear combination of the vectors $b_1, \ldots, b_{k-1}$. This is essentially the same process as Babai's nearest plane algorithm, described below. After that, one tests Lovász's condition. If it is satisfied, then the index k can be incremented. If not, the vectors b_k and b_{k-1} are swapped, and the index k is decremented. The correctness of the LLL algorithm is relatively simple to prove, but the complexity analysis is significantly more involved. We refer the interested reader to [382].

The LLL algorithm has been extensively studied since its invention [324, 544, 576, 463, 465, 457]. Very often in practice, the returned basis is of better quality than the worst-case bound given above and is obtained faster than expected. We refer to [463] for more details about the practical behavior of the LLL algorithm.

Among many important applications of the LLL algorithm, Babai [23] extracted from it a polynomial-time algorithm for solving CVP with an exponentially bounded approximation factor. We present it in Algorithm A.1.

Theorem A.5 (Babai [23]). *Given an arbitrary basis* $(b_1, \ldots, b_d)$ *of a lattice* $L \subseteq \mathbb{Z}^n$*, and a target vector* $t \in \mathbb{Z}^n$*, Babai's nearest plane algorithm (Algorithm A.1) finds a vector* $b \in L$ *such that*

$$\|b - t\|_2 \leq 2^d \cdot \mathrm{dist}_2(t, L).$$

Moreover, it finishes in polynomial time in d*,* n*,* $\ln A$*, and* $\ln \|t\|$*, where* $A = \max_i \|a_i\|$.

Algorithm A.1 Babai's nearest plane algorithm. The inputs are an LLL-reduced basis $(b_i)_{1 \leq i \leq d}$, its Gram–Schmidt orthogonalization $(b_i^*)_{1 \leq i \leq d}$, and a target vector t. The output is a vector in the lattice spanned by the b_i's that is close to t.

$v \leftarrow t$
for ($j = d\,;\, j \geq 1\,;\, j$--) **do**
  $v \leftarrow v - \left\lfloor \frac{\langle v, b_j^* \rangle}{\langle b_j^*, b_j^* \rangle} \right\rceil b_j$
end for
return $t - v$

Babai's algorithm may also be described with the LLL algorithm directly. This may be seen as a particular case of Kannan's embedding technique [326]. This may be simpler to implement, in particular, if one has access to an implementation of LLL. We give that variant in Algorithm A.2.

Algorithm A.2 Babai's nearest plane algorithm, using LLL. The inputs are an LLL-reduced basis $(b_i)_{1 \leq i \leq d}$ and a target vector t. The output is a vector in the lattice spanned by the b_i's that is close to t.

---

for ($j = 1\,;\, j \leq d\,;\, j$++) **do**
  $c_j \leftarrow (b_j, 0)$
end for
$B \leftarrow \max_i \|b_i\|$; $c_{d+1} \leftarrow (t, B)$
$(c'_1, \ldots, c'_{d+1}) \leftarrow \mathrm{LLL}(c_1, \ldots, c_{d+1})$
return $c_{d+1} - c'_{d+1}$

Appendix B

Previous Floating-Point Standards

B.1 The IEEE 754-1985 Standard

The binary part of the IEEE 754-2008 standard is an evolution of IEEE 754-1985. Hence a large part of what we have presented in Chapter 3 was already present in IEEE 754-1985. In the following, we therefore mainly focus on the differences.

B.1.1 Formats specified by IEEE 754-1985

Of course, since IEEE 754-1985 focused on binary floating-point arithmetic, there were no decimal formats. The two basic formats were

- single precision: this is the format now called binary32 in IEEE 754-2008 (same parameters and bit encoding);
- double precision: this is the format now called binary64 (same parameters and bit encoding).

There was no 16-bit or 128-bit format.

To each basic format, the IEEE 754-1985 standard associated an *extended format*. The standard recommended an extended format for the widest basic format supported only. Hence, to the best of our knowledge, the single-extended precision has never been implemented: when double precision was available, it fulfilled all the purposes of a single-extended format. Table B.1 gives the main parameters of the formats specified by IEEE 754-1985.

J.-M. Muller et al., *Handbook of Floating-Point Arithmetic*,
https://doi.org/10.1007/978-3-319-76526-6

Format	Hidden bit	Precision p	$e_{\min}$	$e_{\max}$
Single precision	yes	24	-126	127
Double precision	yes	53	-1022	1023
Single-extended	optional	≥ 32	≤ -1022	≥ 1023
Double-extended	optional	≥ 64	≤ -16382	≥ 16383
Double-ext. (IA32)	no	64	-16382	16383

Table B.1: Main parameters of the four formats specified by the IEEE 754-1985 standard [12] (©IEEE, 1985, with permission). A specific single-extended format has not been implemented in practice. The last line describes the double-extended format introduced by Intel in the 387 FPU, and available in subsequent IA32 compatible processors by Intel, Cyrix, AMD, and others.

B.1.2 Rounding modes specified by IEEE 754-1985

The IEEE 754-1985 standard defined four rounding modes: round toward $-\infty$ (rounding function RD), round toward $+\infty$ (rounding function RU), round toward zero (rounding function RZ), and round to nearest *ties to even* (RN), also called *round to nearest even*. There was no round to nearest *ties to away* (it was brought up by IEEE 754-2008).

B.1.3 Operations specified by IEEE 754-1985

B.1.3.1 Arithmetic operations and square root

The IEEE 754-1985 standard required that addition, subtraction, multiplication, and division of operands of the same format be provided, for all supported formats, with correct rounding (with the four rounding modes presented above). The standard also required a correctly rounded square root in all supported formats. The FMA (fused multiply-add) operator was not specified.

It was also permitted (but not required) that these operations be provided (still with correct rounding) for operands of different formats (in such a case, the destination format had to be at least as wide as the wider operand's format).

B.1.3.2 Conversions to and from decimal strings

At the time the IEEE 754-1985 standard was released, some of the algorithms presented in Section 4.9 were not known. This explains why the requirements of the standard were clearly below what one could now expect. The 2008 version of the standard has much stronger requirements.

	Decimal to binary		Binary to decimal	
	$M_{10,\max}^{(1)}$	$E_{10,\max}^{(1)}$	$M_{10,\max}^{(2)}$	$E_{10,\max}^{(2)}$
Single precision	$10^9 - 1$	99	$10^9 - 1$	53
Double precision	$10^{17} - 1$	999	$10^{17} - 1$	340

Table B.2: The thresholds for conversion from and to a decimal string, as specified by the IEEE 754-1985 standard [12] (©IEEE, 1985, with permission).

The requirements of the IEEE 754-1985 standard were:

- conversions must be provided between decimal strings in at least one format and binary floating-point numbers in all basic floating-point formats, for numbers of the form

$$\pm M_{10} \times 10^{\pm E_{10}}$$

  with $M_{10} \geq 0$ and $E_{10} \geq 0$;

- conversions must be correctly rounded for operands in the ranges specified in Table B.3;

- when the operands are not in the ranges specified in Table B.3:
  - in round-to-nearest mode, the conversion error cannot exceed 0.97 ulp of the target format;
  - in the directed rounding modes, the "direction" of the rounding must be honored (e.g., for round-toward $-\infty$ mode, the delivered result must be less than or equal to the initial value), and the rounding error cannot exceed 1.47 ulp of the target format;

- conversions must be *monotonic* (if $x \leq y$ before conversion, then $x \leq y$ after conversion);

- when rounding to nearest, as long as the decimal strings have at least 9 digits for single precision and 17 digits for double precision, conversion from binary to decimal and back to binary must produce the initial value exactly (Table B.2). This allows one to store intermediate results in files, and to read them later on, without losing any information, as explained in Chapter 2, Section 4.9.

B.1.4 Exceptions specified by IEEE 754-1985

The five exceptions listed in Section 2.5 (invalid, division by zero, overflow, underflow, inexact) had to be signaled when detected, either by taking a *trap* or by setting a *status flag*. The default was not to use traps.

	Decimal to binary		Binary to decimal	
	$M^{(1)}_{10,\max}$	$E^{(1)}_{10,\text{corr}}$	$M^{(2)}_{10,\max}$	$E^{(2)}_{10,\text{corr}}$
Single precision	$10^9 - 1$	13	$10^9 - 1$	13
Double precision	$10^{17} - 1$	27	$10^{17} - 1$	27

Table B.3: Correctly rounded decimal conversion range, as specified by the IEEE 754-1985 standard [12] (©IEEE, 1985, with permission).

B.2 The IEEE 854-1987 Standard

The IEEE 854-1987 standard [13] covered "radix-independent" floating-point arithmetic. This does not mean that all possible radices were considered: actually, that standard only focused on radices 2 and 10. We will just present it briefly (it is now superseded by IEEE 754-2008 [267]).

Unlike IEEE 754-1985, the IEEE 854-1987 standard did not fully specify formats or internal encodings. It merely expressed constraints between the parameters β, $e_{\min}$, $e_{\max}$, and p of the various precisions provided by an implementation. It also said that for each available precision, we must have two infinities, at least one signaling NaN, and at least one quiet NaN (as in the IEEE 754-1985 standard). In the remainder of this section, β is equal to 2 or 10. The same radix must be used for all available precisions: an arithmetic system compliant to IEEE 854-1987 is either binary or decimal, but it cannot mix up the two kinds of representations.

B.2.1 Constraints internal to a format

A balance must be found between the precision p and the value of the extremal exponents $e_{\min}$ and $e_{\max}$. If p is too large compared to $|e_{\min}|$ and $e_{\max}$, then underflows or overflows may occur too often. Also, there must be some balance between $e_{\min}$ and $e_{\max}$: to avoid underflows or overflows when computing reciprocals of normalized floating-point numbers as much as possible, one might want $e_{\min} \approx -e_{\max}$. Since underflow (more precisely, *gradual underflow*, with subnormal numbers available) is less harmful than overflow, it is preferable to have $e_{\min}$ very slightly above[1] $-e_{\max}$. Here are the constraints specified by the IEEE 854-1987 standard.

- We *must* have
$$\frac{e_{\max} - e_{\min}}{p} > 5,$$
and it is *recommended* that
$$\frac{e_{\max} - e_{\min}}{p} > 10.$$

[1] This is why the IEEE 754-2008 standard now requires $e_{\min} = 1 - e_{\max}$ for all formats.

- We must have $\beta^{p-1} \geq 10^5$.
- $\beta^{e_{\max}+e_{\min}+1}$ should be the smallest power of β greater than or equal to 4 (which was a very complicated way of saying that $e_{\min}$ should be $1 - e_{\max}$ in radix 2 and $-e_{\max}$ in radix 10).

For instance, the binary32 format of IEEE 754-2008 satisfies these requirements: with $\beta = 2$, $p = 24$, $e_{\min} = -126$, and $e_{\max} = 127$, we have

$$\begin{cases} \dfrac{e_{\max} - e_{\min}}{p} & = & 10.54\ldots & > & 10; \\ \beta^{p-1} & = & 2^{23} & \geq & 10^5; \\ \beta^{e_{\max}+e_{\min}+1} & = & 2^2 & = & 4. \end{cases}$$

B.2.2 Various formats and the constraints between them

The narrowest supported format was called *single precision*. When a second, wider basic format is supported, it was called *double precision*. The required constraints between their respective parameters $e_{\min s}$, $e_{\max s}$, p_s and $e_{\min d}$, $e_{\max d}$, p_d were:

- $\beta^{p_d} \geq 10\beta^{2p_s}$;
- $e_{\max d} \geq 8e_{\max s} + 7$;
- $e_{\min d} \leq 8e_{\min s}$.

Extended precisions were also possible. For obvious reasons, the only extended precision that was recommended was the one associated with the widest supported basic precision. If $e_{\min}$, $e_{\max}$, and p are the extremal exponents and precision of that widest basic precision, the parameters $e_{\min e}$, $e_{\max e}$, and p_e of the corresponding extended precision had to satisfy:

- $e_{\max e} \geq 8e_{\max} + 7$;
- $e_{\min e} \leq 8e_{\min}$;
- if $\beta = 2$,
$$p_e \geq p + \lceil \log_2 (e_{\max} - e_{\min}) \rceil ; \tag{B.1}$$
- for all β, $p_e \geq 1.2p$.

It was also recommended that

$$p_e > 1 + p + \frac{\log(3\log(\beta)(e_{\max}+1))}{\log(\beta)}. \tag{B.2}$$

The purpose of constraint (B.1) was to facilitate the support of conversion to and from decimal strings for the basic formats, using algorithms that were available at that time. The purpose of (B.2) was to make accurate implementation, in the basic formats, of the power function x^y simpler.

B.2.3 Rounding

The IEEE 854-1987 standard required that the arithmetic operations and the square root be correctly rounded. Exactly as for IEEE 754-1985, four rounding modes were specified: rounding toward $-\infty$, toward $+\infty$, toward 0, and to nearest ties to even.

B.2.4 Operations

The arithmetic operations, the remainder operation, and the square root (including the $\sqrt{-0} = -0$ requirement) were defined very much as in IEEE 754-1985.

B.2.5 Comparisons

The comparisons were defined very much as in IEEE 754-1985. Especially, every NaN compares "unordered" with everything including itself: the test "$x \neq x$" must return **true** if and only if x is a NaN.

B.2.6 Exceptions

The IEEE 754-1985 way of handling exceptions was also chosen for IEEE 854-1987.

B.3 The Need for a Revision

The IEEE 754-1985 standard was a huge improvement. It soon became implemented on most platforms of commercial significance. And yet, 15 years after its release, there was a clear need for a revision.

- Some features that had become common practice needed to be standardized: e.g., the "quadruple-precision" (i.e., 128-bit wide, binary) format, the fused multiply-add operator.

- Since 1985, new algorithms were published that allowed one to easily perform computations that were previously thought too complex. Typical examples are the radix conversion algorithms presented in Section 4.9: now, for an internal binary format, it is possible to have much stronger requirements on the accuracy of the conversions that must be done when reading or printing decimal strings. Another example is the availability of reasonably fast libraries for some correctly rounded elementary functions: the revised standard can now deal with transcendental functions and recommend that some should be correctly rounded.

- There were some ambiguities in IEEE 754-1985. For instance, when evaluating expressions, when a larger internal format is available in hardware, it was unclear in which format the implicit intermediate variables should be represented.

B.4 The IEEE 754-2008 Revision

The IEEE 754-1985 standard has been revised from 2000 to 2006, and the revised standard was adopted in June 2008. Some of the various goals of the working group were as follows (see http://grouper.ieee.org/groups/754/revision.html):

- merging the 854-1987 standard into the 754-1985 standard;
- reducing the implementation choices;
- resolving some ambiguities in the 754-1985 standard (especially concerning expression evaluation and exception handling). The revised standard allows languages and users to focus on portability and reproducibility, or on performance;
- standardizing the fused multiply-add (FMA) operation, and
- including quadruple precision.

Also, the working group had to cope with a very strong constraint: the revised standard would rather not invalidate hardware that conformed to the old IEEE 754-1985 standard.

Bibliography

[1] E. Abu-Shama and M. Bayoumi. A new cell for low power adders. In *International Symposium on Circuits and Systems (ISCAS)*, pages 1014–1017, 1996.

[2] Advanced Micro Devices. 128-bit SSE5 instruction set, 2007. Available at http://pdinda.org/icsclass/doc/AMD_ARCH_MANUALS/AMD64_128_Bit_SSE5_Instrs.pdf.

[3] R. C. Agarwal, F. G. Gustavson, and M. S. Schmookler. Series approximation methods for divide and square root in the Power3™ processor. In *14th IEEE Symposium on Computer Arithmetic (ARITH-14)*, pages 116–123, April 1999.

[4] M. Ajtai. The shortest vector problem in L_2 is NP-hard for randomized reductions. Extended abstract. In *30th Annual ACM Symposium on Theory of Computing (STOC)*, pages 10–19, 1998.

[5] M. Ajtai, R. Kumar, and D. Sivakumar. A sieve algorithm for the shortest lattice vector problem. In *33rd Annual ACM Symposium on Theory of Computing (STOC)*, pages 601–610, 2001.

[6] B. Akbarpour, A. T. Abdel-Hamid, S. Tahar, and J. Harrison. Verifying a synthesized implementation of IEEE-754 floating-point exponential function using HOL. *The Computer Journal*, 53(4):465, 2010.

[7] L. Aksoy, E. Costa, P. Flores, and J. Monteiro. Optimization of area in digital FIR filters using gate-level metrics. In *Design Automation Conference*, pages 420–423, 2007.

[8] J. Alexandre dit Sandretto and A. Chapoutot. Validated Explicit and Implicit Runge-Kutta Methods. *Reliable Computing*, 22:78–103, 2016.

J.-M. Muller et al., *Handbook of Floating-Point Arithmetic*,
https://doi.org/10.1007/978-3-319-76526-6

[9] E. Allen, D. Chase, V. Luchangco, J.-W. Maessen, and G. L. Steele, Jr. Object-oriented units of measurement. In *19th Annual ACM SIGPLAN Conference on Object-Oriented Programming, Systems, Languages, and Applications (OOPSLA)*, pages 384–403, 2004.

[10] Altera Corporation. *FFT/IFFT Block Floating Point Scaling*, 2005. Application note 404-1.0.

[11] B. Amedro, V. Bodnartchouck, D. Caromel, C. Delbé, F. Huet, and G. L. Taboada. Current state of Java for HP. Preprint, Technical Report 0353, INRIA, 2008. Available at http://hal.inria.fr/inria-00312039/en.

[12] American National Standards Institute and Institute of Electrical and Electronic Engineers. *IEEE Standard for Binary Floating-Point Arithmetic*. ANSI/IEEE Standard 754–1985, 1985.

[13] American National Standards Institute and Institute of Electrical and Electronic Engineers. *IEEE Standard for Radix Independent Floating-Point Arithmetic*. ANSI/IEEE Standard 854–1987, 1987.

[14] C. Anderson, N. Astafiev, and S. Story. Accurate math functions on the Intel IA-32 architecture: A performance-driven design. In *Real Numbers and Computers*, pages 93–105, July 2006.

[15] S. F. Anderson, J. G. Earle, R. E. Goldschmidt, and D. M. Powers. The IBM 360/370 model 91: floating-point execution unit. *IBM Journal of Research and Development*, 1967. Reprinted in [583].

[16] M. Andrysco, R. Jhala, and S. Lerner. Printing floating-point numbers: A faster, always correct method. In *43rd Annual ACM SIGPLAN-SIGACT Symposium on Principles of Programming Languages (POPL)*, pages 555–567, 2016.

[17] ARM. *ARM Developer Suite: Compilers and Libraries Guide*. ARM Limited, November 2001. Available at http://infocenter.arm.com/help/index.jsp?topic=/com.arm.doc.dui0067d/index.html or in PDF at http://infocenter.arm.com/help/topic/com.arm.doc.dui0067d/DUI0067.pdf.

[18] ARM. *ARM Developer Suite: Developer Guide*. ARM Limited, 1.2 edition, November 2001. Available at http://infocenter.arm.com/help/index.jsp?topic=/com.arm.doc.dui0056d/index.html or in PDF at http://infocenter.arm.com/help/topic/com.arm.doc.dui0056d/DUI0056.pdf.

[19] ARM. *ARM Compiler: armasm User Guide*. ARM Limited, 6.7 edition, 2017. Available at http://infocenter.arm.com/help/topic/com.arm.doc.100069_0607_00_en/ or in PDF at http://infocenter.arm.com/help/topic/com.arm.doc.100069_0607_00_en/armasm_user_guide_100069_0607_00_en.pdf.

[20] W. Aspray, A. G. Bromley, M. Campbell-Kelly, P. E. Ceruzzi, and M. R. Williams. *Computing Before Computers*. Iowa State University Press, Ames, Iowa, 1990. Available at http://ed-thelen.org/comp-hist/CBC.html.

[21] A. Avizienis. Signed-digit number representations for fast parallel arithmetic. *IRE Transactions on Electronic Computers*, 10:389–400, 1961. Reprinted in [584].

[22] A. Azmi and F. Lombardi. On a tapered floating point system. In *9th IEEE Symposium on Computer Arithmetic (ARITH-9)*, pages 2–9, September 1989.

[23] L. Babai. On Lovász' lattice reduction and the nearest lattice point problem. *Combinatorica*, 6(1):1–13, 1986.

[24] I. Babuška. Numerical stability in mathematical analysis. In *Proceedings of the 1968 IFIP Congress*, volume 1, pages 11–23, 1969.

[25] D. H. Bailey. Some background on Kanada's recent pi calculation. Technical report, Lawrence Berkeley National Laboratory, 2003. Available at http://crd.lbl.gov/~dhbailey/dhbpapers/dhb-kanada.pdf.

[26] D. H. Bailey, R. Barrio, and J. M. Borwein. High precision computation: Mathematical physics and dynamics. *Applied Mathematics and Computation*, 218:10106–10121, 2012.

[27] D. H. Bailey and J. M. Borwein. Experimental mathematics: examples, methods and implications. *Notices of the AMS*, 52(5):502–514, 2005.

[28] D. H. Bailey, J. M. Borwein, P. B. Borwein, and S. Plouffe. The quest for pi. *Mathematical Intelligencer*, 19(1):50–57, 1997.

[29] D. H. Bailey, Y. Hida, X. S. Li, and B. Thompson. ARPREC: an arbitrary precision computation package. Technical report, Lawrence Berkeley National Laboratory, 2002. Available at http://crd.lbl.gov/~dhbailey/dhbpapers/arprec.pdf.

[30] S. Banescu, F. de Dinechin, B. Pasca, and R. Tudoran. Multipliers for floating-point double precision and beyond on FPGAs. *ACM SIGARCH Computer Architecture News*, 38:73–79, 2010.

[31] G. Barrett. Formal methods applied to a floating-point system. *IEEE Transactions on Software Engineering*, 15(5):611–621, 1989.

[32] M. Baudin. Error bounds of complex arithmetic. Technical report, 2011. Available at http://forge.scilab.org/upload/compdiv/files/complexerrorbounds_v0.2.pdf.

[33] P.-D. Beck and M. Nehmeier. Parallel interval Newton method on CUDA. In *International Workshop on Applied Parallel Computing*, pages 454–464, 2012.

[34] A. Becker, L. Ducas, N. Gama, and T. Laarhoven. New directions in nearest neighbor searching with applications to lattice sieving. In *27th Annual ACM-SIAM Symposium on Discrete Algorithms (SODA)*, pages 10–24, 2016.

[35] F. Benford. The law of anomalous numbers. *Proceedings of the American Philosophical Society*, 78(4):551–572, 1938.

[36] M. Bennani and M. C. Brunet. PRECISE: simulation of round-off error propagation model. In *12th World IMACS Congress*, July 1988.

[37] C. Berg. *Formal Verification of an IEEE Floating-Point Adder*. Master's thesis, Universität des Saarlandes, Germany, 2001.

[38] D. J. Bernstein. Multidigit multiplication for mathematicians. Available at https://cr.yp.to/papers/m3.pdf, 2001.

[39] F. Blomquist, W. Hofschuster, and W. Krämer. Real and complex staggered (interval) arithmetics with wide exponent range. Technical Report 2008/1, Universität Wuppertal, Germany, 2008. In German.

[40] A. Blot, J.-M. Muller, and L. Théry. Formal correctness of comparison algorithms between binary64 and decimal64 floating-point numbers. In *10th International Workshop on Numerical Software Verification (NSV)*, pages 25–37, 2017.

[41] G. Bohlender, W. Walter, P. Kornerup, and D. W. Matula. Semantics for exact floating point operations. In *10th IEEE Symposium on Computer Arithmetic (ARITH-10)*, pages 22–26, June 1991.

[42] S. Boldo. Pitfalls of a full floating-point proof: example on the formal proof of the Veltkamp/Dekker algorithms. In *3rd International Joint Conference on Automated Reasoning (IJCAR)*, volume 4130 of *Lecture Notes in Computer Science*, pages 52–66, Seattle, WA, USA, 2006.

[43] S. Boldo. Kahan's algorithm for a correct discriminant computation at last formally proven. *IEEE Transactions on Computers*, 58(2):220–225, 2009.

[44] S. Boldo. How to compute the area of a triangle: a formal revisit. In *21th IEEE Symposium on Computer Arithmetic (ARITH-21)*, pages 91–98, Austin, TX, USA, April 2013.

[45] S. Boldo. Formal verification of programs computing the floating-point average. In *17th International Conference on Formal Engineering Methods (ICFEM)*, volume 9407 of *Lecture Notes in Computer Science*, pages 17–32, Paris, France, November 2015.

[46] S. Boldo and M. Daumas. Representable correcting terms for possibly underflowing floating point operations. In *16th IEEE Symposium on Computer Arithmetic (ARITH-16)*, pages 79–86, Santiago de Compostela, Spain, 2003.

[47] S. Boldo, M. Daumas, and R.-C. Li. Formally verified argument reduction with a fused multiply-add. *IEEE Transactions on Computers*, 58(8):1139–1145, 2009.

[48] S. Boldo, M. Daumas, C. Moreau-Finot, and L. Théry. Computer validated proofs of a toolset for adaptable arithmetic. Technical report, École Normale Supérieure de Lyon, 2001. Available at http://arxiv.org/pdf/cs.MS/0107025.

[49] S. Boldo, S. Graillat, and J.-M. Muller. On the robustness of the 2Sum and Fast2Sum algorithms. *ACM Transactions on Mathematical Software*, 44(1):4:1–4:14, 2017.

[50] S. Boldo, M. Joldeş, J.-M. Muller, and V. Popescu. Formal verification of a floating-point expansion renormalization algorithm. In *8th International Conference on Interactive Theorem Proving (ITP)*, Brasilia, Brazil, 2017.

[51] S. Boldo, J.-H. Jourdan, X. Leroy, and G. Melquiond. Verified compilation of floating-point computations. *Journal of Automated Reasoning*, 54(2):135–163, 2015.

[52] S. Boldo and G. Melquiond. Emulation of FMA and correctly rounded sums: proved algorithms using rounding to odd. *IEEE Transactions on Computers*, 57(4):462–471, 2008.

[53] S. Boldo and G. Melquiond. Flocq: A unified library for proving floating-point algorithms in Coq. In *20th IEEE Symposium on Computer Arithmetic (ARITH-20)*, pages 243–252, Tübingen, Germany, 2011.

[54] S. Boldo and G. Melquiond. *Computer Arithmetic and Formal Proofs*. ISTE Press – Elsevier, 2017.

[55] S. Boldo and J.-M. Muller. Exact and approximated error of the FMA. *IEEE Transactions on Computers*, 60(2):157–164, 2011.

[56] S. Boldo and C. Muñoz. Provably faithful evaluation of polynomials. In *ACM Symposium on Applied Computing*, pages 1328–1332, Dijon, France, 2006.

[57] A. D. Booth. A signed binary multiplication technique. *Quarterly Journal of Mechanics and Applied Mathematics*, 4(2):236–240, 1951. Reprinted in [583].

[58] J. Borwein and D. H. Bailey. *Mathematics by Experiment: Plausible Reasoning in the* 21*st Century*. A. K. Peters, Natick, MA, 2004.

[59] P. Borwein and T. Erdélyi. *Polynomials and Polynomial Inequalities*. Graduate Texts in Mathematics, 161. Springer-Verlag, New York, 1995.

[60] W. Bosma, J. Cannon, and C. Playoust. The Magma algebra system. I. The user language. *Journal of Symbolic Computation*, 24(3–4):235–265, 1997.

[61] N. Boullis and A. Tisserand. Some optimizations of hardware multiplication by constant matrices. *IEEE Transactions on Computers*, 54(10):1271–1282, 2005.

[62] R. T. Boute. The Euclidean definition of the functions div and mod. *ACM Transactions on Programming Languages and Systems*, 14(2):127–144, 1992.

[63] R. P. Brent. On the precision attainable with various floating-point number systems. *IEEE Transactions on Computers*, C-22(6):601–607, 1973.

[64] R. P. Brent. A FORTRAN multiple-precision arithmetic package. *ACM Transactions on Mathematical Software*, 4(1):57–70, 1978.

[65] R. P. Brent, C. Percival, and P. Zimmermann. Error bounds on complex floating-point multiplication. *Mathematics of Computation*, 76:1469–1481, 2007.

[66] R. P. Brent and P. Zimmermann. *Modern Computer Arithmetic*. Cambridge University Press, 2011.

[67] K. Briggs. The `doubledouble` library, 1998. Available at http://www.boutell.com/fracster-src/doubledouble/doubledouble.html.

[68] N. Brisebarre and S. Chevillard. Efficient polynomial L^∞ approximations. In *18th IEEE Symposium on Computer Arithmetic (ARITH-18)*, pages 169–176, Montpellier, France, 2007.

[69] N. Brisebarre, F. de Dinechin, and J.-M. Muller. Integer and floating-point constant multipliers for FPGAs. In *Application-specific Systems, Architectures and Processors*, pages 239–244, 2008.

[70] N. Brisebarre and G. Hanrot. Floating-point L^2-approximations to functions. In *18th IEEE Symposium on Computer Arithmetic (ARITH-18)*, pages 177–186, Montpellier, France, 2007.

[71] N. Brisebarre, G. Hanrot, and O. Robert. Exponential sums and correctly-rounded functions. *IEEE Transactions on Computers*, 66(12):2044–2057, 2017.

[72] N. Brisebarre, M. Joldeş, É. Martin-Dorel, M. Mayero, J.-M. Muller, I. Paşca, L. Rideau, and L. Théry. Rigorous polynomial approximation using Taylor models in Coq. In *4th International Symposium on NASA Formal Methods (NFM)*, volume 7226 of *Lecture Notes in Computer Science*, pages 85–99, Norfolk, VA, USA, 2012.

[73] N. Brisebarre, C. Lauter, M. Mezzarobba, and J.-M. Muller. Comparison between binary and decimal floating-point numbers. *IEEE Transactions on Computers*, 65(7):2032–2044, 2016.

[74] N. Brisebarre and J.-M. Muller. Correctly rounded multiplication by arbitrary precision constants. *IEEE Transactions on Computers*, 57(2):165–174, 2008.

[75] N. Brisebarre, J.-M. Muller, and S.-K. Raina. Accelerating correctly rounded floating-point division when the divisor is known in advance. *IEEE Transactions on Computers*, 53(8):1069–1072, 2004.

[76] W. S. Brown. A simple but realistic model of floating-point computation. *ACM Transactions on Mathematical Software*, 7(4), 1981.

[77] W. S. Brown and P. L. Richman. The choice of base. *Communications of the ACM*, 12(10):560–561, 1969.

[78] C. Bruel. If-conversion SSA framework for partially predicated VLIW architectures. In *4th Workshop on Optimizations for DSP and Embedded Systems (ODES)*, New York, NY, USA, March 2006.

[79] J. D. Bruguera and T. Lang. Leading-one prediction with concurrent position correction. *IEEE Transactions on Computers*, 48(10):1083–1097, 1999.

[80] J. D. Bruguera and T. Lang. Floating-point fused multiply-add: Reduced latency for floating-point addition. In *17th IEEE Symposium on Computer Arithmetic (ARITH-17)*, Cape Cod, MA, USA, June 2005.

[81] M. C. Brunet and F. Chatelin. A probabilistic round-off error propagation model, application to the eigenvalue problem. In *Reliable Numerical Software*, 1987. Available at http://www.boutell.com/fracster-src/doubledouble/doubledouble.html.

[82] N. Brunie. Modified FMA for exact low precision product accumulation. In *24th IEEE Symposium on Computer Arithmetic (ARITH-24)*, pages 106–113, July 2017.

[83] N. Brunie, F. de Dinechin, and B. Dupont de Dinechin. A mixed-precision fused multiply and add. In *45th Asilomar Conference on Signals, Systems, and Computers*, November 2011.

[84] N. Brunie, F. de Dinechin, and B. Dupont de Dinechin. Mixed-precision merged multiplication and addition operator. Patent WO/2012/175828, December 2012.

[85] H. T. Bui, Y. Wang, and Y. Jiang. Design and analysis of low-power 10-transistor full adders using novel XORXNOR gates. *IEEE Transactions on Circuits and Systems II: Analog and Digital Signal Processing*, 49(1), 2003.

[86] R. G. Burger and R. K. Dybvig. Printing floating-point numbers quickly and accurately. In *SIGPLAN'96 Conference on Programming Languages Design and Implementation (PLDI)*, pages 108–116, June 1996.

[87] F. Y. Busaba, C. A. Krygowski, W. H. Li, E. M. Schwarz, and S. R. Carlough. The IBM z900 decimal arithmetic unit. In *35th Asilomar Conference on Signals, Systems, and Computers*, volume 2, pages 1335–1339, November 2001.

[88] P. R. Cappello and K. Steiglitz. A VLSI layout for a pipelined Dadda multiplier. *ACM Transactions on Computer Systems*, 1(2):157–174, 1983. Reprinted in [584].

[89] S. Carlough, A. Collura, S. Mueller, and M. Kroener. The IBM zEnterprise-196 decimal floating-point accelerator. In *20th IEEE Symposium on Computer Arithmetic (ARITH-20)*, pages 139–146, July 2011.

[90] J. W. S. Cassels. *An Introduction to the Geometry of Numbers*. Classics in Mathematics. Springer-Verlag, Berlin, 1997. Corrected reprint of the 1971 edition.

[91] A. Cauchy. Sur les moyens d'éviter les erreurs dans les calculs numériques. *Comptes Rendus de l'Académie des Sciences, Paris*, 11:789–798, 1840. Republished in: Augustin Cauchy, œuvres complètes, 1ère série, Tome V, pages 431–442.

[92] P. E. Ceruzzi. The early computers of Konrad Zuse, 1935 to 1945. *Annals of the History of Computing*, 3(3):241–262, 1981.

[93] P. E. Ceruzzi. *A History of Modern Computing*. MIT Press, 2nd edition, 2003.

[94] K. D. Chapman. Fast integer multipliers fit in FPGAs (EDN 1993 design idea winner). *EDN Magazine*, 1994.

[95] T. C. Chen and I. T. Ho. Storage-efficient representation of decimal data. *Communications of the ACM*, 18(1):49–52, 1975.

[96] S. Chevillard, J. Harrison, M. Joldeş, and C. Lauter. Efficient and accurate computation of upper bounds of approximation errors. *Theoretical Computer Science*, 412(16):1523–1543, 2011.

[97] S. Chevillard, M. Joldeş, and C. Q. Lauter. Sollya: An environment for the development of numerical codes. In *3rd International Congress on Mathematical Software (ICMS)*, volume 6327 of *Lecture Notes in Computer Science*, pages 28–31, September 2010.

[98] S. Chevillard and C. Q. Lauter. A certified infinite norm for the implementation of elementary functions. In *7th International Conference on Quality Software (QSIC)*, pages 153–160, Portland, OR, USA, 2007.

[99] C. W. Chou, D. B. Hume, T. Rosenband, and D. J. Wineland. Optical clocks and relativity. *Science*, 329(5999):1630–1633, 2010.

[100] C. W. Clenshaw and F. W. J. Olver. Beyond floating point. *Journal of the ACM*, 31:319–328, 1985.

[101] W. D. Clinger. How to read floating-point numbers accurately. *ACM SIGPLAN Notices*, 25(6):92–101, 1990.

[102] W. D. Clinger. Retrospective: how to read floating-point numbers accurately. *ACM SIGPLAN Notices*, 39(4):360–371, 2004.

[103] D. Cochran. Algorithms and accuracy in the HP 35. *Hewlett Packard Journal*, 23:10–11, 1972.

[104] W. J. Cody. Static and dynamic numerical characteristics of floating-point arithmetic. *IEEE Transactions on Computers*, C-22(6):598–601, 1973.

[105] W. J. Cody. Implementation and testing of function software. In *Problems and Methodologies in Mathematical Software Production*, volume 142 of *Lecture Notes in Computer Science*, 1982.

[106] W. J. Cody. MACHAR: a subroutine to dynamically determine machine parameters. *ACM Transactions on Mathematical Software*, 14(4):301–311, 1988.

[107] W. J. Cody and W. Waite. *Software Manual for the Elementary Functions*. Prentice-Hall, Englewood Cliffs, NJ, 1980.

[108] T. Coe and P. T. P. Tang. It takes six ones to reach a flaw. In *12th IEEE Symposium on Computer Arithmetic (ARITH-12)*, pages 140–146, July 1995.

[109] S. Collange, M. Daumas, and D. Defour. État de l'intégration de la virgule flottante dans les processeurs graphiques. *Technique et Science Informatiques*, 27(6):719–733, 2008. In French.

[110] S. Collange, M. Daumas, and D. Defour. *GPU Computing Gems Jade Edition*, chapter Interval arithmetic in CUDA, pages 99–107. Morgan Kaufmann, 2011.

[111] J. L. D. Comba and J. Stolfi. Affine arithmetic and its applications to computer graphics. In *VI Simpósio Brasileiro de Computação Gráfica e Processamento de Imagens (SIBGRAPI'93)*, pages 9–18, 1993.

[112] J. H. Conway and N. J. A. Sloane. *Sphere Packings, Lattices and Groups*. Springer-Verlag, New York, 1988.

[113] D. Coppersmith. Finding a small root of a univariate modular equation. In *Advances in Cryptology – EUROCRYPT*, volume 1070 of *Lecture Notes in Computer Science*, pages 155–165, Saragossa, Spain, May 1996.

[114] D. Coppersmith. Finding small solutions to small degree polynomials. In *International Conference on Cryptography and Lattices (CaLC)*, volume 2146 of *Lecture Notes in Computer Science*, pages 20–31, Providence, RI, USA, March 2001.

[115] R. M. Corless and N. Fillion. *A Graduate Introduction to Numerical Methods, From the Viewpoint of Backward Error Analysis*. Springer, 2013.

[116] M. Cornea, C. Anderson, J. Harrison, P. T. P. Tang, E. Schneider, and C. Tsen. A software implementation of the IEEE 754R decimal floating-point arithmetic using the binary encoding format. In *18th IEEE Symposium on Computer Arithmetic (ARITH-18)*, pages 29–37, June 2007.

[117] M. Cornea, J. Harrison, C. Anderson, P. T. P. Tang, E. Schneider, and E. Gvozdev. A software implementation of the IEEE 754R decimal floating-point arithmetic using the binary encoding format. *IEEE Transactions on Computers*, 58(2):148–162, 2009.

[118] M. Cornea, J. Harrison, and P. T. P. Tang. *Scientific Computing on Itanium®-based Systems*. Intel Press, Hillsboro, OR, 2002.

[119] M. Cornea, C. Iordache, J. Harrison, and P. Markstein. Integer divide and remainder operations in the IA-64 architecture. In *4th Conference on Real Numbers and Computers (RNC-4)*, 2000.

[120] M. A. Cornea-Hasegan, R. A. Golliver, and P. Markstein. Correctness proofs outline for Newton–Raphson based floating-point divide and square root algorithms. In *14th IEEE Symposium on Computer Arithmetic (ARITH-14)*, pages 96–105, April 1999.

[121] M. F. Cowlishaw. Decimal floating-point: algorism for computers. In *16th IEEE Symposium on Computer Arithmetic (ARITH-16)*, pages 104–111, June 2003.

[122] M. F. Cowlishaw, E. M. Schwarz, R. M. Smith, and C. F. Webb. A decimal floating-point specification. In *15th IEEE Symposium on Computer Arithmetic (ARITH-15)*, pages 147–154, June 2001.

[123] O. Creţ, F. de Dinechin, I. Trestian, R. Tudoran, L. Creţ, and L. Văcariu. FPGA-based acceleration of the computations involved in transcranial magnetic stimulation. In *Southern Programmable Logic Conference*, pages 43–48, 2008.

[124] X. Cui, W. Liu, D. Wenwen, and F. Lombardi. A parallel decimal multiplier using hybrid binary coded decimal (BCD) codes. In *23rd IEEE Symposium on Computer Arithmetic (ARITH-23)*, pages 150–155, July 2016.

[125] A. Cuyt, B. Verdonk, S. Becuwe, and P. Kuterna. A remarkable example of catastrophic cancellation unraveled. *Computing*, 66:309–320, 2001.

[126] L. Dadda. Some schemes for parallel multipliers. *Alta Frequenza*, 34:349–356, 1965. Reprinted in [583].

[127] L. Dadda. On parallel digital multipliers. *Alta Frequenza*, 45:574–580, 1976. Reprinted in [583].

[128] A. Dahan-Dalmedico and J. Pfeiffer. *Histoire des Mathématiques*. Editions du Seuil, Paris, 1986. In French.

[129] C. Daramy-Loirat, D. Defour, F. de Dinechin, M. Gallet, N. Gast, C. Q. Lauter, and J.-M. Muller. CR-LIBM, a library of correctly-rounded elementary functions in double-precision. Technical report, LIP Laboratory, Arenaire team, December 2006. Available at https://hal-ens-lyon.archives-ouvertes.fr/ensl-01529804.

[130] D. Das Sarma and D. W. Matula. Measuring the accuracy of ROM reciprocal tables. *IEEE Transactions on Computers*, 43(8):932–940, 1994.

[131] D. Das Sarma and D. W. Matula. Faithful bipartite ROM reciprocal tables. In *12th IEEE Symposium on Computer Arithmetic (ARITH-12)*, pages 17–28, June 1995.

[132] D. Das Sarma and D. W. Matula. Faithful interpolation in reciprocal tables. In *13th IEEE Symposium on Computer Arithmetic (ARITH-13)*, pages 82–91, July 1997.

[133] M. Daumas and C. Finot. Division of floating point expansions with an application to the computation of a determinant. *Journal of Universal Computer Science*, 5(6):323–338, 1999.

[134] M. Daumas and G. Melquiond. Certification of bounds on expressions involving rounded operators. *Transactions on Mathematical Software*, 37(1):1–20, 2010.

[135] M. Daumas, G. Melquiond, and C. Muñoz. Guaranteed proofs using interval arithmetic. In *17th IEEE Symposium on Computer Arithmetic (ARITH-17)*, pages 188–195, Cape Cod, MA, USA, 2005.

[136] M. Daumas, L. Rideau, and L. Théry. A generic library of floating-point numbers and its application to exact computing. In *14th International Conference on Theorem Proving in Higher Order Logics (TPHOLs)*, pages 169–184, Edinburgh, Scotland, 2001.

[137] B. de Dinechin. From machine scheduling to VLIW instruction scheduling. *ST Journal of Research*, 1(2), 2004.

[138] F. de Dinechin. The price of routing in FPGAs. *Journal of Universal Computer Science*, 6(2):227–239, 2000.

[139] F. de Dinechin. Multiplication by rational constants. *IEEE Transactions on Circuits and Systems, II*, 52(2):98–102, 2012.

[140] F. de Dinechin and L.-S. Didier. Table-based division by small integer constants. In *Applied Reconfigurable Computing*, pages 53–63, March 2012.

[141] F. de Dinechin, P. Echeverría, M. López-Vallejo, and B. Pasca. Floating-point exponentiation units for reconfigurable computing. *ACM Transactions on Reconfigurable Technology and Systems*, 6(1), 2013.

[142] F. de Dinechin, A. V. Ershov, and N. Gast. Towards the post-ultimate libm. In *17th IEEE Symposium on Computer Arithmetic (ARITH-17)*, pages 288–295, 2005.

[143] F. de Dinechin and M. Istoan. Hardware implementations of fixed-point Atan2. In *22nd IEEE Symposium of Computer Arithmetic (ARITH-22)*, pages 34–41, June 2015.

[144] F. de Dinechin, M. Istoan, and G. Sergent. Fixed-point trigonometric functions on FPGAs. *SIGARCH Computer Architecture News*, 41(5):83–88, 2013.

[145] F. de Dinechin, M. Joldeş, and B. Pasca. Automatic generation of polynomial-based hardware architectures for function evaluation. In *Application-specific Systems, Architectures and Processors (ASAP)*, 2010.

[146] F. de Dinechin, M. Joldeş, B. Pasca, and G. Revy. Multiplicative square root algorithms for FPGAs. In *Field-Programmable Logic and Applications*, pages 574–577, 2010.

[147] F. de Dinechin, C. Q. Lauter, and G. Melquiond. Certifying the floating-point implementation of an elementary function using Gappa. *IEEE Transactions on Computers*, 60(2):242–253, 2011.

[148] F. de Dinechin, C. Q. Lauter, and J.-M. Muller. Fast and correctly rounded logarithms in double-precision. *Theoretical Informatics and Applications*, 41:85–102, 2007.

[149] F. de Dinechin, C. Q. Lauter, J.-M. Muller, and S. Torres. On Ziv's rounding test. *ACM Transactions on Mathematical Software*, 39(4), 2013.

[150] F. de Dinechin and V. Lefèvre. Constant multipliers for FPGAs. In *Parallel and Distributed Processing Techniques and Applications*, pages 167–173, 2000.

[151] F. de Dinechin, C. Loirat, and J.-M. Muller. A proven correctly rounded logarithm in double-precision. In *6th Conference on Real Numbers and Computers (RNC-6)*, 2004.

[152] F. de Dinechin, E. McIntosh, and F. Schmidt. Massive tracking on heterogeneous platforms. In *9th International Computational Accelerator Physics Conference (ICAP)*, October 2006.

[153] F. de Dinechin and B. Pasca. Large multipliers with fewer DSP blocks. In *Field Programmable Logic and Applications*, pages 250–255, August 2009.

[154] F. de Dinechin and B. Pasca. Floating-point exponential functions for DSP-enabled FPGAs. In *Field Programmable Technologies*, pages 110–117, December 2010. Best paper candidate.

[155] F. de Dinechin, B. Pasca, O. Creţ, and R. Tudoran. An FPGA-specific approach to floating-point accumulation and sum-of-products. In *Field-Programmable Technologies*, 2008.

[156] F. de Dinechin and A. Tisserand. Multipartite table methods. *IEEE Transactions on Computers*, 54(3):319–330, 2005.

[157] A. DeHon and N. Kapre. Optimistic parallelization of floating-point accumulation. In *18th Symposium on Computer Arithmetic (ARITH-18)*, pages 205–213, June 2007.

[158] T. J. Dekker. A floating-point technique for extending the available precision. *Numerische Mathematik*, 18(3):224–242, 1971.

[159] M. deLorimier and A. DeHon. Floating-point sparse matrix-vector multiply for FPGAs. In *Field-Programmable Gate Arrays*, pages 75–85, 2005.

[160] J. Demmel. Underflow and the reliability of numerical software. *SIAM Journal on Scientific and Statistical Computing*, 5(4):887–919, 1984.

[161] J. Demmel and H. D. Nguyen. Fast reproducible floating-point summation. In *21th IEEE Symposium on Computer Arithmetic (ARITH-21)*, pages 163–172, April 2013.

[162] J. Demmel and H. D. Nguyen. Parallel reproducible summation. *IEEE Transactions on Computers*, 64(7):2060–2070, 2015.

[163] J. W. Demmel and X. Li. Faster numerical algorithms via exception handling. In *11th IEEE Symposium on Computer Arithmetic*, pages 234–241, June 1993.

[164] J. W. Demmel and X. Li. Faster numerical algorithms via exception handling. *IEEE Transactions on Computers*, 43(8):983–992, 1994.

[165] J. Demmel, P. Ahrens, and H. D. Nguyen. Efficient reproducible floating point summation and BLAS. Technical Report UCB/EECS-2016-121, EECS Department, University of California, Berkeley, June 2016.

[166] J. Demmel and Y. Hida. Accurate and efficient floating point summation. *SIAM Journal of Scientific Computing*, 25(4):1214–1248, 2003.

[167] J. Demmel and Y. Hida. Fast and accurate floating point summation with application to computational geometry. *Numerical Algorithms*, 37(1):101–112, 2004.

[168] A. G. Dempster and M. D. Macleod. Constant integer multiplication using minimum adders. *Circuits, Devices and Systems, IEE Proceedings*, 141(5):407–413, 1994.

[169] R. Descartes. *La Géométrie*. Paris, 1637.

[170] J. Detrey and F. de Dinechin. Table-based polynomials for fast hardware function evaluation. In *Application-Specific Systems, Architectures and Processors*, pages 328–333, 2005.

[171] J. Detrey and F. de Dinechin. Floating-point trigonometric functions for FPGAs. In *Field-Programmable Logic and Applications*, pages 29–34, August 2007.

[172] J. Detrey and F. de Dinechin. Parameterized floating-point logarithm and exponential functions for FPGAs. *Microprocessors and Microsystems, Special Issue on FPGA-based Reconfigurable Computing*, 31(8):537–545, 2007.

[173] J. Detrey and F. de Dinechin. A tool for unbiased comparison between logarithmic and floating-point arithmetic. *Journal of VLSI Signal Processing*, 49(1):161–175, 2007.

[174] J. Detrey, F. de Dinechin, and X. Pujol. Return of the hardware floating-point elementary function. In *18th Symposium on Computer Arithmetic (ARITH-18)*, pages 161–168, June 2007.

[175] W. R. Dieter, A. Kaveti, and H. G. Dietz. Low-cost microarchitectural support for improved floating-point accuracy. *IEEE Computer Architecture Letters*, 6(1):13–16, 2007.

[176] V. Dimitrov, L. Imbert, and A. Zakaluzny. Multiplication by a constant is sublinear. In *18th IEEE Symposium on Computer Arithmetic (ARITH-18)*, pages 261–268, June 2007.

[177] W. S. Dorn. Generalizations of Horner's rule for polynomial evaluation. *IBM Journal of Research and Development*, 6(2):239–245, 1962.

[178] C. Doss and R. L. Riley, Jr. FPGA-based implementation of a robust IEEE-754 exponential unit. In *Field-Programmable Custom Computing Machines*, pages 229–238, 2004.

[179] Y. Dou, S. Vassiliadis, G. K. Kuzmanov, and G. N. Gaydadjiev. 64-bit floating-point FPGA matrix multiplication. In *Field-Programmable Gate Arrays*, pages 86–95, 2005.

[180] T. Drane, W.-C. Cheung, and G. Constantinides. Correctly rounded constant integer division via multiply-add. In *IEEE International Symposium on Circuits and Systems (ISCAS)*, pages 1243–1246, Seoul, South Korea, May 2012.

[181] P. Echeverría and M. López-Vallejo. An FPGA implementation of the powering function with single precision floating-point arithmetic. In *8th Conference on Real Numbers and Computers (RNC-8)*, pages 17–26, 2008.

[182] J.-P. Eckmann, A. Malaspinas, and S. O. Kamphorst. *A Software Tool for Analysis in Function Spaces*, pages 147–167. Springer, 1991.

[183] A. Edelman. The mathematics of the Pentium division bug. *SIAM Review*, 39(1):54–67, 1997.

[184] L. Eisen, J. W. Ward, H. W. Tast, N. Mäding, J. Leenstra, S. M. Mueller, C. Jacobi, J. Preiss, E. M. Schwarz, and S. R. Carlough. IBM POWER6 accelerators: VMX and DFU. *IBM Journal of Research and Development*, 51(6):1–21, 2007.

[185] M. Ercegovac. Radix-16 evaluation of certain elementary functions. *IEEE Transactions on Computers*, C-22(6):561–566, 1973.

[186] M. D. Ercegovac and T. Lang. *Division and Square Root: Digit-Recurrence Algorithms and Implementations*. Kluwer Academic Publishers, Boston, MA, 1994.

[187] M. D. Ercegovac and T. Lang. *Digital Arithmetic*. Morgan Kaufmann Publishers, San Francisco, CA, 2004.

[188] M. D. Ercegovac and J.-M. Muller. Complex division with prescaling of the operands. In *14th IEEE Conference on Application-Specific Systems, Architectures and Processors (ASAP'2003)*, pages 304–314, June 2003.

[189] M. Ercegovac, J.-M. Muller, and A. Tisserand. Simple seed architectures for reciprocal and square root reciprocal. In *39th Asilomar Conference on Signals, Systems, and Computers*, November 2005.

[190] M. A. Erle and M. J. Schulte. Decimal multiplication via carry-save addition. In *Application-specific Systems, Architectures and Processors*, pages 348–355, 2003.

[191] M. A. Erle, M. J. Schulte, and B. J. Hickmann. Decimal floating-point multiplication via carry-save addition. In *18th IEEE Symposium on Computer Arithmetic (ARITH-18)*, pages 46–55, June 2007.

[192] M. A. Erle, M. J. Schulte, and B. J. Hickmann. Decimal floating-point multiplication. *IEEE Transactions on Computers*, 58(7):902–916, 2009.

[193] M. A. Erle, M. J. Schulte, and J. M. Linebarger. Potential speedup using decimal floating-point hardware. In *36th Asilomar Conference on Signals, Systems, and Computers*, volume 2, pages 1073–1077, November 2002.

[194] M. A. Erle, E. M. Schwarz, and M. J. Schulte. Decimal multiplication with efficient partial product generation. In *17th IEEE Symposium on Computer Arithmetic (ARITH-17)*, 2005.

[195] G. Even and W. J. Paul. On the design of IEEE compliant floating-point units. *IEEE Transactions on Computers*, 49(5):398–413, 2000.

[196] G. Even and P.-M. Seidel. A comparison of three rounding algorithms for IEEE floating-point multiplication. *IEEE Transactions on Computers*, 49(7):638–650, 2000.

[197] G. Even, P.-M. Seidel, and W. E. Ferguson. A parametric error analysis of Goldschmidt's division algorithm. *Journal of Computer and System Sciences*, 70(1):118–139, 2005.

[198] H. A. H. Fahmy, A. A. Liddicoat, and M. J. Flynn. Improving the effectiveness of floating point arithmetic. In *35th Asilomar Conference on Signals, Systems, and Computers*, volume 1, pages 875–879, November 2001.

[199] W. E. Ferguson, Jr. Exact computation of a sum or difference with applications to argument reduction. In *12th IEEE Symposium on Computer Arithmetic (ARITH-12)*, pages 216–221, Bath, UK, July 1995.

[200] S. A. Figueroa. When is double rounding innocuous? *ACM SIGNUM Newsletter*, 30(3), 1995.

[201] B. P. Flannery, W. H. Press, S. A. Teukolsky, and W. T. Vetterling. *Numerical Recipes in C*, 2nd edition. Cambridge University Press, New York, NY, 1992.

[202] G. E. Forsythe and C. B. Moler. *Computer Solution of Linear Algebraic Systems*. Prentice-Hall, Englewood Cliffs, NJ, 1967.

[203] P. Fortin, M. Gouicem, and S. Graillat. Towards solving the Table Maker's Dilemma on GPU. In *20th Euromicro International Conference on Parallel, Distributed and Network-Based Processing (PDP)*, pages 407–415, February 2012.

[204] L. Fousse, G. Hanrot, V. Lefèvre, P. Pélissier, and P. Zimmermann. MPFR: A multiple-precision binary floating-point library with correct rounding. *ACM Transactions on Mathematical Software*, 33(2), 2007. 15 pages. Available at http://www.mpfr.org/.

[205] D. Fowler and E. Robson. Square root approximations in old Babylonian mathematics: YBC 7289 in context. *Historia Mathematica*, 25:366–378, 1998.

[206] W. Fraser. A survey of methods of computing minimax and near-minimax polynomial approximations for functions of a single independent variable. *Journal of the ACM*, 12(3):295–314, 1965.

[207] P. Friedland. Algorithm 312: Absolute value and square root of a complex number. *Communications of the ACM*, 10(10):665, 1967.

[208] Fujitsu. *SPARC64^{TM} VI Extensions*. Fujitsu Limited, 1.3 edition, March 2007.

[209] M. Fürer. Faster integer multiplication. In *39th Annual ACM Symposium on Theory of Computing (STOC)*, pages 57–66, June 2007.

[210] P. Gaudry, A. Kruppa, and P. Zimmermann. A GMP-based implementation of Schönhage-Strassen's large integer multiplication algorithm. In *International Symposium on Symbolic and Algebraic Computation (ISSAC)*, pages 167–174, Waterloo, ON, Canada, 2007.

[211] D. M. Gay. Correctly-rounded binary-decimal and decimal-binary conversions. Technical Report Numerical Analysis Manuscript 90–10, ATT & Bell Laboratories (Murray Hill, NJ), November 1990.

[212] W. M. Gentleman and S. B. Marovitch. More on algorithms that reveal properties of floating-point arithmetic units. *Communications of the ACM*, 17(5):276–277, 1974.

[213] G. Gerwig, H. Wetter, E. M. Schwarz, J. Haess, C. A. Krygowski, B. M. Fleischer, and M. Kroener. The IBM eServer z990 floating-point unit. *IBM Journal of Research and Development*, 48(3.4):311–322, 2004.

[214] D. Goldberg. What every computer scientist should know about floating-point arithmetic. *ACM Computing Surveys*, 23(1):5–48, 1991. An edited reprint is available at https://docs.oracle.com/cd/E19059-01/fortec6u2/806-7996/806-7996.pdf from Sun's Numerical Computation Guide; it contains an addendum *Differences Among IEEE 754 Implementations*, also available at http://www.validlab.com/goldberg/addendum.html.

[215] I. B. Goldberg. 27 bits are not enough for 8-digit accuracy. *Commun. ACM*, 10(2):105–106, 1967.

[216] O. Goldreich and S. Goldwasser. On the limits of non-approximability of lattice problems. In *30th Annual ACM Symposium on Theory of Computing (STOC)*, pages 1–9, May 1998.

[217] R. E. Goldschmidt. Applications of division by convergence. Master's thesis, Dept. of Electrical Engineering, Massachusetts Institute of Technology, Cambridge, MA, June 1964.

[218] A. Goldsztejn. Modal intervals revisited, part 1: A generalized interval natural extension. *Reliable Computing*, 16:130–183, 2012.

[219] A. Goldsztejn. Modal intervals revisited, part 2: A generalized interval mean value extension. *Reliable Computing*, 16:184–209, 2012.

[220] J. Gosling, B. Joy, G. Steele, G. Bracha, and A. Buckley. *The Java Language Specification, Java SE 8 edition*. Oracle, 2015.

[221] F. Goualard. Fast and correct SIMD algorithms for interval arithmetic. In *Workshop on State-of-the-Art in Scientific and Parallel Computing (PARA)*, May 2008.

[222] S. Graillat, P. Langlois, and N. Louvet. Algorithms for accurate, validated and fast computations with polynomials. *Japan Journal of Industrial and Applied Mathematics*, 26(2):215–231, 2009.

[223] S. Graillat, V. Lefèvre, and J.-M. Muller. On the maximum relative error when computing integer powers by iterated multiplications in floating-point arithmetic. *Numerical Algorithms*, 70:653–667, 2015.

[224] T. Granlund. The GNU multiple precision arithmetic library, release 6.1.2. Accessible electronically at https://gmplib.org/, September 2016.

[225] B. Greer, J. Harrison, G. Henry, W. Li, and P. Tang. Scientific computing on the Itanium™ processor. In *ACM/IEEE Conference on Supercomputing (Supercomputing '01)*, pages 41–41, 2001.

[226] P. M. Gruber and C. G. Lekkerkerker. *Geometry of Numbers*, volume 37 of *North-Holland Mathematical Library*. North-Holland Publishing Co., Amsterdam, second edition, 1987.

[227] A. Guntoro and M. Glesner. High-performance FPGA-based floating-point adder with three inputs. In *Field Programmable Logic and Applications*, pages 627–630, 2008.

[228] J. L. Gustafson. *The End of Error: Unum Computing*. Chapman & Hall/CRC Computational Science. Taylor & Francis, 2015.

[229] J. L. Gustafson and I. Yonemoto. Beating floating point at its own game: Posit arithmetic. In *Supercomputing Frontiers and Innovations*, pages 71–86, July 2017.

[230] O. Gustafsson, A. G. Dempster, K. Johansson, and M. D. Macleod. Simplified design of constant coefficient multipliers. *Circuits, Systems, and Signal Processing*, 25(2):225–251, 2006.

[231] R. W. Hamming. On the distribution of numbers. *The Bell System Technical Journal*, 49:1609–1625, 1970. Reprinted in [583].

[232] G. Hanrot, V. Lefèvre, P. Pélissier, P. Théveny, and P. Zimmermann. The MPFR library, version 3.1.5, 2016. Available at http://www.mpfr.org/mpfr-3.1.5/.

[233] G. Hanrot and D. Stehlé. Improved analysis of Kannan's shortest lattice vector algorithm. In *CRYPTO*, volume 4622 of *Lecture Notes in Computer Science*, pages 170–186, 2007.

[234] G. Hanrot and P. Zimmermann. A long note on Mulders' short product. *Journal of Symbolic Computation*, 37(3):391–401, 2004.

[235] E. R. Hansen and R. I. Greenberg. An interval Newton method. *Journal of Applied Mathematics and Computing*, 12:89–98, 1983.

[236] E. R. Hansen and W. Walster. *Global optimization using interval analysis*. MIT Press, Cambridge, MA, 2004.

[237] G. H. Hardy and E. M. Wright. *An Introduction to the Theory of Numbers*. Oxford University Press, London, 1979.

[238] J. Harrison. Floating-point verification in HOL light: The exponential function. Technical Report 428, University of Cambridge Computer Laboratory, 1997.

[239] J. Harrison. Verifying the accuracy of polynomial approximations in HOL. In *10th International Conference on Theorem Proving in Higher Order Logics (TPHOLs)*, volume 1275 of *Lecture Notes in Computer Science*, pages 137–152, Murray Hill, NJ, USA, 1997.

[240] J. Harrison. A machine-checked theory of floating point arithmetic. In *12th International Conference in Theorem Proving in Higher Order Logics (TPHOLs)*, volume 1690 of *Lecture Notes in Computer Science*, pages 113–130, Nice, France, September 1999.

[241] J. Harrison. Formal verification of floating-point trigonometric functions. In *3rd International Conference on Formal Methods in Computer-Aided Design (FMCAD)*, number 1954 in Lecture Notes in Computer Science, pages 217–233, Austin, TX, USA, 2000.

[242] J. Harrison. Formal verification of IA-64 division algorithms. In *13th International Conference on Theorem Proving in Higher Order Logics (TPHOLs)*, volume 1869 of *Lecture Notes in Computer Science*, pages 233–251, 2000.

[243] J. Harrison. Floating-point verification using theorem proving. In *Formal Methods for Hardware Verification, 6th International School on Formal Methods for the Design of Computer, Communication, and Software Systems, SFM 2006*, volume 3965 of *Lecture Notes in Computer Science*, pages 211–242, Bertinoro, Italy, 2006.

[244] J. Harrison. Verifying nonlinear real formulas via sums of squares. In *20th International Conference on Theorem Proving in Higher Order Logics (TPHOLs)*, volume 4732 of *Lecture Notes in Computer Science*, pages 102–118, Kaiserslautern, Germany, 2007.

[245] J. Harrison, T. Kubaska, S. Story, and P. T. P. Tang. The computation of transcendental functions on the IA-64 architecture. *Intel Technology Journal*, Q4, 1999. Available at http://developer.intel.com/technology/itj/archive/1999.htm.

[246] J. F. Hart, E. W. Cheney, C. L. Lawson, H. J. Maehly, C. K. Mesztenyi, J. R. Rice, H. G. Thacher, and C. Witzgall. *Computer Approximations*. John Wiley & Sons, New York, 1968.

[247] D. Harvey, J. van der Hoeven, and G. Lecerf. Even faster integer multiplication. *Journal of Complexity*, 36:1–30, 2016.

[248] J. R. Hauser. The SoftFloat and TestFloat Packages. Available at http://www.jhauser.us/arithmetic/.

[249] J. R. Hauser. Handling floating-point exceptions in numeric programs. *ACM Transactions on Programming Languages and Systems*, 18(2):139–174, 1996.

[250] B. Hayes. Third base. *American Scientist*, 89(6):490–494, 2001.

[251] C. He, G. Qin, M. Lu, and W. Zhao. Group-alignment based accurate floating-point summation on FPGAs. In *Engineering of Reconfigurable Systems and Algorithms*, pages 136–142, 2006.

[252] O. Heimlich. Interval arithmetic in GNU Octave. In *Summer Workshop on Interval Methods (SWIM)*, 2016.

[253] T. J. Hickey, Q. Ju, and M. H. van Emden. Interval arithmetic: From principles to implementation. *Journal of the ACM*, 48(5):1038–1068, 2001.

[254] Y. Hida, X. S. Li, and D. H. Bailey. Algorithms for quad-double precision floating-point arithmetic. In *15th IEEE Symposium on Computer Arithmetic (ARITH-15)*, pages 155–162, June 2001.

[255] Y. Hida, X. S. Li, and D. H. Bailey. C++/fortran-90 double-double and quad-double package, release 2.3.17, March 2012. Accessible electronically at http://crd-legacy.lbl.gov/~dhbailey/mpdist/.

[256] D. J. Higham and N. J. Higham. *MATLAB Guide*. Society for Industrial and Applied Mathematics, Philadelphia, PA, USA, third edition, 2017.

[257] N. J. Higham. The accuracy of floating point summation. *SIAM Journal on Scientific Computing*, 14(4):783–799, 1993.

[258] N. J. Higham. *Accuracy and Stability of Numerical Algorithms*. SIAM, Philadelphia, PA, 2nd edition, 2002.

[259] A. Hirshfeld. *Eureka Man, The life and legacy of Archimedes*. Walker & Company, 2009.

[260] E. Hokenek, R. K. Montoye, and P. W. Cook. Second-generation RISC floating point with multiply-add fused. *IEEE Journal of Solid-State Circuits*, 25(5):1207–1213, 1990.

[261] J. E. Holm. *Floating-Point Arithmetic and Program Correctness Proofs*. Ph.D. thesis, Cornell University, 1980.

[262] M. S. Hrishikesh, D. Burger, N. P. Jouppi, S. W. Keckler, K. I. Farkas, and P. Shivakumar. The optimal logic depth per pipeline stage is 6 to 8 FO4 inverter delays. In *29th Annual International Symposium on Computer Architecture (ISCA)*, pages 14–24, 2002.

[263] S.-F. Hsiao, P.-H. Wu, C.-S. Wen, and P. K. Meher. Table size reduction methods for faithfully rounded lookup-table-based multiplierless function evaluation. *Transactions on Circuits and Systems II*, 62(5):466–470, 2015.

[264] T. E. Hull, T. F. Fairgrieve, and P. T. P. Tang. Implementing complex elementary functions using exception handling. *ACM Transactions on Mathematical Software*, 20(2):215–244, 1994.

[265] T. E. Hull, T. F. Fairgrieve, and P. T. P. Tang. Implementing the complex arcsine and arccosine functions using exception handling. *ACM Transactions on Mathematical Software*, 23(3):299–335, 1997.

[266] T. E. Hull and J. R. Swenson. Test of probabilistic models for propagation of round-off errors. *Communications of the ACM*, 9:108–113, 1966.

[267] IEEE Computer Society. *IEEE Standard for Floating-Point Arithmetic*. IEEE Standard 754-2008, August 2008. Available at http://ieeexplore.ieee.org/servlet/opac?punumber=4610933.

[268] IEEE Computer Society. *IEEE Standard for Interval Arithmetic*. IEEE Standard 1788-2015, June 2015.

[269] IEEE Computer Society. *IEEE Standard for Floating-Point Arithmetic (revision of IEEE Std 754-2008)*. 2017.

[270] G. Inoue. Leading one anticipator and floating point addition/subtraction apparatus, August 30 1994. US Patent 5,343,413.

[271] Intel Corporation. *Intrinsics Guide*. Available at https://software.intel.com/sites/landingpage/IntrinsicsGuide/. Accessed in June 2017.

[272] Intel Corporation. *Intel 64 and IA-32 Architectures Software Developer's Manual*, March 2017. Order Number: 325462-062US.

[273] International Organization for Standardization. *Programming Languages – C*. ISO/IEC Standard 9899:1999, Geneva, Switzerland, December 1999.

[274] International Organization for Standardization. *Information technology — Language independent arithmetic — Part 2: Elementary numerical functions*. ISO/IEC standard 10967-2, 2001.

[275] *Rationale for International Standard—Programming Languages—C*, 2003. Revision 5.10. Available at http://www.open-std.org/jtc1/sc22/wg14/www/C99RationaleV5.10.pdf.

[276] International Organization for Standardization. *Programming languages – Fortran – Part 1: Base language*. International Standard ISO/IEC 1539-1:2004, 2004.

[277] International Organization for Standardization. *ISO/IEC/IEEE Standard 60559:2011: Information technology – Microprocessor Systems – Floating-Point arithmetic*. International Electrotechnical Commission, 1st edition, 2011.

[278] International Organization for Standardization. *Programming Languages – C*. ISO/IEC Standard 9899:2011, Geneva, Switzerland, November 2011.

[279] International Organization for Standardization. *Floating-point extensions for C – Part 1: Binary floating-point arithmetic*. ISO/IEC Technical Specification 18661-1:2014, Geneva, Switzerland, July 2014.

[280] International Organization for Standardization. *Floating-point extensions for C – Part 2: Decimal floating-point arithmetic*. ISO/IEC Technical Specification 18661-2:2015, Geneva, Switzerland, May 2015.

[281] International Organization for Standardization. *Floating-point extensions for C – Part 3: Interchange and extended types*. ISO/IEC Technical Specification 18661-3:2015, Geneva, Switzerland, October 2015.

[282] International Organization for Standardization. *Floating-point extensions for C – Part 4: Supplementary functions*. ISO/IEC Technical Specification 18661-4:2015, Geneva, Switzerland, October 2015.

[283] International Organization for Standardization. *Floating-point extensions for C – Part 5: Supplementary attributes*. ISO/IEC Technical Specification 18661-5:2016, Geneva, Switzerland, August 2016.

[284] C. Iordache and P. T. P. Tang. An overview of floating-point support and math library on the Intel XScale architecture. In *16th IEEE Symposium on Computer Arithmetic (ARITH-16)*, pages 122–128, June 2003.

[285] C. Jacobi and C. Berg. Formal verification of the VAMP floating point unit. *Formal Methods in System Design*, 26(3):227–266, 2005.

[286] C. Jacobsen, A. Solovyev, and G. Gopalakrishnan. A parameterized floating-point formalization in HOL Light. In *7th and 8th International Workshops on Numerical Software Verification (NSV)*, volume 317 of *Electronic Notes in Theoretical Computer Science*, pages 101–107, 2015.

[287] C.-P. Jeannerod. A radix-independent error analysis of the Cornea-Harrison-Tang method. *ACM Transactions on Mathematical Software*, 42(3):19:1–19:20, 2016.

[288] C.-P. Jeannerod and J. Jourdan-Lu. Simultaneous floating-point sine and cosine for VLIW integer processors. In *23rd International Conference on Application-Specific Systems, Architectures and Processors (ASAP'12)*, pages 69–76, 2012.

[289] C.-P. Jeannerod, J. Jourdan-Lu, and C. Monat. Non-generic floating-point software support for embedded media processing. In *IEEE Symposium on Industrial Embedded Systems (SIES'12)*, pages 283–286, 2012.

[290] C.-P. Jeannerod, J. Jourdan-Lu, C. Monat, and G. Revy. How to square floats accurately and efficiently on the ST231 integer processor. In *20th IEEE Symposium on Computer Arithmetic (ARITH-20)*, pages 77–81, Tübingen, Germany, July 2011.

[291] C.-P. Jeannerod, H. Knochel, C. Monat, and G. Revy. Faster floating-point square root for integer processors. In *IEEE Symposium on Industrial Embedded Systems (SIES'07)*, pages 324–327, 2007.

[292] C.-P. Jeannerod, H. Knochel, C. Monat, and G. Revy. Computing floating-point square roots via bivariate polynomial evaluation. *IEEE Transactions on Computers*, 60(2):214–227, 2011.

[293] C.-P. Jeannerod, H. Knochel, C. Monat, G. Revy, and G. Villard. A new binary floating-point division algorithm and its software implementation on the ST231 processor. In *19th IEEE Symposium on Computer Arithmetic (ARITH-19)*, June 2009.

[294] C.-P. Jeannerod, P. Kornerup, N. Louvet, and J.-M. Muller. Error bounds on complex floating-point multiplication with an FMA. *Mathematics of Computation*, 86(304):881–898, 2017.

[295] C.-P. Jeannerod, N. Louvet, and J.-M. Muller. Further analysis of Kahan's algorithm for the accurate computation of 2×2 determinants. *Mathematics of Computation*, 82(284):2245–2264, 2013.

[296] C.-P. Jeannerod, N. Louvet, and J.-M. Muller. On the componentwise accuracy of complex floating-point division with an FMA. In *21st IEEE Symposium on Computer Arithmetic (ARITH-21)*, pages 83–90, April 2013.

[297] C.-P. Jeannerod, N. Louvet, J.-M. Muller, and A. Panhaleux. Midpoints and exact points of some algebraic functions in floating-point arithmetic. *IEEE Transactions on Computers*, 60(2), 2011.

[298] C.-P. Jeannerod, N. Louvet, J.-M. Muller, and A. Plet. Sharp error bounds for complex floating-point inversion. *Numerical Algorithms*, 73(3):735–760, 2016.

[299] C.-P. Jeannerod, J.-M. Muller, and A. Plet. The classical relative error bounds for computing $\sqrt{a^2+b^2}$ and $c/\sqrt{a^2+b^2}$ in binary floating-point arithmetic are asymptotically optimal. In *24th IEEE Symposium on Computer Arithmetic (ARITH-24)*, July 2017.

[300] C.-P. Jeannerod and G. Revy. FLIP 1.0: a fast floating-point library for integer processors. http://flip.gforge.inria.fr/, February 2009.

[301] C.-P. Jeannerod and G. Revy. Optimizing correctly-rounded reciprocal square roots for embedded VLIW cores. In *43rd Asilomar Conference on Signals, Systems, and Computers*, pages 731–735, November 2009.

[302] C.-P. Jeannerod and S. M. Rump. Improved error bounds for inner products in floating-point arithmetic. *SIAM Journal on Matrix Analysis and Applications*, 34(2):338–344, 2013.

[303] C.-P. Jeannerod and S. M. Rump. On relative errors of floating-point operations: optimal bounds and applications. *Mathematics of Computation*, 2016. To appear.

[304] F. Johansson. Arb: a C library for ball arithmetic. *ACM Communications in Computer Algebra*, 47(4):166–169, 2013.

[305] K. Johansson, O. Gustafsson, and L. Wanhammar. A detailed complexity model for multiple constant multiplication and an algorithm to minimize the complexity. In *Circuit Theory and Design*, pages 465–468, 2005.

[306] P. Johnstone and F. E. Petry. Rational number approximation in higher radix floating-point systems. *Computers & Mathematics with Applications*, 25(6):103–108, 1993.

[307] M. Joldeş. *Rigorous polynomial approximations and applications*. Ph.D. thesis, École Normale Supérieure de Lyon, Lyon, France, 2011.

[308] M. Joldeş, J.-M. Muller, and V. Popescu. Tight and rigourous error bounds for basic building blocks of double-word arithmetic. *ACM Transactions on Mathematical Software*, 44(2), 2017.

[309] M. Joldeş, J.-M. Muller, V. Popescu, and W. Tucker. CAMPARY: Cuda multiple precision arithmetic library and applications. In *5th International Congress on Mathematical Software (ICMS)*, July 2016.

[310] M. Joldeş, O. Marty, J.-M. Muller, and V. Popescu. Arithmetic algorithms for extended precision using floating-point expansions. *IEEE Transactions on Computers*, 65(4):1197–1210, 2016.

[311] J. Jourdan-Lu. *Custom floating-point arithmetic for integer processors: algorithms, implementation, and selection*. Ph.D. thesis, Université de Lyon - ÉNS de Lyon, France, November 2012.

[312] E. Kadric, P. Gurniak, and A. DeHon. Accurate parallel floating-point accumulation. In *21th IEEE Symposium on Computer Arithmetic (ARITH-21)*, pages 153–162, April 2013.

[313] W. Kahan. Pracniques: further remarks on reducing truncation errors. *Communications of the ACM*, 8(1):40, 1965.

[314] W. Kahan. A more complete interval arithmetic. Lecture notes for the University of Michigan, 1968.

[315] W. Kahan. Why do we need a floating-point standard? Technical report, Computer Science, UC Berkeley, 1981. Available at http://www.cs.berkeley.edu/~wkahan/ieee754status/why-ieee.pdf.

[316] W. Kahan. Minimizing q*m-n. Text accessible electronically at http://http.cs.berkeley.edu/~wkahan/. At the beginning of the file "nearpi.c", 1983.

[317] W. Kahan. Branch cuts for complex elementary functions. In *The State of the Art in Numerical Analysis*, pages 165–211, 1987.

[318] W. Kahan. Lecture notes on the status of IEEE-754. Available at http://www.cs.berkeley.edu/~wkahan/ieee754status/IEEE754.PDF, 1997.

[319] W. Kahan. Matlab's loss is nobody's gain. Technical report, Computer Science, UC Berkeley, 1998. Available at https://people.eecs.berkeley.edu/~wkahan/MxMulEps.pdf.

[320] W. Kahan. How futile are mindless assessments of roundoff in floating-point computation? Available at http://http.cs.berkeley.edu/~wkahan/Mindless.pdf, 2004.

[321] W. Kahan. A logarithm too clever by half. Available at http://http.cs.berkeley.edu/~wkahan/LOG10HAF.TXT, 2004.

[322] W. Kahan and J. Darcy. How Java's floating-point hurts everyone everywhere. Available at http://www.cs.berkeley.edu/~wkahan/JAVAhurt.pdf, 1998.

[323] R. Kaivola and K. Kohatsu. Proof engineering in the large: formal verification of Pentium®4 floating-point divider. *International Journal on Software Tools for Technology Transfer*, 4(3):323–334, 2003.

[324] E. Kaltofen. On the complexity of finding short vectors in integer lattices. In *European Computer Algebra Conference (EUROCAL)*, volume 162 of *Lecture Notes in Computer Science*, pages 236–244, 1983.

[325] R. Kannan. Improved algorithms for integer programming and related lattice problems. In *Annual ACM Symposium on Theory of Computing (STOC)*, pages 193–206, 1983.

[326] R. Kannan. Minkowski's convex body theorem and integer programming. *Mathematics of Operations Research*, 12(3):415–440, 1987.

[327] L. V. Kantorovich. On some new approaches to computational methods and to processing of observations. *Siberian Mathematical Journal*, 3(5):701–709, 1962. In Russian.

[328] A. Karatsuba and Y. Ofman. Multiplication of many-digital numbers by automatic computers. *Doklady Akad. Nauk SSSR*, 145:293–294, 1962. Translation in Physics-Doklady 7, 595–596, 1963.

[329] R. Karpinsky. PARANOIA: a floating-point benchmark. *BYTE*, 10(2), 1985.

[330] E. Kaucher. Interval analysis in the extended interval space IR. In *Fundamentals of Numerical Computation (Computer-Oriented Numerical Analysis)*, pages 33–49. Springer, 1980.

[331] R. B. Kearfott. *Rigorous global search: continuous problems*. Kluwer, Dordrecht, 1996.

[332] R. B. Kearfott, M. T. Nakao., A. Neumaier, S. M. Rump, S. P. Shary, and P. V. Hentenryck. Standardized notation in interval analysis. In *XIII Baikal International School-seminar "Optimization methods and their applications"*, volume 4, pages 106–113, 2005.

[333] B. W. Kernighan and D. M. Ritchie. *The C Programming Language*. Prentice-Hall, Englewood Cliffs, NJ, 1978.

[334] A. Y. Khinchin. *Continued Fractions*. Dover, New York, 1997.

[335] S. Khot. Hardness of approximating the shortest vector problem in lattices. In *FOCS*, pages 126–135, 2004.

[336] Khronos OpenCL Working Group. *The OpenCL SPIR-V Environment Specification, version 2.2*, May 2017. Available at https://www.khronos.org/registry/OpenCL/specs/opencl-2.2-environment.pdf.

[337] N. G. Kingsbury and P. J. W. Rayner. Digital filtering using logarithmic arithmetic. *Electronic Letters*, 7:56–58, 1971. Reprinted in [583].

[338] P. Kirchberger. *Ueber Tchebychefsche Annaeherungsmethoden*. Ph.D. thesis, Gottingen, 1902.

[339] A. Klein. A generalized Kahan-Babuška-summation-algorithm. *Computing*, 76:279–293, 2006.

[340] A. Knöfel. Fast hardware units for the computation of accurate dot products. In *10th IEEE Symposium on Computer Arithmetic (ARITH-10)*, pages 70–74, June 1991.

[341] S. Knowles. A family of adders. In *14th IEEE Symposium on Computer Arithmetic (ARITH-14)*, pages 30–34, April 1999.

[342] D. E. Knuth. *The Art of Computer Programming*, volume 2. Addison-Wesley, Reading, MA, 3rd edition, 1998.

[343] J. Koenig, D. Biancolin, J. Bachrach, and K. Asanovic. A hardware accelerator for computing an exact dot product. In *24th IEEE Symposium on Computer Arithmetic (ARITH-24)*, July 2017.

[344] P. M. Kogge and H. S. Stone. A parallel algorithm for the efficient solution of a general class of recurrence equations. *IEEE Transactions on Computers*, 100(8):786–793, 1973.

[345] I. Koren. *Computer Arithmetic Algorithms*. Prentice-Hall, Englewood Cliffs, NJ, 1993.

[346] P. Kornerup, C. Lauter, V. Lefèvre, N. Louvet, and J.-M. Muller. Computing correctly rounded integer powers in floating-point arithmetic. *ACM Transactions on Mathematical Software*, 37(1):4:1–4:23, 2010.

[347] P. Kornerup, V. Lefèvre, N. Louvet, and J.-M. Muller. On the computation of correctly rounded sums. *IEEE Transactions on Computers*, 61(3):289–298, 2012.

[348] P. Kornerup and D. W. Matula. Finite-precision rational arithmetic: an arithmetic unit. *IEEE Transactions on Computers*, C-32:378–388, 1983.

[349] P. Kornerup and D. W. Matula. Finite precision lexicographic continued fraction number systems. In *7th IEEE Symposium on Computer Arithmetic (ARITH-7)*, 1985. Reprinted in [584].

[350] P. Kornerup and J.-M. Muller. Choosing starting values for certain Newton–Raphson iterations. *Theoretical Computer Science*, 351(1):101–110, 2006.

[351] W. Krandick and J. R. Johnson. Efficient multiprecision floating point multiplication with optimal directional rounding. In *11th IEEE Symposium on Computer Arithmetic (ARITH-11)*, pages 228–233, June 1993.

[352] H. Kuki and W. J. Cody. A statistical study of the accuracy of floating point number systems. *Communications of the ACM*, 16(4):223–230, 1973.

[353] U. W. Kulisch. Circuitry for generating scalar products and sums of floating-point numbers with maximum accuracy. United States Patent 4622650, 1986.

[354] U. W. Kulisch. *Advanced Arithmetic for the Digital Computer: Design of Arithmetic Units*. Springer-Verlag, Berlin, 2002.

[355] U. W. Kulisch. *Computer Arithmetic and Validity: Theory, Implementation, and Applications*. de Gruyter, Berlin, 2008.

[356] M. Kumm, O. Gustafsson, M. Garrido, and P. Zipf. Optimal single constant multiplication using ternary adders. *IEEE Transactions on Circuits and Systems II: Express Briefs*, 2016.

[357] M. Kumm and P. Zipf. Pipelined compressor tree optimization using integer linear programming. In *Field Programmable Logic and Applications*, 2014.

[358] B. Lambov. Interval arithmetic using SSE-2. In *Reliable Implementation of Real Number Algorithms: Theory and Practice*, volume 5045 of *Lecture Notes in Computer Science*, pages 102–113, August 2008.

[359] T. Lang and J. D. Bruguera. Floating-point multiply-add-fused with reduced latency. *IEEE Transactions on Computers*, 53(8):988–1003, 2004.

[360] T. Lang and A. Nannarelli. A radix-10 combinational multiplier. In *40th Asilomar Conference on Signals, Systems, and Computers*, pages 313–317, October/November 2006.

[361] M. Lange and S. M. Rump. Error estimates for the summation of real numbers with application to floating-point summation. *BIT Numerical Mathematics*, 57(3):927–941, 2017.

[362] M. Lange and S. M. Rump. Faithfully rounded floating-point computations. Manuscript available at http://www.ti3.tu-harburg.de/rump/, 2017.

[363] M. Lange and S. M. Rump. Sharp estimates for perturbation errors in summations. Manuscript available at http://www.ti3.tu-harburg.de/rump/, 2017.

[364] M. Langhammer and B. Pasca. Faithful single-precision floating-point tangent for FPGAs. In *ACM/SIGDA International Symposium on Field Programmable Gate Arrays*, pages 39–42, 2013.

[365] M. Langhammer and B. Pasca. Floating-point DSP block architecture for FPGAs. In *ACM/SIGDA International Symposium on Field-Programmable Gate Arrays*, pages 117–125, 2015.

[366] M. Langhammer and B. Pasca. Single precision logarithm and exponential architectures for hard floating-point enabled FPGAs. *IEEE Transactions on Computers*, 66(12):2031–2043, 2017.

[367] P. Langlois. Automatic linear correction of rounding errors. *BIT Numerical Algorithms*, 41(3):515–539, 2001.

[368] P. Langlois and N. Louvet. How to ensure a faithful polynomial evaluation with the compensated Horner algorithm. In *18th IEEE Symposium on Computer Arithmetic (ARITH-18)*, pages 141–149, June 2007.

[369] J. Laskar, P. Robutel, F. Joutel, M. Gastineau, A. C. M. Correia, and B. Levrard. A long term numerical solution for the insolation quantities of the Earth. *Astronomy & Astrophysics*, 428:261–285, 2004.

[370] C. Q. Lauter. Basic building blocks for a triple-double intermediate format. Technical Report 2005-38, LIP, École Normale Supérieure de Lyon, September 2005.

[371] C. Q. Lauter. *Arrondi Correct de Fonctions Mathématiques*. Ph.D. thesis, École Normale Supérieure de Lyon, Lyon, France, October 2008. In French, available at http://www.ens-lyon.fr/LIP/Pub/Rapports/PhD/PhD2008/PhD2008-07.pdf.

[372] J. Le Maire, N. Brunie, F. de Dinechin, and J.-M. Muller. Computing floating-point logarithms with fixed-point operations. In *23rd IEEE Symposium of Computer Arithmetic (ARITH-23)*, July 2016.

[373] G. Lecerf and J. van der Hoeven. Evaluating Straight-Line Programs over Balls. In *23rd IEEE Symposium on Computer Arithmetic*, pages 142–149, 2016.

[374] B. Lee and N. Burgess. Parameterisable floating-point operations on FPGA. In *36th Asilomar Conference on Signals, Systems, and Computers*, volume 2, pages 1064–1068, November 2002.

[375] V. Lefèvre. Multiplication by an integer constant. Technical Report RR1999-06, Laboratoire de l'Informatique du Parallélisme, Lyon, France, 1999.

[376] V. Lefèvre. *Moyens Arithmétiques Pour un Calcul Fiable*. Ph.D. thesis, École Normale Supérieure de Lyon, Lyon, France, 2000.

[377] V. Lefèvre. The Euclidean division implemented with a floating-point division and a floor. Research report RR-5604, INRIA, June 2005.

[378] V. Lefèvre. New results on the distance between a segment and $\mathbb{Z}^2$. Application to the exact rounding. In *17th IEEE Symposium on Computer Arithmetic (ARITH-17)*, pages 68–75, June 2005.

[379] V. Lefèvre. Correctly rounded arbitrary-precision floating-point summation. *IEEE Transactions on Computers*, 66(12):2111–2124, 2017.

[380] V. Lefèvre and J.-M. Muller. Worst cases for correct rounding of the elementary functions in double precision. In *15th IEEE Symposium on Computer Arithmetic (ARITH-15)*, June 2001.

[381] V. Lefèvre and P. Zimmermann. Optimized binary64 and binary128 arithmetic with GNU MPFR. In *24th IEEE Symposium on Computer Arithmetic (ARITH-24)*, July 2017.

[382] A. K. Lenstra, H. W. Lenstra, Jr., and L. Lovász. Factoring polynomials with rational coefficients. *Mathematische Annalen*, 261:515–534, 1982.

[383] X. Leroy. Formal verification of a realistic compiler. *Communications of the ACM*, 52(7):107–115, 2009.

[384] X. Li, J. Demmel, D. H. Bailey, G. Henry, Y. Hida, J. Iskandar, W. Kahan, A. Kapur, M. Martin, T. Tung, and D. J. Yoo. Design, implementation and testing of extended and mixed precision BLAS. Technical Report 45991, Lawrence Berkeley National Laboratory, 2000. https://publications.lbl.gov/islandora/object/ir%3A115848.

[385] X. Li, J. Demmel, D. H. Bailey, G. Henry, Y. Hida, J. Iskandar, W. Kahan, A. Kapur, M. Martin, T. Tung, and D. J. Yoo. Design, implementation and testing of extended and mixed precision BLAS. *ACM Transactions on Mathematical Software*, 28(2):152–205, 2002.

[386] Y. Li and W. Chu. Implementation of single precision floating-point square root on FPGAs. In *FPGAs for Custom Computing Machines*, pages 56–65, 1997.

[387] C. Lichtenau, S. Carlough, and S. M. Mueller. Quad precision floating point on the IBM z13^{TM}. *23rd IEEE Symposium on Computer Arithmetic (ARITH-23)*, pages 87–94, 2016.

[388] G. Lienhart, A. Kugel, and R. Männer. Using floating-point arithmetic on FPGAs to accelerate scientific N-body simulations. In *FPGAs for Custom Computing Machines*, 2002.

[389] W. B. Ligon, S. McMillan, G. Monn, K. Schoonover, F. Stivers, and K. D. Underwood. A re-evaluation of the practicality of floating-point operations on FPGAs. In *FPGAs for Custom Computing Machines*, 1998.

[390] T. Lindholm, F. Yellin, G. Bracha, and A. Buckley. *The Java^{TM}Virtual Machine Specification, Java SE 8 edition*. Oracle, 2015.

[391] S. Linnainmaa. Software for doubled-precision floating-point computations. *ACM Transactions on Mathematical Software*, 7(3):272–283, 1981.

[392] J. Liu, M. Chang, and C.-K. Cheng. An iterative division algorithm for FPGAs. In *Field-Programmable Gate Arrays*, pages 83–89, 2006.

[393] E. Loh and G. W. Walster. Rump's example revisited. *Reliable Computing*, 8(3):245–248, 2002.

[394] F. Loitsch. Printing floating-point numbers quickly and accurately with integers. In *31st ACM SIGPLAN Conference on Programming Language Design and Implementation (PLDI '10)*, pages 233–243, 2010.

[395] A. R. Lopes and G. A. Constantinides. A fused hybrid floating-point and fixed-point dot-product for FPGAs. In *6th International Symposium on Reconfigurable Computing: Architectures, Tools and Applications (ARC)*, volume 5992 of *Lecture Notes in Computer Science*, pages 157–168, Bangkok, Thailand, March 2010.

[396] N. Louvet. *Algorithmes Compensés en Arithmétique Flottante: Précision, Validation, Performances*. Ph.D. thesis, Université de Perpignan, Perpignan, France, November 2007. In French.

[397] L. Lovász. *An Algorithmic Theory of Numbers, Graphs and Convexity*, volume 50 of *CBMS-NSF Regional Conference Series in Applied Mathematics*. Society for Industrial and Applied Mathematics (SIAM), 1986.

[398] Z. Luo and M. Martonosi. Accelerated pipelined integer and floating-point accumulations in configurable hardware with delayed addition techniques. *IEEE Transactions on Computers*, 49(3):208–218, 2000.

[399] D. Lutz. Fused multiply-add microarchitecture comprising separate early-normalizing multiply and add pipelines. In *20th IEEE Symposium on Computer Arithmetic (ARITH-20)*, pages 123–128, 2011.

[400] D. Lutz and N. Burgess. Overcoming double-rounding errors under IEEE 754-2008 using software. In *44th Asilomar Conference on Signals, Systems, and Computers*, pages 1399–1401, November 2010.

[401] N. Macon and A. Spitzbart. Inverses of Vandermonde matrices. *American Mathematical Monthly*, 65(2):95–100, 1958.

[402] K. Makino and M. Berz. Taylor models and other validated functional inclusion methods. *International Journal of Pure and Applied Mathematics*, 6(3):239–312, 2003.

[403] M. A. Malcolm. Algorithms to reveal properties of floating-point arithmetic. *Communications of the ACM*, 15(11):949–951, 1972.

[404] M. V. Manoukian and G. A. Constantinides. Accurate floating point arithmetic through hardware error-free transformations. In *Reconfigurable Computing: Architectures, Tools and Applications (ARC)*, pages 94–101, 2011.

[405] P. Markstein. Computation of elementary functions on the IBM RISC System/6000 processor. *IBM Journal of Research and Development*, 34(1):111–119, 1990.

[406] P. Markstein. *IA-64 and Elementary Functions: Speed and Precision*. Hewlett-Packard Professional Books. Prentice-Hall, Englewood Cliffs, NJ, 2000.

[407] M. Martel. Propagation of Roundoff Errors in Finite Precision Computations: A Semantics Approach. In *ESOP*, pages 194–208, 2002.

[408] É. Martin-Dorel and G. Melquiond. Proving tight bounds on univariate expressions with elementary functions in Coq. *Journal of Automated Reasoning*, 57(3):187–217, 2016.

[409] É. Martin-Dorel, G. Melquiond, and J.-M. Muller. Some issues related to double rounding. *BIT Numerical Mathematics*, 53(4):897–924, 2013.

[410] W. F. Mascarenhas. Floating point numbers are real numbers. Manuscript available at https://arxiv.org/abs/1605.09202, 2016.

[411] D. W. Matula. In-and-out conversions. *Communications of the ACM*, 11(1):47–50, 1968.

[412] D. W. Matula and P. Kornerup. Finite precision rational arithmetic: Slash number systems. *IEEE Transactions on Computers*, 34(1):3–18, 1985.

[413] W. M. McKeeman. Representation error for real numbers in binary computer arithmetic. *IEEE Transactions on Electronic Computers*, EC-16(5):682–683, 1967.

[414] P. Meher, J. Valls, T.-B. Juang, K. Sridharan, and K. Maharatna. 50 years of CORDIC: Algorithms, architectures, and applications. *IEEE Transactions on Circuits and Systems I: Regular Papers*, 56(9):1893–1907, 2009.

[415] G. Melquiond. *De l'arithmétique d'intervalles à la certification de programmes.* Ph.D. thesis, École Normale Supérieure de Lyon, November 2006. In French, available at http://www.ens-lyon.fr/LIP/Pub/PhD2006.php.

[416] G. Melquiond. Proving bounds on real-valued functions with computations. In *4th International Joint Conference on Automated Reasoning (IJCAR)*, volume 5195 of *Lecture Notes in Artificial Intelligence*, pages 2–17, 2008.

[417] G. Melquiond. Floating-point arithmetic in the Coq system. *Information and Computation*, 216:14–23, 2012.

[418] V. Ménissier. *Arithmétique Exacte.* Ph.D. thesis, Université Pierre et Marie Curie, Paris, December 1994. In French.

[419] D. Micciancio. The hardness of the closest vector problem with preprocessing. *IEEE Transactions on Information Theory*, 47(3):1212–1215, 2001.

[420] D. Micciancio and S. Goldwasser. *Complexity of Lattice Problems: a Cryptographic Perspective*, volume 671 of *The Kluwer International Series in Engineering and Computer Science.* Kluwer Academic Publishers, Boston, MA, 2002.

[421] J. H. Min and E. E. Swartzlander. Fused floating-point two-term sum-of-squares unit. In *Application-Specific Systems, Architectures and Processors (ASAP)*, 2013.

[422] O. Møller. Quasi double-precision in floating-point addition. *BIT*, 5:37–50, 1965.

[423] D. Monniaux. The pitfalls of verifying floating-point computations. *ACM TOPLAS*, 30(3):1–41, 2008. A preliminary version is available at http://hal.archives-ouvertes.fr/hal-00128124.

[424] R. K. Montoye, E. Hokonek, and S. L. Runyan. Design of the IBM RISC System/6000 floating-point execution unit. *IBM Journal of Research and Development*, 34(1):59–70, 1990.

[425] P. Montuschi and P. M. Mezzalama. Survey of square rooting algorithms. *Computers and Digital Techniques, IEE Proceedings E.*, 137(1):31–40, 1990.

[426] J. S. Moore, T. Lynch, and M. Kaufmann. A mechanically checked proof of the correctness of the kernel of the AMD5K86 floating point division algorithm. *IEEE Transactions on Computers*, 47(9):913–926, 1998.

[427] R. E. Moore. *Interval arithmetic and automatic error analysis in digital computing*. Ph.D. thesis, Applied Math Statistics Lab., Report 25, Stanford, 1962.

[428] R. E. Moore. *Interval analysis*. Prentice Hall, 1966.

[429] R. E. Moore. *Methods and applications of interval analysis*. SIAM Studies in Applied Mathematics, Philadelphia, PA, 1979.

[430] R. E. Moore, R. B. Kearfott, and M. J. Cloud. *Introduction to Interval Analysis*. SIAM, Philadelphia, PA, 2009.

[431] S. K. Moore. Intel makes a big jump in computer math. *IEEE Spectrum*, 2008.

[432] R. Morris. Tapered floating point: A new floating-point representation. *IEEE Transactions on Computers*, 20(12):1578–1579, 1971.

[433] C. Mouilleron and G. Revy. Automatic generation of fast and certified code for polynomial evaluation. In *20th IEEE Symposium on Computer Arithmetic*, pages 233–242, 2011.

[434] S. Muchnik. *Advanced Compiler Design and Implementation*. Morgan Kaufmann, 1997.

[435] T. Mulders. On short multiplications and divisions. *Applicable Algebra in Engineering, Communication and Computing*, 11(1):69–88, 2000.

[436] J.-M. Muller. *Arithmétique des Ordinateurs*. Masson, Paris, 1989. In French.

[437] J.-M. Muller. Algorithmes de division pour microprocesseurs: illustration à l'aide du "bug" du pentium. *Technique et Science Informatiques*, 14(8), 1995.

[438] J.-M. Muller. A few results on table-based methods. *Reliable Computing*, 5(3):279–288, 1999.

[439] J.-M. Muller. On the definition of ulp(x). Technical Report 2005-09, LIP Laboratory, ENS Lyon, 2005.

[440] J.-M. Muller. Avoiding double roundings in scaled Newton-Raphson division. In *47th Asilomar Conference on Signals, Systems, and Computers*, pages 396–399, November 2013.

[441] J.-M. Muller. On the error of computing $ab + cd$ using Cornea, Harrison and Tang's method. *ACM Transactions on Mathematical Software*, 41(2):7:1–7:8, 2015.

[442] J.-M. Muller. *Elementary Functions, Algorithms and Implementation*. Birkhäuser Boston, MA, 3rd edition, 2016.

[443] J.-M. Muller, V. Popescu, and P. T. P. Tang. A new multiplication algorithm for extended precision using floating-point expansions. In *23nd IEEE Symposium on Computer Arithmetic (ARITH-23)*, pages 39–46, July 2016.

[444] J.-M. Muller, A. Scherbyna, and A. Tisserand. Semi-logarithmic number systems. *IEEE Transactions on Computers*, 47(2):145–151, 1998.

[445] M. Müller, C. Rüb, and W. Rülling. Exact accumulation of floating-point numbers. In *10th IEEE Symposium on Computer Arithmetic (ARITH-10)*, pages 64–69, June 1991.

[446] A. Munk-Nielsen and J.-M. Muller. Borrow-save adders for real and complex number systems. In *Real Numbers and Computers 2*, April 1996.

[447] C. Muñoz and A. Narkawicz. Formalization of Bernstein polynomials and applications to global optimization. *Journal of Automated Reasoning*, 51(2):151–196, 2013.

[448] A. Naini, A. Dhablania, W. James, and D. Das Sarma. 1-GHz HAL SPARC64 dual floating-point unit with RAS features. In *15th IEEE Symposium on Computer Arithmetic (ARITH-15)*, pages 174–183, June 2001.

[449] P. Nataraj and K. Kotecha. Higher Order Convergence for Multidimensional Functions with a New Taylor-Bernstein Form as Inclusion Function. *Reliable Computing*, 9:185–203, 2003.

[450] R. Nathan, B. Anthonio, S.-L. Lu, H. Naeimi, D. J. Sorin, and X. Sun. Recycled error bits: Energy-efficient architectural support for floating point accuracy. In *International Conference for High Performance Computing, Networking, Storage and Analysis (SC '14)*, pages 117–127, 2014.

[451] R. Nave. Implementation of transcendental functions on a numerics processor. *Microprocessing and Microprogramming*, 11:221–225, 1983.

[452] M. Nehmeier. libieeep1788: A C++ implementation of the IEEE interval standard P1788. In *Norbert Wiener in the 21st Century (21CW), 2014 IEEE Conference on*, pages 1–6, 2014.

[453] H. Neto and M. Véstias. Decimal multiplier on FPGA using embedded binary multipliers. In *Field Programmable Logic and Applications*, pages 197–202, 2008.

[454] A. Neumaier. Rundungsfehleranalyse einiger Verfahren zur Summation endlicher Summen. *ZAMM*, 54:39–51, 1974. In German.

[455] A. Neumaier. *Interval methods for systems of equations*. Cambridge University Press, Cambridge, UK, 1990.

[456] A. Neumaier. *Introduction to Numerical Analysis*. Cambridge University Press, 2001.

[457] A. Neumaier and D. Stehlé. Faster LLL-type reduction of lattice bases. In *ACM International Symposium on Symbolic and Algebraic Computation (ISSAC)*, pages 373–380, 2016.

[458] I. Newton. *Methodus Fluxionum et Serierum Infinitarum*. 1664–1671.

[459] K. C. Ng. Argument reduction for huge arguments: Good to the last bit. Technical report, SunPro, 1992.

[460] K. Ng. Method and apparatus for exact leading zero prediction for a floating-point adder, April 20 1993. US Patent 5,204,825.

[461] H. D. Nguyen. *Efficient algorithms for verified scientific computing : Numerical linear algebra using interval arithmetic*. Phd thesis, École Normale Supérieure de Lyon, January 2011.

[462] H. D. Nguyen, B. Pasca, and T. Preusser. FPGA-specific arithmetic optimizations of short-latency adders. In *Field Programmable Logic and Applications*, pages 232–237, 2011.

[463] P. Nguyen and D. Stehlé. An LLL algorithm with quadratic complexity. *SIAM Journal on Computing*, 39(3):874–903, 2009.

[464] K. R. Nichols, M. A. Moussa, and S. M. Areibi. Feasibility of floating-point arithmetic in FPGA based artificial neural networks. In *Computer Applications in Industry and Engineering (CAINE)*, pages 8–13, 2002.

[465] A. Novocin, D. Stehlé, and G. Villard. An LLL-reduction algorithm with quasi-linear time complexity: extended abstract. In *43rd ACM Symposium on Theory of Computing (STOC)*, pages 403–412, 2011.

[466] J. Oberg. Why the Mars probe went off course. *IEEE Spectrum*, 36(12):34–39, 1999.

[467] S. F. Oberman. Floating point division and square root algorithms and implementation in the AMD-K7 microprocessor. In *14th IEEE Symposium on Computer Arithmetic (ARITH-14)*, pages 106–115, April 1999.

[468] S. F. Oberman, H. Al-Twaijry, and M. J. Flynn. The SNAP project: design of floating-point arithmetic units. In *13th Symposium on Computer Arithmetic (ARITH-13)*, 1997.

[469] S. F. Oberman and M. J. Flynn. Division algorithms and implementations. *IEEE Transactions on Computers*, 46(8):833–854, 1997.

[470] S. F. Oberman and M. Y. Siu. A high-performance area-efficient multifunction interpolator. In *17th IEEE Symposium on Computer Arithmetic (ARITH-17)*, June 2005.

[471] T. Ogita, S. M. Rump, and S. Oishi. Accurate sum and dot product. *SIAM Journal on Scientific Computing*, 26(6):1955–1988, 2005.

[472] T. Ogita, S. M. Rump, and S. Oishi. Verified solutions of linear systems without directed rounding. Technical Report 2005-04, Advanced Research Institute for Science and Engineering, Waseda University, Tokyo, Japan, 2005.

[473] V. G. Oklobdzija. An algorithmic and novel design of a leading zero detector circuit: Comparison with logic synthesis. *IEEE Transactions on Very Large Scale Integration (VLSI) Systems*, 2, 1994.

[474] J. O'Leary, X. Zhao, R. Gerth, and C.-J. H. Seger. Formally verifying IEEE compliance of floating point hardware. *Intel Technology Journal*, 3(1), 1999.

[475] F. W. J. Olver and P. R. Turner. Implementation of level-index arithmetic using partial table look-up. In *8th IEEE Symposium on Computer Arithmetic (ARITH-8)*, May 1987.

[476] F. Ortiz, J. Humphrey, J. Durbano, and D. Prather. A study on the design of floating-point functions in FPGAs. In *Field Programmable Logic and Applications*, volume 2778 of *Lecture Notes in Computer Science*, pages 1131–1135, Lisbon, Portugal, September 2003.

[477] M. L. Overton. *Numerical Computing with IEEE Floating Point Arithmetic*. SIAM, Philadelphia, PA, 2001.

[478] K. Ozaki, F. Bünger, T. Ogita, S. Oishi, and S. M. Rump. Simple floating-point filters for the two-dimensional orientation problem. *BIT Numerical Mathematics*, 56(2):729–749, 2016.

[479] K. Ozaki, T. Ogita, F. Bünger, and S. Oishi. Accelerating interval matrix multiplication by mixed precision arithmetic. *Nonlinear Theory and its Applications, IEICE*, 6(3):364–376, 2015.

[480] A. Paidimarri, A. Cevrero, P. Brisk, and P. Ienne. Fpga implementation of a single-precision floating-point multiply-accumulator with single-cycle accumulation. In *17th IEEE Symposium on Field Programmable Custom Computing Machines*, 2009.

[481] A. Panhaleux. Génération d'itérations de type Newton-Raphson pour la division de deux flottants à l'aide d'un FMA. Master's thesis, École Normale Supérieure de Lyon, Lyon, France, 2008. In French.

[482] B. Parhami. On the complexity of table lookup for iterative division. *IEEE Transactions on Computers*, C-36(10):1233–1236, 1987.

[483] B. Parhami. *Computer Arithmetic: Algorithms and Hardware Designs*. Oxford University Press, 2000.

[484] M. Parks. Number-theoretic test generation for directed roundings. *IEEE Transactions on Computers*, 49(7):651–658, 2000.

[485] B. Pasca. Correctly rounded floating-point division for DSP-enabled FPGAs. In *Field Programmable Logic and Applications*, 2012.

[486] D. Patil, O. Azizi, M. Horowitz, R. Ho, and R. Ananthraman. Robust energy-efficient adder topologies. In *18th IEEE Symposium on Computer Arithmetic (ARITH-18)*, pages 29–37, June 2007.

[487] M. Payne and R. Hanek. Radian reduction for trigonometric functions. *SIGNUM Newsletter*, 18:19–24, 1983.

[488] C. Peikert. Limits on the hardness of lattice problems in l_p norms. *Computational Complexity*, 17(2):300–351, 2008.

[489] P. Pélissier and P. Zimmermann. The DPE library. Available at http://www.loria.fr/~zimmerma/free/dpe-1.4.tar.gz.

[490] M. Pichat. Correction d'une somme en arithmétique à virgule flottante. *Numerische Mathematik*, 19:400–406, 1972. In French.

[491] R. V. K. Pillai, D. Al-Khalili, and A. J. Al-Khalili. A low power approach to floating point adder design. In *International Conference on Computer Design*, 1997.

[492] J. A. Pineiro and J. D. Bruguera. High-speed double-precision computation of reciprocal, division, square root, and inverse square root. *IEEE Transactions on Computers*, 51(12):1377–1388, 2002.

[493] S. Pion. *De la Géométrie Algorithmique au Calcul Géométrique*. Ph.D. thesis, Université de Nice Sophia-Antipolis, France, November 1999. In French.

[494] V. Popescu. *Towards fast and certified multiple-precision libraries*. Ph.D. thesis, Université de Lyon, 2017. Available at https://hal.archives-ouvertes.fr/tel-01534090.

[495] D. M. Priest. Algorithms for arbitrary precision floating point arithmetic. In *10th IEEE Symposium on Computer Arithmetic (ARITH-10)*, pages 132–143, June 1991.

[496] D. M. Priest. *On Properties of Floating-Point Arithmetics: Numerical Stability and the Cost of Accurate Computations*. Ph.D. thesis, University of California at Berkeley, 1992.

[497] D. M. Priest. Efficient scaling for complex division. *ACM Transactions on Mathematical Software*, 30(4), 2004.

[498] C. Proust. Masters' writings and students' writings: School material in Mesopotamia. In Gueudet, Pepin, and Trouche, editors, *Mathematics curriculum material and teacher documentation: from textbooks to shared living resources*, pages 161–180. Springer, 2011.

[499] J. D. Pryce and G. F. Corliss. Interval arithmetic with containment sets. *Computing*, 78:251–276, 2006.

[500] S. Putot. *Static Analysis of Numerical Programs and Systems*. Habilitation, Université Paris-Sud, 2012.

[501] E. Quinnell, E. E. Swartzlander, and C. Lemonds. Floating-point fused multiply-add architectures. In *41st Asilomar Conference on Signals, Systems, and Computers*, pages 331–337, November 2007.

[502] S.-K. Raina. *FLIP: a Floating-point Library for Integer Processors*. Ph.D. thesis, École Normale Supérieure de Lyon, September 2006. Available at http://www.ens-lyon.fr/LIP/Pub/PhD2006.php.

[503] J. Ramos and A. Bohorquez. Two operand binary adders with threshold logic. *IEEE Transactions on Computers*, 48(12):1324–1337, 1999.

[504] B. Randell. From analytical engine to electronic digital computer: the contributions of Ludgate, Torres, and Bush. *IEEE Annals of the History of Computing*, 4(4):327–341, 1982.

[505] E. Remez. Sur un procédé convergent d'approximations successives pour déterminer les polynômes d'approximation. *C.R. Académie des Sciences, Paris*, 198:2063–2065, 1934. In French.

[506] N. Revol, K. Makino, and M. Berz. Taylor models and floating-point arithmetic: proof that arithmetic operations are validated in COSY. *Journal of Logic and Algebraic Programming*, 64:135–154, 2005.

[507] N. Revol and F. Rouillier. Motivations for an arbitrary precision interval arithmetic and the MPFI library. *Reliable Computing*, 11:1–16, 2005.

[508] N. Revol and P. Théveny. Numerical reproducibility and parallel computations: Issues for interval algorithms. *IEEE Transactions on Computers*, 63(8):1915–1924, 2014.

[509] G. Revy. *Implementation of binary floating-point arithmetic on embedded integer processors: polynomial evaluation-based algorithms and certified code generation*. Ph.D. thesis, Université de Lyon - ÉNS de Lyon, France, December 2009.

[510] T. J. Rivlin. *Chebyshev polynomials. From approximation theory to algebra.* John Wiley & Sons, New York, 2nd edition, 1990.

[511] J. E. Robertson. A new class of digital division methods. *IRE Transactions on Electronic Computers*, EC-7:218–222, 1958. Reprinted in [583].

[512] E. Roesler and B. Nelson. Novel optimizations for hardware floating-point units in a modern FPGA architecture. In *Field Programmable Logic and Applications*, volume 2438 of *Lecture Notes in Computer Science*, pages 637–646, 2002.

[513] R. Rojas. Konrad Zuse's legacy: the architecture of the Z1 and Z3. *IEEE Annals of the History of Computing*, 19(2):5–16, 1997.

[514] R. Rojas. The Z1: Architecture and algorithms of Konrad Zuse's first computer. Technical report, Freie Universität Berlin, June 2014. Available at https://arxiv.org/abs/1406.1886.

[515] R. Rojas, F. Darius, C. Göktekin, and G. Heyne. The reconstruction of Konrad Zuse's Z3. *IEEE Annals of the History of Computing*, 27(3):23–32, 2005.

[516] P. Roux. Innocuous double rounding of basic arithmetic operations. *Journal of Formalized Reasoning*, 7(1):131–142, 2014.

[517] S. M. Rump. Solving algebraic problems with high accuracy (Habilitationsschrift). In *A New Approach to Scientific Computation*, pages 51–120, 1983.

[518] S. M. Rump. Algorithms for verified inclusions: theory and practice. In *Reliability in Computing, Perspectives in Computing*, pages 109–126, 1988.

[519] S. M. Rump. Fast and parallel interval arithmetic. *BIT*, 39(3):534–554, 1999.

[520] S. M. Rump. INTLAB - INTerval LABoratory. In T. Csendes, editor, *Developments in Reliable Computing*, pages 77–104. Kluwer Academic Publishers, Dordrecht, 1999. http://www.ti3.tuhh.de/rump/.

[521] S. M. Rump. Ultimately fast accurate summation. *SIAM Journal on Scientific Computing*, 31(5):3466–3502, 2009.

[522] S. M. Rump. Error estimation of floating-point summation and dot product. *BIT Numerical Mathematics*, 52(1):201–220, 2012.

[523] S. M. Rump. Fast interval matrix multiplication. *Numerical Algorithms*, 1(61):1–34, 2012.

[524] S. M. Rump. Interval arithmetic over finitely many endpoints. *BIT Numerical Mathematics*, 52(4):1059–1075, 2012.

[525] S. M. Rump. Computable backward error bounds for basic algorithms in linear algebra. *Nonlinear Theory and its Applications, IEICE*, 6(3):360–363, 2015.

[526] S. M. Rump and H. Böhm. Least significant bit evaluation of arithmetic expressions in single-precision. *Computing*, 30:189–199, 1983.

[527] S. M. Rump, F. Bünger, and C.-P. Jeannerod. Improved error bounds for floating-point products and Horner's scheme. *BIT Numerical Mathematics*, 56(1):293–307, 2016.

[528] S. M. Rump and C.-P. Jeannerod. Improved backward error bounds for LU and Cholesky factorizations. *SIAM Journal on Matrix Analysis and Applications*, 35(2):684–698, 2014.

[529] S. M. Rump and M. Kashiwagi. Implementation and improvements of affine arithmetic. *Nonlinear Theory and its Applications, IEICE*, 6(3):341–359, 2015.

[530] S. M. Rump and M. Lange. On the definition of unit roundoff. *BIT Numerical Mathematics*, 56(1):309–317, 2016.

[531] S. M. Rump, T. Ogita, and S. Oishi. Accurate floating-point summation part I: Faithful rounding. *SIAM Journal on Scientific Computing*, 31(1):189–224, 2008.

[532] S. M. Rump, T. Ogita, and S. Oishi. Accurate floating-point summation part II: Sign, K-fold faithful and rounding to nearest. *SIAM Journal on Scientific Computing*, 31(2):1269–1302, 2008.

[533] S. M. Rump, P. Zimmermann, S. Boldo, and G. Melquiond. Computing predecessor and successor in rounding to nearest. *BIT Numerical Mathematics*, 49(2):419–431, 2009.

[534] D. M. Russinoff. A mechanically checked proof of IEEE compliance of a register-transfer-level specification of the AMD-K7 floating-point multiplication, division, and square root instructions. *LMS Journal of Computation and Mathematics*, 1:148–200, 1998.

[535] D. M. Russinoff. A mechanically checked proof of correctness of the AMD K5 floating point square root microcode. *Formal Methods in System Design*, 14(1):75–125, 1999.

[536] D. M. Russinoff. A case study in formal verification of register-transfer logic with ACL2: The floating point adder of the AMD Athlon processor. *Lecture Notes in Computer Science*, 1954:3–36, 2000.

[537] E. Salamin. Computation of π using arithmetic-geometric mean. *Mathematics of Computation*, 30:565–570, 1976.

[538] H. H. Saleh and E. E. Swartzlander. A floating-point fused dot-product unit. In *International Conference on Computer Design (ICCD)*, pages 426–431, 2008.

[539] J.-L. Sanchez, A. Jimeno, H. Mora, J. Mora, and F. Pujol. A CORDIC-based architecture for high performance decimal calculation. In *IEEE International Symposium on Industrial Electronics*, pages 1951–1956, June 2007.

[540] J. Sawada and R. Gamboa. Mechanical verification of a square root algorithm using Taylor's theorem. In *4th International Conference on Formal Methods in Computer-Aided Design (FMCAD)*, volume 2517 of *Lecture Notes in Computer Science*, pages 274–291, 2002.

[541] H. Schichl and A. Neumaier. Interval analysis on directed acyclic graphs for global optimization. *Journal of Global Optimization*, 33:541–562, 2005.

[542] H. Schmid and A. Bogacki. Use decimal CORDIC for generation of many transcendental functions. *EDN*, pages 64–73, 1973.

[543] M. M. Schmookler and K. J. Nowka. Leading zero anticipation and detection – a comparison of methods. In *15th IEEE Symposium on Computer Arithmetic (ARITH-15)*, pages 7–12, June 2001.

[544] C. P. Schnorr. A more efficient algorithm for lattice basis reduction. *J. Algorithms*, 9(1):47–62, 1988.

[545] A. Schönhage. Schnelle Berechnung von Kettenbruchentwicklungen. *Acta Informatica*, 1:139–144, 1971. In German.

[546] A. Schönhage, A. F. W. Grotefeld, and E. Vetter. *Fast algorithms: a Multitape Turing Machine Implementation*. Bibliographisches Institut, Mannheim, 1994.

[547] A. Schönhage and V. Strassen. Schnelle Multiplikation grosser Zahlen. *Computing*, 7:281–292, 1971. In German.

[548] E. M. Schwarz, J. S. Kapernick, and M. F. Cowlishaw. Decimal floating-point support on the IBM system Z10 processor. *IBM Journal of Research and Development*, 53(1):36–45, 2009.

[549] E. M. Schwarz, M. Schmookler, and S. D. Trong. FPU implementations with denormalized numbers. *IEEE Transactions on Computers*, 54(7):825–836, 2005.

[550] P.-M. Seidel. Multiple path IEEE floating-point fused multiply-add. In *46th International Midwest Symposium on Circuits and Systems*, pages 1359–1362, 2003.

[551] P.-M. Seidel and G. Even. How many logic levels does floating-point addition require. In *International Conference on Computer Design*, pages 142–149, 1998.

[552] P.-M. Seidel and G. Even. On the design of fast IEEE floating-point adders. In *15th IEEE Symposium on Computer Arithmetic (ARITH-15)*, pages 184–194, June 2001.

[553] C. Severance. IEEE 754: An interview with William Kahan. *Computer*, 31(3):114–115, 1998.

[554] A. M. Shams and M. A. Bayoumi. A novel high-performance CMOS 1-bit full-adder cell. *IEEE Transactions on Circuits and Systems II: Analog and Digital Signal Processing*, 47(5), 2000.

[555] J. R. Shewchuk. Adaptive precision floating-point arithmetic and fast robust geometric predicates. *Discrete Computational Geometry*, 18:305–363, 1997.

[556] N. Shirazi, A. Walters, and P. Athanas. Quantitative analysis of floating point arithmetic on FPGA based custom computing machine. In *FPGAs for Custom Computing Machines*, pages 155–162, 1995.

[557] V. Shoup. NTL, a library for doing number theory, version 10.5.0. http://shoup.net/ntl/, 2017.

[558] T. Simpson. *Essays on several curious and useful subjects in speculative and mix'd mathematicks, illustrated by a variety of examples*. London, 1740.

[559] D. P. Singh, B. Pasca, and T. S. Czajkowski. High-level design tools for floating point FPGAs. In *ACM/SIGDA International Symposium on Field-Programmable Gate Arrays, Monterey, CA, USA, February 22–24, 2015*, pages 9–12, 2015.

[560] R. A. Smith. A continued-fraction analysis of trigonometric argument reduction. *IEEE Transactions on Computers*, 44(11):1348–1351, 1995.

[561] R. L. Smith. Algorithm 116: Complex division. *Communications of the ACM*, 5(8):435, 1962.

[562] A. Solovyev, C. Jacobsen, Z. Rakamarić, and G. Gopalakrishnan. *Rigorous Estimation of Floating-Point Round-off Errors with Symbolic Taylor Expansions*, pages 532–550. Springer International Publishing, 2015.

[563] A. Solovyev, C. Jacobsen, Z. Rakamarić, and G. Gopalakrishnan. Rigorous estimation of floating-point round-off errors with symbolic Taylor expansions. In *20th International Symposium on Formal Methods (FM)*, June 2015.

[564] E. Sprangle and D. Carmean. Increasing processor performance by implementing deeper pipelines. In *29th Annual International Symposium on Computer Architecture (ISCA)*, pages 25–34, 2002.

[565] H. M. Stark. *An Introduction to Number Theory*. MIT Press, Cambridge, MA, 1981.

[566] G. L. Steele, Jr. and J. L. White. How to print floating-point numbers accurately. *ACM SIGPLAN Notices*, 25(6):112–126, 1990.

[567] G. L. Steele, Jr. and J. L. White. Retrospective: how to print floating-point numbers accurately. *ACM SIGPLAN Notices*, 39(4):372–389, 2004.

[568] D. Stehlé. *Algorithmique de la Réduction de Réseaux et Application à la Recherche de Pires Cas pour l'Arrondi de Fonctions Mathématiques*. Ph.D. thesis, Université Henri Poincaré – Nancy 1, France, December 2005.

[569] D. Stehlé. On the randomness of bits generated by sufficiently smooth functions. In *7th Algorithmic Number Theory Symposium (ANTS)*, volume 4078 of *Lecture Notes in Computer Science*, pages 257–274, 2006.

[570] D. Stehlé, V. Lefèvre, and P. Zimmermann. Worst cases and lattice reduction. In *16th IEEE Symposium on Computer Arithmetic (ARITH-16)*, pages 142–147, June 2003.

[571] D. Stehlé, V. Lefèvre, and P. Zimmermann. Searching worst cases of a one-variable function. *IEEE Transactions on Computers*, 54(3):340–346, 2005.

[572] P. H. Sterbenz. *Floating-Point Computation*. Prentice-Hall, Englewood Cliffs, NJ, 1974.

[573] G. W. Stewart. A note on complex division. *ACM Transactions on Mathematical Software*, 11(3):238–241, 1985.

[574] J. E. Stine and M. J. Schulte. The symmetric table addition method for accurate function approximation. *Journal of VLSI Signal Processing*, 21:167–177, 1999.

[575] J. Stolfi and L. de Figueiredo. *Self-Validated Numerical Methods and Applications*. Monograph for 21st Brazilian Mathematics Colloquium, Rio de Janeiro, Brazil, 1997.

[576] A. Storjohann. Faster algorithms for integer lattice basis reduction. Technical report, ETH Zürich, 1996.

[577] V. Strassen. Gaussian elimination is not optimal. *Numerische Mathematik*, 13:354–356, 1969.

[578] Sun. *Numerical Computation Guide – Sun™ Studio 11*, 2005. Available at http://docs.sun.com/source/819-3693/.

[579] Sun Microsystems. Interval arithmetic in high performance technical computing. Technical report, 2002.

[580] T. Sunaga. Geometry of Numerals. Master's thesis, U. Tokyo, Japan, 1956.

[581] D. A. Sunderland, R. A. Strauch, S. W. Wharfield, H. T. Peterson, and C. R. Cole. CMOS/SOS frequency synthesizer LSI circuit for spread spectrum communications. *IEEE Journal of Solid State Circuits*, SC-19(4):497–506, 1984.

[582] A. Svoboda. Adder with distributed control. *IEEE Transactions on Computers*, C-19(8), 1970. Reprinted in [583].

[583] E. E. Swartzlander. *Computer Arithmetic*, volume 1. World Scientific Publishing Co., Singapore, 2015.

[584] E. E. Swartzlander. *Computer Arithmetic*, volume 2. World Scientific Publishing Co., Singapore, 2015.

[585] E. E. Swartzlander and A. G. Alexpoulos. The sign-logarithm number system. *IEEE Transactions on Computers*, 1975. Reprinted in [583].

[586] N. Takagi. A hardware algorithm for computing the reciprocal square root. In *15th IEEE Symposium on Computer Arithmetic (ARITH-15)*, pages 94–100, June 2001.

[587] N. Takagi and S. Kuwahara. A VLSI algorithm for computing the Euclidean norm of a 3D vector. *IEEE Transactions on Computers*, 49(10):1074–1082, 2000.

[588] P. T. P. Tang. Table-driven implementation of the exponential function in IEEE floating-point arithmetic. *ACM Transactions on Mathematical Software*, 15(2):144–157, 1989.

[589] P. T. P. Tang. Table-driven implementation of the logarithm function in IEEE floating-point arithmetic. *ACM Transactions on Mathematical Software*, 16(4):378–400, 1990.

[590] P. T. P. Tang. Table lookup algorithms for elementary functions and their error analysis. In *10th IEEE Symposium on Computer Arithmetic (ARITH-10)*, pages 232–236, June 1991.

[591] P. T. P. Tang. Table-driven implementation of the expm1 function in IEEE floating-point arithmetic. *ACM Transactions on Mathematical Software*, 18(2):211–222, 1992.

[592] Y. Tao, G. Deyuan, R. Xianglong, H. Limin, F. Xiaoya, and Y. Lei. A novel floating-point function unit combining MAF and 3-input adder. In *Signal Processing, Communication and Computing (ICSPCC)*, 2012.

[593] Y. Tao, G. Deyuan, and F. Xiaoya. A multi-path fused add-subtract unit for digital signal processing. In *Computer Science and Automation Engineering (CSAE)*, 2012.

[594] Y. Tao, G. Deyuan, F. Xiaoya, and J. Nurmi. Correctly rounded architectures for floating-point multi-operand addition and dot-product computation. In *Application-Specific Systems, Architectures and Processors (ASAP)*, 2013.

[595] Y. Tao, G. Deyuan, F. Xiaoya, and R. Xianglong. Three-operand floating-point adder. In *12th International Conference on Computer and Information Technology*, pages 192–196, 2012.

[596] M. B. Taylor. Is dark silicon useful? Harnessing the four horsemen of the coming dark silicon apocalypse. In *49th Design Automation Conference*, pages 1131–1136, 2012.

[597] A. F. Tenca. Multi-operand floating-point addition. In *19th IEEE Symposium on Computer Arithmetic (ARITH-19)*, pages 161–168, 2009.

[598] T. Teufel and M. Baesler. FPGA implementation of a decimal floating-point accurate scalar product unit with a parallel fixed-point multiplier. In *Reconfigurable Computing and FPGAs*, pages 6–11, 2009.

[599] The FPLLL development team. fplll, a lattice reduction library. Available at https://github.com/fplll/fplll, 2016.

[600] P. Théveny. *Numerical Quality and High Performance in Interval Linear Algebra on Multi-Core Processors*. Ph.D. thesis, Université de Lyon, École Normale Supérieure de Lyon, 2014.

[601] D. B. Thomas. A general-purpose method for faithfully rounded floating-point function approximation in FPGAs. In *22nd IEEE Symposium on Computer Arithmetic (ARITH-22)*, pages 42–49, 2015.

[602] K. D. Tocher. Techniques of multiplication and division for automatic binary computers. *Quarterly Journal of Mechanics and Applied Mathematics*, 11(3):364–384, 1958.

[603] A. L. Toom. The complexity of a scheme of functional elements realizing the multiplication of integers. *Soviet Mathematics Doklady*, 3:714–716, 1963.

[604] S. Torres. *Tools for the Design of Reliable and Efficient Function Evaluation Libraries*. Ph.D. thesis, Université de Lyon, 2016. Available at https://tel.archives-ouvertes.fr/tel-01396907.

[605] W. J. Townsend, E. E. Swartzlander, Jr., and J. A. Abraham. A comparison of Dadda and Wallace multiplier delays. In *SPIE's 48th Annual Meeting on Optical Science and Technology*, pages 552–560, 2003.

[606] L. Trefethen. *Approximation Theory and Approximation Practice*. SIAM, 2013.

[607] S. D. Trong, M. Schmookler, E. M. Schwarz, and M. Kroener. P6 binary floating-point unit. In *18th IEEE Symposium on Computer Arithmetic (ARITH-18)*, pages 77–86, 2007.

[608] W. Tucker. The Lorenz attractor exists. *Comptes Rendus de l'Académie des Sciences-Series I-Mathematics*, 328(12):1197–1202, 1999.

[609] W. Tucker. *Validated Numerics – A Short Introduction to Rigorous Computations*. Princeton University Press, 2011.

[610] A. Tyagi. A reduced-area scheme for carry-select adders. *IEEE Transactions on Computers*, 42(10):1163–1170, 1993.

[611] H. F. Ugurdag, A. Bayram, V. E. Levent, and S. Gören. Efficient combinational circuits for division by small integer constants. In *23rd IEEE Symposium on Computer Arithmetic (ARITH-23)*, pages 1–7, July 2016.

[612] P. van Emde Boas. Another NP-complete problem and the complexity of computing short vectors in a lattice. Technical Report 81-04, Mathematisch Instituut, University of Amsterdam, 1981.

[613] A. Vázquez. *High-Performance Decimal Floating-Point Units*. Ph.D. thesis, Universidade de Santiago de Compostela, 2009.

[614] A. Vázquez, E. Antelo, and P. Montuschi. A new family of high performance parallel decimal multipliers. In *18th IEEE Symposium on Computer Arithmetic (ARITH-18)*, pages 195–204, 2007.

[615] L. Veidinger. On the numerical determination of the best approximations in the Chebyshev sense. *Numerische Mathematik*, 2:99–105, 1960.

[616] G. W. Veltkamp. ALGOL procedures voor het berekenen van een inwendig product in dubbele precisie. Technical Report 22, RC-Informatie, Technishe Hogeschool Eindhoven, 1968.

[617] G. W. Veltkamp. ALGOL procedures voor het rekenen in dubbele lengte. Technical Report 21, RC-Informatie, Technishe Hogeschool Eindhoven, 1969.

[618] D. Villeger and V. G. Oklobdzija. Evaluation of Booth encoding techniques for parallel multiplier implementation. *Electronics Letters*, 29(23):2016–2017, 1993.

[619] J. E. Volder. The CORDIC trigonometric computing technique. *IRE Transactions on Electronic Computers*, EC-8(3):330–334, 1959.

[620] J. E. Volder. The birth of CORDIC. *Journal of VLSI Signal Processing Systems*, 25(2):101–105, 2000.

[621] Y. Voronenko and M. Püschel. Multiplierless multiple constant multiplication. *ACM Trans. Algorithms*, 3(2), 2007.

[622] J. E. Vuillemin. Exact real computer arithmetic with continued fractions. *IEEE Transactions on Computers*, 39(8), 1990.

[623] J. E. Vuillemin. On circuits and numbers. *IEEE Transactions on Computers*, 43(8):868–879, 1994.

[624] C. S. Wallace. A suggestion for a fast multiplier. *IEEE Transactions on Electronic Computers*, pages 14–17, 1964. Reprinted in [583].

[625] J. S. Walther. A unified algorithm for elementary functions. In *Joint Computer Conference Proceedings*, 1971. Reprinted in [583].

[626] J. S. Walther. The story of unified CORDIC. *Journal of VLSI Signal Processing Systems*, 25(2):107–112, 2000.

[627] L.-K. Wang and M. J. Schulte. Decimal floating-point division using Newton–Raphson iteration. In *Application-Specific Systems, Architectures and Processors*, pages 84–95, 2004.

[628] L.-K. Wang, M. J. Schulte, J. D. Thompson, and N. Jairam. Hardware designs for decimal floating-point addition and related operations. *IEEE Transactions on Computers*, 58(2):322–335, 2009.

[629] X. Wang, S. Braganza, and M. Leeser. Advanced components in the variable precision floating-point library. In *Field-Programmable Custom Computing Machines*, pages 249–258, 2006.

[630] W. H. Ware, editor. Soviet computer technology—1959. *Communications of the ACM*, 3(3):131–166, 1960.

[631] N. Whitehead and A. Fit-Florea. Precision & performance: Floating point and IEEE 754 compliance for NVIDIA GPUs. Technical report, NVIDIA, 2011. Available at http://developer.download.nvidia.com/devzone/devcenter/cuda/files/NVIDIA-CUDA-Floating-Point.pdf.

[632] Wikipedia. Slide rule — Wikipedia, The Free Encyclopedia, 2017. [Online; accessed 20-November-2017].

[633] Wikipedia. Square root of 2 — Wikipedia, The Free Encyclopedia, 2017. [Online; accessed 20-November-2017].

[634] J. H. Wilkinson. *Rounding Errors in Algebraic Processes*. Prentice-Hall, Englewood Cliffs, NJ, 1963.

[635] J. H. Wilkinson. Some Comments from a Numerical Analyst. *Journal of the ACM*, 18(2):137–147, 1971.

[636] M. J. Wirthlin. Constant coefficient multiplication using look-up tables. *VLSI Signal Processing*, 36(1):7–15, 2004.

[637] W. F. Wong and E. Goto. Fast hardware-based algorithms for elementary function computations using rectangular multipliers. *IEEE Transactions on Computers*, 43(3):278–294, 1994.

[638] S. Xing and W. Yu. FPGA adders: Performance evaluation and optimal design. *IEEE Design & Test of Computers*, 15:24–29, 1998.

[639] T. J. Ypma. Historical development of the Newton-Raphson method. *SIAM Rev.*, 37(4):531–551, 1995.

[640] X. Y. Yu, Y.-H. Chan, B. Curran, E. Schwarz, M. Kelly, and B. Fleischer. A 5GHz+ 128-bit binary floating-point adder for the POWER6 processor. In *European Solid-State Circuits Conference*, pages 166–169, 2006.

[641] N. Zhuang and H. Wu. A new design of the CMOS full adder. *IEEE Journal on Solid-State Circuits*, 27:840–844, 1992.

[642] L. Zhuo and V. K. Prasanna. Scalable and modular algorithms for floating-point matrix multiplication on FPGAs. In *18th International Parallel and Distributed Processing Symposium (IPDPS)*, April 2004.

[643] L. Zhuo and V. K. Prasanna. High performance linear algebra operations on reconfigurable systems. In *Supercomputing*, 2005.

[644] G. Zielke and V. Drygalla. Genaue Loesung Linearer Gleichungssysteme. *GAMM Mitteilungen der Gesellschaft für Angewandte Mathematik und Mechanik*, 26:7–107, 2003.

[645] R. Zimmermann. *Binary Adder Architectures for Cell-Based VLSI and Their Synthesis*. Ph.D. thesis, Swiss Federal Institute of Technology, Zurich, 1997.

[646] A. Ziv. Fast evaluation of elementary mathematical functions with correctly rounded last bit. *ACM Transactions on Mathematical Software*, 17(3):410–423, 1991.

[647] R. Zumkeller. Formal global optimisation with Taylor models. In *3th International Joint Conference on Automated Reasoning (IJCAR)*, volume 4130 of *Lecture Notes in Computer Science*, pages 408–422, August 2006.

[648] D. Zuras. More on squaring and multiplying large integers. *IEEE Transactions on Computers*, 43(8):899–908, 1994.

[649] V. Zyuban, D. Brooks, V. Srinivasan, M. Gschwind, P. Bose, P. N. Strenski, and P. G. Emma. Integrated analysis of power and performance for pipelined microprocessors. *IEEE Transactions on Computers*, 53(8):1004–1016, 2004.

Index

J.-M. Muller et al., *Handbook of Floating-Point Arithmetic*,
https://doi.org/10.1007/978-3-319-76526-6

GPSR Compliance
The European Union's (EU) General Product Safety Regulation (GPSR) is a set of rules that requires consumer products to be safe and our obligations to ensure this.

If you have any concerns about our products, you can contact us on

ProductSafety@springernature.com

In case Publisher is established outside the EU, the EU authorized representative is:

Springer Nature Customer Service Center GmbH
Europaplatz 3
69115 Heidelberg, Germany

www.ingramcontent.com/pod-product-compliance
Lightning Source LLC
Chambersburg PA
CBHW082005130926
51338CB00021B/43
* 9 7 8 3 0 3 0 0 9 5 1 3 0 *